W0112430

CELLS, MEMBRANES, AND DISEASE, Including Renal

METHODOLOGICAL SURVEYS IN BIOCHEMISTRY AND ANALYSIS

Series Editor: Eric Reid

Guildford Academic Associates
72 The Chase
Guildford GU2 5UL, United Kingdom

The series is divided into Subseries A: Analysis, and B: Biochemistry
Enquiries concerning Volumes 1–11 should be sent to the above address.

A Continuation Order Plan is available for this series. A continuation order will bring delivery of each new volume immediately upon publication. Volumes are billed only upon actual shipment. For further information please contact the publisher.

CELLS, MEMBRANES, AND DISEASE, Including Renal

Edited by

Eric Reid
Guildford Academic Associates
Guildford, United Kingdom

and

G. M. W. Cook and
J. P. Luzio
University of Cambridge
Cambridge, United Kingdom

PLENUM PRESS • NEW YORK AND LONDON

Library of Congress Cataloging in Publication Data

International Subcellular Methodology Forum (10th: 1986: University of Surrey)
  Cells, membranes, and disease, including renal.

  (Methodological surveys in biochemistry and analysis; v. 17 (B))
  "Based on the proceedings of the 10th International Subcellular Methodology Forum,
held September 1–4, 1986, at the University of Surrey, in Guildford, United Kingdom"
—T.p. verso.
  Includes bibliographies and index.
  1. Pathology, Cellular—Methodology—Congresses. 2. Physiology, Pathological—
Methodology—Congresses. 3. Kidney—Diseases—Congresses.I. Reid, Eric, 1922–
II. Cook, G. M. W. (Geoffrey Malcolm Weston), 1938–   . III. Luzio, J. P. IV. Series:
Methodological surveys in biochemistry and analysis; v. 17. V. Series: Methodological
surveys in biochemistry and analysis. Subseries B, Biochemistry. [DNLM: 1. Cell Mem-
branes—physiology—congresses. 2. Cells—physiology—congresses. 3. Diseases—
etiology—congresses. W1 ME9612NT v.17 / QH 601 I633 1986c]
RB25.I56   1986   616.07'1   87-22060
ISBN-13: 978-1-4684-1285-7   e-ISBN-13: 978-1-4684-1283-3
DOI: 10.1007/978-1-4684-1283-3

Based on the proceedings of the 10th International Subcellular Methodology
Forum, held September 1–4, 1986, at the University of Surrey,
in Guildford, United Kingdom

© 1987 Plenum Press, New York
Softcover reprint of the hardcover 1st edition 1987
A Division of Plenum Publishing Corporation
233 Spring Street, New York, N.Y. 10013

All rights reserved

No part of this book may be reproduced, stored in a retrieval system, or transmitted
in any form or by any means, electronic, mechanical, photocopying, microfilming,
recording, or otherwise, without written permission from the Publisher

Senior Editor's Preface

This volume widens the topic span of the book series. It focuses on experimental and clinical pathology. It complements Vol. 11 (*Cancer-Cell Organelles*) and pays some attention to the matrix (notably collagen) as well as the cell. It also touches on pharmacological and nutritional abnormalities. Especial attention has been paid to proteolysis - a notable 'growth-area'.

After a somewhat unproductive decade, prospects for elucidating the cellular basis of pathological changes do seem to be improving, as exemplified in this book. Certain kindred books (not methodology-oriented) purport to survey cell pathology but are dominated by normal-cell observations, e.g. on membranes, of potential rather than present-day applicability to cell-pathology studies. This holds to some extent for the present book too. Yet its carefully chosen title is reckoned unlikely to raise false expectations, whilst open to criticism because *Disease* does not self-evidently embrace cell toxicology. At least the title is an improvement on that of the Forum which led to the book, as is amplified on p. xiv ('Forum Co-Organizer's Foreword'). As in the past, the aim of long-term usefulness without the shortcomings of typical 'Proceedings' was an incentive to optimizing the sequence of articles and to thorough editing and indexing. Certain special topics are collated in a Sub-Index.

Appreciation is expressed to authors for 'finding time' to produce publication texts, even if delayed or requiring much editorial rectification - which earned gratifying compliments from some authors. Alterations to authors' texts were mainly for the sake of clarity, consistency or crispness, or of adherence to conventions mentioned below. No attempt was made to standardise terminology in respect of proteinases/proteases/peptidases (see art. #E-6), desirable though this might have been. Nor was action taken on two pet aversions, with no easy remedy, which are now aired (comments invited!).- Where a reaction mixture is kept for a set time at $0°$, 'incubate' seems an unhappy term; likewise the term 'staining' where, in contexts such as electron microscopy, visualization does not involve a stain.

Some authors who 'grew up' with subcellular discoveries (e.g. lysosomes, mid-1950's) and innovations (e.g. zonal rotors, mid-1960's) seemed to assume that younger readers would be equally knowledgeable and wary. For such readers, whose possible unfamiliarity with vital concepts such as enzyme latency warranted some editorial re-phrasings, there are some useful basic sources (besides *Methods in Enzymology*, e.g. Vol. 31A), including: *Subcellular Components* (ed. G.D. Birnie;

2nd edn.; Butterworth), *Density Gradient Centrifugation* (R. Hinton & M. Dobrota, in the North-Holland/Elsevier *Laboratory Techniques* series) and *Centrifugation* (ed. D. Rickwood; IRL Press). E. Reid's advocacy (in 1972) of 'differential pelleting' (counterpart: 'differential banding') as a clearer term than 'differential centrifugation' has gained adherents. Early 'B' volumes (up to Vol. 11) in the present series, as listed opposite the title page, also offer guidance on separating cells and subcellular elements, and on their characterization. The latter, often skimped, relies largely on judicious assays for markers, as amplified in a Forum-derived 'manifesto' by Morré et al. (1979; *Eur. J. Cell Biol. 20*, 195-199) aimed at the community of editors and referees as well as authors.

Acknowledgements. - The Forum had the benefit of planning help from Honorary Advisers including both Co-Editors and Drs. W.H. Evans and T.J. Peters, and of support awards from Beechams Pharmaceuticals, Ciba-Geigy (Advanced Drug Delivery Research, Horsham) and the Cancer Research Campaign. Two academic participants had company support (LKB, Bromma; Nyegaard, Oslo). Certain publishers readily gave permission to reproduce published material, as acknowledged in individual articles. The cover 'logo' was adapted from a diagram in *Membranes and their Cellular Functions* (2nd edn., ed. J.B. Finean, R. Coleman & R.H. Michell; Blackwell, Oxford).

Conventions and abbreviations.- For density (d) values, the unit (g/ml) is not usually stated. For mol. wts., commonly derived by comparison with reference proteins on gels, expression in Daltons was felt less appropriate than M_r (or kM_r, analogous to kDa). All temperatures (°) are in degrees Celsius. Units such as M (molar) have been favoured, although not conforming to SI practices. Abbreviations accepted by the *Biochemical Journal* without definition need no explanation here - e.g. EDTA, Tris and, for absorbance, A (not O.D.). Other abbreviations have been defined in the articles concerned, but some warrant collation here:

Ab, antibody (MAb = monoclonal)	i.p., intraperitoneal
BSA, bovine serum albumin	i.v., intravenous
ELISA, enzyme-linked immuno-absorbent assay	LL, lipid-lowering (hypolipidaemic)
e.m., electron microscopy/micrograph	PAGE, polyacrylamide gel electrophoresis
e.r., endoplasmic reticulum	p.m., plasma membrane
FFE, free-flow electrophoresis	RIA, radioimmunoassay
HPLC, high-pressure liquid chromatography	s.a., specific activity
	s.c., subcutaneous
	SDS, sodium dodecyl sulphate

Guildford Academic Associates
72 The Chase, Guildford
Surrey GU2 5UL, U.K.

ERIC REID

15 May 1987

Contents

Contents ix

List of Authors

Primary author

D. Allan - pp. 211-217
Univ. Coll. London, London

K-J. Andersen - pp. (i) 351-357, (ii) 427-432;
also xiv *& as for* Berg(i) *and* Haga
University of Bergen (Haukeland Sykehus)

P.H. Bach - pp. 417-426
Univ. of Surrey, Guildford

T. Berg - pp. (i) 315-325, (ii) 383-386
Inst. for Nutrition Res., Univ. of Oslo

R.K. Berge - pp. 53-65
Univ. of Bergen (Haukeland Sykehus)

P. Bohley - pp. 299-306
Physiol. Chem. Inst., Univ. of Tübingen

N. Crawford - pp. 219-230
Roy. Coll. of Surgeons, London

M. Davies - pp. 453-464
KRUF Inst., Roy. Infirmary, Cardiff

M. Dobrota pp. 433-438, *and as*
for Andersen (ii) *and* Haga
Univ. of Surrey, Guildford

K. Donaldson - pp. 379-382
Inst. of Occupational Med., Edinburgh

R.L. Dormer - pp. 273-283
Univ. of Wales Coll. of Med., Cardiff

P. Druet - pp. 445-452
INSERM, Hôpital Broussais, Paris

W.H. Evans - pp. 289-297
Nat. Inst. for Med. Res., London

J.T.R. Fitzsimons - pp. 127-132
Univ. of Southampton

B.A. Fowler - pp. 99-108
Nat. Inst. of Environ. Health Sciences, NC

P. Friberger - pp. 327-333
KabiVitrum Haematol., Mölndal, Sweden

Co-authors, with relevant name to be consulted in left column

A. Aarsland - Berge
N. Aarsæther - Berge
A. Abraha - Luzio
G. Adam - Bohley
J.M.F.G. Aerts - Schram
A.R. Al-Mutairy - Dormer
H.J. Ballard - Wilkes
J.A. Barranger - Schram
K. Bartlett - Osmundsen
H. Baum - O'Connell
I.K. Berezesky - Trump
R. Bjerkvig - Andersen (i)
E. Boghaert - Mareel
R.E. Bolton - Donaldson
P.N. Boyd - Price (i)
M. Bracke - Marreel

A.K. Campbell - Luzio
V. Carbonelle - Wattiaux
D. Chescoe - Hinton
J.R. Christensen - Smith
P. Coopman - Mareel
G.M. Cowell - Sjöström
E.M. Danielsen - Sjöström
R.A. Daw - Luzio
V. Dean - Rucklidge
H. Depauw - Hilderson
M. De Wolf - Hilderson
W. Dierick - Hilderson
E. Druet - Druet

W. Eskild - Berg (i)
G. Fox - Hinton
S.U. Friis - Sjöström

Forum Co-Organizer's Foreword

"In all organs of the body whose functions have been investigated by physiologists, it has been found that difference of function is invariably associated with a difference of structure, so that inter-dependence of function and structure has become an axiom. We are therefore justified in founding theories concerning the physiological function of an organ on a purely anatomical study of its structure, although the complete establishment of such theories must ultimately be afforded by physiological investigations" [statement by Starling as cited by G.C. Huber (1909/10) *Harvey Lects.* 5, 100-149].

With the increased use of animal models and tissue culture techniques in combination with the highly sensitive and advanced analytical methodology to study human diseases, it has long been felt that a subcellular approach to organelle changes was needed to reveal mechanisms leading to cellular derangements. The 10th International Subcellular Methodology Forum, 'Investigation of Cellular Derangements'* (Guildford, U.K.; 1-4 September 1986) which gave rise to this book was planned with no attempts made to be comprehensive. The Forum was planned to encourage the 'old' Forum tradition of vigorous debates made possible only by the presence of the very special array of cellular and subcellular experts attending.

It is believed that the book emphasizes the bridging role and also shows the applicability of the subcellular approach to the study of the processes involved in disease.

Medical Department A   KNUT-JAN ANDERSEN
Haukeland Sykehus
University of Bergen
Bergen, Norway
(during sabbatical stay with
Dr. J.K. McDonald - cf. #E-6)

---

* In view of dictionary definitions of 'derange' (e.g. 'throw into confusion or out of gear', 'to make insane'), mild misgivings about the aptness of the term 'Derangements' were expressed to Eric Reid (Chief Organizer), who later sought other opinions, notably from Peter Campbell, and who comments in the Senior Editor's Preface on the aptness of the different title chosen for this Forum-based book.

Section #A

INVESTIGATIVE APPROACHES TO ORGANELLE DISTURBANCES

#A-1

INVESTIGATION OF ISCHAEMIC EFFECTS ON SUBCELLULAR STRUCTURES BY CENTRIFUGATION METHODS

**Robert Wattiaux, Simone Wattiaux-De Coninck
and Véronique Carbonelle**

Laboratoire de Chimie Physiologique
Facultés Universitaires Notre-Dame de la Paix
61, rue de Bruxelles, B-5000 Namur, Belgium

Both differential centrifugation* and isopycnic centrifugation are both useful for investigating membrane pathology. Two abnormal situations may be manifest after differential centrifugation of a homogenate originating from a pathological organ: an increase in the % of marker enzymes recovered in the unsedimentable fraction or a change in distribution curve after incomplete sedimentation. On the other hand, the distribution after isopycnic centrifugation of organelles derived from a pathological organ can be perturbed in two main ways.- The pathological process can cause an increase of the particle permeability to the molecule used to make the gradient, resulting in a density change, or there may be a modification in the particle content causing an increase or decrease of its density. These general considerations are illustrated below by examples, originating from our laboratory and confined to lysosomes and mitochondria, on rat liver subjected to permanent or transitory ischaemia.

Amongst methods used to investigate alterations in subcellular membranes during a pathological process, centrifugation methods are of some help, either analytical or preparative and either differential (particle behaviour depending mainly on size) or isopycnic. To study pathological changes in membrane composition or certain physical properties, it is in general necessary to obtain purified organelles, commonly by preparative centrifugation. On the other hand, analytical centrifugation (which does not furnish purified organelles in bulk) can nevertheless give interesting information on the physicochemical

---

**Editor's note:* Where sedimentation is complete, the term 'differential pelleting' may be apt [Reid, E. (1972) *Subcell. Biochem. 1*, 217]. Some 'basics' in the MS. have been curtailed (consult standard sources).

properties of organelles and therefore on their alteration due to injury of some sort.

Two abnormal situations can arise after differential centrifugation of a homogenate originating from a pathological organ: an increase in % of an enzyme recovered in the unsedimentable fraction or a change in its distribution curve after incomplete sedimentation. The former may result from a true disruption of particles within the cells, caused by the pathological process. It may also originate from an alteration of the particle, which may not be sufficient to allow the enzyme to escape into the cytosol but makes the organelle more susceptible to homogenization and centrifugation procedures. Obviously the distinction between these two situations, both leading to increased unsedimentable enzyme, is very important; the former must be more deleterious to cell function. A change in the sedimentation coefficient distribution curve generally occurs if the size of the particle is modified, particularly if the particle swells – which can likewise lead to its disruption during homogenization.

In isopycnic centrifugation the distribution of organelles originating from the pathological organ can be perturbed as a result of two main modifications. The pathological process can cause an increase of the particle permeability particularly to the molecule used to make the gradient, with a resulting change in equilibrium density. Another possibility is a modification of the particle content causing its density to change, e.g. an accumulation of lipids in the particle would decrease its density.

We now illustrate these general considerations, for lysosomes and mitochondria only, by a few examples from work done in our own laboratory on rat liver subjected to permanent and transitory ischaemia. In brief, the experimental procedure consists in clamping the vascular pedicle of the left lobe of the liver with small forceps that are removed at selected times if re-establishment of the circulation is required. The animals are killed at various times after inducing ischaemia or after re-perfusion; the left lobe is excised and processed for differential or isopycnic centrifugation.

DIFFERENTIAL CENTRIFUGATION

Fig. 1 illustrates how β-galactosidase (a lysosomal hydrolase) distributes after differential centrifugation according to de Duve et al. [1]. Normally (top of Fig.) the hydrolase is recovered mainly in the pellet fractions designated **M** and **L** with a peak of relative specific activity (r.s.a.) in **L**. Little is recovered in the soluble fraction **S**. On the left is illustrated the distribution observed after 2 h of ischaemia and at different times after restoration of blood flow. What is striking is the increase, caused by ischaemia, in the proportion of enzyme located in **S**, indicating that a disruption

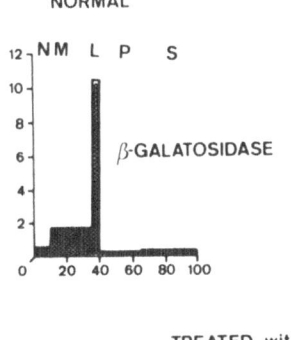

Fig. 1. Distributions of β-galactosidase after differential centrifugation, as amplified in the text. The times refer to re-perfusion, started after 2 h of ischaemia (pedicle ligature). The relative specific activity is the ratio

% of total recovered activity
―――――――――――――――――
% of total recovered proteins

and the abscissa is the relative protein content of fractions (cumulatively from left to right). The chlorpromazine-injected rats received, at the time of ligature, 2 mg/100 g body wt., s.c. *From ref. [3], by permission.*

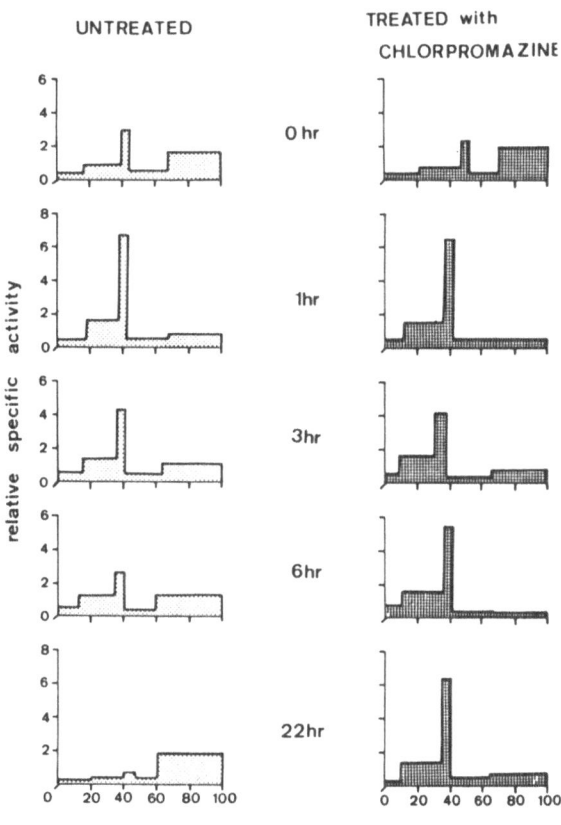

of lysosome membrane occurred *in vivo* or *in vitro*. Note that lysosomes which are not disrupted keep the same distribution pattern as normal lysosomes; therefore the disruption does not concern specifically 'big' or 'small' organelles. Some recovery is observed 1 h after re-perfusion. Later, a progressive loss of β-galactosidase from **M + L** takes place with a parallel increase in **S**. In similar experiments (right) where the rat had been treated with chlorpromazine, which according to Chien et al. [2] prevents cell death caused by ischaemia, the distribution patterns after 2 h are similar to those in the untreated rat. However, under these conditions the quasi-normal distribution re-attained 1 h after re-establishment of the blood flow persists thereafter.

The main conclusion that can be derived from these observations is that 2 h of ischaemia causes a release of lysosomal enzymes. Such a release does not originate from a disruption of the lysosomal membrane within the cell. Indeed, it is difficult to imagine by what mechanism the hydrolases present in the cytosol could be associated with lysosomes no more than 1 h after the re-flow of blood. A plausible explanation for the release of acid hydrolases observed just after ischaemia is that it results from a lysosomal membrane disruption during homogenization owing to increasing fragility of these organelles. However, for the secondary rise of unsedimentable β-galactosidase a true release within the cell cannot be excluded because the process is apparently irreversible. Chlorpromazine does not prevent the first release of hydrolase, but protects the lysosomes from irreversible alterations after restoration of blood flow [3]. As shown by Wattiaux-De Coninck et al. [4], differential centrifugation can also be used to investigate the perturbations of heterophagy caused by ischaemia; their differential centrifugation results suggest that the intracellular traffic of the pinocytosed protein is markedly impaired by ischaemia.

Enzyme latency.- Mention is warranted here of a method which is cognate to differential centrifugation in that it provides similar information on the state of the organelles, viz. assessment of the structure-linked latency of enzymes associated with subcellular structures. — The enzyme release is measured not after separation of sedimentable structures but directly on the homogenate or the granule preparation. Indeed, normally an enzyme that is present in intact granules has no access to external substrate and the reaction cannot take place, whereas free enzyme has access to substrate. Accordingly, under these conditions, the enzyme activity depends on the proportion of altered granules. In general, measured values are similar for the unsedimentable activity and for free activity.

ISOPYCNIC CENTRIFUGATION

When a subcellular structure is centrifuged in a density gradient, its size, shape and density can change during the centrifugation. Advantage can be taken of these changes to analyze some properties of the particles and to investigate whether these properties are modified as a result of a pathological process such as ischaemia.

Let us consider mitochondria. They are surrounded by two membranes. The outer membrane is freely permeable to molecules such as sucrose or metrizamide, whereas the inner membrane is impermeable to such

molecules.   When mitochondria migrate through a sucrose density-
gradient, the inter-membrane space is filled with sucrose solution
of increasing concentration while the matrix compartment shrinks by
losing water.   Therefore it can be expected that if ischaemia alters
the permeability of the inner membrane, it will affect the behaviour
of mitochondria in a density gradient.   The same considerations hold
for lysosomes.

In Fig. 2 we illustrate the distribution of cytochrome oxidase
(mitochondria) and cathepsin C (lysosomes) after isopycnic centrifugation in a Percoll gradient in 0.25 M sucrose medium.   The equilibrium density the particles attain in such a gradient   is the one
that they are endowed with in isoosmotic sucrose.   The experiments
were performed on mitochondrial fractions isolated at different times
after induction of ischaemia.   In the normal rat, mitochondria equilibrate   at a median density of around 1.10; 30 min after ischaemia
a second small peak becomes apparent in a lower density region.
The proportion of mitochondria recovered in that region increases
with the duration of ischaemia, so that after 2 h the cytochrome
oxidase distribution is totally shifted towards lower-density regions
of the gradient.

The effect of ischaemia on cathepsin C distribution is less
apparent.   However, as we have shown, some lysosomes are disrupted
in the homogenate of the ischaemic liver.   It is therefore possible
that most of the altered lysosomes escape analysis by gradient centrifugation because they have been disrupted during homogenization.

What can cause such a distribution change of mitochondria as
ascertained by cytochrome oxidase distribution? There are two possible
explanations: ischaemia either brings about an accumulation of low-
density substances in the organelles, or induces an increase in permeability of the inner membrane to sucrose.   Use of the model proposed
by Beaufay & Berthet [5] can help answer the question. Figs. 3 and 4
show how, according to these authors, the density of mitochondria
changes as a function of the molality of the sucrose solution.   An
increase of sucrose space (Fig. 3) causes a decrease in mitochondrial
density in 0.25 M sucrose.   The decrease is less pronounced when
the sucrose concentration of the medium increases, and indeed when
it is sufficiently high the converse is observed, the density of
mitochondria with a high sucrose space exceeding that with a low
sucrose space.   On the contrary (Fig. 4), the density decrease caused
by an accumulation of low-density substances in the organelles is
greater at high than at low sucrose concentration.

Experimentally, the influence of the sucrose concentration of
the medium on mitochondrial density can be assessed by subjecting
the granules to isopycnic centrifugation in a gradient of a macromolecular substance, glycogen or Percoll, in sucrose medium, and
alternatively in a sucrose gradient.   The equilibrium density that
[ctd. on p. 10

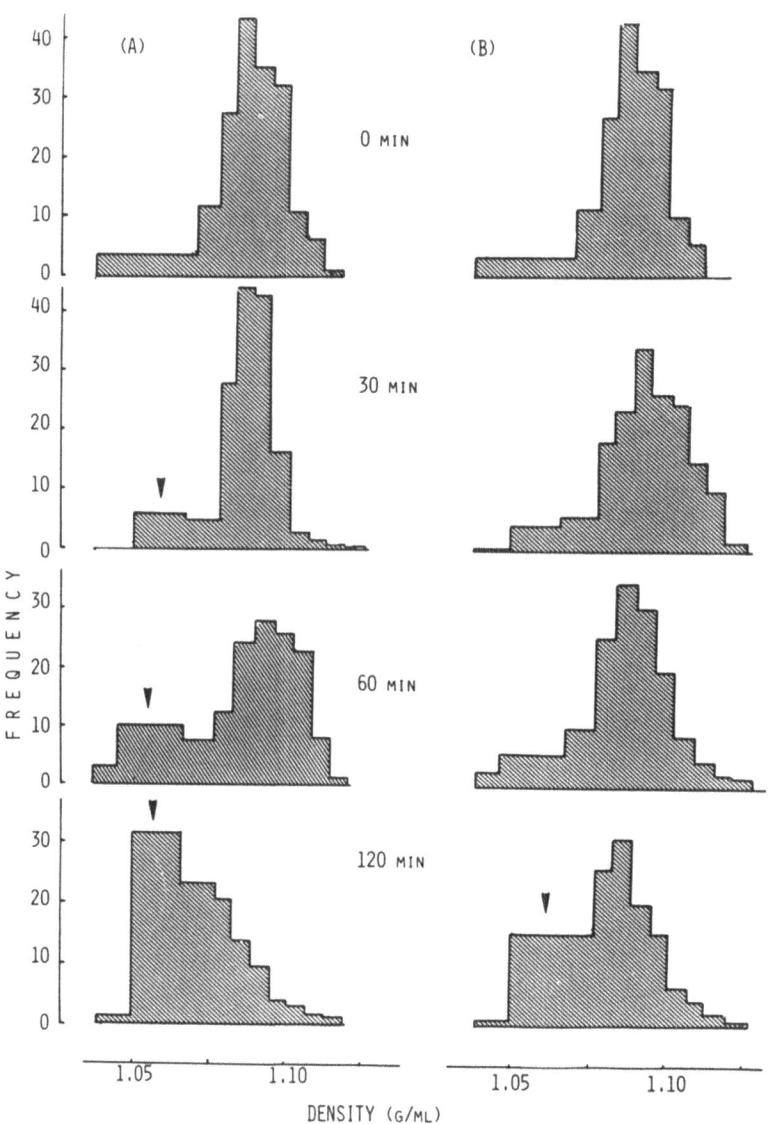

Fig. 2. Distribution of cytochrome oxidase (A) and cathepsin C (B) after isopycnic centrifugation of liver mitochondrial (M + L) fractions in 0.25 M sucrose with Percoll as gradient material. Lobe ligature to produce ischaemic was continued for the time indicated. The time interval of the square angular velocity was 24.6 rad²/n-sec.

Ordinate ('Frequency'): $Q/\Sigma Q.\Delta\rho$ where Q represents the activity found in the fraction, ΣQ the total recovered activity, and $\Delta\rho$ the increment of density from top to bottom of the fraction.
The short *arrows* illustrate the density shift caused by ischaemia.

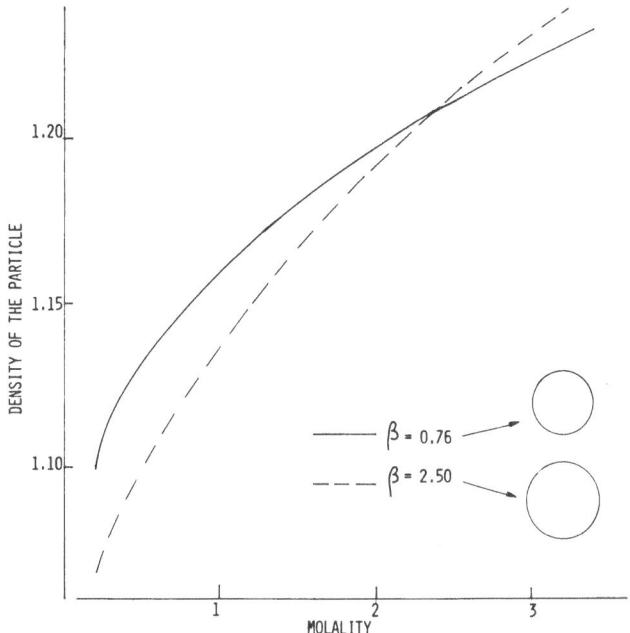

Figs. 3 *(above)* **and 4** *(below).*   Density of rat-liver mitochondria as a function of the molality of the sucrose solution.   The curves were drawn by using the relationship of Beaufay & Berthet [5], supposing (Fig. **3**) a β (sucrose space) of 0.76 or 2.5, **or** (Fig. **4**) a ρ_d (matrix density) of 1.22 or 1.12 g/cm^3 without change in size. The circles in Fig. 3 indicate the particle size difference for β = 0.77 *vs.* 2.5.

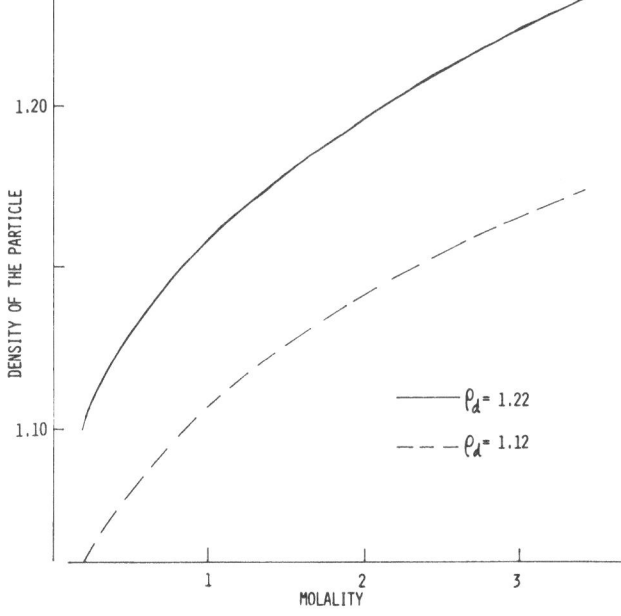

the particles exhibit in a Percoll or glycogen gradient is the density it is endowed with in the sucrose medium, since the macromolecular component does not affect the particle density. By repeated centri-fugation in various sucrose solutions it is possible to infer how the density of the particles changes as a function of the sucrose concentration of the medium. On the other hand, the maximal equilibrium density of the organelle is that which the particle assumes in a sucrose gradient. Fig. 5 allows us to compare the cytochrome oxidase distributions in a Percoll gradient, containing 0.25 or 0.5 M sucrose, and in a sucrose gradient. As expected, the equilibrium density of the particles increases with the sucrose concentration of the medium and is maximal in a sucrose gradient.

The effect of ischaemia on the granule density strikingly depends on the sucrose distribution of the medium. It is less pronounced in 0.5 M than in 0.25 M sucrose. In a sucrose gradient, an increase

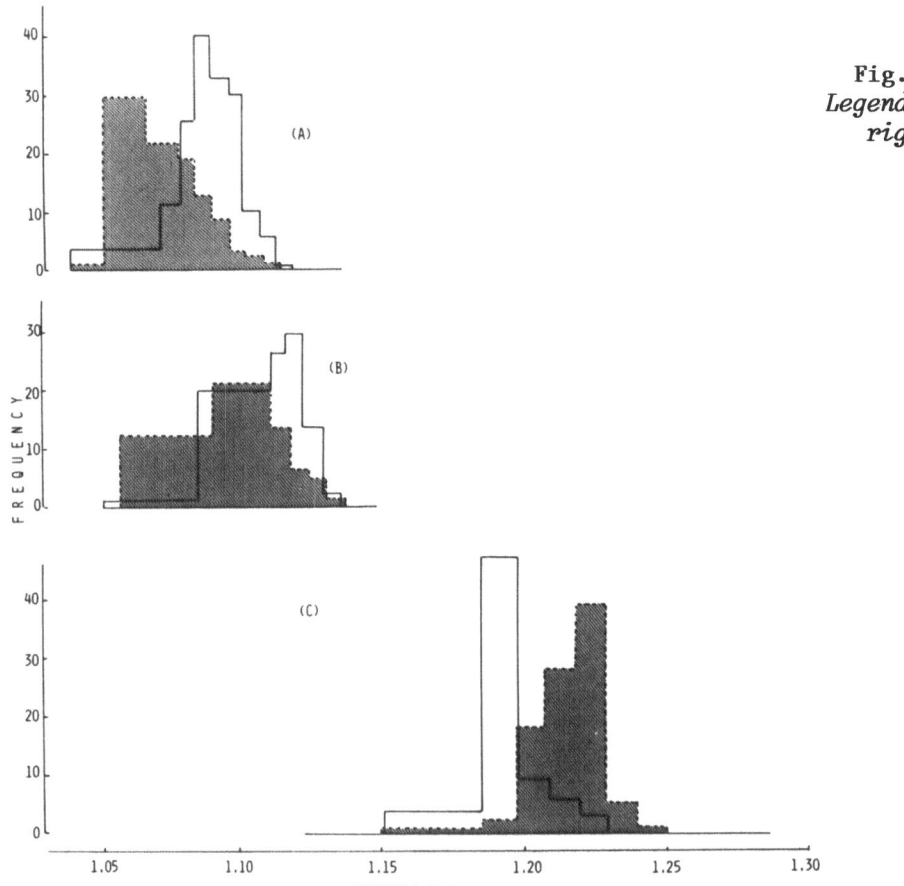

Fig. 5.
Legend on
right.

in equilibrium density of the organelles is observed when they originate
from an ischaemic liver. Such an observation strengthens the hypothesis
that ischaemia causes an increase in the permeability of the inner
membrane to sucrose.

Obviously it is also possible to investigate by isopycnic centrifugation the fate of the structures when the circulation is re-
established. This is exemplified in Fig. 6. We show again the distribution of cytochrome oxidase observed 2 h after ischaemia and in addition
the distribution seen 1 h and 22 h after re-perfusion when animals
have been injected with nembutal. As previously shown [6], nembutal
treatment prevents cell death caused by ischaemia. The treatment
does not prevent the cytochrome oxidase distribution change observed

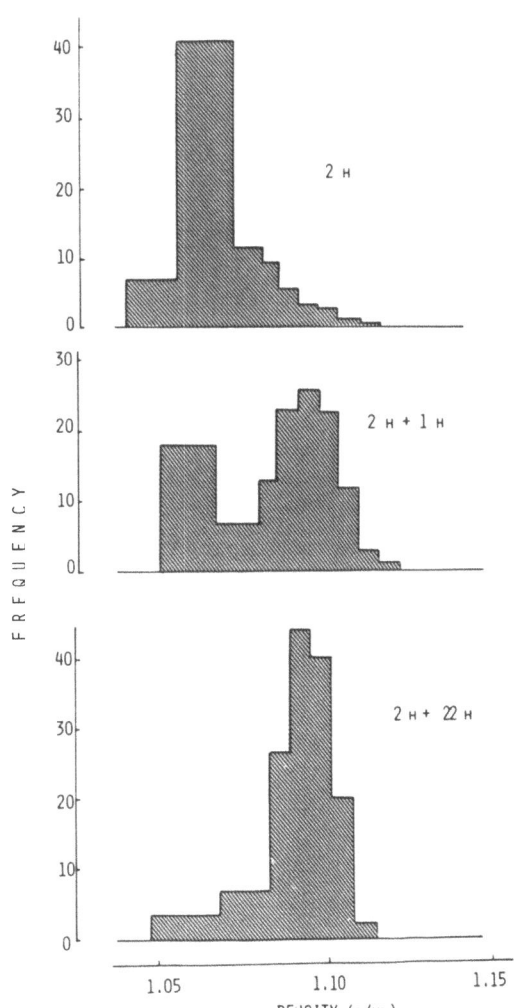

Figs. 5 *(left)* and 6 *(right)*.
Distribution of cytochrome
oxidase after isopycnic
centrifugation of rat-liver
mitochondrial fractions (as
used in the experiments of
Fig. 2; see its legend for
explanation of the graphs)
in density gradients con-
taining sucrose. The liver
lobes had been subjected to
ischaemia for 2 h.

Fig. 5: Percoll gradients
with 0.25 M (A) or 0.5 M (B)
sucrose, **or** (C) a sucrose
gradient. Time interval of
the square angular velocity:
for Percoll and sucrose
gradients, respectively
24.6 and 144 rad^2/n-sec.

Fig. 6: Percoll gradient
with 0.25 M sucrose. The
ischaemic lobes had been re-
perfused for the different
times stated in the Figure.
Initial nembutal treatment:
5 mg/100 g body wt., s.c.

after 2 h of ischaemia. However, when blood is re-supplied there is progressive re-attainment of the normal distribution. Such an observation suggests that the mitochondrial membrane's change in permeability, induced by 2 h of ischaemia, is a reversible phenomenon and, accordingly, cannot be blamed for the cell death resulting from blood deprivation.

CONCLUDING COMMENT

Examples presented in this article show that centrifugation experiments can help elucidate the effect on the cell of a pathological condition such as ischaemia. We think that deeper insight into the behaviour of organelles in pathological cells could be gained if centrifugation methods that we have mentioned could be more frequently applied to the study of physicochemical and functional modifications of subcellular structures.

References

1. de Duve, C., Pressman, B.C., Gianetto, R., Wattiaux, R. & Appelmans, F. (1955) *Biochem. J. 63*, 604-617.
2. Chien, K.R., Abrams, J., Pfau, R.G. & Farber, J.L. (1977) *Am. J. Path. 88*, 539-558.
3. Wattiaux, R. & Wattiaux-De Coninck, S. (1981) *Biochem. J. 196*, 861-866.
4. Wattiaux-De Coninck, S., Dubois, F. & Wattiaux, R. (1982) in *Protection of Tissues against Hypoxia* (Wauquier, A., Borgers, M. & Amery, W.K., eds.), Elsevier, Amsterdam, pp. 139-143.
5. Beaufay, H. & Berthet, J. (1963) *Biochem. Soc. Symp. 23*, 66-85.
6. Wattiaux, R. & Wattiaux-De Coninck, S. (1984) *Int. Rev. Exp. Path. 26*, 85-106.

#A-2

INVESTIGATION OF SUBCELLULAR DERANGEMENTS
IN HUMAN BIOPSY MATERIAL

T.J. Peters

Division of Clinical Cell Biology
MRC Clinical Research Centre
Harrow, Middlesex HA1 3UJ, U.K.

Tissue biopsy samples have routinely been used for diagnostic purposes for nearly a century. More recently, such material has been employed for biochemical investigations. Studies of organelle abnormalities can be quantitatively assessed by marker enzyme analysis (e.g. N-acetyl-β-glucosaminidase and acid phosphatase for lysosomes, lactate dehydrogenase for cytosol) in combination with subcellular fractionation by density-gradient centrifugation. Thereby hitherto unassigned enzymes etc. can be localized.

Functional studies of biopsy fragments can be performed with assessment of protein and enzyme synthesis by cultured tissue fragments. Recent studies have developed micro-techniques for investigating lipid synthesis and metabolism. Similarly, in vitro studies of intestinal fragments can be used to measure the absorptive capacity for various nutrients or for assessing mucosal permeability. Selective application of this approach to a variety of human tissue disorders is discussed.

Since the work of Virchow, histopathological examination of biopsy samples of human organs has been performed increasingly. Currently this represents the major definitive technique used in the investigation of human disease, and in many instances has even made post-mortem examination obsolete. The material is generally placed in fixatives and examined by a variety of morphological techniques at both the light- and the electron-microscopic level. Fixation precludes much biochemical examination of the tissue; but increasingly techniques of cell and molecular biology, as well as more classical biochemical procedures, are applicable to these mg-sized biopsy samples. This review discusses some of the techniques which we have developed to investigate morphological and functional disorders in tissue biopsy samples. Because this review is concerned with techniques and methodology rather than the investigation of a particular disorder, the breadth of topics covered is wide.

SUBCELLULAR FRACTIONATION AND ENZYMIC ANALYSIS OF INFLAMMATORY BOWEL

The equipment and methods used in these studies have been reviewed elsewhere [1].* More recently, these techniques have been applied to the study of inflammatory disorders of the large bowel. The major non-malignant disorders that afflict this organ in Western society are ulcerative colitis - a relapsing inflammatory disease confined entirely to the large bowel - and Crohn's disease, a condition of increasing incidence which patchily may affect the whole gastro-intestinal (G-I) tract, most typically the terminal ileum, but which may cause segmental inflammation of the large bowel. In contrast to ulcerative colitis, Crohn's disease may respond only poorly to a variety of treatments and frequently recurs after apparently successful excision of the affected region of the bowel. Distinction between these disorders is thus important for therapeutic and prognostic reasons. In many cases histological examination of the colonic/rectal biopsies may indicate the nature of the disease. In ulcerative colitis the predominant inflammatory cell is the poly-morphonuclear leucocyte whereas in Crohn's disease chronic inflamma-tory cells such as monocytes and lymphocytes are present in large numbers. However, the lesions may be patchy and in many cases the distinction between the two disorders is not clear.

Morphological examination is subjective, and in an attempt to quantitate the nature of the inflammation, biochemical markers of acute and chronic inflammatory cells were assayed in biopsies from patients with the two diseases and control subjects together with rectal biopsies from patients with Crohn's disease apparently not affecting the rectum. Increased activities of three inflammatory cell markers compared to controls were found in both forms of colitis (Table 1). The results did not distinguish between the two forms of active disease, and in quiescent disease the activities were within the control range. It was concluded that biochemical markers for phagocytic cells were not particularly useful in the differential diagnosis of inflammatory bowel disease, but rather point to a broad overlap between the two conditions [3].

Possible involvement of lysosomes or other organelles

Morphological studies on animals with caragheenin-induced experi-mental ulcerative colitis indicate lysosomal changes in epithelial cells of the large bowel together with accumulation of foreign material

---

*Noted by Ed.: homogenate (Dounce) in isotonic sucrose/EDTA/ethanol to stabilize catalase; isopycnic centrifugation in mini-size zonal rotor (H. Beaufay's design); Triton-X-100 to liberate latent activity.

Table 1. Inflammatory cell marker activities in rectal mucosa from patients with inflammatory bowel disease and controls. From ref.[2]. In this and other Tables, the values are mean ± S.E. for (n) biopsies assayed; statistical analysis by Student's t test: a, $p < 0.05$; b, $p < 0.01$.

Sample type	*Neutrophils* Vitamin B_{12} binding capacity, pg/mg protein	*Neutrophils* Myeloperoxidase, mU/mg protein	*Myelo-monocytes* Lysozyme, µg/mg protein
Controls (15)	88 ±40	42.9 ±5.1	0.66 ±0.10
Ulcerative colitis:			
active (11)	336 ±115[b]	75.6 ±16.1[b]	2.5 ±1.2[b]
quiescent (15)	120 ±40	50.2 ±6.2	0.35 ±0.15
Crohn's colitis (16)	217 ±45[b]	68.5 ±12.1[a]	2.7 ±1.4[b]

in lysosomes [4]. It is thus possible that this form of inflammatory bowel disease in man is due to lysosomal involvement. Organelle markers were therefore assayed in the biopsy homogenates, and subcellular distribution studies undertaken by sucrose density-gradient centrifugation [5]. Table 2 shows these enzyme activities in the rectal biopsy homogenates. Both lysosomal enzymes and β-glucuronidase (not shown) show a significant decrease in activity in the biopsies from patients with ulcerative colitis, both acute and, less markedly, in remission. Activities in Crohn's disease are normal. Lactate and malate dehydrogenase and N α-glucosidase are all within the control range for all disease groups (Table 3).

Assays for the latent *N*-acetyl-β-glucosaminidase activity, expressed as % of total homogenate activity (mean ±S.E.; n observations), show similar values for control and colitic patient tissue: 50.4 ±5.5 (8) and 50.4 ±6.1 (8), respectively. Subcellular fractionation by sucrose density-gradient centrifugation (Fig. 1) confirmed that the proportion of soluble and particulate (lysosomal) activity in acute ulcerative colitis was similar to that in control tissue although the total activity (mU/mg protein) was reduced to half of

Fig. 1. Subcellular fractionation of rectal mucosa from control subjects (———) and patients with acute ulcerative colitis (▬▬▬): mean distribution of lysosomal marker enzyme. Activity at low d's is due to soluble (non-latent) enzyme. ['Frequency': fractional activity ÷ fractional density span (g/ml)]. From ref. [5], by permission.

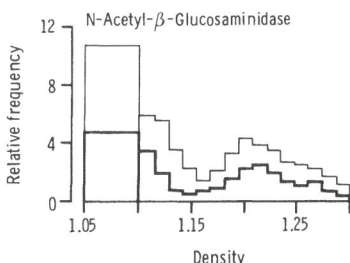

Table 2. Organelle marker enzyme activities in rectal mucosa from patients with inflammatory bowel disease and controls. The values (see Table 1 heading) are mU/mg protein. Data from ref. [6].

Sample type	N-Acetyl-β-glucosaminidase (lysosomes)	Acid phosphatase (lysosomes)	Lactate dehydrogenase (cytosol)
Controls (24)	7.00 ±0.84	11.1 ±1.53	74 ±12
Ulcerative colitis:			
active (16)	3.52 ±0.78[a]	4.20 ±1.20[a]	78 ±13
quiescent (24)	4.80 ±0.84[a]	7.32 ±1.20	86 ±12
Crohn's colitis (22)	6.56 ±0.84	9.45 ±1.86	85 ±16

Table 3. As for Table 2. Data from refs. [2, 6].

Sample type	5'-Nucleotidase (plasma membrane)	N α-Glucosidase (endoplasmic reticulum)[⊗]	Malate dehydrogenase (mitochondria/cytosol)
Controls (22)	7.29 ±0.99	0.55 ±0.01	919 ±76
Ulcerative colitis:			
active (29)	8.70 ±1.48	0.45 ±0.20	606 ±88
quiescent (21)	8.48 ±0.99	0.48 ±0.11	682 ±98
Crohn's colitis (21)	15.4 ±2.71[a]	0.73 ±0.09	792 ±206

[a] $p < 0.05$ (both Tables)   [⊗] Prefix N denotes 'neutral'.

control values. Reduced activity of this enzyme in rectal biopsies from patients with acute colitis (nature unspecified) has been reported previously [7-9].

The significance of these lysosomal changes is not clear. Increased activities of certain lysosomal enzymes are seen in conditions associated with intra-lysosomal accumulation of undegradable material. If this is accompanied by enhanced lysosomal fragility, cell damage occurs; this is clearly not the situation in ulcerative colitis and thus the carragheenin model, although showing some similarities to the human disorder, clearly has a different pathogenesis and the significance of the lysosomal enzyme deficiency in ulcerative colitis remains to be determined.

INTESTINAL PERMEABILITY IN INFLAMMATORY BOWEL DISEASE

An important question is whether Crohn's disease, unlike ulcerative colitis, affects the small as well as the large bowel even

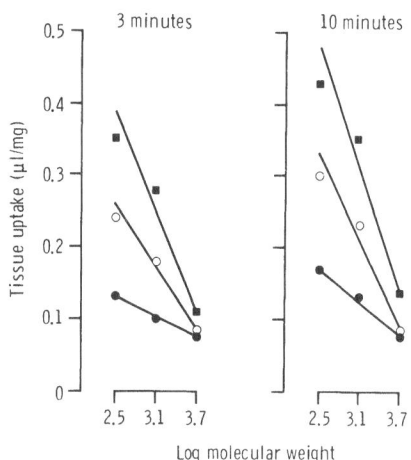

Fig. 2. *In vitro* determination of intestinal permeability: tissue uptake mean plotted against log mol. wt. of the 3 probes. Control subjects, ●; coeliac disease in remission, ○; coeliac disease in relapse, ■. From ref. [12], by permission.

without apparent clinical involvement of small bowel. In addition, the underlying cause of this and other small-bowel disorders remains in dispute. A popular hypothesis suggests that many G-I diseases represent an abnormal inflammatory response to certain absorbed antigens, e.g. gluten in coeliac disease, bacterial products in Crohn's disease. It was therefore important to determine the permeability of the small-bowel mucosa. Previous methods for determining intestinal permeability in man involved oral administration of a range of poly-ethylene glycols (PEG's) or various combinations of non-metabolizable sugars, e.g. lactulose or mannitol, and measuring their urinary excretion [10, 11]. Results and conclusions from studies with these probes were conflicting, and it was claimed that in small-bowel disorders permeability was increased to large probes but decreased to small - a paradoxical conclusion. Moreover, there was conflict as to whether abnormalities in permeability returned to normal following successful treatment of the gut disease.

In order to resolve this paradox, *in vitro* methods were established to determine the integrity of the mucosal permeability barrier in small-bowel biopsy specimens. Use of an *in vitro* technique will avoid such variables as gastric emptying, G-I transit, mucosal blood flow, blood volume and renal excretion of absorbed probes [11]. Biopsy fragments, 1-5 mg wet wt., were incubated in oxygenated physiological buffer containing three radio-labelled permeability probes, selected to cover a range of mol. wts. In addition, probes were selected to which the mucosa was normally highly permeable so that small increases in permeability could readily be detected. Procedural details and results in normal subjects and patients are given elsewhere [12].

Fig. 2 compares the apparent permeability of normal and coeliac mucosa both in relapse and in remission. In all patient groups permea-bility is inversely proportional to log mol. wt., and permeability

is clearly increased to a greater degree for the small than for the
large probes.   These results thus accord with recognized physical
principles.   There is an apparently paradoxical result related to
differences in uptake of the various probes by normal mucosa.   Thus
the small-sized probes, e.g. mannitol, rhamnose and PEG 300, are
significantly absorbed by normal mucosa (up to 20%), whereas the
larger probes, e.g. lactulose and PEG 600, resemble the probes used
in the present study by being only slightly absorbed (<5%). In contrast, damaged mucosa showed increased permeability to all probes,   but
for small probes the changes in mucosal architecture leads to a major
decrease in surface area and thus uptake is reduced to a greater
extent than would be expected from the mucosal damage.   The resolution
of the paradox is discussed in detail elsewhere [10].

     The *in vitro* studies, besides rationalizing the previous results,
have led to the introduction of ^{51}Cr-EDTA as the definitive method
for assessing small-bowel permeability *in vivo* in man.   This probe
has been used widely in a variety of disorders [10, 11, 13-15] and
clearly exemplifies the value of *in vitro* studies in man in the development of clinically valuable *in vivo* methods.

IN VITRO METHOD FOR ASSESSING SMALL-BOWEL MUCOSAL FUNCTION

     Current techniques for assessing absorption of various nutrients
in man involves lengthy, complex and expensive tests on actual patients.
Methods for assessing nutrient absorption in man are as follows.-
(1) Balance studies.
(2) Isotope studies - whole-body retention; plasma studies; urinary
and faecal excretion.
(3) Intestinal perfusion studies.
(4) *In vitro* techniques.

     (1) Balance studies involve prolonged in-patient collections
with accurate assessment of dietary intake of the appropriate nutrient.
They do not provide detailed information on the site, mechanism or
kinetics of intestinal absorption.   In addition, although if carefully
performed they give an accurate measure of overall absorption, this
approach does not correct for loss of the nutrient into the gut lumen,
e.g. biliary and pancreatic secretions, or desquamated epithelial
cells.   Besides, it is an expensive, time-consuming and (for the
biochemist) unpleasant procedure.   To overcome some of these difficulties, particularly to avoid careful stool collections and prolonged
in-patient stays, other techniques have been developed, usually
involving radioisotopes.

     (2) If a whole-body counter is available, retention techniques can
be used, suitable for studying absorption of γ-emitting isotopes.
The techniques are accurate and relatively convenient, although serial
studies are very time-consuming.   It is not possible to assess

absorption of such nutrients as carbohydrates, fats and proteins by this technique. In addition, this approach provides limited kinetic information on the site or nature of the absorptive process. An alternative approach is to administer the isotope or test substance orally and assay the plasma, urine or faecal levels at various time periods. Such techniques are usually more convenient and often can be conducted on an out-patient basis. Some are reasonably reliable, but additional problems of assaying plasma levels of an orally adminis-tered substance include influences of G-I transit, plasma distribution volumes, tissue uptake and renal function.

(3) In order to assess intestinal absorption directly, perfusion techniques have been introduced in which a segment of intestine, isolated by intra-luminal balloons, is perfused with varying concentra-tions of the nutrient. A non-absorbable marker is used to assess loss from, or leakage into, the perfused segment. This approach, which is almost exclusively a research procedure, is very demanding to both the patient and the experimentalist. Information can be obtained on the kinetics of nutrient absorption, i.e. K_m and v_{max} can be calculated. However, absorption is assessed by measuring the decrease in nutrient concentration along the perfused segment, and at low perfusate concentrations it is difficult to measure absorption rates accurately. Absorption is also very dependent on perfusate flow rates, and the isolating balloons cause a varying degree of obstruction with consequent disturbance of intestinal function.

(4) In order to overcome these many difficulties, we have developed an *in vitro* technique for assessing intestinal absorption. A fragment of jejunal mucosa, collected by a simple biopsy technique, is incubated in oxygenated medium for up to 10 min in the presence of the labelled nutrient and an extracellular fluid marker. The tissue fragments are removed, briefly washed in nutrient medium and counted for radio-activity. Net uptake is determined by correcting for adherent medium with extracellular fluid markers [16]. Several different nutrients including Ca^{2+}, Fe^{2+}, Fe^{3+}, P_i and 3-O-methylglucose have been studied by this technique. The technique is clearly suited for a variety of substances including amino acids, sugars, vitamins, drugs and ions [17-20].

Application of the approach to studying calcium absorption

Ca^{2+} absorption has been investigated in normal subjects and patients with renal stones. Fig. 3 shows diagrammatically the technique used to study $^{45}Ca^{2+}$ absorption *in vitro*. The apparatus is simply constructed and is readily available. Tissue uptake is linear with respect to time and is not energy-requiring. It is, however, effectively and specifically inhibited by ruthenium red, a selective reagent for calcium pores in the plasma membrane.

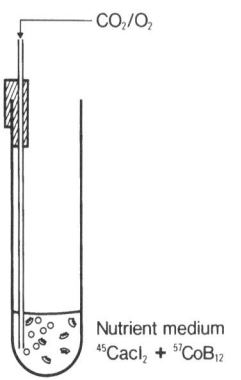

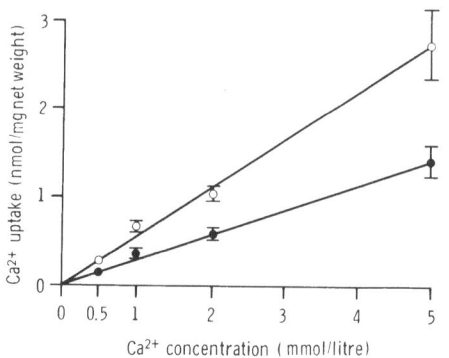

Fig. 3. *In vitro* measurement of calcium absorption: duodenal fragments in oxygenated medium containing $^{45}Ca^{2+}$ with ^{57}Co-vitamin B_{12} as an extracellular fluid marker.

Fig. 4. Mean (±S.E.) uptake of $^{45}Ca^{2+}$ by duodenal mucosa, corrected for adherent medium with ^{57}Co-vitamin B_{12}. Control subjects, •; patients with idiopathic hypercalciuria, o. From ref. [21].

Fig. 4 shows the *in vitro* uptake of $^{45}Ca^{2+}$ by biopsies from control subjects and patients with idiopathic hypercalciuria. In both groups the uptake is linearly related to the Ca^{2+} concentration in the medium. However, uptake by the hypercalciuric patients is significantly greater than controls at all medium concentrations. Patients having renal calculi without hypercalciuria show similar uptake to controls [21]. It is suggested that these results reflect increased selective permeability of the brush-border membrane to Ca^{2+}, possibly due to increased plasma levels of, or tissue sensitivity to, 1,25-dihydroxy-vitamin D_3. Subcellular fractionation techniques are now being applied to follow the labelled $^{45}Ca^{2+}$ across the cell and, in particular, to investigate the transfer across the isolated brush-border membrane.

IN VITRO METHODS FOR ASSESSING BIOPSY FUNCTION

Modern clinical investigational procedures have made it possible to biopsy safely virtually all organs and tissues. Although primarily intended for diagnostic purposes, portions of the tissue can be readily used for all types of biochemical, cellular and molecular analyses with the obvious restrictions associated with the limited amount of tissue available, heterogeneity of human samples, non-representative sampling, and difficulties in obtaining comparable control tissue. Using intestinal biopsies in culture [22], measurements can be made of protein [23], DNA [24] and specific enzyme synthesis [25], and important differences have been reported in various intestinal disorders. Methods have recently been developed for investigating disorders of fatty acid metabolism in patients with fatty liver, particularly that due to alcohol abuse.

Fig. 5. Lipogenesis by liver biopsies. Statistical analysis: controls *vs*. mild and moderate alcoholic fatty liver: $p < 0.001$; controls *vs*. diabetic fatty liver: $p > 0.01$; mild *vs*. moderate alcoholic fatty liver: $p > 0.05$. From **ref**. [27]. *Courtesy of Oxford University Press.*

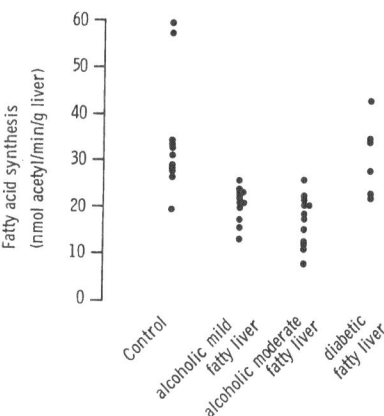

Lipid metabolism in alcoholic fatty liver

The nature and subcellular localization of the accumulated lipid is discussed in a companion article (#A-3), but two aspects of fatty liver are considered here. The first is the evaluation of the widely held view [26] that alcoholic fatty liver is due to enhanced lipogenesis consequent upon the increased redox status following hepatic ethanol oxidation. Using incorporation of ^3H from ^3H_2O as a measure of absolute lipogenesis rates, fragments of needle biopsy specimens were incubated with ^3H_2O in Krebs-bicarbonate buffer containing glucose and amino acids. After 2 h the lipid fraction of the tissue was isolated, saponified and the free fatty acids assayed for radioactivity [27]. Lipogenesis rates were calculated according to Jungas [28], correcting for the tritium isotope effect. Lipogenesis was linear over a 4 h incubation period, and was clearly reduced (Fig. 5) in patients with fatty liver. Addition of 50 mmol/l ethanol to the incubation medium did not affect the lipogenesis rates in either control or fatty liver tissue. It was therefore concluded that lipogenesis rates were reduced – not increased – in fatty liver. Recent studies with animals fed an alcohol-rich diet showed that lipogenesis rates decreased *after* the accumulation of triglyceride (TG) had occurred [29].

The second aspect relates to the evident need to seek other explanations for the accumulation of TG. The metabolism of palmitate by human biopsies was studied. Liver biopsy fragments were incubated with [1-^{14}C]palmitate in physiological buffer, and CO_2, ketone body and TG formation were measured [30]. The results are summarized in Table 4. Overall metabolism was similar in all patient groups, but the relative contribution of the three products differed. All groups were similar in respect of ketone body formation, which comprised

T.J. Peters [#A-2

Table 4. Palmitic acid metabolism by human liver biopsies. The values are mean ±S.E. pmol metabolized/h/mg wet wt. for (n) biopsies assayed; results calculated from data in ref. [30]. For patient group compared with controls: a, $p < 0.05$; b, $p < 0.02$; c, $p < 0.01$; d, $p < 0.0025$.

	Controls (7)	Fatty liver mild (15)	Fatty liver severe (10)
Ketone body production	440 ±123	375 ±61	380 ±74
Esterification to TG	420 ±81	570 ±106	580 ±83[a]
CO_2 production	88.5 ±17.5	60.0 ±10.9[b]	53.1 ±13.3[d]
Total metabolism	949 ±221	1010 ±180[c]	1010 ±170[c]

~40% of the metabolized palmitate. This is an important observation, as it clearly confirms our previous reports that mitochondrial function is not, contrary to popular belief, impaired in alcoholic fatty liver [31]. Esterification of palmitate into TG was increased in the fatty liver biopsies, significantly so in the biopsies with severe fatty change. In contrast, $^{14}CO_2$ production, although responsible for only 5-10% of the overall palmitate metabolism, is progressively reduced in fatty liver of increasing severity.

The cause of this impaired CO_2 production is not clear. It has been observed previously [32], and postulated to be a direct inhibitory effect of ethanol on Krebs cycle activity [33]. The present studies, where the incubations were carried out in the absence of ethanol, clearly show that impaired TCA cycle activity still occurs. Similarly impaired $^{14}CO_2$ production from palmitate occurs even after one month's abstention in patients with alcoholic fatty liver. Addition of malate increases $^{14}CO_2$ production in fatty liver biopsies, whereas a small reduction occurs with control tissue [30]. It is therefore suggested that there is a functional disorder of TCA activity probably due to a deficiency of a cycle intermediate.

These studies clearly highlight the importance of human, rather than experimental animal, studies in investigating the pathogenesis of alcoholic liver disease. Results to date, including studies discussed elsewhere [34], suggest that the initial accumulation of TG in alcoholic fatty liver occurs in the Golgi and that there is a small but significant shift in free fatty acid metabolism from oxidation to esterification to yield TG. However, the mechanisms of these changes remain to be determined. Contrary to popular belief, lipogenesis is impaired probably as a consequence of the TG accumulation.

References

1. Peters, T.J. (1981) *J. Clin.Path. 34*, 1-12.
2. O'Morain, C., Smethurst, P., Levi, A.J. & Peters, T.J. (1983) *J. Clin. Path. 36*, 1312-1316.
3. Krawisz, J.E., Sharon, P. & Stenson, W.F. (1984) *Gastroenterology 87, 1344-1350.*
4. Abraham, R., Fabian, R.J., Goldberg, L. & Coulston, F. (1974) *Gastroenterology 67*, 1169-1181.
5. O'Morain, C., Smethurst, P., Levi, A.J. & Peters, T.J. (1985) *Scand. J. Gastroenterology 20*, 209-214.
6. O'Morain, C., Smethurst, P., Levi, A.J. & Peters, T.J. (1984) *Gut 25*, 455-459.
7. Danovitch, S.H., Gallucci, A. & Shora, W. (1972) *Am. J. Digest. Dis. 17*, 977-992.
8. Binder, V., Soltoft, J. & Gudmand-Hoyer, E. (1974) *Scand. J. Gastroenterology 9*, 293-297.
9. Kane, S.P. & Vincent, A.C. (1979) *Clin. Sci. 57*, 295-303.
10. Peters, T.J. & Bjarnason, I. (1985) in *Food and the Gut* (Hunter, J.O. & Alun Jones, V., eds.), Baillière Tindall, London, pp. 30-44.
11. Bjarnason, I., Peters, T.J. & Levi, A.J. (1986) *Digest. Dis. 4*, 83-92.
12. Bjarnason, I. & Peters, T.J. (1984) *Gut 25*, 145-150.
13. Bjarnason, I., O'Morain, C., Levi, A.J. & Peters, T.J. (1983) *Gastroenterology 85*, 318-322.
14. O'Morain, C., Peters, T.J. & Veall, N. (1983) *Lancet i*, 323-325.
15. Bjarnason, I., Ward, K. & Peters, T.J. (1984) *Lancet i*, 179-182.
16. Cox, T.M. & Peters, T.J. (1979) *J. Physiol. 289*, 469-478.
17. Cox, T.M. & Peters, T.J. (1978) *Lancet i*, 123-124.
18. Duncombe, V.M., Watts, R.W.E. & Peters, T.J. (1980) *Lancet ii*, 1334-1336.
19. Scott, J. & Peters, T.J. (1983) *Am. J. Physiol. 244*, G532-540.
20. Hamilton, J.W., Li, B., Shug, A.L. & Olsen, W.A. (1986) *Gastroenterology 91*, 10-16.
21. Duncombe, V.M., Watts, R.W.E. & Peters, T.J. (1984) *Quart. J. Med. 54*, 69-79.
22. L'Hirondel, C., Doe, W.F. & Peters, T.J. (1976) *Clin. Sci. Mol. Med. 50*, 425-429.
23. Jones, P.E., L'Hirondel, C. & Peters, T.J. (1981) *Gut 22*, 623-627.
24. Jones, P.E. & Peters, T.J. (1977) *Br. Med. J. i*, 1130-1134.
25. Jones, P.E., L'Hirondel, C. & Peters, T.J. (1982) *Gut 23*, 108-114.
26. Lieber, C.S. (1985) *Acta Med. Scand. (Suppl.) 703*, 11-55.
27. Venkatesan, S., Leung, N.W.Y. & Peters, T.J. (1986) *Clin. Sci. 71*, 723-728.
28. Jungas, R.L. (1968) *Biochemistry 7*, 3708-3717.
29. Venkatesan, S., Ward, R.J. & Peters, T.J. (1987) *Clin. Sci.*, in press.

30. Leung, N.W.Y. & Peters, T.J. (1986) *Clin. Sci. 71*, 253-260.
31. Jenkins, W.J. & Peters, T.J. (1978) *Gut 19*, 341-344.
32. Fischel, P. & Oette, K. (1974) *Res. Exp. Med. 163*, 1-16.
33. Grunnet, N., Kondrup, J. & Dich, J. (1985) *Biochem. J. 228*, 673-681.
34. Peters, T.J. & Cairns, S.R. (1985) *Alcohol 2*, 447-451.

#A-3

ANALYTICAL SUBCELLULAR STUDIES IN ALCOHOLIC LIVER DISEASE

T.J. Peters

MRC Clinical Research Centre
Harrow, Middlesex HA1 3UJ, U.K.

Morphological studies have disclosed a variety of organelle abnormalities in liver biopsy samples from patients with alcoholic liver disease. These include mitochondrial, lysosomal, peroxisomal and e.r.* changes; but whether these alterations are adaptive or pathological remains to be determined. In addition, lipid accumulation is a major feature of alcohol toxicity; but the nature and, in particular, the subcellular localization of the lipid and the mechanism of its accumulation are ill-understood.

Analytical subcellular fractionation, in conjunction with enzymic microanalysis, can be used to investigate the significance of the organelle changes. For these studies sucrose density-gradient centrifugation in the Beaufay automatic zonal rotor is ideal. However, this rotor is unsatisfactory for studying the localization of lipid. Differential flotation methods have been developed for isolating macro- and micro-lipid droplets from liver samples of needle-biopsy size (10-20 mg). Use of the vertical pocket reorienting rotor is particularly useful in the separation of cellular lipid droplets and for characterization of the Golgi complex, a major site of TG accumulation in alcoholic fatty liver.

Alcoholism has been described as the syphilis of the 20th century. It is estimated that there are at least one-million alcohol-dependent people and an equal number of alcohol abusers in the U.K. and the vast majority of these will have liver disease of varying severity. The liver damage is progressive with increasing duration and severity of abuse, as well as with other synergistic factors. There are, however, wide variations in individual susceptibility dependent on sex, race, age, etc. The initial effects, i.e. cellular hyperplasia

---

* *Editor's abbreviations:* e.r., endoplasmic reticulum; TG, triglyceride; PL, phospholipid; p.m., plasma membrane.

and fatty liver, are reversible with abstinence, but lesions that
have progressed to hepatitis, fibrosis, cirrhosis and hepatocarcinoma
will not, even with abstinence, return to normal. Although progression
is related to the accumulated dose of alcohol consumed, only 10-20%
of even heavy alcohol abusers develop cirrhosis. Thus there remain
several unanswered questions relating to the pathogenesis of the
liver injury due to alcohol abuse.

Morphological studies, including electron microscopy, have
disclosed diverse abnormalities in hepatocytes in patients with alcoholic liver disease of varying severity; but whether these changes are
pathological or are adaptive responses is uncertain. Thus, hyper-
trophied (mega) mitochondria are frequently reported as a characteristic of alcoholic liver disease [1-3] and are assumed to represent
a pathological change [4]. However, ethanol metabolism by the hepatocyte induces striking cytoplasmic redox changes [5], and conceivably
the mitochondrial hypertrophy represents an adaptive response.
Ultrastructural changes have been reported in a variety of other
organelles including lysosomes, peroxisomes and e.r. It is not,
however, certain whether the alterations are representative or reflect
a subjective morphological opinion.

For these reasons a biochemical approach to the study of organelle
alterations in alcoholic liver disease has been undertaken. The
strategy has been to assay organelle marker enzymes and to study
the properties of the organelles, e.g. latent enzyme activities and
centrifugation behaviour, and to use these results as a basis for
future studies. The most characteristic lesion of alcoholic liver
disease is fatty liver, and recently this condition has been studied
in detail. In particular, the subcellular distribution of the accumulated lipid has been investigated in order to determine the initial
site of lipid accumulation and thus the pathogenesis of the lesion.

An important technical problem in these studies has been the
very limited amount of tissue available from a percutaneous needle-
biopsy specimen of human liver (10-20 mg). This has meant that micro-
methods for subcellular fractionation and enzymic analysis have had
to be developed [6]. Subcellular fractionation has entailed single-
step zonal centrifugation techniques, initially with the Beaufay
zonal rotor [see #A-2, earlier in this vol.- *Ed*.], but more recently
with a vertical pocket reorientating rotor. Enzyme analysis has
entailed use of fluorigenic and radiolabelled substrates [7].

ENZYMIC ANALYSIS OF ALCOHOLIC LIVER DISEASE

Table 1 shows representative organelle marker enzyme activities
in biopsy samples from control subjects and patients with varying
degrees of alcoholic liver disease. Lysosomal, peroxisomal and e.r.
markers show no significant changes between the various patient groups.

Table 1. Enzymic analysis of alcoholic liver disease. Values are mUnits/mg protein: mean ±S.E. (with no. of samples). *Data from [8-10].*

Patient group	*N*-Acetyl-β-glucosamininidase (lysosomes)	5'-Nucleo-tidase (p.m.)	N α-gluco-sidase[⊗] (e.r.)	Glutamate dehyd'ase (mito'dria)	Catalase (peroxi-somes)
Controls	2.03 ±0.28 (37)	13.7 ±2.4 (37)	0.60 ±0.06 (37)	103 ±12 (11)	239 ±2 (13)
Fatty liver	1.99 ±0.31 (21)	16.3 ±2.9 (21)	0.58 ±0.16 (21)	214 ±15[b] (12)	199 ±29 (5)
Alcoholic hepatitis	3.07 ±0.94 (7)	21.2 ±5.4 (7)	0.76 ±0.14 (7)	–	–
Cirrhosis	2.66 ±0.27 (16)	22.3 ±3.6[a] (16)	0.63 ±0.06 (16)	–	–

[a] $p < 0.05$, [b] $p < 0.01$, for patient group compared with controls (t-test)
[⊗] N denotes neutral.

5'-Nucleotidase activity increases with increasing severity of liver damage, but the increase becomes significant only with cirrhotic liver. Similar conclusions were reached for other p.m. enzymes, viz. alkaline phosphatase, leucyl-2-naphthyl-amidase and γ-glutamyl transferase [9]. In addition, the mitochondrial enzyme glutamate dehydrogenase showed increased activity (Table 1), and similar conclusions were reached for other matrix marker enzymes [8]. Markers for the outer membrane (monoamine oxidase) and inner membrane (succinate dehydrogenase, cytochrome **c** oxidase) showed similar activities in tissue from control and alcoholic fatty liver disease patients. Centrifugation studies confirmed these results and showed normal mitochondrial integrity in alcoholic liver disease. It is concluded that mitochondria are not damaged in alcoholic liver disease, at least during the early stages, and that the increased enzyme activities, particularly affecting enzymes of the malate shuttle, represent an adaptive response to ethanol metabolism, i.e. to the increased redox state. Thus mega mitochondria reflect a functional hypertrophy and are not diseased *per se* [11].

In view of recent interest in lysosomal changes manifested by animal models of alcoholic liver disease [12-14], these organelles have been studied in greater detail. Assays of several lysosomal enzymes have been made in biopsies from various patient groups [9]. In addition, measurements of lysosomal integrity were made by assaying latent and sedimentable enzyme activities.[⊗] The results are shown in Table 2. Apart from a small increase in latent and sedimentable *N*-acetyl-β-glucosaminidase in biopsies from patients with fatty liver, lysosomal integrity as well as total enzyme activities were normal. The increased latent activity in the patients with fatty liver probably represents the stabilizing effects of certain lipids on lysosomes

[⊗]Cf. R.Wattiaux et al., #A-1, this vol.: 'free' ≅ non-sedimentable activity.

Table 2. Hepatic lysosomal integrity in alcoholic liver disease.
Values are % of total activity (& see Table 1 heading). *Data from [9].*

| Patient group | *N*-Acetyl-β-glucosaminidase | | Acid phosphatase | |
	Latent	*Sedimentable*	*Sedimentable*	
Controls	64.7 ±1.7 (23)	56.1 ±2.8 (9)	50.6 ±2.4 (9)	a, $p<0.01$;
Fatty liver	71.8 ±1.7 (13)[a]	69.4 ±1.0 (8)[b]	52.7 ±4.5 (7)	b, $p<0.001$
Alcoholic hepatitis	58.8 ±4.4 (10)	–	38, 45 (2)	
Cirrhosis	57.5 ±2.6 (11)	57.1 ±5.3 (7)	46.0 ±3.0 (6)	

[15, 16]. In addition, there may be lipid accumulation within lysosomes
– resulting in so-called lipolysosomes which have been demonstrated
by centrifugal flotation studies [17].

SUBCELLULAR FRACTIONATION IN ALCOHOLIC LIVER DISEASE

Liver needle-biopsy homogenates were fractionated as described
previously [7]. Typical results from patients with alcoholic cirrhosis
are shown in Fig. 1. The p.m. marker enzymes show little change
in distributions. Similar conclusions were reached for mitochondria
and peroxisomes (not shown). One lysosomal marker (*N*-acetyl-β-
glucosaminidase) shows a decrease in median density . Acid phosphatase
shows a small shift, but β-glucuronidase is unaffected. This probably
reflects the lipid associated with hepatocyte lysosomes, reducing
their equilibrium density in the sucrose gradients. As the latent
or sedimentable activity was unaffected, this increased *N*-acetyl-β-
glucosaminidase recovered in the sample layer does not reflect
increased cytosolic enzyme released from disrupted lysosomes. Similar
conclusions were reached for smooth muscle cell lysosomes [16].

β-Glucuronidase and, to a lesser extent, acid phosphatase are
localized in non-hepatocyte cells, notably Kupffer cells which clearly
do not participate in the cirrhotic process. The other noteworthy
alteration in gradient distributions of organelle marker enzymes
was the decrease in median density of neutral α-glucosidase (Fig. 1).
The enzyme, an integral marker for the e.r., accurately reflects
the relative proportions of smooth and rough e.r. [18]. This result
indicates that there is a relative increase in the proportion of
smooth e.r. in alcoholic liver disease, a finding consistent with
morphological studies [19] and in agreement with the enzyme-inducing
effects of ethanol on the smooth e.r.

The enzyme and centrifugation studies in alcoholic liver disease
indicate little if any change in p.m. or peroxisomes. Mitochondria
show favourable adaptive alterations, and a recent report of clinico-
pathological correlations in patients with alcoholic hepatitis shows

Fig. 1. Isopycnic centrifugation of post-nuclear supernatant from control (·····) and cirrhotic (—I—) liver biopsies. Results show distributions as mean ± S.D. where frequency is defined as fractional activity divided by fractional density span (g/ml). For further details see ref. [9].

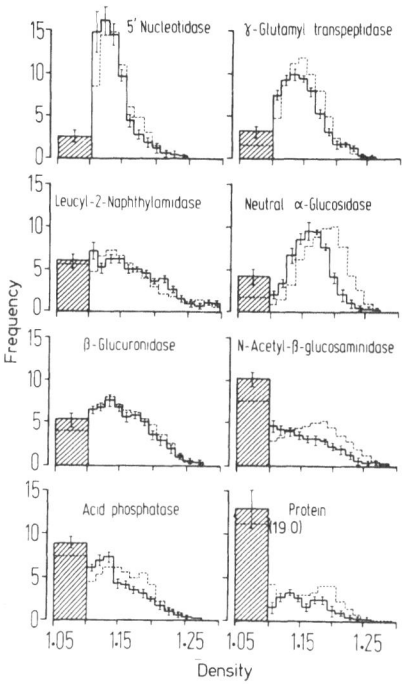

that patients with so-called mega mitochondria have a less serious form of liver disease with a better prognosis and less cirrhosis [20], a result entirely consistent with the above conclusions. Lysosomes become associated with some of the accumulated lipid, but there is no consistent change in their properties. Lysosomal disruption is not a feature of uncomplicated alcoholic liver disease. The e.r. changes are also probably an adaptive response to the chronic ethanol load. No clear pathogenic mechanism for the development of cirrhosis has been identified as yet.

NATURE AND SUBCELLULAR LOCALIZATION OF LIPID IN ALCOHOLIC LIVER DISEASE

Fatty liver is a characteristic of alcoholic liver disease although the degree of fatty change varies widely between patients. Because uncomplicated fatty liver is an early but reversible feature of alcohol-mediated damage to the liver, this has been the subject of considerable study. Table 3 shows the nature of the accumulating lipids in alcoholic fatty liver and cirrhosis compared to controls. In normal liver ~50% of the lipid is PL with a PL:free cholesterol ratio of nearly 8. Free fatty acids and TG each contribute ~20% of the lipid. In fatty liver there is an ~10-fold increase in TG which now comprises about two-thirds of the total lipid. The other lipid classes are relatively unchanged and thus contribute relatively small proportions to the total lipid. The PL:free cholesterol ratio is unchanged.

Table 3. Major lipid classes in alcoholic liver disease. Values are μmol/mg DNA, mean ±S.E. (with no. of patients, and % lipid is tabulated thus, (..%). *From ref. [21], by permission.*

Patient group	Free cholesterol	Cholesteryl ester	Phospho- lipid, PL	Free fatty acids	Trigly- ceride, TG	TOTAL LIPID
Control (7)	3.0 ±0.6 (7%)	0.4 ±0.1 (1%)	23.5 ±3.6 (52%)	8.7 ±0.8 (20%)	8.9 ±2.0 (20%)	44.5 ±5.9
Fatty liver (10)	5.8 ±1.4 (4%)	1.4 ±0.3[b] (1%)	30.8 ±5.7 (28%)	8.6 ±1.8 (8%)	96.5 ±22.5[b] (56%)	143 ±40[b]
Cirrhosis (6)	4.4 ±1.4 (7%)	0.51 +0.11 (1%)	17.4 ±2.6 (28%)	5.0 ±0.5[a] (8%)	34.8 ±11.9[a] (56%)	62.1 ±15.5

a, $p < 0.05$; b, $p < 0.01$ (patient compared with control group)

In cirrhosis the total lipid is increased (Table 3), but in view of the wide scatter of results the increase is not statistically significant: TG comprises ~50% of the total lipid. The PL:free cholesterol ratio was significantly reduced to 5, clearly causing reduced membrane fluidity. The level of free fatty acids was reduced but the other lipids were relatively unaffected.

More detailed lipid analyses including proportions of individual PL's and profiles of fatty acids in the individual lipids are reported elsewhere [22]. The important finding is that the alterations in PL fatty acids, like the PL:cholesterol ratio, indicate a reduced membrane fluidity. This conclusion was confirmed by studies of plasma PL fatty acids [23, 24] and appears to be a consistent abnormality in chronic alcohol toxicity, both in man and in the experimental animal.

Several possible mechanisms for these changes have been suggested. Ethanol itself has a direct fluidizing effect on various cell membranes, and the apparent decrease in lipid membrane fluidity may represent a compensatory effect [25]. There is currently considerable interest in the role of free radical-mediated membrane damage in alcohol toxicity [26], particularly in the presence of iron excess as frequently found in alcoholic liver disease [27] (see M.J. O'Connell et al., #A-9, this vol.-*Ed.*). There is a reduction in the proportion of polyunsaturated fatty acids that may be due to enhanced lipid peroxidation, and there is also the possibility that this reduction is a consequence of reduced fatty acid desaturase activities [28-30].

The subcellular distribution of the accumulating TG has been determined by centrifugation procedures. Firstly, by analogy with the separation of plasma lipoprotein classes by differential flotation [cf. art. by A. Mallinson & R.H. Hinton in Vol. 3, this series -*Ed.*],

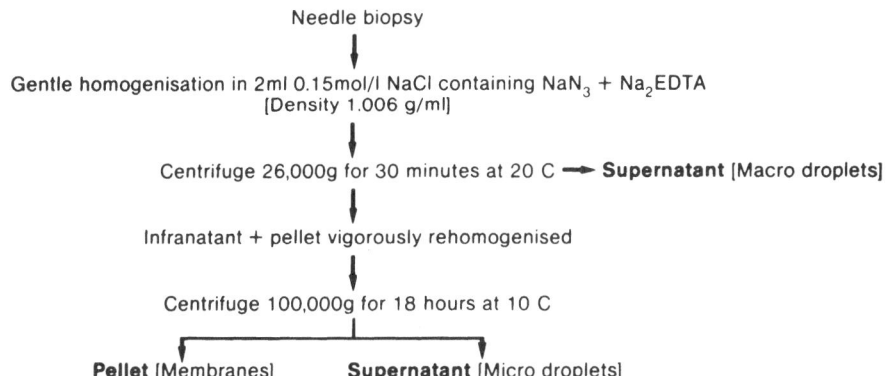

Scheme 1. Flow diagram for fractionation of liver biopsy homogenates by differential centrifugation/flotation. See [21] for more detail.

Table 4. Distribution of lipid in liver biopsy fractions. Values are mg total lipid/mg wet wt. tissue, mean ±S.E. (with no. of samples analyzed). Distribution of recovered lipid shown thus, (..%). For fatty liver compared with controls, $p < 0.05$. *From ref. [21].*

Tissue sample	Membranes	Macrodroplets	Microdroplets
Controls (6)	37.1 ±8.2 *(84%)*	4.5 ±1.7 *(9%)*	3.3 ±1.2 *(7%)*
Fatty liver (6)	67.6 ±17.0 *(76%)*	10.3 ±5.4 *(12%)*	10.1 ±4.1 *(12%)*

liver biopsy samples have been fractionated into macro- and micro-lipid droplet fractions and into membrane-bound lipid [31]. This procedure is illustrated in Scheme 1. Detailed morphological and biochemical analyses of these fractions have been reported. Table 4 shows the lipid content of the three fractions isolated from both control and fatty liver samples. In normal liver >80% of the lipid is membrane-bound. Surprisingly, even in severe fatty liver, although there was a 3- to 4-fold increase in lipid in the floating lipid fractions, there is a considerable amount of lipid associated with membranes. Analyses of these fractions indicate that much of the accumulating TG is membrane-bound [31].

In order to study further the subcellular localization of the accumulating TG, homogenates of liver biopsies were subjected to analyltical subcellular fractionation by sucrose density-gradient centrifugation. Initial experiments with the Beaufay automatic zonal centrifuge were unsatisfactory, with poor recoveries of TG from the gradient due to loss of the floating lipid fractions. This problem was overcome with a vertical pocket reorientating rotor [32] as illustrated in Scheme 2. Analysis of the fractions for TG (Fig. 2) shows a distinct peak at d = 1.13. This corresponds to Golgi marker enzymes.

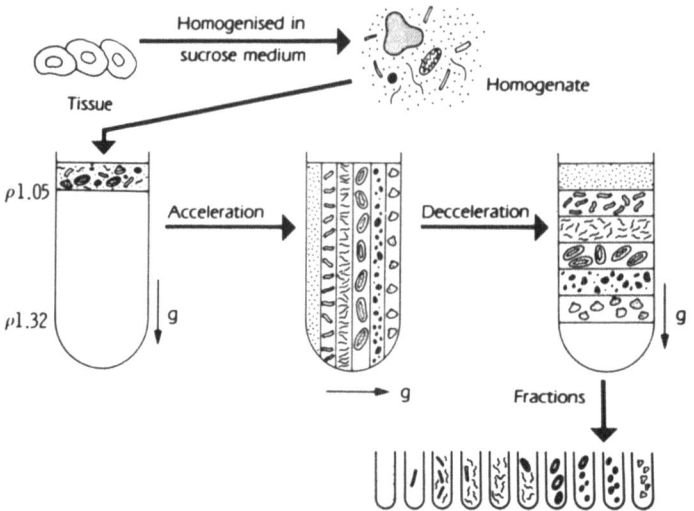

Scheme 2. Flow diagram for fractionation of liver biopsy homogenate by sucrose density gradient centrifugation in a vertical pocket reorientating rotor. *From ref. [21], by permission.*

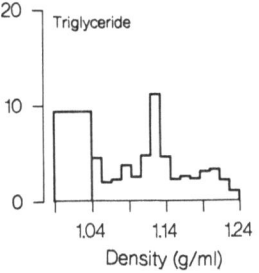

Fig. 2 *(right).* Isopycnic centrifugation of liver biopsy homogenate from patients with alcoholic fatty liver. Material at d <1.04 represents floating TG fraction. *From [31].*

It is thus concluded that a significant proportion of the accumulating lipid in alcoholic fatty liver is membrane–bound and probably associated with the Golgi. This accords with recent results showing impaired VDL secretion, an important Golgi function, in experimentally induced fatty liver [33, 34]. Other workers have postulated impaired secretory function of the hepatocyte in alcoholic liver disease [35, 36]. The basis for this defect, whether it reflects a cytoskeletal abnormality or perhaps impaired Golgi formation of exocytic vesicles, possibly consequent upon the changes in membrane fluidity referred to above, remains to be determined.

References

1. Petersen, P. (1977) *Acta Path. Microbiol. Scand. 85A*, 421–427.
2. Stewart, R.V. & Dincsoy, H.P. (1982) *Am. J. Clin. Path. 78*, 293–298.
3. Uchida, T., Kronborg, I. & Peters. R.L. (1984) *Liver 4*, 29–38.
4. Arai, L., Leo, M.A., Nakano, M., Gordon, E.R & Lieber, C.S. (1984) *Hepatology 4*, 165–174.
5. Peters, T.J. (1982) *Br. Med. Bull. 38*, 17–20.
6. Peters, T.J. (1981) *J. Clin. Path. 34*, 1–12.

7. Peters, T.J. & Seymour, C.A. (1978) *Biochem. J. 174*, 435–446.
8. Jenkins, W.J. & Peters, T.J. (1978) *Gut 19*, 341–344.
9. Seymour, C.A. & Peters, T.J. (1978) *Clin. Sci. Mol. Med. 55*, 383–389.
10. Selden, C., Seymour, C.A. & Peters, T.J. (1980) *Clin. Sci. 58*, 211–219.
11. French, S.W., Ruebner, B.H., Mezey, E., Tamura, T. & Halsted,C.H. (1983) *Hepatology 3*, 34–40.
12. Platt, D., Stein, U. & Heissmeyer, H. (1971) *Z. Klin. Chem. Klin. Biochem. 9*, 126–129.
13. Mezey, E., Potter, J.J. & Ammon, R.A. (1976) *Biochem. Pharmacol. 25*, 2663–2667.
14. Mezey, E., Potter, J.J., Slusser, R.J., Brandes, G., Romero, J., Tamura, T. & Halsted, C.H. (1980) *Lab. Invest. 43*, 88–93.
15. de Duve, C., Wattiaux, R. & Wibo, M. (1962) *Biochem. Pharmacol. 9*, 97–116.
16. Peters, T.J. & de Duve, C. (1974) *Exp. Mol. Path. 20*, 228–256.
17. Shinoda, T. & Hayashi, H. (1980) *J. Ultrastruct. Res. 72*, 235–244.
18. Smith, G.D. & Peters, T.J. (1980) *Eur. J. Biochem. 104*, 305–311.
19. Ishii, H., Joly, J-G. & Lieber, C.S. (1973) *Biochim. Biophys. Acta 291*, 411–420.
20. Chedid, A., Mendenhall, C.L., Tosch, T., Chen, T., Rabin, L., Garcia-Pont, P., Goldberg, S.J., Kiernan, T., Seeffl, B., Sorrell, M., Tamburro, C., Weesner, R.E. & Zetterman, R. (1986) *Gastroenterology 90*, 1858–1864.
21. Peters, T.J. & Cairns, S.R. (1985) *Alcohol 2*, 447–451.
22. Cairns, S.R. & Peters, T.J. (1983) *Clin. Sci. 65*, 645–652.
23. Alling, C., Aspenström, G., Dencker, S.J. & Svennerholm, L. (1979) *Acta Med. Scand. Suppl. 631*, 4–38.
24. Johnson, S.B., Gordon, E., McClain, C., Low, G. Holman, & R.T. (1985) *Proc. Nat. Acad. Sci. 82*, 1815–1818.
25. Sun, G.Y. & Sun, A.Y. (1985) *Alcohol Clin. Exp. Res. 9*, 164–179.
26. Peters, T.J. & Cairns, S.R. (1984) in *Pharmacological Treatments for Alcoholism* (Edwards, E. & Littleton, J., eds.), Methuen, New York, pp. 87–109.
27. Peters, T.J., O'Connell, M.J. & Ward, R.J. (1985) in *Free Radicals in Liver Injury* (Poli, G., Cheesman, K.H., Dianzani, M.V. & Slater, T.F., eds.), IRL Press, Oxford, pp. 107–115.
28. Nervi, A.M., Peluffo, R.O., Brenner, R.R. & Lieken, A.I. (1980) *Lipids 15*, 263–268.
29. Rao, G.A., Lew, G. & Larkin, E.C. (1984) *Lipids 19*, 151–153.
30. Umeki, S., Shiojiri, H. & Nozawa, Y. (1984) *FEBS Lett. 169*, 274–278.
31. Cairns, S.R. & Peters, T.J. (1984) *Clin. Sci. 67*, 337–345.
32. Rickwood, D. (1982) *Anal. Biochem. 122*, 33–40.
33. Baraona, E., Leo, M.A., Barowsky, S.A. & Lieber, C.S. (1977) *J. Clin. Invest. 60*, 546–554.
34. Sabesin, S.M., Frase, S. & Ragland, J.B. (1977) *Lab. Invest. 37*, 127–135.

35. Matsuda, Y., Baraona, E., Salaspuro, M. & Lieber, C.S. (1979)
    Lab. Invest. 41, 455-463.
36. Volentine, G.D., Tuma, D.J. & Sorrell, M.F. (1986)
    Gastroenterology 90, 158-165.

#A-4

METHODOLOGICAL APPROACHES IN STUDIES OF
ALTERED MITOCHONDRIAL AND PEROXISOMAL β-OXIDATION*

[1]Harald Osmundsen, [2]Kim Bartlett and
[3]H. Stanley A. Sherratt

[1]Institute of Physiology and Biochemistry
Dental School, Blindern
University of Oslo, Oslo 3, Norway

[2]Department of Child Health   [3]Department of Pharmacological
  and Clinical Biochemistry   Sciences
        Medical School, University of Newcastle upon Tyne
        Newcastle upon Tyne, NE2 4HH, U.K.

The study of **Mt** *and* **Px** *β-oxidation (which diseases or xenobiotics may affect) needs proper methodology. This is considered, for liver and other samples including biopsies, under the following heads:*
#Substrate preparation (incl. synthesis of acylCn's and acylCoA's).
#Preparation and use of **Mt** *fractions (incl. 'coupling'; media).*
- Measurement of β-oxidation by **Mt** *(incl. chain-length aspects; assay conditions; polarographic measurement; radiolabelled substrates; spectrophotometric assays).*
#Preparation and use of **Px** *fractions.*
- Isolation of **Px** *fractions. - Measurement of* **Px** *β-oxidation (incl. photometric assays; radiolabelled substrates). - Assay of* **Px** *β-oxidation with isolated fractions (incl. media; factors influencing rate; FFA* vs. *acylCoA as substrate; CoA and BSA roles; post-incubation gradient-centrifugation* **Px** *patterns).*
#Radio-HPLC analysis of acylCoA metabolic intermediates (incl. monitoring; metabolite recovery).
For **Px** *the assay recommendations are reinforced by Figs.*

Refs. list abridged by Editor, and abbreviations introduced:
  Mt, mitochondria(l); **Px**, perixosome(s)/peroxisomal; **de**'ase, dehydrogenase (but LDH if lactate); **tr**'ase, transferase; synthase ≡ synthetase.
  Cn, carnitine; FA, fatty acid; FFA, free FA; PL, phospholipid(s); py, palmitoyl (hence 'palmitoyl-CoA ester' in MS. has become pyCoA); BHT, butylated hydroxytoluene; BSA, bovine serum albumin; DTT, dithiothreitol; PCA, perchloric acid; RCR/RCI, respiratory control ratio/index; s.a., specific (radio)activity; THF, tetrahydrofuran.

Mt and/or **Px** β-oxidation (as reviewed:- [1]), whose character-istics somewhat differ, are impaired in acquired or genetic diseases, and may be altered by drugs and toxins. Acquired disorders can be studied using subcellular fractions prepared from animal models, or the effects of xenobiotics may be determined *in vitro* using sub-cellular fractions from normal animals. However, investigations of β-oxidation defects in patients are severely limited by the minuteness of the biopsy samples usually available. [See T.J. Peters, #A-2 in this vol., for lipid and other studies on biopsies.-*Ed.*]

The proper measurement of β-oxidation is not as simple as often supposed. Some techniques applicable to subcellular fractions will be described, and the choice of substrate considered - this being determined by the fraction and the metabolic problem in question. Many substrates of interest are commercially unavailable or prohibi-tively costly, particularly CoA- or Cn-esters of radiolabelled FA's. With minimal flair for organic synthesis this problem can be circum-vented. Similarly the selection of subcellular fractions requires some thought. Preparative complexity is minimal for **Mt** fractions - although great care has to be taken (during incubation also) if respiratory coupling is to be maintained - but is high for isolating **Px** fractions, although eased by the advent of vertical-tube rotors and modern density-gradient media. **Px** fractions can now be prepared and used for subsequent incubations within a normal working day.

SUBSTRATE PREPARATION

FFA's are frequently used, e.g. with tissue homogenates. However, with defined subcellular fractions FFA's are often unsatisfactory because it may be both difficult and tedious to define optimal condi-tions for their oxidation and to avoid the acylCoA synthase or Cn-py-**tr**'ase reactions being rate-limiting. Using acylCoA's or acylCn's as substrates, this problem is circumvented. With isolated **Mt** the acylCn's are usually preferable, unless the study concerns effects of changes in outer Cn-py-**tr**'ase activity on β-oxidation - in which case an acylCoA would be preferred.

Synthesis and preservation of acylcarnitines (acylCn's)

Conventionally acylCn's are synthesized by use of acyl chlorides corresponding to FA's ('FA chlorides'). Many are available commerci-ally; otherwise they can be generated from oxalyl chloride. This procedure sometimes works with polyunsaturated FA's [2; cf. below] as well as saturated FA's. It has been used too with radiolabelled FA's; thus, with a procedural modification, [16-^{14}C]pyCn of high s.a. was prepared [3].

For synthesizing Cn esters of polyunsaturated FA's, FA chlorides may be too reactive, and a method using the considerably less reactive acyl-imidazole has been developed [4]. It involves condensing

N,*N*-carbonyldiimidazole with FA in dried benzene for 30 min, and
reaction with Cn-perchlorate in dry acetonitrile medium. When purification of unsaturated acylCn's is attempted by a crystallization
procedure that works well for saturated acylCn's, an oil is invariably
obtained; instead, a simple chromatographic procedure with RP-SepPak
cartridges (Waters Assoc.) may be used [4]. Polyunsaturated acylCn's
are very susceptible to peroxidation, prevented by including BHT
(0.005% w/w) in all solvents used. They are best stored at -70° as
methanolic solutions (of defined concentrations) containing BHT.
On the day of use, small aliquots are withdrawn, methanol is evaporated
with a N_2 stream, and the residue dissolved in a set volume of water;
5 or 10 mM is a convenient working strength. Saturated or mono-
unsaturated esters can be stored as solids.

Synthesis of acylCoA's

The classical approach with use of FA chloride at mildly alkaline
pH has been improved to obtain 1-^{14}C-labelled fatty acylCoA's of
high s.a. [5], but typically yields unpredictable results, due probably
to attack on other functional groups in CoA by the very reactive
acid chlorides. Gentler acylation of CoA is achievable with FA anhydrides [6]: yields are low with long-chain FA's, but excellent (40-60%)
with short-chain (<10-C) FA's, for which this is the method of choice.
Most of these FA anhydrides are commercially available, as is diketene
which serves to prepare acetoacetyl-CoA. AcylCoA's of medium-chain
FA's have also been prepared from corresponding thioglycollates by
ester exchange [7].

An alternative approach, extensively used, that gives high-
quality acylCoA's for all types of FA's utilizes the mixed anhydride
of FA and ethyl chloroformate [8]. However, yields are poor (10-25%),
and the customary use of a 10-fold excess of the anhydride is a drawback
with expensive labelled FA's. This problem was overcome by use of
N-acylsuccinimide, readily prepared from FA and *N*-hydroxysuccinimide
using dicyclohexylcarbodiimide as dehydrating agent [9]. The reagent
is readily isolated, and may be stored at -20° for several months.
In 10-20% molar excess, reaction with CoA in a water/THF medium (e.g.
1:3 by vol.) can give 20-40% yield, provided that care is taken to
ensure that all CoA is present in reduced form, suitably by pre-incuba-
ting it with 50 mM DTT for ~30 min; the DTT is then removed by passage
through a Sephadex G-10 or G-15 column. This approach has furnished
^{14}C-labelled py-, erucoyl- and elaidoyl-CoA of sufficient s.a. for
metabolic work [10, 11].

More recently, the above-mentioned acylimidazole approach has
been used [12], with notable effectiveness reflecting the greater
reactivity of the CoA -SH compared with the Cn -OH group. Both acyl-
imidazole and CoA are each dissolved in water/THF (3:1) and immediately
mixed; as for acylCn, acylation (3-4 h usually suffices) is accelerated
by weakly basic conditions (0.5 M triethylamine present). A yield

of 60-80% is readily obtainable, and quantities can be equimolar
(even 10% molar excess of CoA); hence this procedure suits well for
labelled FA's.   It gives excellent yields with saturated, mono- and
poly-unsaturated FA's, also with benzoic acid and clofibric acid.
It is the present-day prime candidate for synthesizing long-chain
acylCoA's, in terms of yield and convenience.

Purification of acylCoA's merely involves precipitation at pH
<2 (e.g. by HCl) unless the FA is <10-C. Repeated solvent-extraction
steps lead to a white solid.   Where the FA is <10-C, any desired
purification is usually achieved by ion-exchange chromatography and
desalting with a gel column [13].

PREPARATION AND USE OF MITOCHONDRIAL (Mt) FRACTIONS

Potter-Elvehjem homogenates of liver readily and reproducibly
furnish **Mt** fractions by differential centrifugation. Other tissues
may present difficulties, e.g. because of connective tissue - which
may be removed initially by forcing the tissue through a hand-press.
With skeletal or cardiac muscle, initial use (~5 sec only) of an
Ultra-Turrax homogenizer is very effective. If the tissue contains
abnormal amounts of lipid, or active PLases, surface-active lysoleci-
thins may be formed and impair **Mt** function, circumventable by including
at least 10 mg/ml of defatted BSA in the homogenization medium.
In general, obtaining reputable **Mt** calls for practice (not neglecting
purity validation - *Ed.**).   Removal of contaminating microsomes and
Px, however, requires centrifugation in a density-gradient, such
as can be self-generated with Percoll ([14]; cf. H. Pertoft et al.,
in Vol. 8, this series).

Respiratory coupling.- Mt intactness is conventionally assessed
by the RCR or RCI, entailing oxygen-electrode recording of the rate
of oxidation of a suitable substrate   (e.g. 10 mM glutamate + 1 mM
malate). The RCR is the ratio of the rate with ADP present in limiting
amount (usually ~10 mM; state 3) to the final slower rate observed
when all ADP has been converted to ATP (state 4). Values in the
range 3-8 are considered acceptable, depending on the tissue. Higher
values occasionally reported may be partly artefactual. **Mt** gradually
become uncoupled when stored (even on ice), and should be used within
hours of preparation. Best stability is achieved with thick suspensions
(60-100 mg protein/ml).

Maintaining respiratory coupling is crucial: **Mt** must be incubated under conditions that maintain defined energy status, vital for
satisfactory **Mt** redox status and hence β-oxidation rates. This can
be particularly important for many polyunsaturated FA's whose β-oxidation needs the NADPH-dependent 2,4-dienoyl-CoA reductase [15]. Even
slight uncoupling can depress intra-**Mt** NADPH and hence the oxidation
rate, e.g. of γ-linolenic acid [16].

*See Preface (also for sources of basic guidance on cell fractionation).

Media suitable for Mt fractions.- Mt fractions from most tissues are typically, although not invariably, prepared with near-isotonic sucrose (which supposedly does not enter the Mt matrix) or mannitol, buffered with Tris or ·preferably Hepes or Mops and containing a trace of EGTA to chelate Ca^{2+} released during homogenization: e.g. 300 mM mannitol, 10 mM Hepes, 0.1 mM EGTA, pH 7.2 [17]. The Mt isolated in hypertonic sucrose (0.6-1 M) as in density gradients have hopelessly impaired respiratory activity, even if then incubated under conditions facilitating reswelling of the matrix.

Hepatic Mt isolated in sucrose (or mannitol, although disagreement exists) from glucagon-injected rats show up to 50% stimulation of oxidation of various substrates including FA's - perhaps reflecting an increased volume of the matrix compartment.

The tonicity of the incubation medium is important [18]. For incubation of liver Mt, a sucrose compared with a KCl isotonic medium gives very inferior respiratory rates - particularly for β-oxidation [9], attributable to dehydration of the matrix; cardiac or skeletal muscle Mt hardly show this effect (H.O., unpublished) [but see 19]. When the osmolarity in Mt incubations is increased by increasing the sucrose concentration, the oxidation rate of pyCn + malate declines quicker with increasing osmolarity than for other substrates [18].

Measurement of β-oxidation by Mt

Chain-length considerations in substrate selection.- In the intact cell, pyCoA synthase in the outer Mt membrane, the endoplasmic reticulum or the Px membrane converts long-chain FA's to acylCoA's whose acyl, if destined for β-oxidation by Mt, is shifted to L-carnitine by Cn-py-tr'ase I on the outer face of the inner membrane. Once transported across the membrane by the Cn-acylCn translocator, the acyls are shifted to intra-Mt CoA by the Cn-py-tr'ase II on the inner surface of the inner membrane. Physiologically the oxidation of medium- and short-chain FA's is Cn-independent since they diffuse directly into the Mt matrix, where butyryl-CoA synthase converts them to acylCoA's [1]. *In vitro*, however, entry into Mt via the Cn-translocator occurs with acylCn's from FA's of all chain lengths, including medium-chain FA's resulting from chain-shortening of long-chain FA's in Px (review: [20]).

Achieving valid assay conditions.- Care is required in measuring FA oxidation rates. The Cn-dependence of long-chain FA's and their acylCoA's is observable only with a low substrate concentration, achieved by including defatted BSA (suitably 2 mg/ml) in all incubations so as to buffer their free concentrations although the free levels are uncertain.- BSA has 2 high-affinity and 5 lower-affinity sites for palmitate [21], while values are lacking for BSA binding of acylCoA's and acylCn's. Unbound substrate is in lower proportion to total substrate at low than at high BSA concentrations, and is not a linear function of total substrate + BSA concentrations even when these are varied while maintaining a constant molar ratio [22].

AcylCn's (10-40 μM) will be β-oxidized directly, while acylCoA's also need added Cn (L, to 0.5 mM) in the incubations. With long-chain FFA's (10-20 μM), Cn, CoA-SH (0.1 mM) and ATP (1 mM) are all needed, and the ADP/ATP ratio may require adjustment to get maximal β-oxidation rates. A **Mt** preparation exhibiting CoA-independent oxidation of long-chain FA's is likely to contain a high proportion of damaged **Mt** (and **Px**) from which CoA-SH is escaping during incubation. For short- and medium-chain FA's, β-oxidation only required ADP (+ phosphate) to stimulate the flux.

Polarographic measurements of β-oxidation.- Direct polarographic measurements of β-oxidation are feasible with liver **Mt** in state 3 or state 4 (uncoupled) when the citrate cycle is inoperative (e.g. with 5 mM malonate or 50 μM fluorocitrate present): then the resulting acetyl-CoA is essentially converted to acetoacetate and, due to short-chain acylCoA hydrolase in the matrix, small amounts of acetate. In the presence of exogenous Cn, acylCn will also be formed, and the oxygen uptake rate is now an unambiguous measure of the flux through β-oxidation [23]. In state 3 some of the NADH formed during β-oxidation reduces acetoacetate to 3-hydroxybutyrate, so that the β-oxidation rate slightly exceeds that indicated by recorded oxygen uptake [24]. If malonate is replaced by 2 mM malate the citrate cycle will operate and so augment the recorded respiratory rate. In the presence of fluoroacetate, citrate accumulates. **Mt** from non-ketogenic tissues, e.g. muscle, may need added malate for β-oxidation to operate [1, 25].

Ideally all β-oxidation products should be measured to get precise oxidation rates, but this is rarely feasible. With liver **Mt**, or hepatocytes, ketone-body formation alone is a fair index of β-oxidation. CO_2 formation may represent barely 5% of the degradation products [26] and can often be disregarded.

Useful information may be obtained from the stoichiometry of acylCn-dependent pulses of respiration (the acylCn concentration being limiting for respiration). Thus, the active metabolite of hypoglycin methylenecyclopropylacetyl-CoA, inactivates both short- and medium-chain acylCoA **de'**ase, but not pyCoA **de'**ase; with liver **Mt** from poisoned animals both the rate and extent of β-oxidation is decreased, since the oxidation proceeds only as far as butyryl-CoA [27].

Use of radiolabelled substrates.- FA's labelled with ^{14}C in various positions are often used[⊗] as substrates for β-oxidation by **Mt**. Many authors measure only the $^{14}CO_2$ formed without apparently realizing that this is only a minor degradation product (5-20%) in

⊗ ^3H label also serves

in vitro preparations from all tissues or in cultured fibroblasts [25, 28, 29]. The main degradation products are acid-soluble, e.g. short-chain acylCoA's and acylCn's, citrate-cycle intermediates (also: with liver, ketone bodies; with isolated hepatocytes, glucose). β-Oxidation can therefore be followed merely by measurement of the radioactivity in the supernatant obtained by adding ice-cold PCA (to 2.5% w/w) to the incubation mixture and centrifuging at 20,000 **g** for 10 min; unchanged long-chain FA and FA's incorporated into complex lipids are precipitated. Acid-soluble radioactivity plus $^{14}CO_2$ comprise the total flux through β-oxidation.

   PCA precipitation is non-quantitative for FA's below 12-C, for which the acylCoA **de'**ases may be genetically deficient such that fibroblasts oxidize the FA's poorly. Only $^{14}CO_2$ can therefore be measured in simple experiments with these FA's as substrates. Using [^{14}C]octanoate, this limitation has been circumvented by finally removing unreacted substrate with Lipidex 1000 and measuring the remaining acid-soluble radioactivity [29].

   Spectrophotometric assays.- It is feasible to trap the electrons derived from oxidation of all substrates at the level of cytochrome **c**, using an electron acceptor which can be chosen to manifest an absorbance change when reduced. Thereby, using ferricyanide with isolated **Mt** fractions, a direct-reading β-oxidation assay was introduced which is both sensitive and may be applicable to many tissues as citrate-cycle interference is probably minimal; the ferricyanide traps electrons flowing to cytochrome **c** from the acylCoA **de'**ase reaction [30]. A high concentration of oxaloacetate (10 mM) is present, functioning as a sink for NADH generated by the 3-hydroxyacyl-CoA **de'**ase and for acetyl-CoA (by the action of citrate synthase). Therefore the rate of ferricyanide reduction is a direct measure of β-oxidation. With skeletal muscle **Mt**, recent studies (N.J. Watmough, unpublished) suggest that the assay also records some flow of electrons to ferricyanide from succinate formed from citrate.

   AcylCn's are the substrates usually employed, although acylCoA's can be used in the presence of Cn. Allowance must, however, be made for the formation of CoA-SH as acylCn is formed from the acylCoA, because CoA-SH will spontaneously reduce ferricyanide. Free octanoate has also been used as substrate, with 10 mM ATP present [28]. With skeletal muscle **Mt** the reaction tends to accelerate before reaching a steady state. When β-oxidation is deficient, it may be useful to measure also the oxidation of 10 mM succinate (omit oxaloacetate) or of 10 mM glutamate + 1 mM malate (omit oxalacetate and rotenone), thereby verifying that the amount of **Mt** protein present is adequate.

   The ferricyanide assay procedure has been used to detect an impaired rate of β-oxidation in skeletal muscle **Mt** from a patient with a defect of butyryl-CoA **de'**ase activity, apparently confined to muscle [28]. The assay is more sensitive than polarographic assay, requiring only 50-110 µg of protein/assay; but a dual-wavelength spectrophotometer is essential for following ferricyanide reduction at 425-475 nm in turbid **Mt** suspensions.   Increased sensitivity is obtainable by replacing ferricyanide with tetracyano-2,2-bispyridine-iron (III) [not commercially available] whose absorption coefficient is ~2.6 times that of ferricyanide (D.M. Turnbull et al.; see [28]).

Cytochrome **c** (III) can also be used as an exogenous electron acceptor at the level of endogenous cytochrome **c**.   Prior swelling of the **Mt** in a hypotonic medium (essentially 50 mM K phosphate, 5 mM $MgCl_2$, pH 7.4; 10 min at 37°) is required, to render the outer **Mt** membrane permeable to exogenous cytochrome **c**. This approach where reduction of cytochrome **c** is followed at 550 nm has been applied to skeletal muscle homogenates oxidizing pyCn [31]; but the extent of non-specific reduction is unclear.   There is also uncertainty because cytochrome **c** appears to accept electrons at a rate which is only about one-third of the rate observed with ferricyanide as the acceptor [see 32].

PREPARATION AND USE OF PEROXISOMAL (Px) FRACTIONS

Isolation of Px fractions

In differential centrifugation, **Px** (which comprise <6% of total cell protein even after drug-induced proliferation in rats) largely co-sediment with **Mt** and lysosomes.   Few laboratories have practised **Px** purification by density-gradient centrifugation, since this has necessitated hours of ultracentrifugation or use of a Beaufay rotor (not commercially available).   The advent of suitable gradient media and new rotors has altered this situation.   With a vertical-tube rotor, **Px** can now be isolated in a self-generated Percoll gradient after ultracentrifugation at speed for 30 min [33], or high-speed centrifugation (e.g. Sorvall RC 5B) for 60 min.   Success has also been claimed with a fixed-angle head (e.g. type Ti 60), although invariably with poorer resolution and **Px** purity [e.g. 34].

The Percoll procedure served for study of β-oxidation by **Px** after high-fat diets [32], in genetically obese mice (see [35]) and in riboflavin-deficient rats [35].   Iodinated gradient-media (Metrizamide and Nycodenz) have proved eminently suitable for isolating very pure **Px** fractions, comprising ~95% **Px** protein [36, 37].   With these media, self-generation of the gradient is not possible and an ultracentrifuge is necessary.

Most work with **Px** fractions (review: [38]) has involved liver or, more rarely, kidney cortex, largely in the rat or mouse although **Px** fractions have also been prepared from liver and kidney of beef, sheep and cat [39] and human liver [3].

Measurement of peroxisomal (Px) β-oxidation

Photometric assays.- The classical assay where acylCoA-dependent NAD$^+$ reduction is followed spectrophotometrically has been applied to homogenates of liver, heart and intestinal mucosa (e.g. 40, 41]. There can be interference when activity is low and crude homogenates are used, due to acyl-CoA oxidase action and to NAD$^+$ reduction resulting from the action of β-hydroxyacyl-CoA de'ase on an intermediate formed by this oxidase [see 1]. This problem can be overcome by assaying the acylCoA-dependent H$_2$O$_2$ generation by acyl-CoA oxidase activity directly. Thus, this reaction has been coupled to horseradish peroxidase and a chromogen, the oxidized (coloured) form of which is measured [42]. This assay works well for tissue homogenates [43], with the drawback that the reaction remains linear only during the initial 3-4 min. A fluorimetric acyl-CoA oxidase assay [44] is notably sensitive, and suitable for small biopsies.

Assay with radiolabelled substrates.- With acylCoA's such as [1-^{14}C]- or [U-^{14}C]-pyCoA, β-oxidation is easily followed by the increase in acid-soluble (acetyl-CoA) radioactivity [45]. This convenient assay is unfortunately of limited use in preparations which are heavily contaminated by **Mt**, e.g. a tissue homogenate. It is of course possible to block **Mt** β-oxidation with a respiratory inhibitor, suitably rotenone or antimycin (not KCN, which can inhibit other enzymes, e.g. catalase), but incompletely [30]. Even with 99% blockage the remaining activity will be enough to cause severe errors in estimates of **Px** activity. The use of antimycin plus myxothiazol (which decreases the antimycin-resistant electron flow by 96%; see [46]) may be invaluable. In general the assay procedure can provide meaningful results where, as after hypolipidaemic drugs [cf. R.K. Berge, #A-5 in this vol.] there is a large change in β-oxidative activity [47].

A similar but more precise approach [48] entails measuring the ratio of oxidation rates obtained with [1-^{14}C]- compared with [16-^{14}C]-palmitate in the presence and absence of antimycin A. Especially in its presence, the ratio should be high in a tissue having a high rate of **Px** β-oxidation - which will usually not degrade the FA's sufficiently to generate acid-soluble ^{14}C from [16-^{14}C]-palmitate as occurs with **Mt** β-oxidation. Use of the latter therefore serves to correct for remaining antimycin-sensitive **Mt** β-oxidation. This approach has been used to get an estimate of **Px** β-oxidative activity in liver, kidney, heart and skeletal muscle from rat and man, following high-fat diets, clofibrate treatment, and in hyper- or hypo-thyroid animals [48].

Assay of Px β-oxidation with isolated fractions

Media.- In classical studies such as those of C. de Duve, **Px** β-oxidation was assayed after detergent solubilization of membrane-enclosed enzymes; but to get a true picture of physiological **Px** function as intact organelles, their integrity before and during assay should be safeguarded. Damage to fragile **Px** was hardly avoidable as long as isolation necessitated sucrose density gradients with their high osmolarity, especially in view of the transfer into an iso-osmotic medium for incubation. The problem has been solved by the advent of media such as Percoll (gradients iso-osmotic) and iodinated compounds (mildly hyperosmotic).

It remains to devise assay conditions suitable for maintaining **Px** intactness. With BSA in the incubation to largely bind FA's ([21], & see above), membrane damage by FFA's or acylCoA's is avoided. Such conditions have been developed, with isotonic KCl medium [11].

β-Oxidation rates as influenced by substrate and conditions.- Compared with the rate under these conditions, the rate under solubilizing conditions and with no BSA present was ~3-fold higher (Fig. 1), largely attributable to the BSA variable (absent in the detergent-stimulated assay), as evidenced by Fig. 2.- Here, with BSA present in both incubations, the rate difference was smaller although still manifest, confirming an earlier finding from spectrophotometric assay that **Px** β-oxidative activity exhibits latency. (The difference in substrate between Fig. 1 and Fig. 2 is not pertinent.)

FFA *vs.* acylCoA as substrate.- With FFA, conversely to the above effect, β-oxidation is slower with solubilizing than with non-solubilizing conditions (Fig. 3). Seemingly with FFA the acylCoA synthase reaction becomes rate-limiting under solubilizing conditions, although not with non-solubilizing conditions where ^{14}C-FA (oleate) is similar in rate to that of the acylCoA (Fig. 4). However, the latter is oxidized somewhat faster than the FFA in the first 10 min of incubation (Fig. 4), probably because the FA activation step is circumvented. Taking account of similar findings obtained with palmitate and pyCoA, evidently radiolabelled FFA's are satisfactory substrates for **Px** β-oxidation, particularly where the **Px** are intact. Their use is more convenient and saves the expense of radiolabelled acylCoA's.

Roles of CoA and BSA.- With acylCoA's the CoA addition requisite for optimal rates under solublizing conditions [11] is not needed under non-solubilizing conditions. Presumably a CoA pool shown (in our laboratory, & see [33]) to be associated with liver **Px** is diluted when **Px** membrane is solubilized. Since the **Px** acylCoA is associated with the membrane's cytosolic (outer) face [33], a general CoA-dependence is to be expected using FFA's as substrate. CoA now has to be added whether the conditions are solubilizing or non-solubilizing (Fig. 5). FA's must be added as BSA complexes, since commonly used organic solvents inhibit **Px** β-oxidation (Fig. 6). FFA's incubated with isolated **Px** under non-solubilizing conditions are chain-shortened by 2-3 cycles of β-oxidation [14], analogous to findings using acyl-CoA's [10, 45].

Fig. 1. Effect of incubation conditions on acylCoA β-oxidation by a **Px** fraction (from liver of rats fed for ~10 days on a diet containing 0.5% w/w clofibrate; isolated in a Percoll gradient [see 11]; 1.5 mg protein /assay), with 50 μM (~5000 dpm/nmol) [9,10-(n)³H]pyCoA in 2 ml of a medium containing:
o: 130 mM KCl, 10 mM Hepes, 0.1 mM EGTA, 2 mg defatted BSA/ml, 0.5 mM pyruvate, 1 mM NH₄Cl, 0.5 mM NAD⁺, 0.1 mM NADP⁺, 1 mM DTT; 10 μg antimycin and 2 mU LDH/ml; pH 7.2; **or** □, 30 mM K phosphate; NAD⁺, NADP⁺, DTT, LDH, NH₄Cl and antimycin as in o; 0.2 mM CoA-SH, 0.005% (v/v) Triton X-100; pH 7.5.
During the incubation (37°, shaking water bath) 200 μl aliquots were added to 200 μl ice-cold 5% (w/v) PCA and centrifuged; the radioactivity in the supernatant was measured.

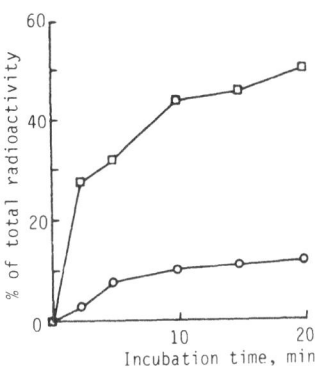

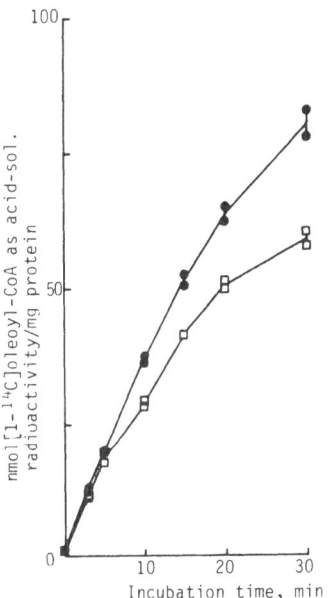

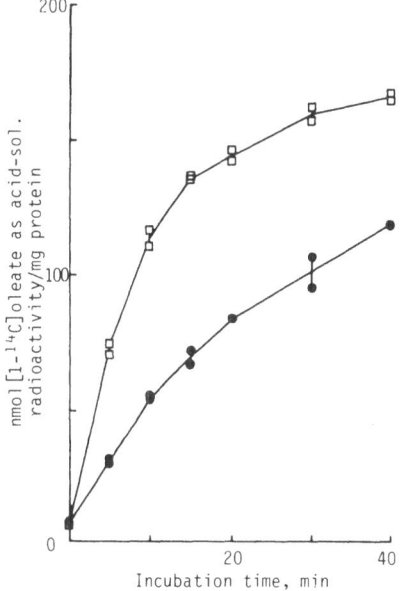

Fig. 2. AcylCoA β-oxidation rates: •, non-solubilizing conditions as for o in Fig. 1, *vs.* □, solubilizing conditions as for □ in Fig. 1 but with 2 mg BSA/ml. Other details as in Fig. 1 legend, but with [1-¹⁴C]-oleyl-CoA (50 μM, 6700 dpm/nmol) and 0.6 mg **Px** protein.

Fig. 3. FA β-oxidation by **Px**: □, non-solubilizing conditions, *vs.* •, solubilizing conditions but with 2 mg BSA/ml. Other details: Fig. 1 legend. Mg-ATP (5 mM) present to aid FA activation by the acylCoA synthase. Substrate: [1-¹⁴C]linolenic acid (50 μM, 4500 dpm/nmol); very similar results found with [1-¹⁴C]oleic acid.

Fig. 4. β-Oxidation by **Px** fractions (1.4 mg protein), under non-solubilizing conditions, of [1-^{14}C]-oleic acid (●) and -oleoyl-CoA (□), each 50 μM (6500 and 6700 dpm/nmol respectively). Incubates as in Fig. 1 (2 ml vol.), with 5 mM Mg-ATP in ●, as in Fig. 3 legend.

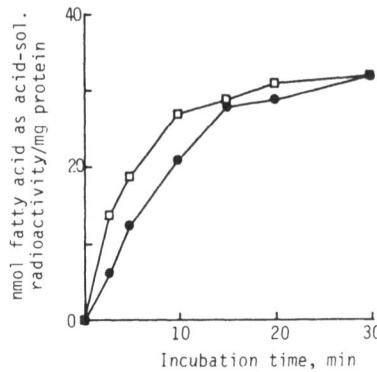

Fig. 5, *below.* Effect of added CoA-SH on β-oxidation by **Px** fractions (1.1 mg protein), under non-solubilizing conditions, of [1-^{14}C]palmitic acid (50 μM; 3400 dpm/nmol). Incubates as in Fig. 4, with Mg-ATP. No CoA, □; 0.2 mM CoA, Δ. With Triton X-100 (0.005% v/v): no CoA, o; 0.2 mM CoA, ◖.

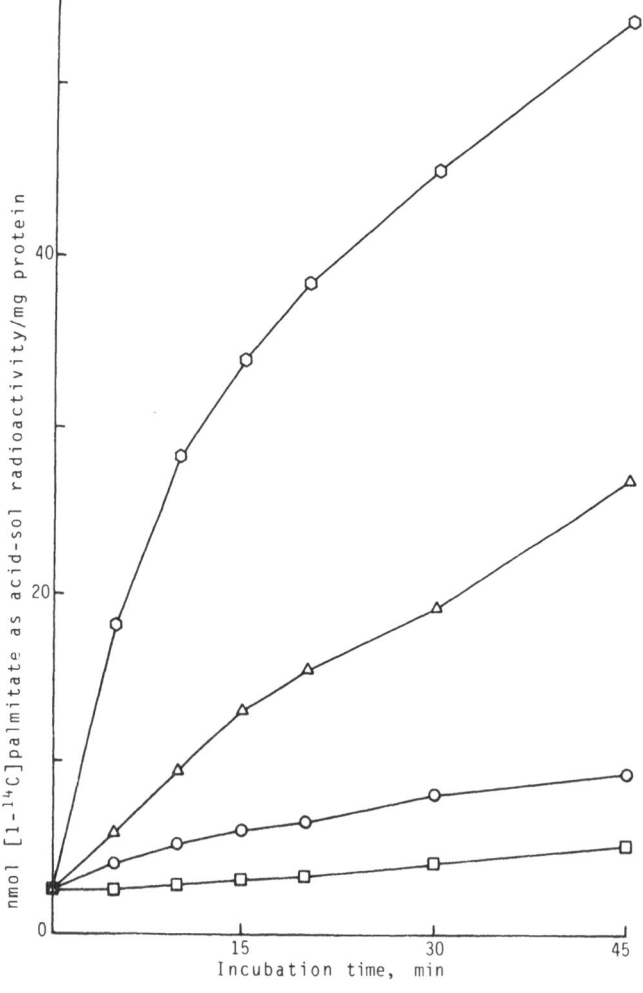

Fig. 6. Effects of various
solvents on β-oxidation by
Px fractions (180 µg protein),
under non-solubilizing conditions,
of 50 µM (*solid* symbols) or 10 µM
(*open symbols*) [1-¹⁴C]linoleic
acid (4500 dpm/nmol). MgATP (5 mM)
present; incubation in 200 µl, for
10 min; 200 µl ice-cold 5% PCA
added for termination. Details
otherwise as in Fig. 1 legend.
Solvents present (concn. as given
in the Fig.):
○,●: ethanol;
□,■: dimethylsulphoxide;
△,▲: dimethylformamide.

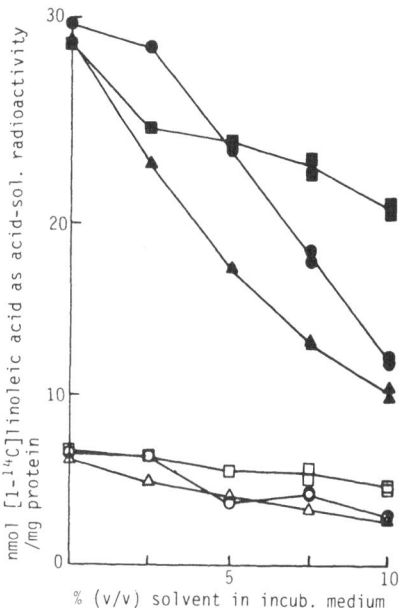

Post-incubation gradient centrifugation.- Percoll gradient runs
with marker enzyme assays were performed on iso-osmotic KCl incubates
(Fig. 7). A **Px** fraction incubated without detergent manifested ~10%
of its catalase as soluble activity (in the 2 fractions at top of
gradient), and a peak (fraction 7) coinciding with that of uricase activity. With Triton X-100 present in the incubation, the catalase peak
shifted virtually to the top of the gradient, indicative of catalase
solubilization, whereas uricase - a particulate activity - shifted
only marginally.

RADIO-HPLC ANALYSIS OF ACYL-CoA METABOLIC INTERMEDIATES

The methods discussed above enable **Mt** or **Px** β-oxidation fluxes
to be determined. However, overall rates may not throw light on
sites at which drugs or disease may impair β-oxidation. A complementary
approach with **Mt** and **Px**, potentially very useful, is to identify
and measure all acylCoA intermediates, looking for abnormalities
in organelles from tissues with impaired function.

Complete β-oxidation of pyCoA to acetyl-CoA involves a total
of 28 acylCoA intermediates. Their identification has presented
an intractable analytical problem. Early attempts, with limited
success, involved their saponification followed by radio-GC of the
methylated FFA's (no attempt was made to conserve keto-acids) [3].
Since the scale of **Mt** incubations is small and the GC carrier-gas
flow rates are 15-30 ml/min giving a brief residence time in the
radioactivity detector, radiolabelled substrates of high s.a. are
required to achieve the necessary specificity of detection. Peak
trapping is a cumbersome and time-consuming alternative.

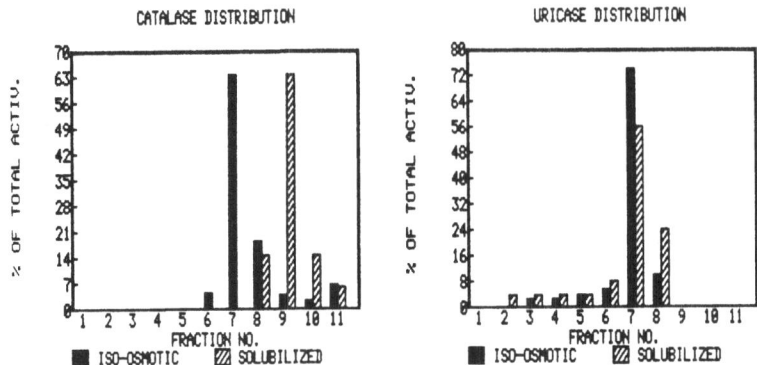

Fig. 7. Effects of incubation conditions on the distribution of catalase and uricase in a Percoll gradient. A **Px** fraction (1.6 mg protein) was incubated under non-solubilizing conditions at 37° in 2 ml, as in Fig. 1 legend, in the absence *(solid bars; some NIL values)* and presence *(hatched bars)* of 0.005% (v/v) Triton X-100. After 10 min the incubate was placed on Percoll (50% v/v) and centrifuged (conditions: ref. [32]). The resulting gradient was divided into 11 fractions which were assayed for the enzymes [32].

    An alternative strategy involves the analysis of intact acylCoA's by HPLC. There are reported methods for short- and medium-chain acylCoA's [e.g. 49, 50] and, to analyze freeze-clamped liver, for long-chain acylCoA's [51]. The only method for resolving both long- and short-chain acylCoA's in one chromatogram [52] involved use of tetrabutylammonium phosphate, which is both corrosive and expensive. The best method for analyzing acylCoA's in biological samples uses phosphate buffers, as in a modification [53] of a published gradient-elution method [51]: a homologous series of saturated acylCoA's up to C-16 was resolved within 30 min, or 50 min with an extended gradient that enhances resolution of long-chain CoA's (A.G. Causey, unpub-lished).

    Effluent monitoring.- For biological samples, UV monitoring is insufficiently selective and sensitive. This problem has been overcome [e.g. 49] by using a ^{14}C- or ^{3}H-labelled FA as substrate and on-line radiochemical effluent monitoring.

    Acid-soluble metabolite recovery.- Although 3-methyl-2oxo-pentanoate gave ~95% recovery of metabolites [49], recovery is prob-lematical for long-chain acylCoA's (e.g. derived from palmitate cata-bolism), which are not only precipitated by acid but form complexes with denatured proteins. In a methodological variant [54] that gives good recoveries of acylCoA's even of long chain-length, the incubation (in 1 ml) is quenched by glacial acetic acid (100 µl) and washed with diethyl ether (3×5 ml); saturated $(NH_4)_2SO_4$ (100 µl) is added, then chloroform/methanol (1:2 by vol.; 6 ml, added slowly). After

Fig. 8. HPLC, with radio-
monitoring *(lower panel)*, of
rat-liver **Mt** incubated with
120 μM [U-^{14}C]palmitate
(5 μCi/μmol), in the presence
of rotenone (10 μg) [49] and
also 5 mM ATP, 1 mM L-Cn and
0.1 mM CoA-SH; the FA was a
complex (5:1) with BSA.
Elution positions of standard
saturated acylCoA's:
1, acetyl-; 2, butyryl-;
3, hexanoyl-; 4, octanoyl-;
5, decanoyl-; 6, dodecanoyl-;
7, tetradecanyl-; 8, hexadec-
anoyl-. **Mt** acylCoA peaks:
A = 1; B, acetoacetyl-; C-F
are 2,3-enoyl derivatives of
5-8 respectively.
Incubate extraction: see
text.
HPLC: C-18 column; methanol/
phosphate buffer gradient.
Detection at 280 nm; lower
trace = radioactivity
elution profile.
For details see [53].

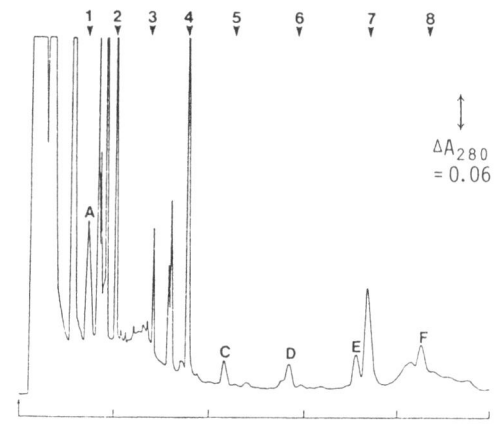

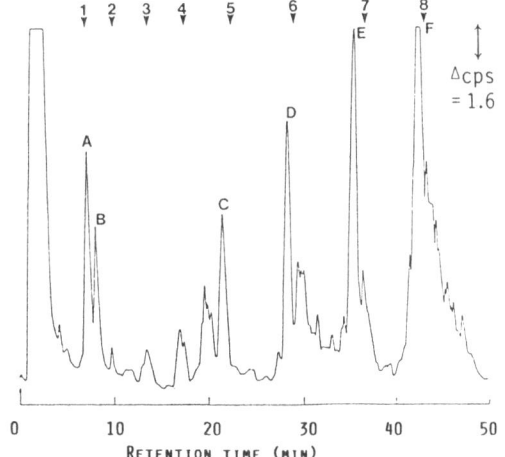

RETENTION TIME (MIN)

20 min, the salt-protein complex is centrifuged down, washed with
chloroform/methanol (2 ml), re-centrifuged, and the supernatants combined. The residue after solvent evaporation at 50° with a slow
N$_2$ stream is suspended in 50 mM K phosphate (pH 5.3; 400 μl), and
200 μl is chromatographed. Recoveries (mean ±S.E.M.; n = 3) as established with authentic acylCoA's were 99 ±3, 100 ±34 and 68±5% for
butyryl-CoA, octanoyl-CoA and tetradecanoyl CoA respectively. The
essentiality of the (NH$_4$)$_2$SO$_4$ treatment was evidenced by the lower
recoveries if it were omitted: 74 ±2, 12 ±4 and 8 ±1% respectively.
In preliminary studies, liver **Mt** were incubated with [U-^{14}C]palmitate
and the incubates treated and analyzed as above. The accumulation
of U-^{14}C-intermediates was satisfactory, especially in the presence
of rotenone to slow the flux through β-oxidation. Fig. 8 shows a
representative run.

References

1. Bremer, J. & Osmundsen, H. (1984) in *Fatty Acid Metabolism and its Regulation* (Numa, S., ed.), Elsevier, Amsterdam, pp. 113-154.
2. Christophersen, B.O. & Bremer, J. (1972) *Biochim. Biophys. Acta 260*, 515-526.
3. Stanley, K.K. & Tubbs, P.K. (1975) *Biochem J. 150*, 77-78.
4. Cervenka, J. & Osmundsen, H. (1982) *J. Lipid Res. 23*, 1243-1246.
5. Bishop, J.E. & Hajra, A.K. (1980) *Anal. Biochem. 106*, 344-350.
6. Simon, E.J. & Shemin, D. (1953) *J. Am. Chem. Soc. 75*, 2520-2524.
7. Chase, J.F.A. & Tubbs, P.K. (1972) *Biochem. J. 129*, 55-65.
8. Goldman, P. & Vagelos, P.R. (1961) *J. Biol. Chem. 236*, 2620-2625.
9. Al-Ahrif, A. & Blecher, M. (1969) *J. Lipid Res. 10*, 344-345.
10. Osmundsen, H., Neat, C.E. & Norum, K.R. (1979) *FEBS Lett. 99*, 292-296.
11. Osmundsen, H. (1982) *Int. J. Biochem. 14*, 905-914.
12. Kawaguci, A., Yoshimura, T. & Okuda, S. (1981) *J. Biochem. (Tokyo) 89*, 337-339.
13. Bartlett, K. & Gompertz, D. (1974) *Biochem. Med. 10*, 15-20.
14. Hiltunen, K., Karki, T., Hassinen, I.E. & Osmundsen, H. (1986) *J. Biol. Chem. 261*, 16484-16493.
15. Osmundsen, H., Cervenka, J. & Bremer, J. (1982) *Biochem. J. 208*, 749-757.
16. Osmundsen, H. & Bjørnstad, K. (1985) *Biochem J. 230*, 329-338.
17. Billington, D., Osmundsen, H. & Sherratt, H.S.A. (1978) *Biochem. Pharmacol. 27*, 2879-2890.
18. Halestrap, A.P. & Dunlop, J.L. (1986) *Biochem. J. 239*, 559-565.
19. Osmundsen, H. & Bremer, J. (1976) *FEBS Lett. 69*, 221-224.
20. Bremer, J. & Norum, K.R. (1982) *J. Lipid Res. 23*, 243-256.
21. Krag-Hansen, U. (1981) *Pharmacol. Rev. 33*, 17-53.
22. Bartlett, K., Bartlett, P., Bartlett, N., Sherratt, H.S.A. (1985) *Biochem. J. 229*, 559-560.
23. Sherratt, H.S.A. & Osmundsen, H. (1976) *Biochem. Pharmacol. 25*, 743-750.
24. Garland, P.B. (1968) in *The Metabolic Roles of Citrate* (Goodwin, T.J., ed.), Academic Press, London, pp. 41-60.
25. Veerkamp, J.H. (1981) in *Mitochondria and Muscular Diseases* (Busch, H.F.M., Jennekens, F.G.I. & Scholte, H.R., eds.), Mefar, Beesterzwoog, pp. 29-50.
26. Bremer, J. & Wojtczak, A.B. (1972) *Biochim. Biophys. Acta 280*, 515-530.
27. Osmundsen, H. & Sherratt, H.S.A. (1975) *FEBS Lett. 55*, 38-41.
28. Turnbull, D.M., Bartlett, K., Younan, S.I.M. & Sherratt, H.S.A. (1984) *Biochem. Pharmacol. 33*, 475-481.
29. Veerkamp, J.H., van Moerkerk, H.T.B. & Bakkern, J.A.J.M. (1986) *Biochim. Biophys. Acta 876*, 133-137.
30. Osmundsen, H. (1981) *Meths. Enzymol. 72*, 306-314.

31. Gohil, A.G., Jones, D.A. & Edwards, R.H.T.(1984) *Clin. Sci.* *66*, 173-178.

32. Turnbull, D.M., Sherratt, H.S.A., Davies, D.M. & Sykes, A.G. (1982) *Biochem. J. 206*, 511-516.

33. Neat, C.E., Thomassen, M.S. & Osmundsen, H. (1981) *Biochem. J. 196*, 149-159.

34. Kolvraa, S. & Gregersen, N. (1986) *Biochim. Biophys. Acta 876*, 515-526.

35. Brady, L.J., Brady, P.S., Romson, D.R. & Hoppel, M.D. (1985) *Biochem. J. 231*, 439-444 (also *229*, 717-721).

36. Osmundsen, H. (1983) in *Iodinated Density Gradient Media* (Rickwood, D., ed.),   IRL Press, Oxford, pp. 139-146.

37. Hajra, A.K. & Wu, D. (1985) *Anal. Biochem. 148*, 233-244.

38. Osmundsen, H., Thomassen, M.S., Hiltunen, K. & Berge, R.K. (1987) *Eur. J. Cell Biol.*, in press.

39. Zaar, K., Volkl, A. & Fahimi, H.D. (1986) *Eur. J. Cell Biol. 40*, 16-24.

40. Norseth, J. & Thomassen, M.S. (1983) *Biochim. Biophys. Acta 751*, 312-320.

41. Thomassen, M.S., Helgerud, P. & Norum, K.R. (1985) *Biochem. J. 255*, 301-306.

42. Hryb, D.J. & Hogg, J.F. (1979) *Biochem. Biophys. Res. Comm. 87*, 1200-1206.

43. Eliassen, K.J. & Osmundsen, H. (1984) *Biochem. Pharmacol. 33*, 1023-1031.

44. Walusimbi-Kisitu, M. & Harrison, E.H. (1983) *J. Lipid Res. 24*, 1077-1084.

45. Osmundsen, H., Neat, C.E. & Borrebaek, B. (1980) *Int. J. Biochem. 12*, 625-630.

46. West,I.D., Mitchell, R.A., Moody, A.J. & Mitchell, P. (1986) *Biochem. J. 236*, 15-21.

47. Lazarow, P.B. (1977) *Science 197*, 580-581.

48. Veerkamp, J.H. & van Moerkerk, H.T.B. (1986) *Biochim. Biophys. Acta 875*, 301-310.

49. Causey, A.G., Middleton, B. & Bartlett, K. (1986) *Biochem. J. 235*, 343-350.

50. Hosokawa, Y., Shimomura, Y., Harris, R.A. & Osawa, T. (1986) *Anal. Biochem. 153*, 45-49.

51. Woldegiorgios, G., Spennetta, T., Corkey, B.E. & Shrago, E. (1985) *Anal. Biochem. 150*, 8-12.

52. Baker, F.C. & Schooley, D.A. (1979) *Anal. Biochem. 94*, 417-424.

53. Bartlett, K. & Causey, A.G. (1987) *Meths. Enzymol.*, in press.

54. Causey, A.G. & Bartlett, K. (1986) *Biochem. Soc. Trans, 14*, 1175-1176.

#A-5

PHARMACOLOGICAL AND TOXICOLOGICAL ASPECTS OF PEROXISOME PROLIFERATORS

R.K. Berge, N. Aarsæther, A. Aarsland and T. Ghezai

Laboratory of Clinical Biochemistry
University of Bergen
N-5016 Haukeland Sykehus, Bergen, Norway

In male rats, LL* drugs including clofibrate, tiadenol, niadenate, nicotinic acid and cholestyramine differed in their effects on hepatic enzymic activities involved in long-chain fatty acyl-CoA (pyCoA) formation and breakdown, and on peroxisome $\bar{s}$ values that could explain some of the observed changes in s.a.'s of fatty acid metabolizing enzymes with multiple subcellular localization. Long-chain acyl-CoA content was strongly correlated with pyCoA hydrolase and peroxisomal β-oxidation activity (which bring about peroxisomal proliferation), pointing to a common regulation mechanism for the enzymes.

A microsomal carboxylesterase acts on administered clofibrate to give clofibric acid and clofibroyl-CoA. Hence the acyl-CoA pattern as well as level may trigger the proliferation. The proliferation may also be related to cancer, insofar as 2-stage transformation experiments with mouse embryo fibroblasts showed a tumour-promoting action of clofibrate.

The range of bioactive compounds possessing or readily furnishing a carboxyl group includes LL* agents. Different LL drugs are now considered in relation to enzymes and metabolites involved in fatty acid metabolism. LL agents can be roughly classified into those which are absorbed and those which act within the lumen of the ·GI tract and are poorly absorbed. Besides clofibrate, now widely used, absorbable drugs include tiadenol (structurally unrelated to clofibrate), nicotinic acid, and niadenate - a pro-drug of nicotinic acid

---

Abbreviations - by author:* see text, e.g. **M = a centrifugal fraction; *editorial:* antihyperlipidaemic/lipid-lowering/hypolipidaemic, LL; palmitoyl-CoA, pyCoA; specific activity, s.a.; centrifugal integral ('effect') as measure of *g*-min, $c_i = {_o}\!\int^t \mathrm{rpm}^2 dt \times 10^{-8}$ min.

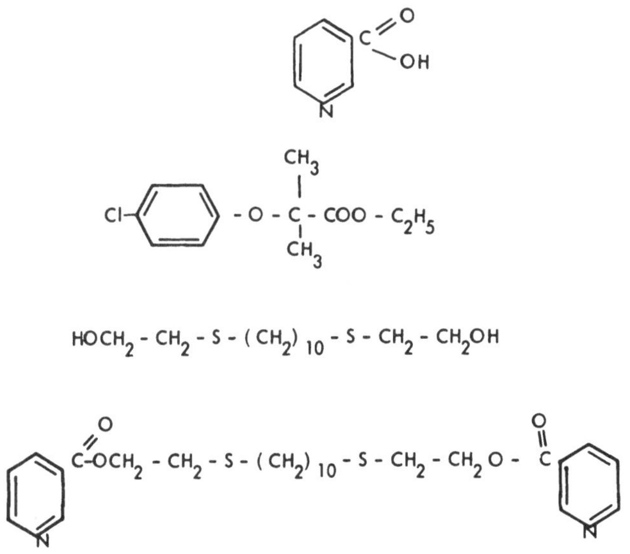

Fig. 1. LL drug structures.

and tiadenol (Fig. 1). Clofibrate and tiadenol are highly effective
in lowering serum cholesterol, triglycerides (TG), and the ratio
VLDL + LDL/HDL in different types of hyperlipoproteinaemias [1, 2].
Nicotinic acid represents a different class of LL drug and is thought
to act mainly by inhibiting peripheral lipolysis, thereby making
less free fatty acids available for the synthesis of TG, VLDL and
eventually LDL [3, 4].   Cholestyramine, an anionic resin which is
not absorbed, traps bile acids and other acidic sterols in the intestinal lumen, and reduces serum cholesterol and LDL by reducing the
negative feedback on cholesterol 7-β-hydroxylase [5, 6]; it does
not lower serum TG.

The absorbable LL drugs cause various changes in the histology,
enzyme activities and the content of metabolites, viz. free CoA-SH
and long-chain acyl-CoA in the liver of animals [7, 8]. Pronounced
hepatomegaly develops, and hepatocyte hypertrophy, proliferation of
peroxisomes - with increased peroxisomal β-oxidation - and e.r.,
and an increase in mitochondrial number and size [9-11]. Hepatocellular
carcinoma appears long-term [12, 13].

Similarly to LL drugs, high-fat diets, starvation and diabetes
entail enhanced fatty acid oxidation and induction of long-chain
acyl-CoA hydrolase (pyCoA hydrolase activity) and the peroxisome
β-oxidation system [14, 15]. Hepatic acyl-CoA hydrolases comprise
a series varying in chain-length specificity [16]. Conceivably their
function may be to ensure that free CoA-SH is always available for
cellular metabolism. Moreover, as LL drugs, starvation and high-fat
diets increase the hepatic long-chain acyl-CoA content and fatty

acid β-oxidation, the hydrolase and β-oxidation augmentations may represent detoxication pathways, induced when the organism is faced with a high influx of fatty acids which are poorly oxidized by mito-chondria.

Concerning regulation, enzymes confined to one compartment are of course influenced only by regulators present therein. Hence sub-cellular fractionation of cells and tissues is informative concerning identification, characterization, biosynthesis, turnover and function of enzymes involved in fatty acid metabolism. Such approaches have now been chosen to investigate compartmentalization and how the distri-bution of pyCoA hydrolase and other peroxisomal enzymes is influenced nutritionally and after treatment with LL drugs, absorbable or non-absorbable. [The author's text and ref. list have been slightly curtailed. See #A-4 by H. Osmundsen et al. for some pertinent assay methodology.- *Ed.*]

CELLULAR LOCALIZATION OF LONG–CHAIN ACYL–CoA HYDROLASE

Differential centrifugation approach

Fractions were prepared from liver homogenates (0.25 M sucrose, 10 mM pH 7.4 Hepes) at 0-4° by a procedure [17] based on that of de Duve et al. [18], using a Sorvall RC-5 centrifuge and a HB-4 rotor (rad. 4.8 cm min., 14.6 cm max.). The following pellets were obtained, adding two washings to the supernatant except for the final pellet ('ci' definition: title p., footnote): **N**, crude nuclear (ci 0.63; 10 min); **M**, mitochondrial (ci 6.4; 10 min); **L₁**, light mitochondrial (ci 43; 30 min); and **P**, microsomal (35,000 rpm, ci 735; 60 min).

This classical approach having shown that **M** (with a mitochondrial matrix localization) and **P** each have a pyCoA hydrolase, both enzymes were purified and characterized [16, 19]. Evidently they differ, the estimated values being as follows for the mitochondrial and micro-somal hydrolases respectively.- M_r: 19,000, 59,000 (1 subunit in each); $s_{20,w}$ (S): 2.1, 4.3; Stoke's radius: 19, 31 Å; pI: 6.0, 6.9; susceptible substrates: C-10 to C-18 and C-7 to C-18 (C-16 maximal for each). Complex kinetics were found for hydrolysis of pyCoA: micelle formation governs its availability as a substrate, and the hydrolase shows a different reaction behaviour towards monomeric and micellar forms [19, 20].

Observed influences.- Dietary administration at a high concentrat-ion of an absorbable LL drug (clofibrate, tiadenol, niadenate; 0.3%) significantly increased the specific pyCoA hydrolase recovery and s.a. in the **M** and cytosol fractions, but conversely for the **P** fraction; moreover, the recovery in the peroxisome–enriched fraction (**L₁**) rose slightly. Nicotinic acid failed to affect the enzyme activities in daily doses up to 200 mg/kg/day; but when given by stomach tube, twice daily for 10 days, it slightly enhanced pyCoA s.a. in **L**, **P**

Fig. 2. Marker enzyme s.a.'s in
L_1 and L_2 fractions obtained
from liver homogenates by differential centrifugation (see
text, & [17]). The ci
values to furnish L_2 (which
was not re-spun) were lower
than, but overlapped with,
those for L_1.
The s.a. values shown for
drug-treated rats are relative to
normal-liver s.a.'s taken as 1.0.
a, pyCoA hydrolase; b, catalase;
c, urate oxidase; d, NADPH-
cytochrome c reductase; e, cytochrome c oxidase; f, acid phosphatase.

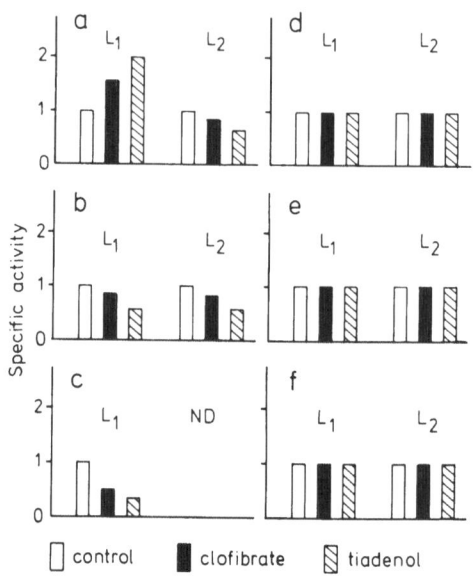

and (1.3-fold) **M** fractions; the cytosol fraction showed no change
in s.a. [21].   Cholestyramine similarly intubated (1000 mg/kg/day)
did not significantly affect the activity in any fraction [21].

The clofibrate-like LL drugs and high-fat diets or starvation
had a remarkably similar effect on fatty acid oxidation and induction
of long-chain acyl-CoA hydrolyase activity [14, 22]. After feeding
male rats a 20% (w/w) partially hydrogenated fish-oil diet for 29 days
an adaption in lipid metabolism was observed as indicated by changes
in the s.a.'s of peroxisomal β-oxidation enzymes, pyCoA hydrolase,
carnitine py-transferase and pyCoA synthetase (to be published).
Subcellular fractionation of liver homogenates revealed that pyCoA
s.a. increased 2.4-, 1.3- and 2.0-fold in the mitochondrial, peroxisomal and cytosolic fractions respectively.   Fasting increased the
mitochondrial pyCoA hydrolase activity 1.5- to 1.8-fold and the cytosolic activity 1.5-fold, but hardly affected the **L**-fraction activity.

Approaches entailing isopycnic equilibration

The foregoing classical centrifugation approach gives insuffici-
ent resolution to determine whether acyl-CoA hydrolase activity is
also present in the peroxisomes.   As an alternative to L_1, a fraction
designated L_2 enriched in mitochondria and peroxisomes was prepared
(between ci = 2.1 and ci = 12).   In order to prepare pure peroxisomes, L_2 was further fractionated in a linear gradient of 38–49%
(w/w) sucrose, using a vertical rotor, Sorvall TV850 [17].   It should
be noted that the amounts and, as shown in Fig. 2, the s.a.'s as
well as recoveries of peroxisomal enzymes in L-fractions (L_1 and
L_2) after subcellular fractionation depended on the centrifugal cut
taken.

Observed influences.- With both L_1 and L_2 the recovery and (see Fig. 2) the s.a. of catalase fell after clofibrate or tiadenol adminis- tration, and the s.a. of peroxisomal β-oxidation activity rose. L_1 also had decreased urate oxidase s.a. Both recovery and s.a. of pyCoA hydrolase rose in L_1 and fell in L_2 following treatment with LL agents, indicating that most of the extra activity after administration of peroxisomal proliferators is localized in particles that have $\bar{s}$ values [average sedimentation coefficient for group of particles] <10,000 S. The density gradient distribution profile for this activity was similar to that of acid phosphatase [17]; thus the possibility of a lysosomal origin had to be considered.

Sedimentation profile of pyCoA hydrolase and some marker enzymes

The approach adopted to learn more about the sedimentation char- acteristics of these activities, using the post-nuclear supernatant ('100% homogenate'), was analytical differential centrifugation with parallel sedimentation in swinging bucket tubes for whose rate equation a logarithmic expression applies (the spins being monitored by an integrator) [17], enabling $\bar{s}$ to be estimated [17] for peroxisomes, lysosomes and mitochondria.-

$$('Z' =) \ \log_{10}\{1 - [1 - (R_{min}/R_{max})(Y/100)]\} = \frac{-s_o \int^t rpm^2 \ dt}{3.5 \times 10^{13}}$$

where R_{max} and R_{min} are the distances (cm) from the axis of rotation to the bottom and the surface of the fluid column respectively; Y = % of particles that have sedimented. The plot of 'Z' against rpm^2 gives a straight line for a single population of particles [17].

The sedimentation profile of pyCoA hydrolase activity was discon- tinuous in normal and especially in the clofibrate-treated animals (Fig. 3), indicating the presence of more than a single population of particles associated with this activity. A log plot of the data (Fig. 4) clearly demonstrated a deviation from the simple linear rela- tionship expected for a single population of particles. The data for pyCoA hydrolase appear to fit two linear functions corresponding to $\bar{s}$ values of 20,650 S and 4400 S for clofibrate-treated animals and 19,500 S and 6740 S for normal animals.

The acid phosphatase profile was likewise discontinuous, but identical in the control and clofibrate-treated animals (Fig. 3). Evidently LL drugs do not change lysosomal polydispersity. A log plot of the data for the two groups appeared to fit 3 linear functions, whose equations were calculated using linear regression analysis by the method of least squares [17]. Using convergence theory, the phosphatase $\bar{s}$ values for the 3 populations in normal rats were found to be 33,500 ±5,600, 14,000 ±3,200 and 5000 ±400.

Based on malate dehydrogenase and catalase profiles, a single population of particles was evident for mitochondria in normal rats

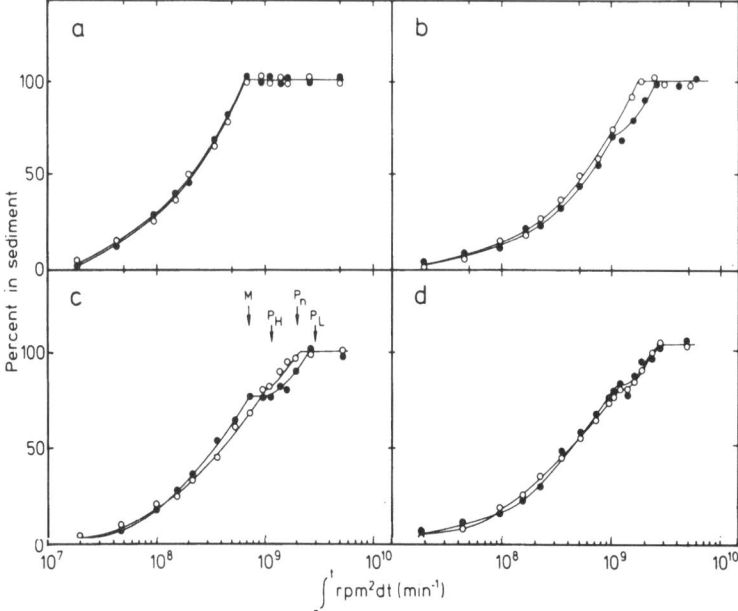

Fig. 3. Sedimentation profiles, normal (o) and after clofibrate (●), for pyCoA hydrolase (c) and marker enzymes – malate dehydrogenase (a), catalase (b) and acid phosphatase (d). **Arrows** indicate the convergence level for sedimentation of mitochondria (M) and peroxisomes – normal (P_n), heavy (P_H) and light (P_L). *Figs. 3 (in part) & 4 are from ref. [17], by permission.*

or after clofibrate ($\bar{s}$ = 18,750) and for peroxisomes in normal rats ($\bar{s}$ = 7,420). For catalase, however, clofibrate led – maybe resulting from peroxisomal proliferation – to heavy and light populations, $\bar{s}$ = 11,860 and 4,240; the latter was notable in having a high s.a. for pyCoA-dependent dehydrogenase activity compared with urate oxidase and catalase [23]. Peroxisomal β-oxidation surpassed catalase in polydispersity: clofibrate gave $\bar{s}$ = 11,860 and a range from 4,200 to 850, compared with a single $\bar{s}$ value (6,160) in normal rats.

Lysosomes show no change in acid phosphatase s.a. or polydispersity after peroxisomal proliferator (Figs. 2–4), and it further appears from the post-drug profiles that acid phosphatase and pyCoA hydrolase are not attributable to the same subcellular particles. One line of evidence for a peroxisomal location of pyCoA hydrolase is the clofibrate-induced enrichment in an appropriate centrifugal fraction (Fig. 2), Further evidence that the post-mitochondriaL activity, in two populations of particles, is partly peroxisomal comes from the estimated $\bar{s}$ values, matching those for pyCoA-dependent dehydrogenase in the two groups of animals. This localization was most evident for the peroxisomal light population after clofibrate treatment; it sediments only partly in preparing the L_2 fraction, which

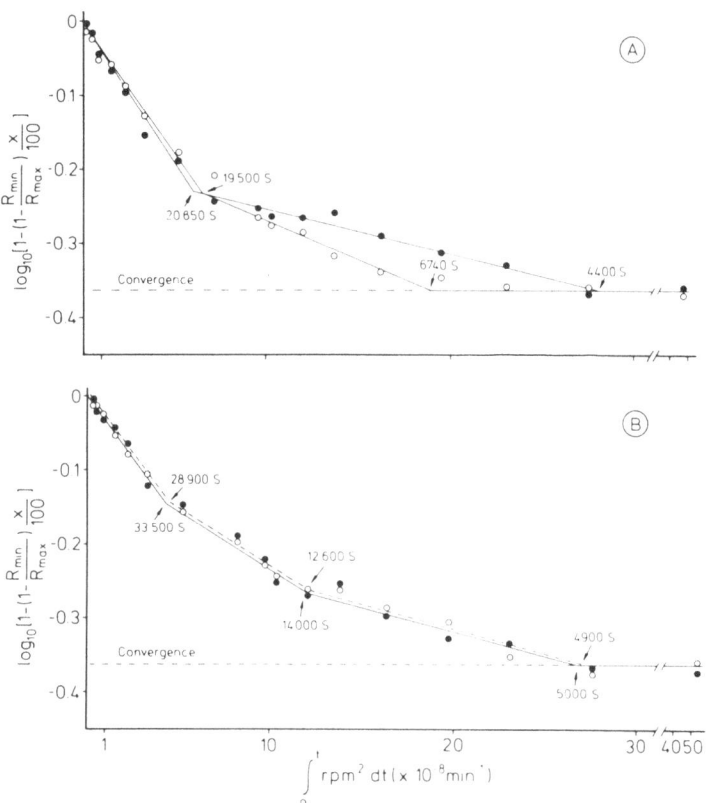

Fig. 4. Determination of $\bar{s}$, normal (o) or after clofibrate (●; interrupted line in case of B), for (A) pyCoA hydrolase (mitochondrial and peroxisomal), and (B) acid phosphatase. Data from Fig. 3.

accordingly shows no increased activity. Our finding of high pyCoA hydrolase activity in the cytosolic fraction after LL drug treatment may be due to induced peroxisomal heterogeneity and not only to increased peroxisomal fragility and hence leakage of peroxisomal matrix enzymes.

RELATIONSHIP BETWEEN PEROXISOMAL β-OXIDATION, LONG-CHAIN ACYL-CoA HYDROLASE ACTIVITY AND PEROXISOME PROLIFERATION

With all four of the absorbable drugs tested, there are increases in peroxisomal β-oxidation and in pyCoA hydrolase activity [9]; for both (correlation r = 0.96, P <0.01) the potency order for enzyme induction was niadenate > tiadenol > clofibrate > nicotinic acid. Also well correlated (r = 0.94) were the increases produced in the two activities by diets containing 5–30% (w/w) of partially hydrogenated fish oil [14].

For the two activities, in whole homogenates or cytosol or a combined **LP** fraction, tiadenol showed sigmoidal dose-dependency with increasing doses above the near-ineffective dose, ~90 mg/kg/day. For urate oxidase (latter dose ineffective) the effect of tiadenol markedly differed from that on these two activities: at the highest dose levels the homogenate and **LP** activity decreased (no activity detectable in the cytosol).

Morphometric analysis has shown that the frequency of peroxisomes with dense core, and the area of dense core within each peroxisome, decreased as a result of peroxisome proliferation [23]. The observation that this LL drug gives rise to new populations of peroxisomes with altered enzyme content and membrane characteristics, and the observation of marked dissociation of the activities of urate oxidase and peroxisomal pyCoA oxidation and pyCoA hydrolase together with a shift of the latter two enzymes to the cytosol, suggest that they are located in induced anucleoid peroxisomes. Hence decreased urate oxidase after LL drug administration may be a consequence of proliferation of peroxisomes. Another biochemical parameter which may well be related to peroxisomal proliferation is an increased pyCoA hydrolase activity in the cytosol.

Administration of nicotinic acid and cholestyramine slightly but significantly enhanced the activities of peroxisomal β-oxidation and of catalase. Amongst the enzyme activivities measured, urate oxidase showed the greatest increase - ~2-fold in the L-fraction [21] – as also obtainable by acetylsalicylic acid treatment [24, 25]. As neither LL drug increased cytosolic pyCoA hydrolase activity, the induction data suggest that these drugs, unlike other LL drugs, increase the core-containing peroxisomes. Since the changes observed here with high doses of nicotinic acid and cholestyramine were small, it is unlikely that peroxisomal proliferation and increased fatty acyl-CoA oxidation represent the primary mechanism of action of these LL drugs. The structural similarity of nicotinic acid, acetylsalicylic acid and other aromatic carboxylic acids which cause peroxisome proliferation, possibly by a mechanism involving formation of CoA-esters, may have some bearing on interpretation of the present findings.

MECHANISM OF INDUCTION OF PEROXISOMAL β-OXIDATION AND pyCoA HYDROLASE

In our experiments niadenate and tiadenol were considerably more potent than clofibrate in inducing enlargement of the liver and in inducing lipid-metabolizing enzymes associated with the peroxisomes and other subcellular organelles. The findings were qualitatively similar with all the drugs, but some of the phenomena observed with clofibrate were more clearly discernible in experiments with tiadenol and niadenate. Since tiadenol and clofibrate are structurally unrelated and since with high fat-feeding there is a similar correlation between pyCoA hydrolase activity and peroxisomal β-oxidation, a common regulatory mechanism for these two enzyme systems is plausible.

Fig. 5. The concentrations of
long-chain acyl-CoA and of
free CoA-SH, and their ratio
(---------), after administration
of different doses of tiadenol.

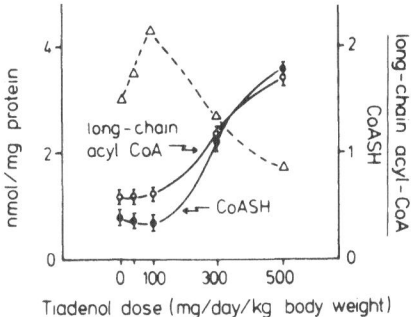

Tiadenol dose (mg/day/kg body weight)

Long-chain acyl-CoA, the substrate for both pyCoA hydrolase
and peroxisomal β-oxidation, may regulate the enzyme system by a
substrate-induced mechanism.   Long-chain acyl-CoA and free CoA-SH
have now been measured in whole homogenates and in subcellular fractions. Linear regression analysis of the long-chain acyl-CoA content,
but not the free CoA-SH, *vs.* pyCoA hydrolase and peroxisomal β-oxidation
activities showed highly significant linear correlations both in
total liver homogenates and in the L-fraction.   Interestingly, the
ratio of long-chain acyl-CoA to free CoA-SH was enhanced by low doses
of tiadenol, maximally with 90 mg/kg/day where induction of the
two activities was seen (Fig. 5). If the ratio of long-chain acyl-CoA
to free CoA-SH is not changed during the cell-fractionation procedure,
the overall results provide support for the hypothesis that a common
induction mechanism exists for pyCoA hydrolase and peroxisomal β-oxidation activities, possibly exerted through an increased cellular level
of long-chain acyl-CoA.

Clofibrate is rapidly hydrolysed by tissue and serum esterases
to clofibric acid, which circulates in the blood and may be the pharmacologically active form. Recently (to be published) we have observed
that bis(hydroxyethylthio)-1,10-decane, a possible   dicarboxylic
metabolite of tiadenol, triggers the induction of peroxisomal fatty
acid oxidation and pyCoA hydrolase activity to the same extent as
tiadenol. Of the 4 distinct carboxylesterases in rat liver microsomes,
the dominant one (pI 6.0) cleaves clofibrate; but clofibrate hydrolase
s.a. was hardly affected by clofibrate administration, indicating
that this esterase was not induced by its own substrate [26]. Because
of such enzymic activity, clofibric acid as well as clofibroyl-CoA
appears in rat liver after clofibrate administration [27]. The CoA
derivative of the dicarboxylic acid of tiadenol can be formed in
liver by an acyl-CoA synthetase (work to be published). Normal β-
oxidation does not occur with this derivative (there are β-located S
atoms in the chain), nor with similar derivatives of clofibrate or of
clofibrate-like LL drugs. As we have found a clofibroyl-CoA hydrolase activity in liver, seemingly not merely the acyl-CoA level but
also the acyl-CoA pattern may be the trigger of peroxisomal proliferation accompanying increased peroxisomal fatty acid oxidation and
pyCoA hydrolase and/or xenobiotic acyl-CoA hydrolase activities.

For rat hepatocytes treated with LL drugs *in vitro* the earliest reported change is increased triglyceride synthesis [28]. Furthermore, we have observed that long-term exposure of mouse-embryo fibroblasts to clofibrate, tiadenol or niadenate stimulated the formation of cell foci which on maturation contained adipocytes [29]. As these drugs mediated induction of peroxisomal enzymes and pyCoA hydrolase in these cells [30], fat accumulation therein may itself be the stimulus for induction of peroxisomal β-oxidation. Lipid accumulation in the livers of chlorpromazine-treated rats is well documented, but is not accompanied by peroxisomal proliferation ([31]; cf. #A-1, this vol.). Thus lipid accumulation appears not to be a trigger for peroxisomal proliferation.

MICROSOMAL pyCoA HYDROLASE AND CARBOXYLESTERASE IDENTITY

Besides the above-mentioned clofibrate-cleaving enzyme [26], an esterase with pI 6.2-6.4 and of M_r 60,000 is reported to hydrolyse propanidid and aspirin [32]. Recently we have shown biochemical identity between this esterase and microsomal (hepatic) pyCoA hydrolase: (1) antisera against the two purified enzyme preparations were cross-reactive; (2) they co-migrate in SDS-PAGE; (3) they have identical inhibition characteristics with cations and different substrates [33]. However, immunoprecipitation and inhibition experiments confirm that these microsomal hydrolase/esterase findings from the pyCoA hydrolases of rat-liver cytosol and mitochondria [26, 33].

THE LINK BETWEEN CANCER AND PEROXISOMAL PROLIFERATION

It is known that long-term treatment of rats and mice with different peroxisomal proliferators, viz. LL drugs and the plasticizer DEHP, induces hepatocellular tumours (questionably in exposed humans), although the agents seem non-mutagenic; possibly they comprise a distinct class of carcinogens, but conceivably they may be merely tumour-promoters. Promoters are not themselves carcinogenic or genotoxic, and their effectiveness depends on the timing of the exposure in relation to that for the initiator. That carcinogenesis is a multi-step process has been established by diverse approaches.

In recent 2-stage transformation experiments with mouse embryo fibroblasts (C3H/10T1/2 C18), we have looked for 'Type III foci', which are of 2-10 mm diam. and have, at their circumference, a characteristic criss-cross cell-pattern [34] which matured into foci containing adipocytes. The following results represent no. of dishes with foci/no. scored (*N.B.: the* [...] *entries are **not** refs.; defined below):*
#One treatment only
  Acetone, 0.5% - 1/35; MCA, 3.7 μM - 8/36; niadenate, 5μM - 0/37, 20 μM - 0/38; clofibrate, 1 μM - 0/39, 5 μM - 0/27; tiadenol, 2 μM - 0/12, 5 μM - 0/12.
#Two treatments
  2nd with 0.5% acetone; first: same - 1/54 [relative plating efficiency taken as 100%]; MCA, 3.7 μM - 10/42 [69]; 0.37 μM - 2/41 [87];

First with MCA, 0.37 μM; second with -
#12-0-tetradecanoylphorbol-13-acetate (TPA; a known promoter): 24/50
[68];
#niadenate, 5 μM – 4/29 [112], 20 μM – 9/35 [94]; clofibrate, 5 μM –
8/23 [55]; tiadenol 2 μM – 2/24 [96], 5 μM – 0/11 [not determined],
20 μM – 0/17 [74].

Evidently MCA-treated C18 cells developed more foci if exposed long-
term (for 6 weeks) to clofibrate or niadenate, although not tiadenol.

There is much evidence that reactive oxygen, especially free
radicals, is important in tumour promotion by TPA. The occurrence
of DNA breaks in TPA-exposed cells co-cultured with phagocytes suggests
indirect as well as direct mechanisms. Moreover, many general characteristics of initiation and promotion as established for skin appear
to hold for the 2-stage model of hepatocarcinogenesis also. Recently
(unpublished) we have observed that the peroxisomal β-oxidation system
is increased in C18 cells by TPA, clofibrate or tiadenol, and similarly
in liver by TPA administration. Fahl et al. [35] have observed a
link between DNA damage and H_2O_2 generation by LL drug-induced liver
peroxisomes. There are, then, similarities between the classical
co-carcinogen, TPA, and a potent LL drug, clofibrate: (1) both are
tumour-promoting but not mutagenic: (2) both induce peroxisomal
β-oxidation *in vivo* and *in vitro*; (3) both lead to lipid accumulation.

For LL drugs, increased synthesis of the peroxisomal β-oxidation
enzyme, which generates H_2O_2, has been postulated as the link between
peroxisomal proliferation and hepatocarcinogenesis. Increased lipid
peroxidation has been observed in the liver of rats treated with
peroxisome proliferators [36]. One manifestation of membrane damage
is lipid peroxidation. Free (oxygen) radicals initiate this process,
and their generation has been postulated as a step in tumour promotion.
A very pertinent recent finding is that peroxisomes can produce the
most potent oxygen radical, viz. the toxic hydroxyl radical ($^.$OH).
As clofibrate and niadenate are tumour-promoting, an excess production
of H_2O_2 and $^.$OH conceivably contribute to hepatocarcinogenesis, by
mechanisms unrelated to direct DNA damage.

Acknowledgements

The authors are grateful for grant support from the Norwegian
Society for Fighting Cancer, the Norwegian Research Council for Science
and the Humanities, and the Norwegian Cancer Society.

References

1. Baggio, C., Briani, G., Fellin, R., Martini, S., Baiocchi, M.R.
    & Crepaldi, C. (1979) *Artery 5*, 486-494.

2. Saba, P., Paglial, E., Scalabrino, A., Galeone, F.,
    Giuntoli, F., Guidi, G., Morini, S. & Lavoratti, G.   (1980)
    Curr. Ther. Res. 27, 677-683.

3. Miettinen, T.A. (1976) in *Lipid Pharmacology* (Paoletti, R. &
    Glueck, C.I., eds.), Vol. 2, Academic Press, New York, pp. 83-87.

4. Grundy, S.M., Mok, H.Y.I., Zech, L. & Berman, M. (1983)
    J. Lipid Res. 22, 24-30.

5. Gibbons, G.F., Mitropoulos, K.A. & Myant, N.B. (1982)
    Biochemistry of Cholesterol, Elsevier, Amsterdam.

6. Dietschy, I.M. & Brown, M. (1974) *J. Lipid Res. 15*, 508-516.

7. Berge, R.K. & Bakke, O.M.(1981) *Biochem. Pharmacol. 30*, 2251-
    2256.

8. Berge, R.K. & Aarsland, A. (1985) *Biochim. Biophys. Acta
    837*, 141-151.

9. Svoboda, D.I. & Azarnoff, D.L. (1971)   *Cancer Res. 39*, 3419-
    3428.

10. Cohen, A.I. & Grasso, P. (1981)   *Food Cosmet.. Toxicol. 19*,
    585-605.

11. Lalwani, N.D., Reddy, J.K., Gureshi, S.A., Reddy, M.K. & Moehle, C.M.
    (1984) *Am. J. Path. 114*, 171-174.

12. Hashimoto, T. (1982) *Ann. N.Y. Acad. Sci. 386*, 5-13.

13. Reddy, J.K., Azarnoff, D.L. & Hignite, C.E. (1980) *Nature 283*,
    397-398.

14. Berge, R.K. & Thomassen, M.S. (1984) *Lipids 20*, 49-52.

15. Thomassen, M.S., Christiansen, E.N. & Norum, K.R. (1982)
    Biochem. J. 206, 195-202.

16. Berge, R.K. (1980) *Biochim. Biophys. Acta   574*, 321-323.

17. Berge, R.K., Flatmark, T. & Osmundsen, H. (1984) *Eur. J.
    Biochem., 141*, 637-644.

18. de Duve, C., Pressman, B.C., Gianetto, R., Wattiaux, R. &
    Appelmans, F. (1955) *Biochem. J. 60*,   604-617.

19. Berge, R.K. & Farstad, M. (1981) *Meths. Enzymol. 71*, 234-242.

20. Berge, R.K., Slinde, E. & Farstad, M. (1980) *Biochim. Biophys.
    Acta 666*, 25-35.

21. Bakke, O.M. & Berge, R.K. (1984) *Biochem. Pharmacol.33*, 3077-3080.

22. Berge, R.K., Hosøy, L.H. & Farstad, M. (1984) *Int. J. Biochem.
    16*, 403-410.

23. Flatmark, T., Christiansen, E.N. & Kryvi, H. (1981) *Eur. J.Cell
    Biol. 24*, 62-69.

24. Sakurai, T., Miyazawa, S., Osumi, T., Furuta, S. & Hasimoto, T.
    Toxicol. Appl. Pharmacol. 59, 8-18.

25. Ishii, H. & Suga, T. (1979) *Biochem. Pharmacol. 28*, 2829-2834.

26. Mentlein, R., Lembke, B., Vik, H. & Berge, R.K. (1986)
    Biochem. Pharmacol. 35, 2727-2730.

27. Lygre, T., Aarsæther, N., Stensland, E. & Berge, R.K. (1986)
    J. Chromatog. 381, 95-105.

28. Cappuizzi, P.M., Intenzo, C.M., Lackman, R.D., Whereat, A.F. &
    Scott, D.M. (1985) *Biochem. Pharmacol. 32*, 2195-2203.

29. Lillehaug, I.R., Aarsæther, N. Berge, R.K., & Male, R.
    (1986)   *Int. J. Cancer 39*,   97-100.

30. Berge, R.K. & Lillehaug, I.R. (1985) *Int. J. Cancer 36*, 489- 494.
31. Price, S.C., Hall, D.E. & Hinton, R.H. (1985) *Toxicol. Lett. 25*, 11- 13.
32. Mentlein, R., Schumann, M. & Heymann, E. (1985) *Arch. Biochem. Biophys. 234*, 612-621.
33. Mentlein, R., Berge, R.K. & Heymann, E. (1985} *Biochem. J. 232*, 479-483.
34. Mondal, S., Brankow, D.W. & Heidelberger, C. (1975) *Cancer Res. 36*, 2254-2260.
35. Fahl, W.E., Lalwani, N.D., Watanabe, T., Goel, S.K. & Reddy, I.K. (1984) *Proc. Nat. Acad. Sci. 81*, 7827-7831.

#A-6

STUDIES ON THE MECHANISMS OF CHANGES PRODUCED IN THE LIVER, THYROID, PANCREAS AND KIDNEY BY HYPOLIPIDAEMIC DRUGS AND DI-(2-ETHYLHEXYL) PHTHALATE

Shirley C. Price, Willis Ochieng, Richard Weaver, Glen Fox, Fiona E. Mitchell, *Dawn Chescoe, *Jenny Mullervy and Richard H. Hinton

Robens Institute of Industrial and Environmental Health and Safety, and *Microstructural Studies Unit University of Surrey Guildford, Surrey GU2 5XH, U.K.

With emphasis on methodology, this article concerns effects, in the rat, of LL⊗ drugs (clofibrate, fenofibrate) and DEHP, administered for up to 18 months. Short-term alterations were also studied in hepatocytes, both freshly isolated and cultured. Alterations were found in the liver, as expected, and in several other tissues. Similarly to liver, the kidney proximal tubule showed an initial increase in peroxisomes and then an enlargement of lysosomes which were filled with material staining as for lipofuscin. The pancreas showed exocrine-cell hyperactivity, and with prolonged treatment the islets did not show the hypertrophy normally observed in ageing rats. All the alterations may well relate to alterations in fat metabolism.

*There are also thyroid changes. As early as 3 days the serum shows reduced T4 although not T3. Acinar cells show, by e.m., hyperactivity which persists with continued treatment. These changes are similar to those produced by polyhalogenated hydrocarbons, but the peroxisome-proliferating agents do not induce hepatic GA **tr**'ase.*

The LL drugs related to clofibrate have been intensively studied over recent years. They are not mutagenic in a wide variety of prokaryotic and eukaryotic test systems [1], yet cause a marked increase in cancer in the liver of rats [1, 2] and in the exocrine pancreas. Short-term alterations in the kidney similar to those in liver have also been reported [1], although there have, as yet, been no published

---

⊗*Abbreviations*: LL, lipid-lowering (hypolipidaemic); DEHP, as in title; e.m., electron microscopy/micrograph; GA **tr**'ase, UDP–glucuronyl transferase; αGP **de**'ase, α–glycerophosphate dehydrogenase.

reports of an increase in kidney tumours. Furthermore, a kindred compound, DEHP, causes a marked increase in testicular tumours [3], while we have observed thyroid changes indicative of hyperactivity [4, 5] and similar to those produced by agents such as polyhalogenated hydrocarbons which increase the incidence of thyroid adenomas after prolonged treatment [6].

There is thus evidence that, in rats, LL drugs and related compounds are carcinogenic in many tissues. This does not appear to be the case in humans. Although the highest therapeutic dose of clofibrate is as high as 7% of the lowest dose reported to produce an increase in cancer in rats, clinical evidence gained during 25 years rules out any association between clofibrate treatment and an increase in cancer in man [7].

Thus man and rats appear to differ markedly in the long-term effects of clofibrate treatment. There are also marked differences in the short-term effects, at least on the liver. In rats, treatment with clofibrate causes a massive increase in liver weight and in the concentrations of peroxisomes in the liver ([1], & G.G. Gibson, #A-7, this vol.), while changes in human liver are minimal [8]. However, organs besides the liver are affected by long-term treatment with LL agents, and while the difference in the short-term hepatic response between humans and rats may be related to the difference in long-term response, there is no evidence whatsoever on whether other human tissues respond to LL drugs. However, it is by no means certain that the responses of different tissues are in fact independent. Here we review the experimental procedures which we have adopted to investigate the changes in the different tissues in the rat, and summarize our conclusions on the mechanisms of toxicity.

MATERIALS AND METHODS

Wistar albino rats, usually of initial body wt. 80-100 g, were obtained from the University's Animal House or from ICI Rodent Breeding Unit. No strain differences have been observed to date. Resins for e.m. were obtained from TAAB Ltd. (Reading, Berks.), and other chemicals from Sigma or BDH Chemicals. The phthalic acid esters were kindly donated by BP Chemicals (Sulley, Penarth) and the LL drugs by Dr. G. Blane (Dijon, France). Radiochemicals were obtained from Amersham International, and cell-culture materials from Flow Laboratories.

In vivo experiments (details in [5, 9-11]).— In summary, rats were fed a powdered diet, initially as supplied and after 1 week, where appropriate, containing the compound. Young rats readily accepted diets containing very high concentrations of LL drugs or phthalates, but older (9 month) rats showed a marked aversion to some of the compounds [5]. Finally a lethal dose of pentobarbital was given, and the rats were bled by cardiac puncture while unconscious, then fully autopsied; any abdominal or thoracic organs that showed signs of gross toxicity were further examined [9-11]. A limited autopsy was performed on rats killed at intermediate times.

Tissue preparation for microscopy (& see legends to Figs. 2 & 4).- For light microscopy the tissues, fixed for at least 14 days, were usually embedded in paraffin wax (7 µm sections), but frozen fixed material was used for Oil Red O staining. Besides H & E, the following stains were used.- Periodic acid-Schiff for polysaccharide with and without diastase digestion to remove glycogen, Oil Red O for neutral lipid [12]; and Nile Blue for acidic lipid, Schmorl's stain for lipofuscin, Perl's Prussian Blue for iron, Masson-Fontana stain for melanin [13]. For e.m., tissue cubes (~0.5 mm, cut with 2 disposable microtome blades) were fixed for 6-8 h, washed overnight in the buffer, and counter-fixed for 2 h; OsO_4 solutions were more reliable if prepared from crystals than if bought as vials. After ethanol dehydration and exposure to epoxypropane, the samples were embedded in Epon 812 for 48 h at 60°.

Hepatocytes.- The isolation method, based on that of Rao et al. [14], entailed use of Ca^{2+}-free pH 7.4 buffers - Krebs-Ringer phosphate (**PBS**), and HCO_3^-/95% O_2-5% CO_2 (**C**). The rats (male) were anaesthetized with pentobarbital. The portal vein was cannulated 22 swg × 32 mm catheter), and the liver perfused with **PBS**, the inferior vena cava being cut just below an untied ligature which had been pre-inserted just above the renal veins. Then this ligature was tightened after carefully cutting open the rib cage and cannulating (18 swg × 32 mm) the inferior vena cava via the right atrium, for further perfusion. Following **PBS** (15 min) and then **C** (15 min), not re-circulated, perfusion with re-circulation was carried out for up to 20 min with 150 ml of **C** containing 100 mg collagenase and 5 mM $CaCl_2$. The liver was then rapidly excised and, in **PBS** at 37°, sub-capsular tissue was teased out. It was passed through bolting cloth, and the volume made up to 200 ml with **PBS**. Washing and re-suspension (× 3) were performed either with settling under gravity or by centrifuging at 400 rpm for 1 min. The cell yield was routinely ~200 × 10^6 with viabilities >90% by Trypan Blue exclusion testing. For hepatocyte culture conditions, see [15].

RESULTS

In probing the effects of LL drugs and DEHP we used light microscopy to identify the main tissues affected, e.m. to determine which cells and cell organelles were affected, and biochemical measurements to quantitate the changes. We found changes in the thyroid, pancreas, kidney and testes as well as in liver, and in the concentrations of certain proteins in plasma. We then sought to identify the underlying mechanism and, in particular, the connection between hepatic and thyroid changes, by experimental approaches which are outlined below following description of the changes occurring in each tissue.

Changes in the liver

We have already described the hepatic changes induced by LL drugs and phthalate esters in some detail [5, 9-11]. They accord with those mentioned in accompanying articles [#A-4, #A-5, #A-7]. The initial changes can be reproduced in cultured hepatocytes exposed to the agent [5, 15]; hence freshly prepared hepatocytes serve to investigate very short-term induced changes [15, 16]. The development of alterations may be summarized as follows.

- (a) Immediately after addition of phthalate esters [15] or LL drugs [16, 17] to hepatocytes isolated from fasted rats there is an increase in triglyceride synthesis. It is also observed in hepatocytes from fed rats killed in the afternoon but not at 9 a.m. This indicates that the effect depends on the animal's nutritional state and is consistent with an allosteric effect of the agents on key enzymes in fatty acid metabolism.

- (b) Within 24 h of commencing treatment of intact animals or of adding the agents to isolated hepatocytes there is accumulation of lipid in small droplets. With low doses and with very short treatment times this effect is seen in all parts of the liver lobule, but at high doses or after longer treatment the lipid clears from the liver's centrilobular zone.

- (c) Following rapidly on the hepatic lipid accumulation there is proliferation of hepatic peroxisomes and induction of the P-450 iso-enzyme responsible for ω-oxidation of fatty acids. This effect is observed both in intact animals and in cultured hepatocytes.

- (d) Soon after commencing treatment of intact animals with peroxisome-proliferating agents there is a burst of mitosis in the liver. This is not observed in cultured hepatocytes.

- (e) Following these changes there are alterations indicative of mild hepatotoxicity. There is centrilobular loss of glycogen and of glucose-6-phosphatase activity and a fall in non-protein reducing constituents. These alterations are observed in intact animals but not in cultured hepatocytes. They are first seen 3 days after commencing treatment but increase in magnitude up to 3 weeks, after which time they stabilize.

- (f) One month after commencing treatment there is enlargement of lysosomes. Various histochemical stains gave the following evidence on the nature of the accumulating material.-

i) The lysosomes, as in normal cells, showed periodic acid-Schiff's staining which was not digestible by diastase and most probably connotes glycoprotein.

ii) The lysosomes stained strongly with Schmorl's stain and with Alternative Nile Blue, indicating the presence of acidic lipids.

iii) The lysosomes stained weakly with Nile Red, indicating the presence of a small amount of neutral lipid.

iv) The lysosomes were not stained by Masson & Fontana's stain or by Perl's Prussian Blue, indicating absence of melanin and iron.

These results are consistent with accumulation of peroxidized lipid within the lysosomes. This conclusion is supported by e.m. which showed that the accumulating lipid was not extracted by ethanol or by propylene oxide, suggesting a degree of cross-linking. Of several enzymes examined as possible markers for this lysosomal enlargement, β-D-galactosidase was found most satisfactory.

- (g) Autofluorescent lipofuscin deposits were seen in the livers after 6 months of treatment, identical in staining properties to the enlarged lysosomes seen earlier.

- (h) In the second year of treatment there was evidence for increased cell turnover in the liver.

- (i) After ~18 months of treatment foci of altered cells appear in the liver [18]. They initially disappear on cessation of treatment, but with more prolonged treatment (113 weeks) they remain even at 4 weeks after withdrawal of the compounds.

- (j) Although most of the alterations affect the hepatocytes it is noticeable that the bile duct proliferation characteristic of ageing rats is not found in rats treated long-term with LL agents.

Changes in the thyroid

- (a) Within 3 days of commencing treatment with LL drugs or DEHP, serum thyroxine (T4) shows a fall (Fig. 1) which persists through the period of treatment and is dose-dependent.

- (b) The thyroid at this time shows e.m. alterations indicative of hyperactivity (Fig. 2). There is enlargement of lysosomes and the Golgi apparatus, myelination of mitochondria and an increase in resorption droplets on the apical face of the follicular cells.

- (c) After treatment for 9 months or more, calcified deposits are visible in the colloid. In some cases cast cells are also visible. These alterations are also consistent with persistent hyperactivity.

Fig. 1. Effect of fenofibrate on serum T4 activity (by RIA). Taking account of later experiments, the apparent change in T3 was insignificant. Lower or higher doses gave the same trends of T4 change. It was also seen with clofibrate (400 mg/kg per day).

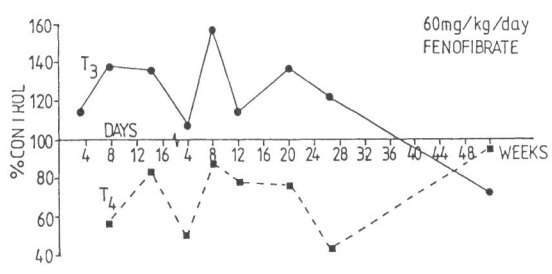

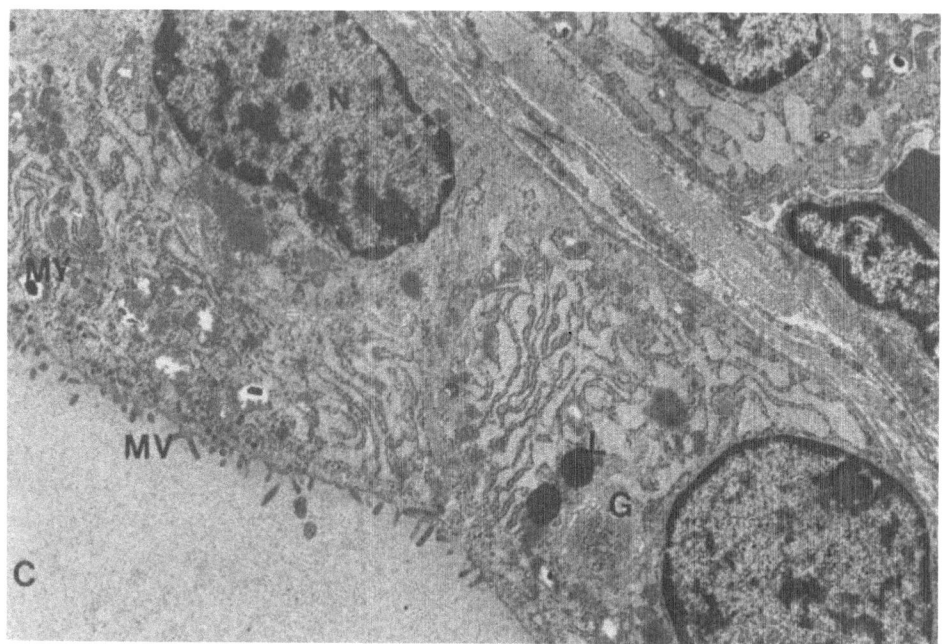

Fig. 2.   Thyroid e.m.'s: *above:* control; *opposite p.:* after DEHP
treatment (1 g/kg per day for 3 days), showing hyperactivity.
Fixation in 4% glutaraldehyde buffered with 0.1 M cacodylate, pH 7.4.
Post-fixation with 2% OsO_4/0.1 M cacodylate.   Counterstaining with
uranyl acetate and Pb citrate.   Bars = 1 μm.   C, colloid; G, Golgi;
L, lysosomes/granules; MV, microvilli; MY, myelin body; N, nucleus.

Changes in the pancreas

- (a) Shortly after commencing treatment the pancreas shows acinar
ultrastructural alterations (Fig. 3), viz. marked Golgi hypertrophy
and an overall reduction in secretion granules although individual
cells vary markedly.

- (b) Prolonged treatment did not result in any major changes in
the acinar cells although there was, frequently, a clear zone just
to the apical side of the nucleus consistent with persistent Golgi
hypertrophy.

- (c) The islet cell hyperplasia typical of ageing rats was reduced.

Changes in the kidney

- (a) Other workers have shown increased peroxisomes in the proximal
tubule cells of rats treated with LL agents [1].

- (b) After 10 days of DEHP (1 g/kg) kidney weight shows a small rise,
but much less marked than for liver.

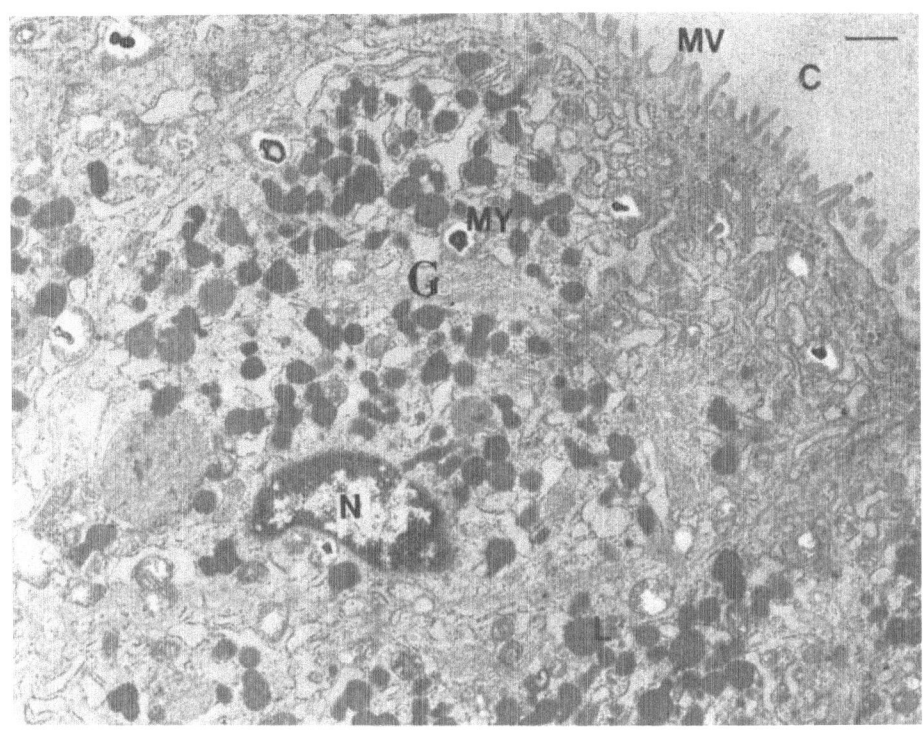

- (c) We have shown proximal tubules to have enlarged lysosomes (Fig. 4), whose content stained as for equivalent structures in liver.

Changes in the testes

In agreement with other authors we found testicular atrophy in rats given DEHP (2 g/kg). There was no effect with 1 g/kg even over 9 months, nor with LL drugs, but testicular atrophy occurred with straight-chain phthalate esters that caused no peroxisomal proliferation; hence the effect appears independent of hepatic effects. Interstitial cell carcinomas were observed in 2 out of 3 rats treated for 18 months with a LL drug; but published data and our past experience led us [11] to conclude that this was a chance occurrence.

Changes in plasma proteins

Effects of proliferating agents have been fully reported elsewhere [19]. They differ both from those associated with inflammation (the acute-phase response) and from those caused by other hepatotoxins [20].

Connection between hepatic and thyroidal changes

An earlier theory that altered thyroid-hormone metabolism explained the LL action of clofibrate is now discredited; but high

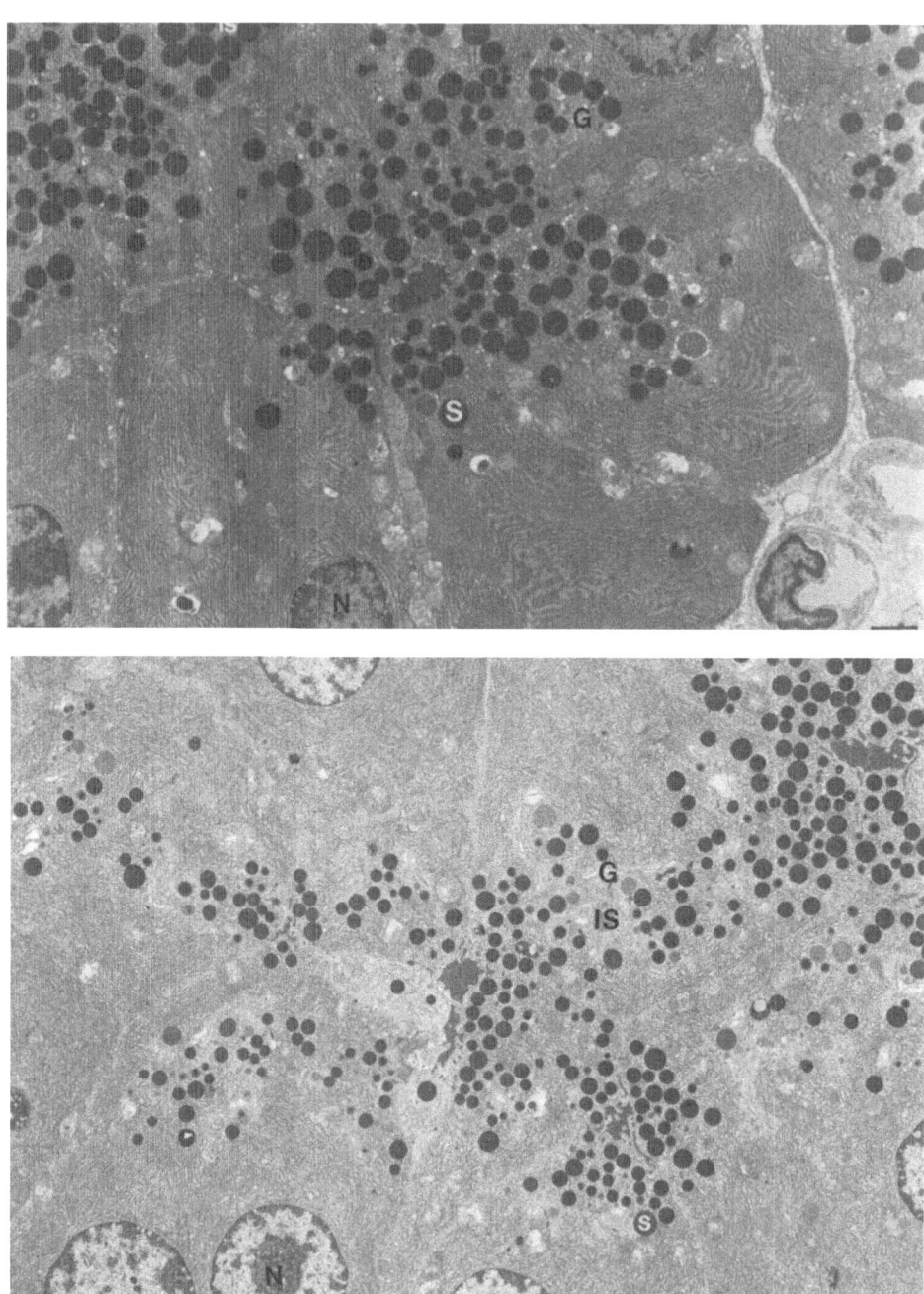

Fig. 3. Pancreas (exocrine) e.m.'s: *upper:* control; *lower:* after clofibrate treatment (400 mg/kg per day for 21 days). Processed as for Fig. 2; bar = 1 µm. G, Golgi zone; SG, secretion granules (IS, immature); N, nucleus.

Fig. 4. Photomicrographs
of kidney sections: *right*
control; *below, right:*
after fenofibrate treat-
ment (200 mg/kg per day
for 6 months). Fixation
in 10% neutral buffered
formalin, pH 7.4;
staining with Schmorl's
stain for lipofuscin.
L, enlarged lysosomes
with the staining pro-
perties of lipofuscin.
× 420.

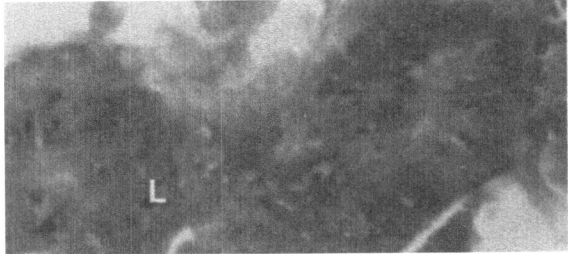

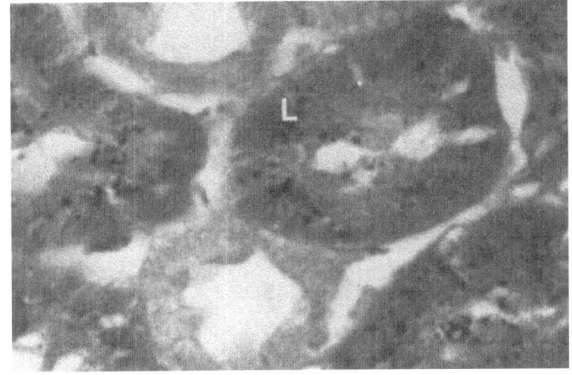

doses of thyroxine are known to induce peroxisome proliferation.
We accordingly wondered whether clofibrate could in fact bind to
a subset of thyroxine receptors such as plasma membrane (p.m.), cytosol
and the nucleus are known to possess; thyroxine is also reported
to affect mitochondria directly. We therefore investigated, but
without success, whether clofibrate could compete with thyroxine
for any of its receptors and whether thyroxine affected mitochondrial
fatty acid oxidation similarly to clofibrate. Thus, using published
procedures [21], we found binding of ^{125}I-T3 to the nuclear receptor
to be only slightly affected by clofibrate: there was ~15% displacement
at 0.1 μM but no increase at levels up to 1 mM. There was saturable
binding of T3 but no displacement by clofibrate with liver cytosol
or, in a preliminary experiment, with p.m. prepared [22] from the
basolateral face of hepatocytes.

Since clofibrate is evidently not thyromimetic and the action
of LL agents on cultured cells rules out the liver action being secondary to thyroxine changes, we examined whether thyroxine changes
could be due to alterations in thyroxine metabolism in the liver.
Polychlorinated hydrocarbons cause thyroidal alterations very like
those found with LL agents, possibly attributable largely to increased
excretion of thyroxine due to GA tr'ase induction [6]. We found,
however (Table 1), no induction of GA **tr'**ase by DEHP, but marked
induction of αGP **de'**ase, an enzyme which supposedly reflects intra-
hepatic thyroxine levels and which is not induced by polyhalogenated
hydrocarbons. In a few preliminary experiments we have found an

Table 1. Enzyme activities in the livers of rats fed for 7 days with diets containing either 1% DEHP or 0.015% Arochlor 1254. α-Glycero-phosphate dehydrogenase (αGP **de**'ase) activity was measured ([23], adapted) on a large-particle fraction (10,000 **g**, 15 min). UDP-gluc-uronyl transferase (GA **tr**'ase) was measured on a microsomal fraction with published conditions for incubation [24] and colorimetry [25]. Protein was measured by the Lowry procedure. Results are presented as μmol/min per mg protein: mean ±S.E.M. (& no. of animals).

	GA **tr**'ase	αGP **de**'ase	
Control	4.02 ±1.60 (4)	45.6 ±5.9 (8)	
DEHP	3.57 ±1.22 (4)	74.6 ±4.3 (4)	} each p <0.05
Arochlor	7.01 ±0.75 (4)	24.9 ±3.5 (4)	} *vs.* control

apparent fall in the plasma half-life of T4, but saw no significant increase in biliary excretion of ^{125}I-T4 or any increase in its hepatic retention.

DISCUSSION

Our results show that LL drugs and DEHP can affect many tissues. Except for the testicular effects, the changes produced by the different agents are remarkably uniform, suggesting a common pathogenesis. Although there is, as yet, no definite proof, the connection would appear to lie in the effect the compounds have on lipid metabolism. It is clear that both LL drugs and mono-2-ethylhexyl phthalate (MEHP) have an immediate effect on triglyceride metabolism in hepatocytes from fasted animals, and as discussed elsewhere [15] this effect is consistent with the compounds mimicking fatty acids and binding to the regulatory site of key enzymes. It would therefore appear likely that the induction of peroxisomal and microsomal fatty acid oxidases is due to binding of the compounds to the regulator site of the appropriate genes. As neither MEHP nor the LL agents will be metabolized by the enzymes induced, the result is sustained induction of the enzymes. The connection between peroxisome proliferation and the remaining part of the hepatic response [5] lies outside the scope of this article.

Concerning effects on non-hepatic tissues, the effect on kidney are most easily explained. The cells of the proximal tubule have many resemblances to hepatocytes. They possess glucose-6-phosphatase activity [26] and are rich in drug-metabolizing enzymes. The induction of peroxisomal enzymes in these cells is not, therefore, surprising, and the subsequent changes in the kidney resemble those in the liver although there have been no published reports of kidney tumours arising from treatment with LL agents. The changes in the pancreatic acinar cells may also be associated with the fatty acid-like behaviour of peroxisome proliferating agents. The changes are indicative of hyperactivity.   Prolonged administration of corn oil to rats is known to increase focal acinar cell hyperplasia and acinar cell adenoma in the pancreas of rats treated over a 2-year period [27].   This suggests the possibility that the pancreatic tumours observed after prolonged treatment with clofibrate [1] may also be associated with a lipid-like action of the drug.

The connection between alterations in fat metabolism and those in the other tissues affected by LL agents is less clear-cut.   It appears likely that the absence of pancreatic islet hyperplasia in treated rats is connected with alterations in lipid metabolism; but the mechanism is uncertain.   There is no obvious explanation for the reduction of bile-duct proliferation in treated animals. Finally it would appear likely that changes in the thyroxine are secondary to alterations in the liver. Studies on the mechanism are continuing in our laboratory [28], but so far we have not been able to explain all the mechanisms that result in there being such a close association between liver and thyroxine toxicity.   We do, however, hope that we have already obtained enough data to show that an integrated multi-disciplinary approach is required for the understanding of the overall effects of foreign compounds on the body.

Acknowledgements

We thank Miss Julie Howarth for skilled assistance in preparing sections for histological examination, and Mr. D.E. Hall, Prof. P. Grasso, Prof. J. Faccini and Prof. J.W. Bridges for useful discussions. We thank Dr. G. Blane, BP Chemicals and ICI Pharmaceuticals for test materials.   Financial support was provided by the Cancer Research Campaign and ECETOC.   Glen Fox thanks Environment Canada for granting Study Leave.

References

1. Reddy, J.K. & Lalwani, N.D. (1984) *CRC Crit. Rev. Toxicol. 12*, 1-58.
2. Cohen, A.J. & Grasso, P. (1981) *Food Cosmet. Toxicol. 19*, 585-605.
3. Gray, T.J.B., Butterworth, K., Gaunt, I.F., Grasso, P. & Gangoli, S.D. (1977) *Food Cosmet. Toxicol. 15*, 389-399.
4. Price, S.C., Chescoe, D. & Hinton, R.H. (1984) *Biochem. Soc. Trans. 12*, 1034-1035.
5. Hinton, R.H., Mitchell, F.E., Mann, A.H., Chescoe, D., Price, S.C., Nunn, A., Grasso, P. & Bridges, J.W. (1986) *Environ. Health Perspect. 70*, 195-205.
6. Cavalieri, R.P. & Pitt-Rivers, R. (1981) *Pharmacol Rev. 33*, 55-80.
7. Canner, P.L., Berge, K.G., Wenger, A.K., Stamler, J., Friedman, L., Princas, R.J. & Friedewald, U. (1986) *J. Am. Coll. Cardiol. 8*, 1245-1255.

8. Hanefeld, M.,   Kemmer, C. & Kadner, E. (1983) *Atherosclerosis*
    46, 239-246.
9. Mann, A.H., Price, S.C., Mitchell, F.E., Grasso, P., Hinton, R.H.
    Bridges, J.W. (1984) *Toxicol. Appl. Pharmacol.* 77, 116-132.
10. Mitchell, F.E., Price, S.C., Hinton, R.H., Grasso, P. &
    Bridges, J.W. (1985) *Toxicol. Appl. Pharmacol. 81*, 371-392.
11. Price, S.C., Hinton, R.H., Mitchell, F.E., Hall, D.E., Grasso, P.
    & Bridges, J.W. (1986) *Toxicology 41*, 169-191.
12. Culling, C.F.A. (1975) *Handbook of Histopathological and*
    Histochemical Techniques, 3rd edn., Butterworth, London.
13. Bancroft, J.D. (1974) *Histochemical Techniques*, 2nd edn.,
    Butterworth, London.
14. Rao, M.L., Rao, J.S., Holler, M., Brener, H., Schattenberg, P.J.
    & Stein, W.D. (1976) *Hoppe-Seyler's Z. Physiol. Chem. 357*, 573-584.
15. Mitchell, F.E., Bridges, J.W. & Hinton, R.H. (1986) *Biochem.*
    Pharmacol. 35, 2941-2947.
16. Price, S.C., Mitchell, F.E. & Hinton, R.H. (1986) *Biochem. Soc.*
    Trans. 14, 636-637.
17. Cappuzzi, D.M., Intenzo, C.M., Lachman, T.D., Whereat, A. &
    Scott, D.H. (1983) *Biochem. Pharmacol. 22*, 2191-2203.
18. Trisarri, E. (1985) *Ph.D. Thesis*, University of Surrey.
19. Hinton, R.H., Price, S.C., Mitchell, F.E., Mann, A., Hall, D.E.
    & Bridges, J.W. (1985) *Human Toxicol. 4*, 261-271.
20. Makarananda, K., Fox, G.A., Price, S.C. & Hinton, R.H. (1987)
    Human Toxicol. 6, 121-126.
21. De Groot, L.J. & Torresani, J. (1975) *Endocrinology 96*, 357-
    368.
22. Mullock, B.M., Luzio, J.P. & Hinton, R.H. (1983) *Biochem. J.*
    214, 823-827.
23. Lee, Y.P. & Lardy, H.A. (1965) *J. Biol. Chem. 240*, 1427-1436.
24. Motohashi, M., Tori, H., Doi, T. & Tanayama, S. (1982) *J.*
    Chromatog. 253, 129-132.
25. Castren, M. & Oikari, A. (1983) *Comp. Biochem. Physiol. 760*,
    365-369.
26. Andersen, K-J., Haga, H.J. & Dobrota, M. (1987) *Kidney Internat.*
    31, 886-897.
27. Eustin, S.L. & Boorman, G.A. (1985) *J. Nat. Cancer Inst. 75*,
    1067-1071.
28. Ozalp, S., Weaver, R., Price, S.C. & Hinton, R.H. (1987)
    Biochem. Soc. Trans., in press.

#A-7

THE USE OF CYTOCHROME P-450 ISOENZYMES AS AN INDEX OF ENDOPLASMIC RETICULUM CHANGES

G. Gordon Gibson and Raj Sharma

Department of Biochemistry
Division of Pharmacology and Toxicology
University of Surrey, Guildford GU2 5XH, U.K.

An outline is given of general properties of cytochrome P-450 and of its spectral determination - which does not distinguish amongst the e.r.* isoenzymes. These may differ in specificity, manifest by measuring a range of catalytic activities as exemplified by findings in clofibrate-treated rats. Ab's that recognize different P-450 apoenzymes can help in isoenzyme identification and quantitation. It has, for example, been shown by matrix analysis that the ability of hypolipidaemic agents to elevate immunochemically determined P-452 correlates with their hepatomegalic action. Induction of these iso- enzymes also correlates well with increases in peroxisomal parameters, especially volume. Consideration is given to important points of interpretation and technique, especially in the immunochemical approach.

Cytochrome P-450 is a collective name for a family of haemoprotein isoenzymes, primarily localized in the e.r. of liver, whose function is to oxidize literally hundreds of endogenous compounds, drugs and other xenobiotic chemicals. It is well established that exposure of animals or man to diverse xenobiotics can induce a population of e.r. cytochrome P-450 isoenzymes, resulting in profound changes in the pharmacological and toxicological properties of either the inducing agent itself or other co-administered drugs.

Accordingly the problem arises what is the best method to identify and quantitate these subtle changes in the e.r. P-450 profile. One approach is to spectrally determine the total amount .of P-450 as

* *Abbreviations:* e.r., endoplasmic reticulum; Ab, antibody; PAGE, polyacrylamide gel electrophoresis; MC, 3-methylcholanthrene; PB, phenobarbital.

the CO-complex - which is informative but does not quantitatively furnish the isoenzyme pattern. A second approach is to measure changes in the substrate metabolism and specificity of the e.r. to oxidize a broad spectrum of structurally diverse substrates; but problems arise because many of the isoenzymes overlap in specificity. A third approach is to use Ab's to probe e.r. P-450 changes. Factors affecting the success of this approach include the nature and specificity of the Ab, the specific P-450 isoenzyme under scrutiny, and the existence of common epitopes between different isoenzymes. The latter method suffers from the disadvantage that the Ab recognizes only the apoprotein and hence gives no indication of enzyme functionality because the holoenzyme (apoprotein plus haem prosthetic group) is absolutely required for catalysis.

It has become clear that the P-450 isoenzymes are a good index of e.r. changes, with the caveat that close attention must be paid to methodology and hence data interpretation.

GENERAL PROPERTIES OF CYTOCHROME P-450

Before discussing in detail the role of cytochrome P-450 analysis in the monitoring of subcellular derangements, it is informative to consider some of the enzyme's general properties in order that the reader can fully appreciate the analytical problems concerned.

- The enzyme is a haemoprotein of monomeric kM_r ~45-55.
- Ubiquitous distribution, occurring in prokaryotic and eukaryotic cells. Detectable in almost all mammalian tissues, the liver being the richest source, and localized in the e.r.
- Terminal haemoprotein component of the mixed-function oxidase system and catalytically functions as a hydroxylase enzyme. Additionally has NADPH-oxidase and peroxidase activity.
- Responsible for the metabolism of hundreds of drugs and chemicals and endogenous substrates including steroids, fatty acids, vitamins, prostaglandins and leukotrienes.
- Exists as isoenzymes (multiple forms), the relative proportions of which may be modulated by xenobiotics.
- Plays a role in both the pharmacological deactivation and toxicological activation of drugs and chemicals.

SPECTRAL DETERMINATION OF CYTOCHROME P-450

The original method of Omura & Sato [1] takes advantage of the fact that cytochrome P-450, when in the reduced ferrous state, complexes CO resulting in the formation of a strong 'soret' band at 450 nm. As the extinction peak for this soret peak is known [1], the gross cytochrome P-450 content is readily determined. However, this technique suffers from several disadvantages including the following.

- The assay method only quantitates the holoenzyme (i.e. incorporating the CO-reactive haem prosthetic group) and does not take into account

any apoprotein present. This is an important point to note regarding
this method as many drugs and xenobiotics can induce the level of
apoprotein synthesis without the matching and necessary increase
in haem incorporation. Accordingly, total cytochrome P-450 content
may be under-estimated.

- The spectral analysis does not differentiate or identify the isoenzymes of cytochrome P-450, and is reactive with all its forms known
to date. This lack of isoenzyme differentiation is important when
analyzing subcellular changes in cytochrome P-450 in response to
xenobiotics or physiological and environmental challenges. For
example, after drug challenge, the complement of cytochrome P-450
isoenzymes may drastically alter without any changes in the total
absolute amount of haemoprotein. One could therefore draw the false
conclusion, based on spectral data, that the drug in question had
no influence on cytochrome P-450 content, whereas drug induction
of individual isoenzymes paints an entirely different picture [2, 3].

CATALYTIC ACTIVITY OF CYTOCHROME P-450

If each isoenzyme had a unique and characteristic substrate
specificity, then simply by observing the catalytic activity we could
unequivocally identify which isoenzyme(s) is present in a particular
biochemical environment. Unfortunately this is not the case, and
P-450 isoenzymes exhibit a similar although overlapping substrate
specificity, resulting in the contribution of more than one isoenzyme
to the metabolism of a particular substrate. However, some drug
inducers will preferentially induce a particular isoenzyme of cytochrome P-450, resulting in the preferential metabolism of a substrate.
For example, as shown in Table 1, PB induces benzphetamine-*N*-
demethylase activity and other inducers will preferentially bring
about metabolism of different substrates. One must, then, recognize
the limitations of such catalytic data in that several different
isoenzymes could conceivably contribute to the metabolism of a
particular substrate.

A striking example of the utility of determining cytochrome
P-450 catalytic activity as an index of subcellular liver changes
is seen with the hypolipidaemic drug clofibrate. Clofibrate (and
related drugs) produces three characteristic liver changes in rodents,
viz. proliferation of the e.r., peroxisomal proliferation and hepatocellular carcinomas on chronic exposure [4]. In addition, proliferation of the e.r. results in the induction of a specific form of
cytochrome P-450 (termed P-452) responsible for the hydroxylation
of fatty acids (Table 2).

This Table shows the complexity of the cytochrome P-450 system
in that although cytochrome P-450 levels are induced by clofibrate
the metabolism of both aminopyrine and benzphetamine is approximately
halved. This is rationalized by the ability of clofibrate to somehow

Table 1. Drug metabolism by hepatic microsomes as influenced by inducers – PB, phenobarbitone; PCN, pregnenolone-16α-carbonitrile; ARO, Arochlor 1254; MC, 3-methylcholanthrene. Activities expressed as nmol product/min per nmol cytochrome P-450; ± values are S.E.M. (n = 4).   *Adapted from ref. [5].*

Substrate	Control	PB	PCN	ARO	MC
Ethylmorphine	13.7 ±0.8	16.8 ±4.3	24.9 ±3.9	9.5 ±1.2	6.4 ±0.5
Benzphetamine	12.5 ±1.2	45.7 ±14.0	6.6 ±0.7	15.8 ±2.7	5.7 ±1.1
Benzo(a)pyrene	0.14	0.14	0.14	not assayed	0.33

Table 2. Induction of rat-liver microsomal constituents – cytochrome P-450, demethylases and fatty acid hydroxylase – by clofibrate (given as stated in Table 4 heading, below; n = 4).

Parameter	Control	Induced	
Cytochrome P-450	1.14	1.90	nmol/mg protein
N-Demethylases:			
benzphetamine	12.00 ±0.01	4.90 ±0.09	nmol HCHO product formed
aminopyrine	11.87 ±1.26	5.93 ±0.10	/nmol cytochrome P-450/min
Lauric acid			total 11-OH + 12-OH formed
hydroxylase	1.53 ±0.25	7.55 ±0.20	/mg protein/min

switch off the genes that code for the isoenzymes that metabolize these two substrates, concomitant with the induction of the fatty acid hydroxylase isoenzyme. Thus it is important to be fully aware of the existence of multiple forms and to carefully choose the 'marker' substrate.

Clofibrate appears to be a useful tool in studying subcellular liver changes, particularly in view of the fact that the induced microsomal enzyme (P-452) catalyzes only the ω-hydroxylation of fatty acids [6, 7]. When the cytochrome P-452 isoenzyme is isolated and purified to electrophoretic homogeneity from the liver microsomes of rats pre-treated with clofibrate, the absolute substrate specificity can be determined in a reconstituted system. As shown in Table 3, cytochrome P-452, unlike the other isoenzymes, exhibits a very narrow and specific substrate range; so far no other hepatic isoenzyme has been demonstrated to catalyze this activity (contrast cytochrome P-452 substrate specificity with that of cytochrome P-450$_b$).

As mentioned previously, hypolipidaemic drugs also produce significant proliferation of peroxisomes, evident both from peroxisomal volume and from the representative catalytic activity of the

Table 3. Substrate specificity of highly purified cytochrome P-450 isoenzymes in a reconstituted system. P-450$_b$ = a major PB-induced isoenzyme, purified to electrophoretic homogeneity. The tabulated activities are nmol product formed/nmol P-450/min. *Derived from [7].*

Substrate		P-452 isoenzyme	P-450$_b$ isoenzyme
Lauric acid (ω-hydroxylation)		43.0	6.7
Benzphetamine (*N*-demethylation)		not detectable	281.0
Testosterone (hydroxylation)	7α	" "	not detectable
	16α	" "	5.2
	6β	" "	4.6
	2β	" "	3.0
Ethoxyresorufin (*O*-deethylase)	"	"	0.4

fatty acid β-oxidation system. We have shown (unpublished work) that peroxisomal volume can increase up to 5-fold after challenge by potent hypolipidaemics such as benzafibrate, concomitant with a 20-40 fold increase in the specific activity of the β-oxidation system. Sustained proliferation of peroxisomes in rodents almost always results in hepatocellular carcinomas [8], and the above short-term changes in both cytochrome P-450 and peroxisomal β-oxidation activities are therefore an important 'marker' for the long-term toxicity of this class of compounds. This is particularly true because amongst compounds studied to date most of those that produce peroxisomal proliferation and hepatocellular carcinomas also induce the activity of the micro-somal cytochrome P-452 hydroxylase [9]. Although the precise inter-relationship between these diverse hepatic responses to hypolipidaemic agents is not clear at present, it is certainly a warning signal that more pronounced and toxicologically significant subcellular derangements will almost definitely follow.

IMMUNOCHEMICAL APROACHES FOR CYTOCHROME P-450 ISOENZYME STUDIES

The early research emphasis on isolating and purifying multiple forms of cytochrome P-450 has enabled the pure proteins to be used as antigens to raise polyclonal Ab's to the haemoproteins. These Ab's have been used in a variety of immunochemical techniques, including Ouchterlony double immunodiffusion, rocket immunoelectrophoresis, radial immunodiffusion, ELISA and Western blotting, to answer such varied questions as the role of cytochrome P-450 isoenzymes in foreign compound metabolism, identifying subcellular fractions that contain the isoenzymes, and the distribution of the multiple forms across the liver lobule; the Ab's have also served for screening of cDNA expression vectors. As in any other immuno-based technique, the specificity and lack of non-specific cross-reactivity of Ab's to cytochrome P-450 isoenzymes is of prime importance. Lack of attention

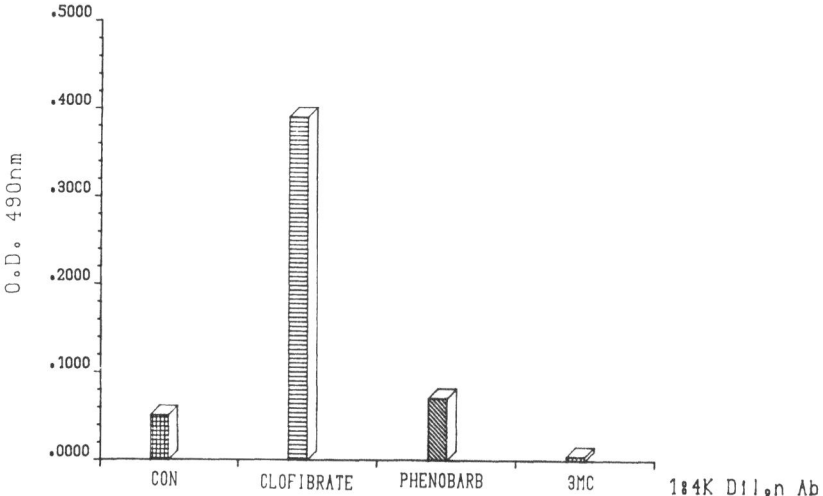

Fig. 1. Specificity of a cytochrome P-452 Ab. Equivalent amounts of CO-discernible cytochrome P-450 (1 pmol) were subjected to the ELISA procedure in the presence of the P-452 Ab (1:4,000 dilution) and the A_{490} determined. The microsomes came from control rats (CON) or rats treated with the agents indicated.

to Ab specificity can often lead to misinterpretation of data and result in false conclusions concerning the antigen.

This laboratory has raised polyclonal Ab's to rat-liver cytochrome P-450 in both rabbit and sheep. In our experience, the immunization schedule is very important. For example, if low initial doses (~100 µg antigen) are given followed by infrequent boosts, then a specific Ab will result. However, if high initial doses (~500-1000 µg) are given followed by frequent boosting, then the specificity is lost and the resulting Ab cross-reacts with additional isoenzymes. With these principles in mind, we have raised to rat-liver cytochrome P-452 a specific polyclonal Ab that recognizes its homologous antigen but does not cross-react with purified cytochrome P-450 isoenzymes induced by either PB or β-naphthoflavone [7]. Furthermore, this Ab recognizes a single protein of the correct mol. wt. for cytochrome P-452 in a Western blot procedure and gives one reactive spot in 2-D electrophoresis, viz. electrofocusing followed by SDS-PAGE (unpublished observations).

With these observations in mind, we have developed a quantitative ELISA procedure for cytochrome P-452. As shown in Fig. 1, the Ab preferentially reacts with clofibrate-induced rat-liver microsomes, in keeping with the induced 12-hydroxylation of lauric acid mediated by cytochrome P-452 as mentioned above. It is evident (Fig. 1) that the Ab reacts to some extent with uninduced, PB-induced or MC-induced microsomes. This observation can be rationalized in one of two ways.

Table 4. Effect of hypolipidaemic agents on liver size and cytochrome P-452 content (nmol/mg of microsomal protein. Compounds were
administered by gavage for 3 days (in parentheses: dose as mg/kg;
controls received 5 ml/kg). Values are means ±S.D. (6 individual
animals). Significant differences (Student's t test) *vs.* control:
[1] = $p < 0.05$, [2] = $p < 0.01$, [3] = $p < 0.001$.

Treatment	Liver wt., % body wt.	Total cyt. P-450, nmol/mg	Specific cyt. P-452	P-452, % of total P-450
Peanut oil	5.34±0.23	1.07±0.17	0.040±0.013	3.73±0.83
WY.14-643 (250)	7.41±0.38[3]	1.48±0.31[1]	0.297±0.060[3]	20.20±2.29[3]
DEHP (1200)	7.10±0.38[3]	1.54±0.14[3]	0.280±0.045[3]	18.30±3.14[3]
MEHP (100)	6.30±0.54[2]	1.27±0.09[1]	0.108±0.030[3]	8.47±2.15[3]
Aspirin (500)	6.56±0.61[2]	1.54±0.39[1]	0.148±0.058[3]	9.55±2.54[3]
Bezafibrate (200)	7.80±0.70[3]	1.21±0.25	0.347±0.107[3]	30.3 ±8.79[3]
Nafenopin (180)	7.45±0.59[3]	1.27±0.14	0.340±0.094[3]	26.53±5.48[3]
Clofibrate (250)	6.80±0.40[3]	1.43±0.16[1]	0.317±0.063[3]	22.2 ±4.55[3]
Clobuzarit (50)	6.95±0.18[3]	1.55±0.26[1]	0.317±0.053[3]	20.63±2.26[3]

Either the Ab recognizes other cytochrome P-450 isoenzymes in the
different preparations, or the other liver samples contain residual
cytochrome P-452, recognized by its homologous Ab. In view of both
the specificity indicated above and the fact that the control, PB-
and MC-induced liver preparations exhibit a low but detectable lauric
acid 12-hydroxylase activity, the latter possibility is the more
plausible.

As mentioned above, hypolipidaemic agents induce both hepatomegaly and lauric acid 12-hydroxylase activity that depends on cytochrome P-452. The ability of various hypolipidaemics to cause liver
changes is shown in Table 4. Evidently the hypolipidaemic agents
that produce the greatest extent of hepatomegaly also induce the
highest amounts of immunochemically determined cytochrome P-452.
At the tabulated dose levels the weak inducers, viz. aspirin and
monoethylhexylphthalate (MEHP), also produced the weakest hepatomegaly
and subcellular liver changes.

More detailed study was warranted to establish any inter-
relationships between the induction of microsomal and peroxisomal
parameters by hypolipidaemic agents. Accordingly, groups of rats
were pre-treated with the hypolipidaemic agents listed in Table 4
and the responses of several liver parameters were determined. These
parameters were then subjected to a correlation matrix analysis in

Table 5. Correlation matrix to show the inter-relationship between hypolipidaemic-induced subcellular liver changes. (Enzyme activities: x, oxidase or hydroxylase; t, transferase; dm & de, desalkylases.)

	Total P-450 cyt	P-452 cyt'	lauric acid hydroxylase 11-OH xL^{11}	lauric acid hydroxylase 12-OH xL^{12}	palmitoylCoA oxidase xP	carnitine acetyl CoA transferase tCA	carnitine palmitoyl CoA transferase tCP	enoyl-CoA total ecA	enoyl-CoA peroxisomal ecA'	benzphetamine-N-demethylase dmB	ethoxyresorufin-O-deethylase deE	Peroxisomal volume
cyt	0.391	0.311	0.482	0.268	0.121	0.198	0.301	0.280	-0.299	-0.405	0.186
cyt'		0.690	0.776	0.945	0.922	0.907	0.739	0.747	-0.731	-0.932	0.973
xL^{11}			0.894	0.736	0.778	0.725	0.544	0.559	-0.309	-0.549	0.646
xL^{12}				0.768	0.741	0.752	0.450	0.464	-0.512	-0.656	0.750
xP					0.964	0.949	0.837	0.845	-0.630	-0.920	0.970
tCA						0.919	0.779	0.795	-0.624	-0.867	0.962
tCP							0.792	0.794	-0.667	-0.884	0.938
ecA								0.999	-0.398	-0.794	0.741
ecA'									-0.398	-0.795	0.741
dmB										-0.844	-0.746
deE											-0.937

order to highlight any relationship (or lack thereof) that may exist between the various liver parameters assayed.   Table 5 summarizes the outcome.

In this matrix for 9 observations and 7 degrees of freedom, the correlation coefficient r in a linear regression analysis is significantly related (95% probability) if $r > 0.666$, and hence causal relationships between two experimentally determined parameters can be inferred.   Thus the usefulness of analyzing a cytochrome P-450 isoenzyme (in this case P-452) in relation to other liver changes can be assessed. As Table 5 clearly shows, induction of the specific cytochrome P-452 by hypolipidaemic drugs shows a negative correlation when compared to either of two dealkylation activities, indicating a specific gene switch-on for cytochrome P-452 concomitantly with the gene switch-off for these two activities.   This conclusion is in complete accord with the substrate specificity data for highly purified, hypolipidaemic-induced cytochrome P-452 (Table 3).

In addition, Table 5 shows that induction of the microsomal cytochrome P-452 isoenzyme correlates very well with the peroxisomal parameters, particularly peroxisomal volume.   This observation highlights the use of cytochrome P-450 isoenzymes in studying drug-

Table 6. Influence of boiling period on the quantitation of a cytochrome P-450 isoenzyme by a Western blot procedure. Tests performed (near neutral pH) with rabbit material by Domin & Philpot; *from [10]*.

Boiling time, min	Relative extent of detection		Relative apparent microsomal concentration
	pure form	microsomal form	
0	*1.00*	*1.00*	*1.00*
1	0.98	1.00	1.02
3	0.58	0.97	1.65
5	0.27	0.84	3.31
10	0.08	0.51	6.40

induced hepatocyte derangements, as pronounced and prolonged peroxisomal proliferation in rodent liver has been causally and mechanistically linked with hypolipidaemic-induced hepatocellular carcinomas [4]. Although there is much further information to be gained from Table 5, it is outside the scope of this discussion; we plan to study, in more detail, the precise temporal relationship between e.r. induction and peroxisomal proliferation in response to hypolipidaemic agents.

Finally, it is vital to consider further the validity of using Ab's in cytochrome P-450 isoenzyme analysis and, in particular, the importance of sample preparation prior to ultimate analysis. The latter point is clearly exemplified in the quantitative Western blot procedure for cytochrome P-450 isoenzyme analysis. As an integral part of this procedure the isoenzymes (either purified or in the membrane-bound form) are solubilized and monomerized by boiling briefly in SDS, prior to PAGE and electro-transfer to nitrocellulose filters for quantitation. As shown in Table 6, the quantitation of isoenzyme 6 is critically dependent on the length of the boiling period. If the basic sample preparation protocols are not fully worked out, it is clear from the above example that the microsomal isoenzyme 6 content can apparently vary dramatically.

Acknowledgements

Part of the work presented was supported by an M.R.C. Studentship (R.S.) and an M.R.C. Project Grant (G.G.G.).

References

1. Omura, T. & Sato, R. (1964) *J. Biol. Chem.* *239*, 2370-2378.
2. Nebert, D.W. & Negishi, M. (1982) *Biochem. Pharmacol.* *31*, 2311-2317.
3. Guengerich, F.P. (1982) in *Principles and Methods in Toxicology* (Hayes, A.W., ed.), Raven Press, New York, pp. 609-634.

4. Reddy, J.K. & Lalwani, N.D. (1983) *CRC Crit. Rev. Toxicol. 12*, 1-58.
5. Powis, G., Talcott, R.A. & Schenkman, J.B. (1977) in *Microsomes and Drug Oxidations* (Ullrich, V., Roots, A., Hildebrandt, A., Estabrook, R.W. & Conney, A.H., eds.), Pergamon, Oxford, pp. 127-135.
6. Gibson, G.G., Orton, T.C. & Tamburini, P.P. (1982) *Biochem. J. 203*, 161-168.
7. Tamburini, P.P., Masson, H., Bains, S.K., Makowski, R., Morris, B. & Gibson, G.G. (1984) *Eur. J. Biochem. 139*, 235-246.
8. Reddy, J.K., Azarnoff, D.L. & Hignite, C.E. (1980) *Nature 283*, 397-398.
9. Lake, B.G., Gray, T.J.B., Pels Rijcken, W.R., Beamand, J.A. & Gangolli, S.D. (1984) *Xenobiotica 14*, 269-276.
10. Domin, B.A. & Philpot, R.M. (1986) *Arch. Biochem. Biophys. 246*, 128-142.

#A-8

DRUG-INDUCED MITOCHONDRIAL PROLIFERATION

William Lijinsky

NCI-Frederick Cancer Research Facility
BRI-Basic Research Program
Frederick, MD 21701, U.S.A.

Although proliferation of mitochondria has been noted in the liver of patients suffering from a number of illnesses, the relation of this phenomenon to the disease is not known. Several liver carcinogens are known to induce mitochondrial proliferation shortly after administration; but again, the relationship to hepatocarcinogenesis is not known. Several carcinogens having this effect are not detectably mutagenic, suggesting that their effect on mitochondria might be related to carcinogenesis. The most notable mitochondrial-proliferation carcinogens are methapyrilene (formerly a commonly used antihistaminic), diethylhexylphthalate (DEHP; a widely used plasticizer), nitrosodiethanolamine and nitrosomethylethanolamine (both contaminants of cutting oils and cosmetics). Methapyrilene induces liver tumours in rats, but not in other species, and several structurally close analogues are not carcinogenic. These analogues do not induce mitochondrial proliferation, nor does methapyrilene in species other than the rat.

Proliferation of mitochondria is not a very common situation in cells, but has been noted in the liver of various hospitalized patients: in about one-third there was an association with chronic hepatitis and cirrhosis [1], but in the others there was no association with a particular disease. The analysis, incompletely described, was based on assessment of the numbers of mitochondria in serial sections examined by electron microscopy (e.m.). The mitochondria were normal in size and in cristal configuration, whilst greatly increased in number. The authors discussed the association between hypoxia and mitochondrial proliferation, as also manifest in nutritional deficiencies of vitamin E and copper. The mechanism of this phenomenon is not known, nor its significance in disease.

Scrutiny of the pathological effects of carcinogens, e.g. azo dyes, has often revealed ultrastructural changes. The non-carcinogenic azo dye, 2-methyl-4-dimethylaminoazobenzene, a close analogue of 3'-methyl-4-dimethylaminoazobenzene which induces liver tumours in rats, caused an increase in the mitochondrial content of rat liver cells, besides other changes [2]. More recently, an ultrastructural study of the action of methapyrilene hydrochloride showed that it produced extensive proliferation of mitochondria in the liver cells of rats [3], the target cells of this carcinogen.

Methapyrilene was one of several antihistaminic drugs that were tested for carcinogenic activity because they were capable of reacting with nitrosating agents to form carcinogenic nitrosamines [4]. It induced a very high incidence of hepatocellular and cholangio-cellular neoplasms in rats, when fed for a year or more at 1,000 ppm; the concurrent feeding of sodium nitrite did not change the results, indicating that the formation of liver tumours was not related to nitrosation of methapyrilene [5]. This compound appears to act as a carcinogen through an unconventional mechanism, since it is not a mutagen in bacteria [6] nor in other mutagenic systems, even when activated by a rat liver microsomal preparation. Nor is it active in the Syrian hamster embryo transformation assay, even when activated [7]. The compound does not appear to be 'genotoxic', i.e. it does not cause a structural change in cellular DNA. There is no suggestion of chromosome damage induced by methapyrilene: searches for sister-chromatid exchanges in rat liver, *in vitro* or *in vivo*, have failed [8].

POSSIBLE BINDING OF METHAPYRILENE TO CELLULAR MACROMOLECULES

In accord with the lack of chromosome damage, a study of the interaction of radiolabelled methapyrilene with cellular macro-molecules showed negligible binding to DNA and RNA in the liver of rats given a single dose of 20 mg (^{3}H ~ 1 mCi) [9]. There was extensive binding to liver proteins, but the specific activity of the nucleic acids was infinitesimal in liver, as for other rat organs - e.g. lung and kidney - which are not target organs for the carcinogenic action of methapyrilene. In the same experiment, radioautography of the livers at 1, 6, 14, 24 and 44 h after treatment showed that the maximum radioactivity was at 6 h [3], coinciding with the maximum binding observed chemically [9]; binding remained substantial 44 h after treatment. At each time, 50 hepatocytes from the periportal region and 50 from the centrilobular region were examined; most radio-activity was in the former. Other cells in the liver were not labelled.

Radioautography by e.m. gave values (as % of total grains/cell) for radiolabel distribution within periportal hepatocytes at 6 h:
- nucleus (10% of cell area) and nucleolus (0.5%) each nil;
- mitochondria (7.5% of cell area), 67%;
- endoplasmic reticulum (e.r.): rough (22%), 9%; smooth (46%), 23%;
- lipid (2.0%), 0.9%; lysosomes (1.4%), 0.2%.

Evidently the highest concentration of radiolabel, two-thirds of the cell total, was in the mitochondria although they occupied <10% of the cell area. Notably, nuclei had no detectable radioactivity. The other major binding site of radioactivity, one-quarter of the cell total, was the smooth e.r., which comprised almost half of the cell area.

Progressive changes were revealed by e.m. examination of liver after 1 and 2 weeks' feeding of a methapyrilene-containing diet (1,000 ppm), a treatment which if continued gives rise to liver tumours in every animal after 1 year.   There was a reduction in the rough and smooth e.r. and a considerable increase in the number of mitochond- ria per cell [10].   Mitochondria in process of division were not uncommon.   At 2 weeks mitochondria occupied as much as one-third of the cross-sectional area of the cytoplasm (Fig. 1).   Serial sectioning was not carried out, so the proportion occupied by volume must be assumed.

It would be reasonable to conclude that in rat liver methapyrilene binds to mitochondria and induces them to divide by a still unknown mechanism.   This in turn could lead to induction of neoplasia by a non-mutagenic mechanism, likewise unknown at present. The mitochond- rial proliferation is not transient: it persists if the treatment is continued, and is seen in the hepatocellular carcinomas which ensue. These neoplasms are as packed with mitochondria as were the liver cells of the rats after 2 weeks' treatment (Fig. 2), leading to their classification as oncocytomas [10].

TESTING OF METHAPYRILENE IN VARIOUS SPECIES

Various substances such as the hypolipidaemic agent clofibrate are non-mutagenic carcinogens and presumably act through their induc- tion of peroxisomal proliferation [11].   In our studies of some unusual types of carcinogen, we have found several compounds that are inducers of mitochondrial proliferation. Probably there are substances that induce this proliferation but are not carcinogenic, but none has achieved prominence.   Our interest in the possible risk to humans of exposure to methapyrilene, which was commonly used as a sleep-aid until discovery of its carcinogenic properties led to its withdrawal, prompted us to test it for chronic toxicity in other species.

In hamsters, methapyrilene caused convulsions, limiting the amount that could be administered at one time.   Nevertheless, after administration of 30 mg/week for 60 weeks no tumours were seen that could be attributed to the treatment [12].   Treatment of guinea pigs with higher doses (200 mg/kg body wt., twice a week) similarly failed to induce tumours [12].   In a chronic toxicity study of methapyrilene in mice, no tumours related to the treatment were seen, but unfortun- ately proof was inadequate because the mice were killed after 7 months' treatment, which was far too small a proportion of the life-span.

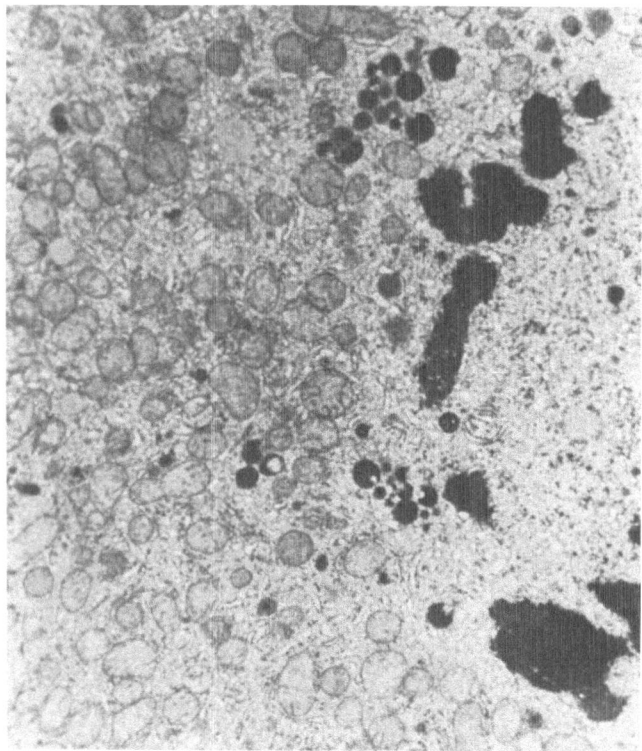

Fig. 1. Electron micrograph of mitotic liver cell from portal area of F344 rat treated with methapyrilene hydrochloride for 2 weeks. × 4500.

Paralleling this lack of carcinogenic activity, e.m. studies of the livers of Syrian hamsters and guinea pigs following chronic treatment with the drug showed no increase in the proportion of mitochondria in the cells. It appears, then, that both hepatocarcinogenesis and induction of mitochondrial proliferation by methapyrilene are restricted to rats.

STUDIES WITH PYRILAMINE AND OTHER COMPOUNDS

An analogue of methapyrilene, the drug pyrilamine (Fig. 3), also a commonly used antihistaminic, was a weak liver carcinogen in rats, many times less effective than methapyrilene and at considerably higher doses [13]. Pyrilamine caused some mitochondrial proliferation in rat liver [14], but much less than with lower doses of methapyrilene. On the other hand, no carcinogenic activity was found with a number of very close analogues of methapyrilene, all antihistaminics, given to rats at doses equal to or greater than those of methapyrilene which had caused a high incidence of liver tumours

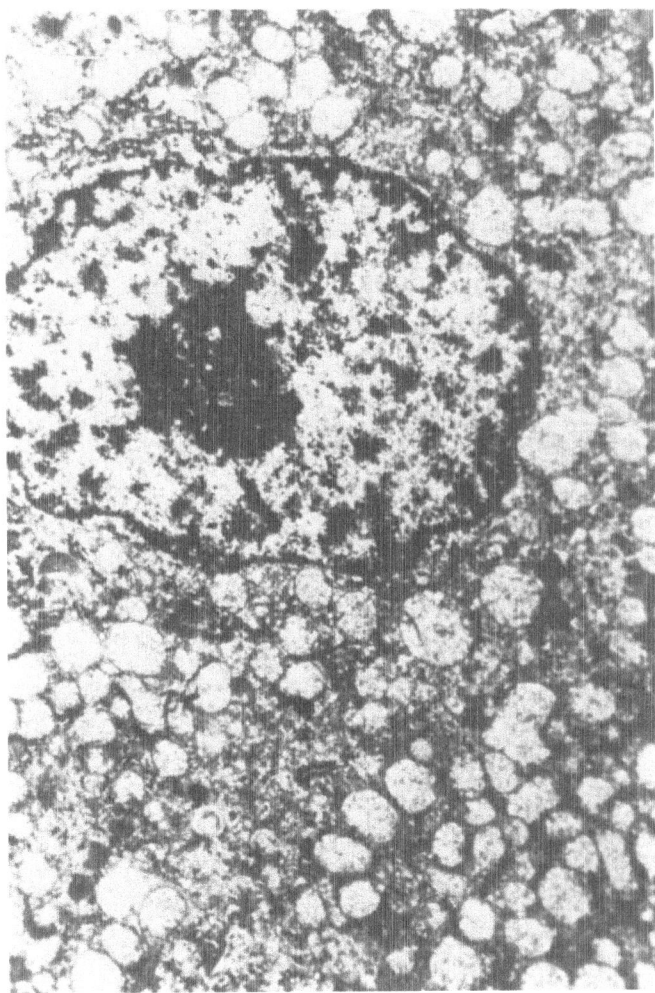

Fig. 2. Cell from hepatocellular carcinoma in rat treated with methapyrilene hydrochloride for 1 year. ×6400.

[15]. These compounds were thenyldiamine, chlorothen, methafurylene and methaphenilene (Fig. 3), each differing from methapyrilene in only one atom or, in the case of thenyldiamine, only the position of substitution in the thiophen ring. Such specificity is, of course, not rare in carcinogenesis among a group of chemically similar substances, but it increases the difficulty of understanding why only the molecule with the unique structure of methapyrilene should be a potent carcinogen, and close analogues inactive.

These four compounds had no significant effect on mitochondrial numbers, contrasting with the large increase caused by methapyrilene

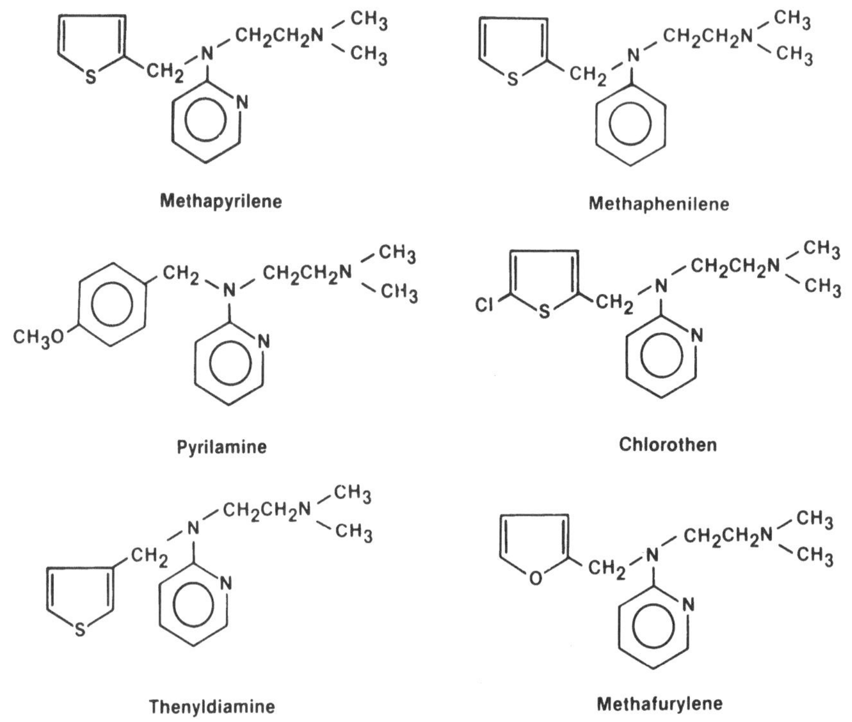

Fig. 3. Structures of methapyrilene and its analogues.

under the same experimental conditions [14]. Unlike the other three non-carcinogenic methapyrilene analogues, methaphenilene did cause cellular derangements in rat liver following several weeks' treatment. However, it induced peroxisome proliferation [16], much as does clofibrate and similar drugs, although, unlike them, it was not a hepatocarcinogen in rats. It is not known why methapyrilene and methaphenilene, differing chemically only in replacement of a pyridine ring in the former by a benzene ring in the latter, differ in effects. Presumably there is subtle specificity for certain receptors related to mitochondria and to peroxisomes respectively.

EXPOSURE OF CULTURED LIVER CELLS TO METHAPYRILENE

In further study of the unusual biological properties of methapyrilene, it was added to liver-cell cultures (25 μg/ml medium). Within 2 h lamellar inclusion bodies were observed, which persisted as long as drug was present. Within 24 h there was an increase in liver-cell mitochondrial content, as measured with a fluorescence-activated cell sorter after staining with Rhodamine 123 [17]. The mitochondrial change appeared to be a change in number, not size, and was the same in the liver of orally dosed rats.

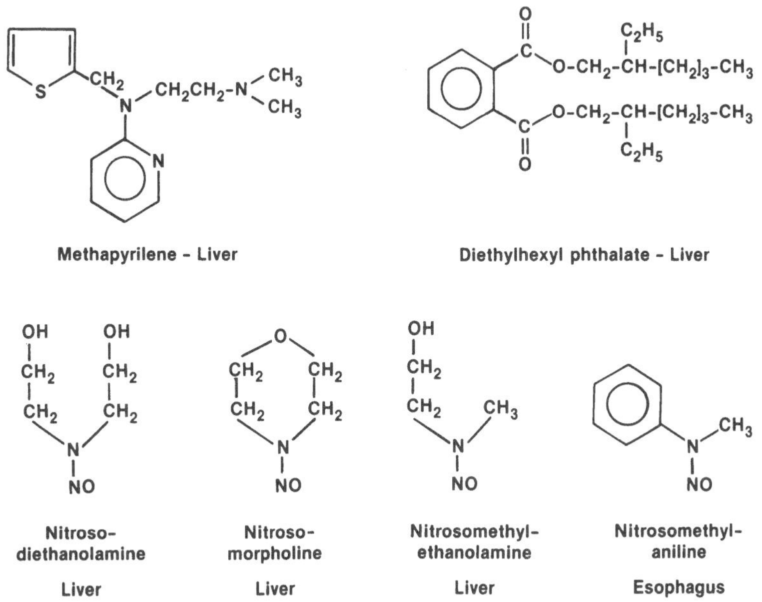

Methapyrilene - Liver

Diethylhexyl phthalate - Liver

Nitroso-diethanolamine

Liver

Nitroso-morpholine

Liver

Nitrosomethyl-ethanolamine

Liver

Nitrosomethyl-aniline

Esophagus

Fig. 4. Structures of carcinogens which induce mitochondrial proliferation, and organs in which tumours appear.

The mitochondrial content of liver cells increased with time
of exposure in culture, but tended to revert to normal if the
methapyrilene were removed. This need for sustained exposure
reinforces the conclusion that the effect of methapyrilene is reversible for a considerable time, and possibly becomes permanent only
when neoplastic transformation of the cells has occurred.

STUDIES WITH PLASTICIZERS AND NITROSO COMPOUNDS

Although mostly we have studied methapyrilene, there are other
compounds of interest that induce mitochondrial proliferation, and
almost all are carcinogenic. The plasticizers diethylhexyl-phthalate
(DEHP) and -adipate induce liver tumours in rats, although only after
administration of very large doses [18]. These compounds also induce
changes in cellular organelles, particularly increases in peroxisomes.
However, DEHP also induced a considerable increase in liver-cell
mitochondrial content [19]. The focus of investigation of the carcinogenic activity of this compound has usually been on the proliferation
of peroxisomes, although the increase in mitochondria might be equally,
or more, important, based on our results with methapyrilene.

Other compounds examined include a number of nitrosamines
(Fig. 4), which are particularly potent carcinogens. As with
methapyrilene, rat liver-cell mitochondrial content was analyzed
by examining the e.m.'s of 5 portal and 5 centrilobular cells per

animal, and estimating the volume fractions of each organelle per hepatocyte (H. Reznik-Schüller & C.J. Michejda, pers. comm.).

The following values were obtained for mitochondria as % of cell area (controls in parentheses):
- methapyrilene, 1000 ppm in food: 28 (8);
- pyrilamine, 2000 ppm in food: 12 (8);
- DEHP, 20000 ppm in food: 35 (16);
- nitrosodiethanolamine, 70 ppm in water: 32 (20);
- nitrosomethylethanolamine, 54 ppm in water: 29 (20);
- nitrosomorpholine, 60 ppm in water: 32 (20);
- nitrosomethylaniline, 71 ppm in water: 34 (20).

Evidently these nitrosamines enhanced mitochondrial content. (Some differences in measurement method from that in the methapyrilene studies gave higher basal values.) Mitochondrial content was not increased by two nitrosamines (not listed) that have been intensively studied, nitrosodimethylamine and nitrosodiethylamine. On the other hand, the content was raised by the oxygen-containing nitrosamines nitrosomethylethanolamine, nitrosomorpholine and nitrosodiethanol-amine.

Because all of these nitrosamines are potent hepatocarcinogens in the rat, the doses used in these experiments were much smaller than for methapyrilene or DEHP. This may explain why the liver-cell mitochondrial increase was less spectacular. The positive result listed for nitrosodiethanolamine, which is also a non-mutagenic carcino-genic nitrosamine, contrasts strongly with the results of an earlier study where the mitochondrial content was sub-normal [14]. The reason for the discrepancy is not clear. One anomaly in these experiments is the positive increase caused by administering nitrosomethylaniline. This compound is a potent carcinogen for rat oesophagus, but has never induced tumours of the liver. It, too, is a non-mutagenic carcinogenic nitrosamine and, as would be predicted from its chemical structure, has been shown not to alkylate DNA [20].

CONCLUSIONS

It appears, therefore, that there is some correlation between the induction of mitochondrial proliferation in rat liver cells by a number of apparently non-mutagenic carcinogens and their induction of tumours in rats, particularly in the liver. Whether the effect on mitochondria is related to the induction of neoplastic transfor-mation in these cells, or is another manifestation of a common bio-chemical lesion, is not known. There seems to be little common chemically between many of these agents: some are simple nitrosamines, and some are complex molecules of unexplored chemistry. Nevertheless, it seems likely that studies of the mechanisms by which these compounds cause the proliferation of mitochondria in liver cells will shed light on the complex problem of carcinogenesis by them, and moreover be an interesting and important addition to our knowledge of the controls and effectors of proliferation of cellular organelles. Investigations are in progress, in rats and other species, of the biochemical transformations of some of these molecules, particularly methapyrilene and its analogues.

Acknowledgments and disclaimer

I thank Drs. H.M. Reznik-Schüller and C.J. Michejda for some of the information presented here. Dr. Reznik-Schüller also provided some photomicrographs.

The research was sponsored by the National Cancer Institute, DHHS, under contract no. N01-CO-23909 with Bionetics Research Inc. The contents of this publication do not necessarily reflect the views or policies of the Dept. of Health & Human Services, nor does mention of commercial products imply endorsement by the U.S. Government.

References

1. Lefkowitch, J.H., Arborgh, B.A.M. & Scheuer, P.J. (1980) *Am. J. Clin. Path. 74*, 432-441.
2. Lafontaine, J.G. & Allard, C. (1964) *J. Cell Biol. 22*, 143-172.
3. Reznik-Schüller, H.M. & Lijinsky, W. (1981) *Arch. Toxicol. 49*, 79-83.
4. Lijinsky, W. (1974) *Cancer Res. 34*, 255-258.
5. Lijinsky, W., Reuber, M.D. & Blackwell, B.N. (1980) *Science 209*, 817-819.
6. Andrews, A.W., Fornwald, J.A. & Lijinsky, W. (1980) *Toxicol. Appl. Pharmacol. 52*, 237-244.
7. Pienta, R.J. (1980) in *The Predictive Value of Short Term Screening Tests in Carcinogenicity Evaluation* (Williams, G.M., et al., eds.), Elsevier/N. Holland, Amsterdam, pp. 149-169.
8. Iype, P.T., Ray-Chaudhuri, R., Lijinsky, W. & Kelley, S.P. (1982) *Cancer Res. 42*, 4614-4618.
9. Lijinsky, W. & Muschik, G.M. (1982) *J. Cancer Res. Clin. Oncol. 103*, 69-73.
10. Reznik-Schüller, H.M. & Gregg, M. (1983) *J. Natl. Cancer Inst. 71*, 1021-1031.
11. Reddy, J.K., Azarnoff, D.L. & Hignite, C.E. (1980) *Nature 283*, 397-398.
12. Lijinsky, W., Knutsen, G. & Reuber, M.D. (1983) *J. Toxicol.Enviro Health 12*, 653-657.
13. Lijinsky, W. (1984) *Food Chem. Toxicol. 22*, 27-30.
14. Reznik-Schüller, H.M. & Lijinsky, W. (1982) *Ecotoxicol. Environ. Safety 6*, 328-335.
15. Lijinsky, W. & Kovatch, R.M. (1986) *J. Cancer Res. Clin. Oncol. 112*, 57-60.

16. Reznik-Schüller, H.M. & Lijinsky, W. (1983) *Arch. Toxicol.* 52, 165-166.
17. Iype, P.T., Bucana, C.D. & Kelley, S.P. (1985) *Cancer Res. 45,* 2184-2191.
18. Kluwe, W.M., McConnell, E.E., Huff, J.E., Haseman, J.K., Douglas, J.F. & Hartwell, W.V. (1982) *Environ. Health Perspect. 45,* 129-133.
19. Ganning, A.W., Brunk, U. & Dallner, G. (1983) *Biochim. Biophys. Acta 763,* 72-82.
20. Lijinsky, W. (1984) in *Genotoxicology of N-Nitroso Compounds* (Rao, T.K., Lijinsky, W. & Epler, J.L., eds.), Plenum, New York, pp. 189-231.

#A-9

ULTRASTRUCTURAL MORPHOMETRIC/BIOCHEMICAL APPROACHES TO ASSESSING METAL-INDUCED CELL INJURY

Bruce A. Fowler

National Institute of Environmental Health Sciences
P.O. Box 12233, Research Triangle Park, NC 27709, U.S.A.

*Correlated approaches (as in title) have been used to assess mechanisms of metal- or metalloid-induced cell injury following in vivo (usually rat) exposure to these agents. Relationships have been studied between **indium-** and **thallium**-induced perturbations of hepatic haem metabolism and associated changes in e.r.* and mitochondrial membrane structure. Seemingly such structural changes largely mediate the inhibitions of cytochrome P-450-dependent enzyme activities observed following acute administration of various metals.*

*With these techniques we have also examined quantitative relationships between formation of **lead**-induced intranuclear inclusion bodies in kidney proximal tubule cells, alterations in their organelle systems, and changes in renal gene expression. The apparently reversible formation of these inclusion bodies seems to be associated with highly specific changes in gene expression following a lead dose which does not produce cellular necrosis or overt structural changes in the organelles. These results accord with already postulated mediation of the initial intranuclear movement of lead by high-affinity receptor-like proteins, target tissue-specific, which regulate lead effects on sensitive target molecules, e.g. ALA hydratase.*

***Arsenicals** administered chronically have reported hepatic effects (e.g. on respiration and ultrastructure) which have now been studied by ^{31}P-NMR in vivo, in an attempt to delineate the temporal relationships between mitochondrial swelling and decreases in hepatic ATP, NAD and phosphorylation of other chemical species. Tentatively it appears that mitochondrial structural damage in periportal areas may play an important early role in subsequent loss of biochemical functionality.*

---

* *Editor's abbreviations:* e.r., endoplasmic reticulum; ALA, δ-amino-
levulinic acid; MT, metallothionein; NMR, nuclear magnetic resonance.

The physical relationships which exist between intracellular organelles and biochemical processes localized within these structures is of ever-growing importance in understanding biochemical mechaniams of cell injury. Methodological approaches for assessing alterations in these relationships and evaluating how these changes relate to injury have led to the use of ultrastructural morphometry in combination with a variety of biochemical techniques for evaluating early metal- or metalloid-induced perturbations of key systems in target-cell populations. This review is intended as an outline of our experience with these techniques. [They feature (not in the injury context) in earlier vols., e.g. D.J. Morré in Vol. 4 and R.L. Deter in Vol. 6.- Ed.] It will hopefully give the reader insight into the value of this integrated approach with regard to other biological investigations in normal animals and under conditions of toxicological stress.

An attempt will be made to consider both the morphological and the biochemical aspects of this evolving approach, in an effort to focus on those areas where these two distinct sets of research techniques have been successfully integrated to yield more definitive answers than could be obtained by more reductionistic approaches. Technical limitations in combining these techniques will also be discussed, since both morphological and biochemical approaches have inherent methodological problems which must be considered before effective integration can be achieved.

ULTRASTRUCTURAL MORPHOMETRY

The theory and practical aspects of ultrastructural morphometry have been extensively described [1], and this discussion will focus on the practical aspects of applying this set of techniques to elucidating mechanisms of toxicity. As previously discussed [1, 2], the procedures are highly labour-intensive, and anyone considering using them should have a clear question or hypothesis delineated before beginning a study. The approach is, then, not readily applicable to large-scale screening for chemical effects. This consideration is particularly true for pharmacology/toxicology studies where dose-response and time-course effects of a given drug or chemical are under investigation. Such studies necessitate the use of multiple sampling groups which greatly increase the number of micrographs that must be evaluated.

On the positive side, previous studies using these techniques have yielded data providing fresh insights [3, 4] into the importance of the mitochondrial and e.r. membranes in mediating the indium- and thallium-mediated disturbances of the hepatic haem biosynthetic pathway and attendant cytochrome P-450-dependent enzyme activities. Data from these studies [3, 4] have quantitatively shown ([†]Tables 1a, 1b) structural changes in e.r. membranes (Fig. 1) which are correlated with induction of microsomal haem oxygenase, depletion of microsomal cytochrome P-450, and the inhibition of attendant microsomal mono-

[†]Tables are near end of article

oxygenase and acid hydrolase activities (Table 1c) [5, 6]. Similar results emerged from evaluation of mitochondrial membrane struct-ure *vs.* synthesis of membrane structural proteins following *in utero* exposure to methylmercury [7] or *in vivo* exposure to arsenate [8].

Time-course studies in the kidneys of rats exposed to cadmium as the MT complex (CdMT) ([9]; cf. M. Dobrota in this vol., # F-5) or lead [10] have yielded similar results with respect to delineating toxic mechanisms for these ubiquitous toxic elements in this organ. Administration of CdMT produces a characteristic vesiculation of the kidney proximal tubule cells (Fig. 2). This follows degradation of the CdMT complex which is associated with marked inhibition of cathepsin D activity and development of a tubular proteinuria [9, 11] similar to that observed in occupationally exposed workers or environmentally exposed persons [12]. Morphometric studies [9] demonstrated that these effects were preceded by a marked reduction in lysosome diameter and more numerous small lysosomes within these tubule cells, suggesting that Cd^{2+} disrupts normal lysosome biogenesis. Further evidence in support of this idea came from the finding that cathepsin D activity was highly resistant to inhibition by either Cd^{2+} or CdMT *in vitro* at concentrations in excess of those measured in kidney lysosome pellets from *in vivo* treatment studies [9].

More recent time-course studies [10] have been conducted to characterize the apparently reversible formation of lead intranuclear inclusion bodies (Fig. 3) within kidney proximal tubule cell nuclei in relation to specific changes in renal gene expression. These studies were conducted as part of an overall study to evaluate the hypothesis that kidney-specific, cytosolic high-affinity lead-binding proteins [13-19], which are the initial binding sites for lead in this tissue, act in a manner analogous to 'receptors' by facilitating the intranuclear movement of this element and its subsequent inter-action with sensitive target molecules or the genetic machinery (Fig. 4) of the target-cell population [10, 14, 15].

Ultrastructural morphometry played a key role in these studies by providing quantitative evidence for the formation and dissolution of these structures within proximal tubule cell nuclei over time. Such data documented the indubitable presence of lead within this target organelle *in vivo* and provided a quantitative time-course basis for biochemical evaluation of the reversible effects of lead on renal gene expression during this period. In addition, the demons-trated absence of quantitative morphological changes in other sensitive organelle systems such as the mitochondria supported the idea that the observed changes in gene expression were the direct result of lead interaction with renal genetic machinery and not secondary to a cell-death and replacement phenomenon.

Overall results of the above studies suggest that ultrastructural morphometry may provide useful data for interpreting biochemical

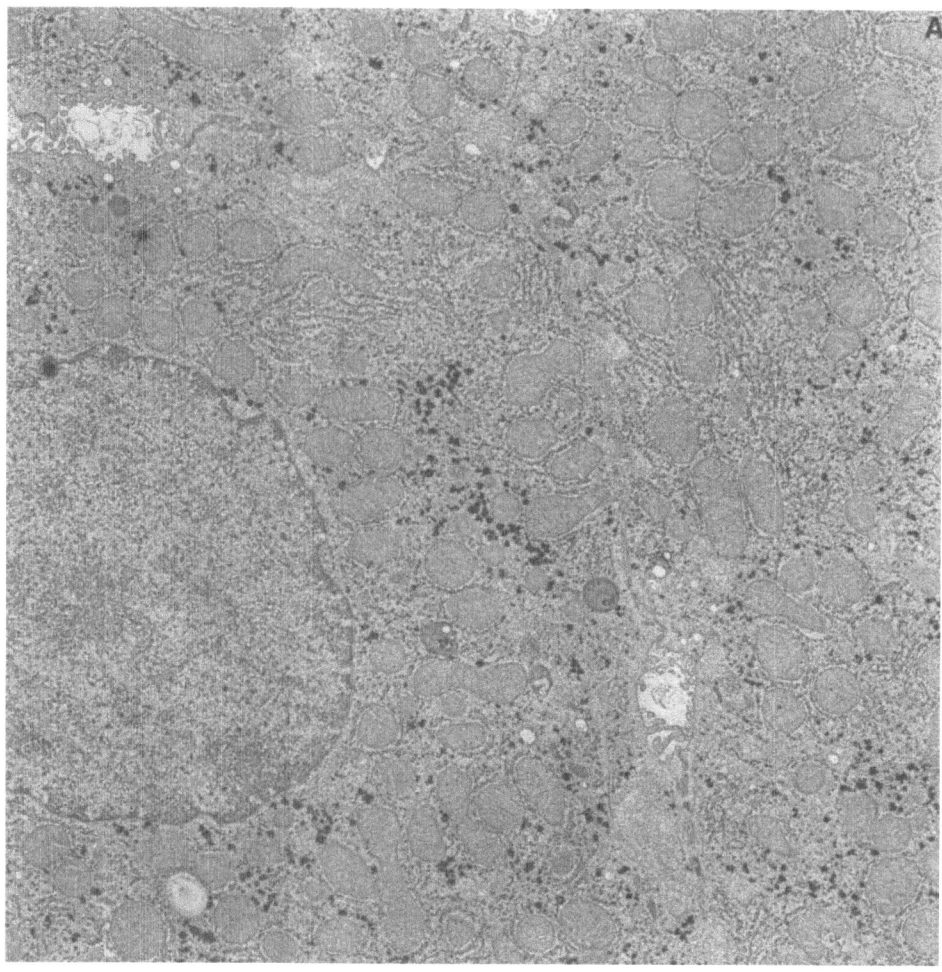

Fig. 1. Hepatocyte from a control rat (A, *above*), and (B, *opposite*)
a rat injected with indium chloride (20 mg/kg, i.p.) 16 h earlier
showing e.r. dilatation and degranulation.   ×11,200.

changes in subcellular elements, by indicating which organelles or
components thereof are physically altered within a target cell
population as a function of time or metal/metalloid dose level.
Such data are of particular value in interpreting results of biochemical
studies conducted on subcellular fractions from whole organs derived
from organisms exposed to these agents *in vivo*. This information
has proved quite useful in 'bridging the gap' between *in vivo* effects
and mechanistic data derived from *in vitro* systems. As discussed
below, the degree of correlation between these ultrastructural changes
and biochemical functionality also depends upon which biochemical
parameters are evaluated, since not all biochemical processes are

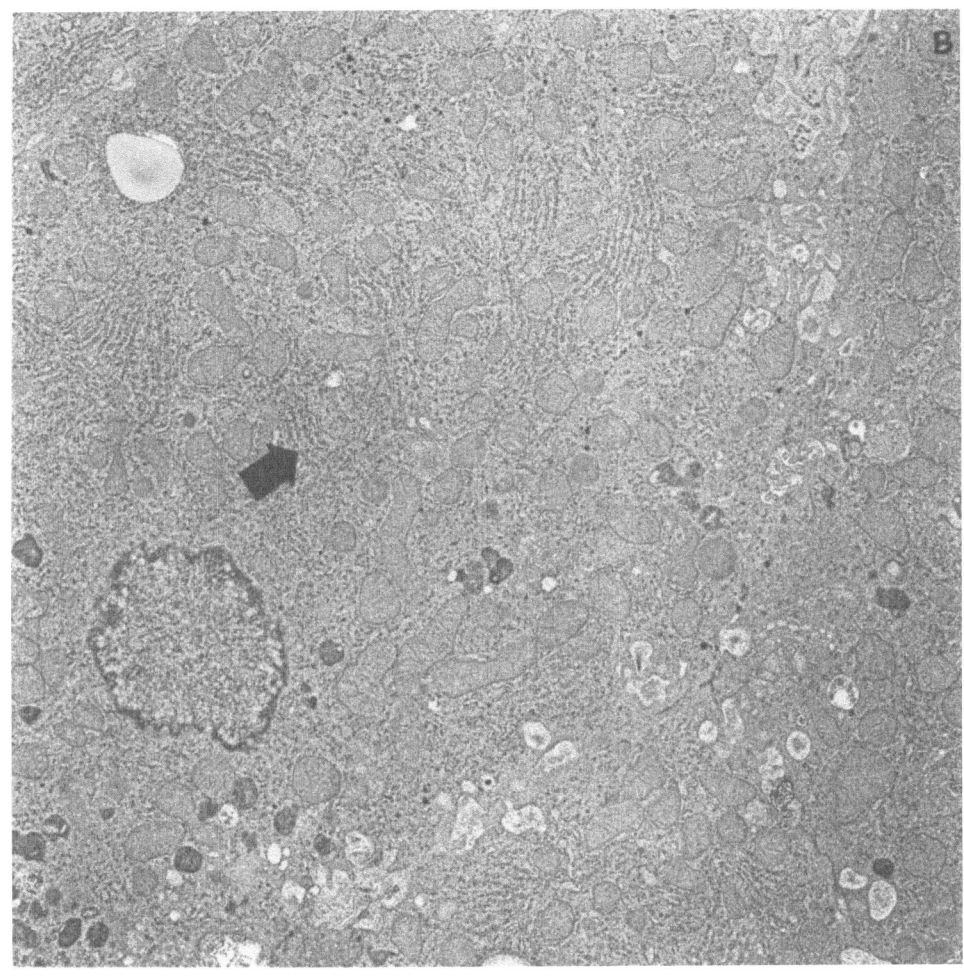

[*Fig. legend opposite*

equally dependent upon organelle structure or equally affected by an administered agent.

BIOCHEMICAL PARAMETERS

The chief advantage of ultrastructural morphometry is that it provides quantitative data on which cells, which organelles therein, and which intraorganelle compartments are affected by administration of a toxic element. Like all morphological techniques, however, it does not directly delineate changes in biochemical function but rather indicates where structural changes within cells/organelles are occurring. This necessitates biochemical measurements to provide functional data of use in interpreting the observed morphological phenomena. In searching for correlative functional indices, it is

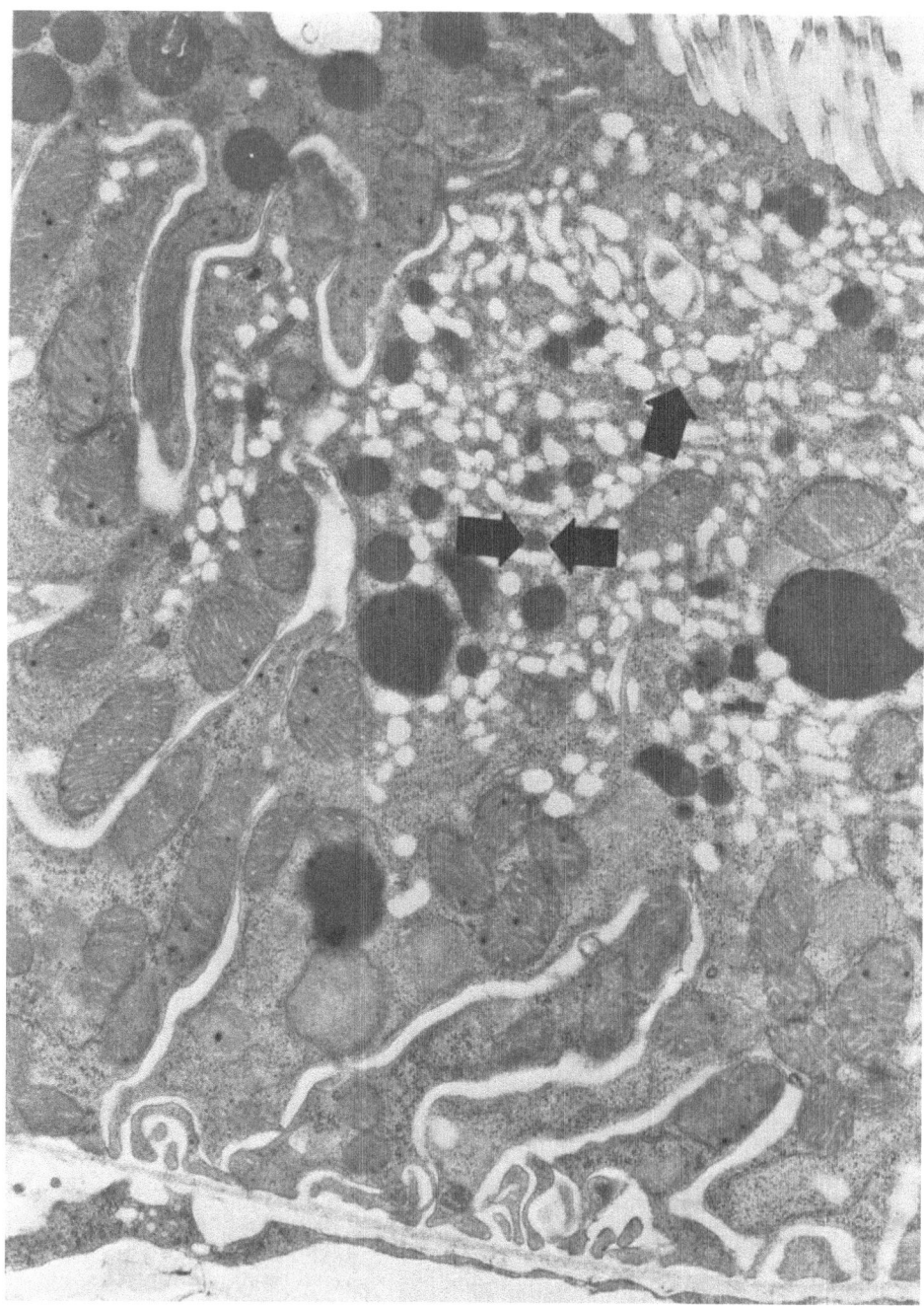

Fig. 2. Kidney proximal tubule cell from a rat injected with CdMT, showing apical vesiculation *(arrow)* and the presence of small, dense lysosomes *(double arrow).* ×20,200.

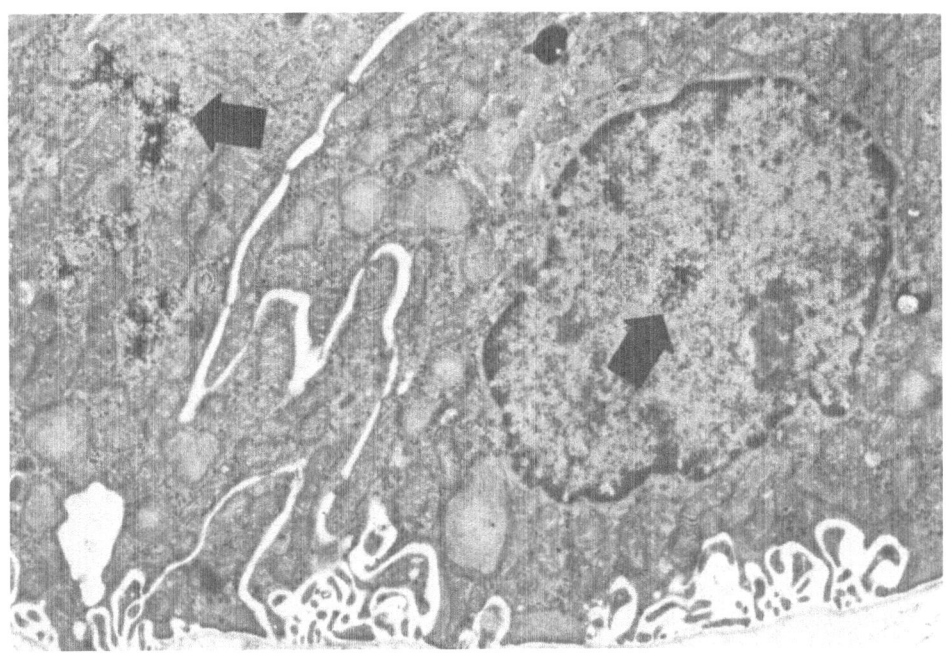

Fig. 3. Kidney proximal tubule cell from a rat at 16 h after i.v. injection of lead acetate, showing both cytoplasmic and intranuclear lead inclusion bodies *(arrows).*   ×9,400. (Dose: 3 mg Pb/kg.)

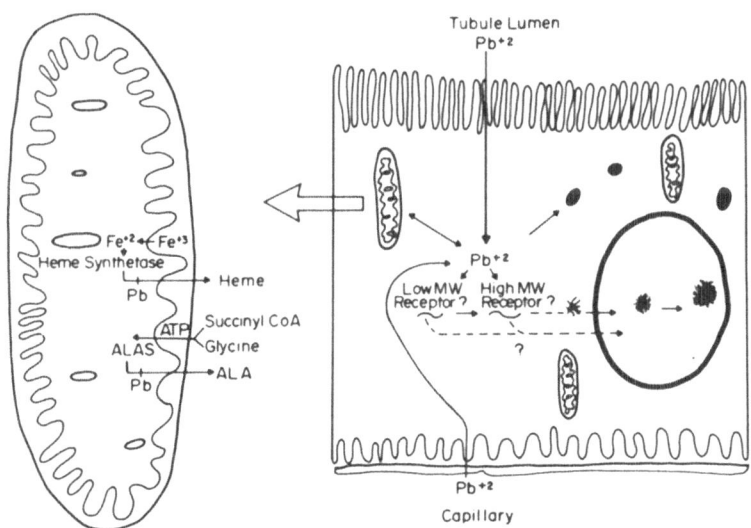

Fig. 4.
Hypothesized roles for Pb-binding proteins in mediating Pb effects in a renal proximal tubule cell in respect of both the haem biosynthetic pathway and intranuclear Pb movement.

important to evaluate a number of parameters since all are not equally
dependent upon organelle structure. Thus, mitochondrial swelling
may more severely influence the activity of a membrane marker enzyme
such as cytochrome oxidase than that of a matrix enzyme such as malate
dehydrogenase.

Obviously there are many biochemical approaches that may be
utilized for evaluating these considerations. In our experience
it is important to examine a number of interrelated parameters since
each may offer a partial insight into the overall question under
study. Some studies [3-9] have utilized marker enzymes to provide
functional information and correlative biochemical data about the
intraorganelle site of metal/metalloid action. The value of these
measurements lies in localizing effects to a particular membrane
compartment (e.g. mitochondrial inner membrane). On the other hand,
without protein synthesis or quantitative immunological studies,
one would not know whether these changes were the result of direct
chemical effects or secondary to changes in the synthesis of the
marker enzyme.

Organelle dysfunction has in fact been examined in other studies
by measuring changes in protein synthesis or, more recently, by use
of 2-D gel electrophoresis to elucidate specific patterns of change
in tissue gene expression [e.g. 10, 20]. Obviously the parameter
selected for evaluation will depend on the question being asked or
the hypothesis under evaluation. The crux of this discussion is
that for integrative studies of this kind, it is important to measure
a number of biochemical parameters.

Finally, it should be noted that the biochemical parameters
discussed above all depend upon subcellular fractionation techniques
to elucidate correlative functional mechanisms. While such approaches
have yielded invaluable information, it should also be remembered
that these procedures need appropriate control measures. Mixtures
of organelles from different cell types within an intact organ are
also a potential problem [21]. More recent studies from a number
of laboratories [22-27] have suggested the possibility of utilizing
non-invasive techniques such as surface-coil NMR *in vivo* to gather
correlative real-time biochemical information without subjecting the
tissue under study to homogenization procedures which may produce
artefacts in affected organelles.[†] Yet NMR techniques, whilst informative, are limited to certain isotopes that have a sufficient
abundance/NMR activity to be readily detected. The vast potential
of this technique needs further exploration. In summary, the value
and potential limitations of integrating different ultrastructural
and biochemical techniques for probing chemically induced cell injury
has been considered. Biochemical regulation within normal intact
cells can be illuminated by such study of relationships between biochemical functionality and organelle structure.

[*†Footnote opposite*

Table 1. Hepatic parameters in control and indium-treated rats – *adapted from Tables in ref. [3], omitting the S.E.M.'s.* Dose-response trends were assessed by Jonckheere's test; NS = not significant. The ANOVA test of significance was used for particular dose levels compared with controls; [1] = $p < 0.06$, [2] = $p < 0.05$, [3] = $p < 0.01$. For **1a** and **1b** (morphometry) a test grid was printed onto each micrograph; assessments were based on line intersections [3, 6].

Dose, mg/kg:	0	10	20	40	Dose–response
1a: Organelle volume densities, *1000 × % of unit volume*					
Mitochondria	198	205	189	194	NS
Nuclei	85	77	77	79	NS
Lysosomes	1.7	5.3	7.0[2]	7.3[2]	$p < 0.01$
Vacuoles	12	9[2]	6[2]	5[3]	$p < 0.05$
1b: Surface densities of different membrane types, *m^3/cm^3 cytoplasm*					
Mitochondrial outer	2.23	2.34	2.23	2.18	NS
Mitochondrial inner	2.22	2.32	2.23	2.20	NS
Cristae	4.46	4.72	4.86	6.54[3]	$p < 0.01$
Rough e.r.	2.92	3.22	3.81[2]	4.25[3]	$p < 0.01$
Smooth e.r.	0.06	0.22	0.21	0.41[2]	$p < 0.05$
1c: Microsomal enzyme activities, *nmol/substrate converted/min/mg protein*					
Aminopyrine demethylase	5.2	5.5	5.8	5.3	NS
Aniline hydroxylase	47.5	34.9[2]	30.1[2]	30.7[3]	$p < 0.01$
Ethoxyresorufin-*O*-deethylase	23.6	21.4	16.5	17.0[2]	$p < 0.05$
NADPH cyt. **c** reductase	105	96	96	86	NS
Acid phosphatase	27.5	23.8[2]	22.7[2]	20.3[3]	$p < 0.001$
β-Glucuronidase	7.0	8.8	9.2	6.7	NS

References

1. Weibel, E. & Bolender, R.P. (1973) in *Principles and Techniques of Electron Microscopy: Biological Applications*, Vol. 3 (Hyat, M.A., ed.), Van Nostrand Reinhold, New York, pp. 237–296.
2. Fowler, B.A. (1983) *Fed. Proc. 42*, 2957–2964.
3. Fowler, B.A., Kardish, R. & Woods, J.S. (1983) *Lab. Invest. 48*, 471–478.
4. Woods, J.S. & Fowler, B.A. (1986) *Toxicol. Appl. Pharmacol. 83*, 218–229.
5. Woods, J.S., Carver, G.T. & Fowler, B.A. (1979) *Toxicol. Appl. Pharmacol. 49*, 455–461.

[†] *(referring to opposite p.)* Recent studies [27] have demonstrated *in vivo* decreases in hepatic ATP content in real-time with concomitant increases in P_i and phosphorylation of other chemical species.

6. Woods, J.S., Fowler, B.A. & Eaton, D.L. (1984) *Biochem. Pharmacol.* *33*, 571-576.

7. Fowler, B.A. & Woods, J.S. (1977) *Lab. Invest. 36*, 122-130.

8. Fowler, B.A., Woods, J.S. & Schiller, C.M. (1979) *Lab. Invest.* *41*, 313-320.

9. Squibb, K.S., Pritchard, J.B. & Fowler, B.A. (1984) *J. Pharmacol. Exp. Ther. 229*, 311-321.

10. Fowler, B.A., Mistry, P. & Victery, W.W. (1985) *The Toxicologist* *5*, 53.

11. Squibb, K.S. & Fowler, B.A. (1984) *Environ. Hlth. Perspect. 54*, 31-35.

12. Kjelstrom, T. (1986) in *Cadmium and Health* (Friberg, L. et al., eds.), CRC Press, Boca Raton, pp. 47-86.

13. Oskarsson, A., Squibb, K.S. & Fowler, B.A. (1982) *Biochem. Biophys. Res. Comm. 104*, 290-298.

14. Mistry, P., Lucier, G.W. & Fowler, B.A. (1985) *J. Pharmacol. Exp. Ther. 232*, 462-469.

15. Mistry, P., Mastri, C. & Fowler, B.A. (1986) *Biochem. Pharmacol.* *35*, 711-713.

16. Goering, P.S. & Fowler, B.A. (1984) *J. Pharmacol. Exp. Ther.* *231*, 66-71.

17. Goering, P.S. & Fowler, B.A. (1985) *J. Pharmacol. Exp. Ther.* *234*, 365-371.

18. Goering, P.S., Mistry, P. & Fowler, B.A. (1986) *J. Pharmacol. Exp. Ther. 237*, 220-225.

19. Gilg, D.E.O., Pentecost, B., Duval, G.E. & Fowler, B.A. (1987) *Fed. Proc.*, in press.

20. Shelton, K.R., Todd, J.M. & Egle, P.M. (1986) *J. Biol. Chem.* *261*, 1935-1940.

21. Blouin, A., Bolender, R.P. & Weibel, E.R. (1977) *J. Cell Biol.* *72*, 441-455.

22. Burt, C.T., Glonek, T. & Barany, M. (1976) *J. Biol. Chem. 251*, 2584-2591.

23. Kopp, S.J., Krieglstein, A., Freidank, A., Rachman, A., Seibert, A. & Cohen, M.M. (1984) *J. Neurochem. 46*, 1716-1731.

24. Bailey, I.A., Williams, S.R. Radda, G.K. & Gadian, D.G. (1981) *Biochem. J. 196*, 171-178.

25. Stubbs, M., Freeman, D. & Ross, B.D. (1984) *Biochem. J. 224*, 241-246.

26. Iles, R.A., Stevens, A.N., Griffiths, J.R. & Morris, P.G. (1985) *Biochem. J. 229*, 141-151.

27. Chen, B., Burt, C.T., Goering, P.S., Fowler, B.A. & London, R.E. (1986) *Biochem. Biophys. Res. Comm. 39*, 228-234.

#A-10

IRON OVERLOAD, LYSOSOMES AND FREE RADICALS

M.J. O'Connell, R.J. Ward, *H. Baum and T.J. Peters

Division of Clinical   *Department of Biochemistry
 Cell Biology   King's College (KQC)
MRC Clinical Research   Kensington Campus
 Centre, Harrow   Campden Hill Road
Middlesex HA1 3UJ, U.K.   London W8 7AH, U.K.

Tissue damage due to iron excess is an important clinical problem. Ferritin is the main iron protein in normal liver. However, in iron overload haemosiderin is predominant. It accumulates in lysosomes and is associated with increased fragility of these organelles [1]. Ferritin and haemosiderin will stimulate free radical reactions in vitro [2, 3] and such species may therefore mediate lysosomal damage in vivo. Methodology for the preparation and analysis of haemosiderin and lysosomal lipids is discussed with reference to evidence for oxidative free radical damage.

Haemosiderin is an aggregate largely comprised of iron oxide cores and incomplete protein shells which appear to have been derived from ferritin [4], together with lipid and small amounts of carbohydrate [5]. Damage to lysosomes in iron-overload may result from mechanical distension as haemosiderin accumulates. Similar pathology is thought to occur in other secondary lysosomal storage diseases, e.g. silicosis [6]. Free radical generation catalyzed by iron-protein may however be important [2, 3] since hydroxyl radical has been shown to increase the permeability of lysosomal membranes *in vitro* [7]. Evidence for the occurrence of oxidative free radical processes was sought by analysis of iron-loaded lysosomes. Since many products of free radical reactions are transitory *in vivo*, emphasis was placed on changes in composition and properties of certain biomolecules, particularly the protein components of haemosiderin and lipid extracts of lysosomes.

PREPARATION AND ANALYSIS OF APOHAEMOSIDERIN AND APOFERRITIN

Human haemosiderin [5] and ferritin [8] were isolated from iron-overloaded human spleens which had been removed from patients with

Table 1. Comparison of amino acid analysis of haemosiderin and ferritin. Values are mol/100 mol total residues and % difference.

	Ferritin	H'siderin	DECREASE		Ferritin	H'siderin	INCREASE
Met	2.3	0.1	-96	Glx	12.8	13.2	+3
Tyr	5.8	0.9	-84	Thr	4.1	5.1	+24
Leu	15.7	4.9	-69	Val	3.5	4.8	+37
Phe	4.7	2.2	-53	Ile	1.7	2.5	+47
His	3.5	2.8	-20	Asx	11.0	16.3	+48
Lys	7.6	6.4	-16	Gly	7.0	12.9	+84
Ala	8.7	8.1	-7	Pro	2.3	5.3	+130
Arg	5.2	5.2	0	Ser	4.1	9.6	+134

β-thalassaemia. Apoferritin and apohaemosiderin were prepared by dialysis against two changes of 0.1 M thioglycollic acid in 0.1 M Na acetate and two changes of 0.1 M NaCl. Each dialysis was for 2 h at room temperature and with continuous bubbling of oxygen-free nitrogen. Dialyzed samples contained a precipitate which was cleared by adding tetramethylammonium hydroxide (to 20 mM final concn.).

Amino acid analyses [5, 9] show that, of the residues detected in haemosiderin, those most likely to be susceptible to oxidation are depleted when compared to ferritin (Table 1), viz. methionine, tyrosine, phenylalanine and histidine.

The iron storage proteins were further compared by fluorescence spectroscopy using the apoproteins to avoid quenching effects of iron. In the UV range, two preparations when examined as 100 mg/ml aqueous solutions gave the following mean value for excitation (ex) and emission (em) peaks and relative fluorescence intensity (RFI):
- apoferritin: ex 283 nm, em 340-357 nm; RFI 643 U;
- apohaemosiderin: same ex & em; RFI 250 U.
The peaks for apoferritin are consistent with the recognized properties of tryptophan-containing proteins [10]. Apohaemosiderin samples prepared from the same spleen had similar UV fluorescence but of intensity ~40% that of protein-matched apoferritin, This is consistent with oxidative loss of aromatic residues in haemosiderin compared to ferritin.

Fluorescence characteristics in the visible region were as follows, for aqueous solutions:
- apoferritin: ex 402, em 500; RFI 6.0 U.;
- apohaemosiderin: same ex & em; RFI 90.2 U.
Evidently apohaemosiderin has substantial fluorescence in the visible region, consistent with the presence of conjugated Schiff's base adducts of the type found in lipofuscin. These compounds are thought to be formed by reaction of malondialdehyde (a product of free radical-

mediated lipid peroxidation) with primary amino groups of proteins. Compared to apohaemosiderin, protein-matched apoferritin samples from the same spleen had very little 'visible' fluorescence (see above). Thus, haemosiderin formation in lysosomes is associated with free radical reactions; indeed, the conversion of ferritin to haemosiderin may be free radical-mediated [11].

PREPARATION OF IRON-LOADED RAT LIVER LYSOSOMES AND LIPID ANALYSIS

Male Sprague-Dawley rats of starting body wt. 50 g were fed 2% reduced pentacarbonyl iron in powdered diet [12] for 129 days. Control rats were injected with Dextran 500 (1 g/100 g body wt.) for 3 days before sacrifice to increase the density of hepatic lysosomes. Liver homogenates (5% w/v in 0.25 M sucrose) from iron-loaded and control animals were centrifuged (1000 g, 10 min); the supernatants were further centrifuged (10,000 g, 10 min) and the high-speed pellets were washed in 0.25 M sucrose with two further centrifugation steps.

The washed pellets were layered onto discontinuous sucrose gradients (5 ml each of d = 1.17, 1.22 and 1.28) and centrifuged (100,000 g, 2 h). The iron- or dextran-laden lysosomes formed a pellet and the mitochondria equilibrated at the 1.17/1.22 interface. These fractions, which contained >70% and >95% of the recovered activity of two marker enzymes - N-acetyl-β-glucosaminidase and succinate dehydrogenase respectively, were extracted for fatty acid analysis with 8 vol. of chloroform/methanol (2:1 by vol.) containing 10 mg/ml butylated hydroxytoluene. Methyl esters of the fatty acids were prepared and analyzed with a Pye-Unicam 204 gas chromatograph [13]. Peak areas representing individual fatty acids were expressed as % of total fatty acid recovery. The ratio double-bond index/saturated fatty acid was calculated by dividing the sum of % proportions of individual unsaturated fatty acids (each multiplied by the no. of double bonds in the fatty acid) by the % of saturated fatty acids.

The iron content of iron-overloaded rat liver lysosomes determined by atomic absorption spectroscopy was increased 6-fold compared to those of control animals, whereas mitochondrial iron content was increased only 2-fold (Table 2). Greater losses in arachidonate were also associated with lysosomes than with mitochondria, and this was reflected in an overall reduction in the ratio double-bond index/ saturated fatty acid. Loss of unsaturated fatty acids is consistent with peroxidative damage at the lysosomal site of iron accumulation.

CONCLUSIONS

The data presented here are indicative of free radical damage in iron-overloaded lysosomes. Experiments in which antioxidants are administered *in vivo* are required to determine whether it is possible to protect against this damage and the increased fragility of lysosomes.

Table 2. Iron content (µg/mg protein) and fatty acid composition of hepatic lysosomes and mitochondria: arachidonate as % of total fatty acid, and DBI/S = double-bond index/saturated fatty acid ratio. The means ± S.E.M. are based on 8 observations; * signifies $p < 0.05$.

	Lysosomes		*Mitochondria*	
	Control	Iron-loaded	Control	Iron-loaded
Total iron	4.3 ±1.5	25.4 ±16.7	0.36 ±0.10	0.61 ±0.20
Arachidonate	16.2 ±2.2	10.2 ±1.7*	20.8 ±0.8	17.6 ±1.2*
DBI/S	2.13 ±0.32	1.49 ±0.18*	3.61 ±0.23	3.09 ±0.41*

Acknowledgements

We are grateful to Mrs. H. Lewis for preparing the typescript and to the Medical Research Council for financial support (M.J.O'C.).

References

1. Seymour, C.A. & Peters, T.J. (1979) *Br. J. Haematol. 40*, 239–253.
2. O'Connell, M.J., Ward, R.J., Baum, H. & Peters, T.J. (1985) *Biochem J. 229*, 135–139.
3. O'Connell, M.J., Halliwell, B., Moorhouse, C.P., Aruoma, O.I., Baum, H. & Peters, T.J. (1986) *Biochem. J. 234*, 727–731.
4. Iancu, T.C. & Neustein, H.B. (1977) *Br. J. Haematol. 37*, 527–535.
5. Weir, M.P., Gibson, J.F. & Peters, T.J. (1984) *Biochem. J. 223*, 31–38.
6. Allison, A.C., Harrington, S.S. & Birbeck, M. (1966) *J. Exp. Med. 124*, 141–154.
7. Fong, K.L., McCay, P.B. & Payer, J.L. (1973) *J. Biol. Chem. 248*, 7792–7797.
8. Cham, B.E., Roeser, H.P., Nikles, A. & Ridgway, K. (1985) *Anal. Biochem. 151*, 561–565.
9. Wustefield, C. & Crichton, R.R. (1982) *FEBS Lett. 150*, 43–48.
10. Weber, G. (1960) *Biochem. J. 75*, 345–352.
11. O'Connell, M.J., Baum, H. & Peters, T.J. (1986) *Biochem. J. 240*, 297–300.
12. Bacon, B.R., Tavill, A.S., Brittenham, G.M., Park, C.H. & Recknagel, R.O. (1983) *J. Clin. Invest. 71*, 429–439.
13. Cairns, S.R. & Peters, T.J. (1983) *Clin. Chim. Acta 127*, 373–382.

#A-11

GLUCOCEREBROSIDASE, A MEMBRANE-ASSOCIATED LYSOSOMAL ENZYME DEFICIENT IN GAUCHER DISEASE

[1]A.W. Schram, [1]J.M.F.G. Aerts, [1]S. van Weely,
[2]J.A. Barranger and [1]J.M. Tager

[1]⊗Laboratory of Biochemistry   [2]Division of Human Genetics
   University of Amsterdam   Children's Hospital of Los Angeles
   P.O. Box 20151   University of Southern California
   1000 HD Amsterdam   4650 Sunset Boulevard
   The Netherlands   Los Angeles, CA 90054, U.S.A.

Lysosomal storage diseases are characterized by a massive intra-lysosomal accumulation of undegraded material, causing dysfunction of the cells involved [1]. The storage can have different causes, viz. (1) a generalized deficiency of one or more enzymes in the lysosomes, (2) deficiency of a cofactor required for the stability of one or more enzymes, or (3) deficiency of a cofactor required for expression of activity in vivo [2-5].

To obtain information on the molecular basis of lysosomal storage diseases it is essential to study enzymological and cell-biological properties of the enzyme(s) involved in tissues and cells from control subjects and from affected patients. This article provides an overview of studies and procedures which have been used to obtain such information for Gaucher disease, where massive amounts of glucosylceramide, the substrate for the deficient enzyme GCase*, accumulate [6].

GCase (EC 3.2.1.45) is a lysosomal enzyme responsible for the hydrolysis of the glycolipid glucosylceramide to glucose and ceramide [6,7]. Deficiency of this enzyme is the metabolic basis of Gaucher disease, a group of autosomal recessive disorders characterized by the accumulation of massive amounts of glucosylceramide in cells of the RES [6-8]. The disease has been divided clinically into three subclasses: type 1 with no neurological involvement; type 2, the

---

⊗ Address for correspondence.
* *Editor's abbreviations:* GCase, glucocerebrosidase; (M)Ab, (monoclonal) antibody; IE, isoelectric; PAGE, polyacrylamide gel electrophoresis; (r.)e.r., (rough) endoplasmic reticulum; s.a., specific activity; SAP-2, TC, etc. - *overleaf.*

acute form with neurological involvement; and type 3, the subacute form with neurological involvement [6]. In patients, glucocerebrosidase has been shown to be deficient in a number of tissues including brain, spleen and liver and also in leucocytes and cultured cells such as skin fibroblasts and amniocytes. No correlation has been observed between clinical symptoms and the amount of stored lipid or the residual enzymic activity [6].

The enzyme has been purified to homogeneity by a variety of procedures [9-14], including hydrophobic chromatography and affinity chromatography. The enzymic properties [6, 15] as well as the structure of the enzyme have been closely examined. The amino acid sequence [16] and the composition of the oligosaccharide moieties of the enzyme have been described [17]. Multiple molecular forms of GCase can be detected in extracts of cells and tissues from control persons and Gaucher patients, both by IE focusing and by PAGE [18]; they are due to heterogeneity in the oligosaccharide moiety. Studies which have elucidated the properties of this and other β-glucosidases are discussed in part below.

1. β-GLUCOSIDASES IN TISSUE

Using the artificial substrate 4-methyl-umbellipheryl-β-glucoside (4MU-β-glu), one can discriminate between 3 different β-glucosidases. The first is a cytosolic hydrolase with maximal activity at neutral pH [19,20]. It is inhibited by taurocholate (TC) but not by the epoxide conduritol-β-epoxide (CBE), and is unable to hydrolyze the glycolipid glucosylceramide. The enzyme does not cross-react with Ab's raised against GCase and is not deficient in tissues obtained from Gaucher patients [21]. It represents a gene product different from GCase. Upon flat-bed IE focusing this isoenzyme migrates at a distinct pH of 4.5 [19, 21]. Gel-permeation HPLC has given a native kM_r value of ~60 [22].

Table 1 summarizes the properties of this enzyme, and of a second β-glucosidase, membrane-associated, which occurs in human spleen – undiminished in Gaucher patients – and is rapidly inactivated upon extraction by detergents [19]. It is not inhibited by CBE and is not lysosomal (unpublished results).

A third enzyme that can hydrolyze artificial substrates is GCase, which can also act on the glycolipid glucosylceramide. Besides detergent to solubilize the lipid substrate, full activity needs TC [23, 24]; thus detergents activate artificial substrate hydrolysis. Phospholipids [25] and so-called activator proteins including Sphingolipid Activating Protein-2 (SAP-2) [26] can also activate (see 4. below). Upon immunoprecipitation of the enzyme present in a human-spleen homogenate, two GCase's can be distinguished [21]. Part of the GCase activity is immunoprecipitated (~80%, form 1) whereas the remaining activity (form 2) is not recognized by either polyclonal or monoclonal

Table 1. Properties of β-glucosidases present in human spleen, including a membrane-associated (memb.) enzyme; n.d. = not determined, inhibn. = inhibition, activn. = activation; GC = glucosylceramide.

Property	β-Glucosidase		Glucocerebrosidase (GCase)	
	soluble	memb.	form 1	form 2
Activity with 4MU-β-glu	+	+	+	+
GC	−	n.d.	+	+
Effect of CBE	none	none	inhibn.	inhibn.
SAP-2	none	none	activn.	none
TC	inhibn.	inactivn.	activn.	none
Reaction with anti-GCase	−	−	+	−
kM_r: in SDS gels	−	n.d.	59 (mature)	59 (mature)
native enzyme	~50–70	n.d.	~60	~200
IE point	~4.5	n.d.	~4.7 & ~6.0	~4.9
Gaucher tissue presence	+	+	−	(−)

anti-GCase Ab's. Both forms are irreversibly inhibited by CBE and are deficient - especially form 2 - in spleen from Gaucher patients [21].

The occurrence of these two forms is explained by the association of GCase with SAP-2, leading to loss of epitopes recognizable by the Ab's probably due to steric hindrance [27]. Upon treatment of the preparation with ethylene glycol, the complex dissociates and immunoprecipitability is restored [27]. Besides loss of immunoreactivity the association with the activator in form 2 has three other consequences for the enzyme (Table 1). (a) The enzyme occurs in a high mol. wt. complex, of kM_r ~200 by gel-permeation HPLC, consisting of GCase (monomeric kM_r ~60), activator protein, and possibly other proteins [27, 28]. (b) Added activator or detergent is without effect [27, 28], indicating that the enzyme is already in a maximally activated configuration. (c) IE focusing shows a pH difference from form 1 (Table 1) [21].

2. β-GLUCOSIDASE IN URINE

Besides being found in tissue, GCase is (like many lysosomal enzymes) also present in urine [29, 30]. This is a surprising finding because in tissue the enzyme is membrane-associated, whereas in urine it behaves as a soluble protein. Moreover, the kinetic parameters, the activity per enzyme molecule, the native mol. wt. and the IE point are indistinguishable from those for the tissue enzyme [31]. Urinary GCase is present mainly as the immunoprecipitable form and is deficient in urine from Gaucher patients. This allows one to use urine to diagnose the disease [29].

3. PURIFICATION OF THE ENZYME

GCase is present in many tissues and cells, and can be efficiently extracted from tissues with detergents [32-34] such as Triton X-100 [21]. It is advisable to first prepare an aqueous extract from tissue and to extract the centrifuged residue (membrane fraction) with detergents: this procedure gives a higher s.a. The subsequent chromatographic steps to purify the enzyme in general consist of hydrophobic chromatography [9-14], gel-permeation chromatography and lectin affinity chromatography [35] to separate glycoproteins from non-glyco-proteins. For efficient elution of GCase from these columns the presence of apolar reagents such as ethylene glycol and detergents is often required. Because of the hydrophobic nature of the enzyme, association of protein with (phospho)lipid material easily occurs during extraction.

It is, moreover, to be expected that *in vivo* the enzyme will interact with (phospho)lipids. To obtain a preparation which is devoid of lipid contaminants, n-butanol extraction of the partially purified enzyme is an adequate procedure (unpublished results). These more conventional procedures to purify the enzyme lead to an apparently homogeneous protein preparation as judged by SDS-PAGE. These published procedures, however, give a low recovery and are time-consuming. Another disadvantage is that they are restricted to a few tissues.

Newer procedures seem to have at least partly overcome these problems. Purified enzyme of high s.a. is rapidly attainable with substrate-analogue affinity resins [10]. Aerts et al. recently described a method using MAb's covalently immobilized to a solid support [14]. Although this immunoaffinity column binds not only GCase but also a considerable amount of other proteins, these can be effectively removed by washing with different buffers containing detergents or low concentrations of ethylene glycol, subsequently used in high concentration to elute the enzyme. The purity of the preparation equals or surpasses that obtained by the more conventional methods, and recovery is much superior. Applicability is not restricted to a few tissues: besides control and Gaucher spleen, placenta, kidney, liver, brain, fibroblasts and urine have been used as sources.

4. ACTIVATION OF ENZYMIC ACTIVITY

Artificial substrates and the natural one may both be used for activity measurement. In all cases, however, the assay system must be supplemented. Firstly, with the glycolipid substrate a detergent is necessary to solubilize it and make it available for enzymic digestion. With artificial substrate the detergent TC is also requisite for expression of full activity [24] or, alternatively, certain low mol. wt. proteins purified from spleen (activator proteins), particularly in the presence of phospholipids [36]. These cofactors are thought to associate with the GCase, forming in the case of SAP-2 a high mol. wt. aggregate for which evidence has been adduced [27, 28]. Sucrose gradient centrifugation showed an altered behaviour of GCase in the presence of purified activator [28]. This finding was confirmed and, using HPLC and specific Ab's raised against activator and enzyme, activator protein was shown to be present in the high mol. wt. aggregate of GCase [27]. Furthermore, no additional activation either by TC or by added activator could be achieved for the aggregate, wheareas the low mol. wt. form of the enzyme was devoid of activator and could be activated by the activator and detergent.

The physiological significance of the activator has been much discussed [36]. Christomanou [37] recently identified a patient in whom an activator protein normally present in spleen from Gaucher patients could not be detected. This patient accumulates glucosyl-ceramide in the spleen and has normal levels of enzyme activity *in vitro*. Although these findings do indeed suggest a Gaucher phenotype due to the absence of activator protein, additional information is required before a definite conclusion can be reached. In particular, data on the kinetic properties and on the subcellular localization of the apparently normal enzyme are lacking.

5. GLUCOCEREBROSIDASE IN CULTURED FIBROBLASTS

Protein biosynthesis in cultured cells is usually monitored by the use of specific Ab's directed towards the protein. Cells are metabolically labelled with a radioactive amino acid and, after a suitable time of pulse or chase with non-radioactive medium, harvested and homogenized. The final extract is incubated with the Ab's immobilized on protein A-Sepharose beads or protein A-containing membranes. Protein A has a high affinity for IgG and allows an efficient washing of the immunocomplex. These washing steps, including the use of detergent-containing buffers, should be carried out extensively to avoid aspecific interaction between non-related proteins and the immunobeads. After washing, the immunocomplex is eluted from the beads using a low-pH buffer and the immunocomplex is subjected to SDS-PAGE. After separation, the gel is incubated in a PPO-containing medium to intensify the radioactive signal. The separated immuno-precipitated radioactive proteins are visualized using fluoro-graphy [38].

In general, *N*-linked glycoproteins are synthesized on e.r.-associated ribosomes and are co-translationally inserted into the e.r. lumen*. Proteins become glycosylated by the transfer of an oligosaccharide having 2 *N*-acetylglucosamines, 9 mannoses and 3 glucose residues from the dolichol-phosphate carrier to specific asparagine residues in the polypeptide backbone (Fig. 1). Upon

---

* *Editor's note:* For pertinent background consult earlier vols., particularly D.J. Morré arts. and 'Glyco....' Index entries; O. Touster in Vol. 4 touches on Gaucher disease.

Lipid-P-P-[GlcNAc]$_2$-[Man]$_9$-[Glc]$_3$

```
            |              |
         Thr or Ser     Thr or Ser
            |              |
   +        X     ——→      X                    + Lipid-P-P
            |              |
          Asn            Asn-[GlcNAc]₂-[Man]₉-[Glc]₃
            |              |
```

Fig. 1. Transfer of the oligosaccharide moiety from dolichol (*lipid*) phosphate to specific Asn residues in the polypeptide backbone. Abbreviations as in (e.g.) *Biochemical Journal*.

subsequent transport through the e.r. and the Golgi apparatus, several modifications occur involving the removal of the terminal glucoses and some mannoses and the addition of *N*-acetylglucosamines, galactoses and sialic acids (Fig. 2). This leads to the formation of glycoproteins with a complex oligosaccharide moiety [39]. This conversion of high-mannose oligosaccharides to complex structures does not occur with all newly synthesized glycoproteins. Many lysosomal proteins, for example, remain in the high-mannose configuration to a significant extent and are phosphorylated at mannose residues [38, 40] (Fig. 2), the significance of which is discussed below.

GCase, like other lysosomal proteins, is also synthesized on e.r.-associated ribosomes and is co-translationally inserted into the e.r. lumen [41]. The first detectable form of the enzyme after immunoprecipitation from labelled fibroblasts is a high-mannose species of kM_r 62 [41, 42]. By use of an endoglycosidase specific for this type of oligosaccharide moiety, its nature has been identified [41] (see below).

Upon transport to the Golgi apparatus, forms with higher kM_r (65–68) are generated [41, 42], through hydrolysis of mannose linkages in the enzyme's oligosaccharide moiety and subsequent addition of *N*-acetylglucosamine, galactose and sialic acid groups. The diffuse appearance of higher M_r material suggests a considerable heterogeneity within this moiety at this stage of transport to the lysosome. Enzyme of kM_r 59 (by SDS-PAGE) then appears in considerable amount, arising from the high M_r material by partial removal of some of the added saccharide groups [41, 42]. Evidencing this glycosidic degradation, GCase of kM_r ~67 occurs in cells lacking sialidase and of kM_r ~63 in cells deficient in galactosidase (S. van Weely et al., unpublished). This suggests that lysosomal sialidase and galactosidase are involved in the conversion of the large GCase (kM_r 65–68) to the mature form (kM_r 59), and also that the highly sialylated form is a Golgi→lysosome transit form.

GCase maturation does not involve proteolytic degradation of precursor proteins to mature forms as is often observed with lysosomal proteins (see below) [38], except for removal of the signal peptide upon entry of the newly synthesized protein into the e.r. lumen.

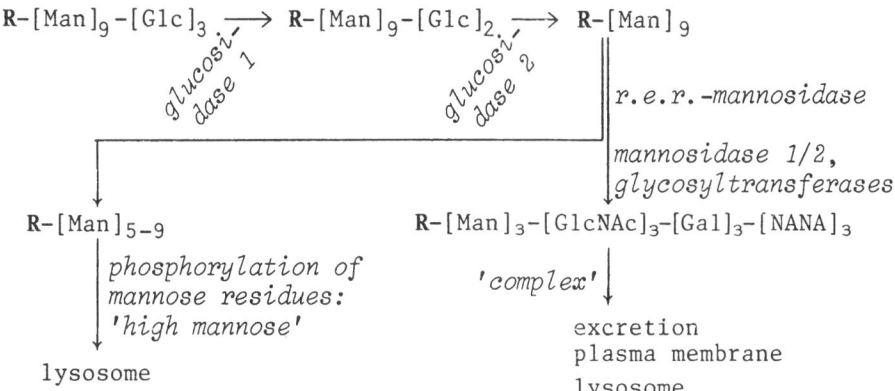

Fig. 2. Two of the possible trimming pathways for oligosaccharide processing of glycoproteins. R = Asn-[GlcNAc]$_2$; NANA = neuraminic (sialic) acid.

For transport to the lysosome many lysosomal enzymes rely on Golgi-located Man-6-P receptors [40, 43], which recognize high-Man chains that have become attached to newly synthesized glycoproteins and phosphorylated (Fig. 2). This binding sorts the lysosomal enzymes from proteins being secreted or transported to other locations in the cell. The receptor loses ligand affinity at a low pH. This dissociation is essential for efficient transport of proteins to the lysosome [43]; it is blocked in the presence of weak bases, which accumulate in acid organelles and raise their internal pH, whereby receptor is depleted. Hence proteins normally transported to the lysosome are excreted [44]. For GCase, however, weak bases do not lead to excretion (J.M.F.G. Aerts & S. van Weely, unpublished), which suggests a different mechanism of intracellular transport. The following observations confirm this.-

(a) The enzyme is normally present in a membrane-associated state in patients suffering from mucolipidosis II [45, 46], who cannot phosphorylate high-mannose glycoproteins: thus proteins normally transported to the lysosome are excreted into the fibroblast culture medium [47]. It is of interest that SAP-2, which can activate and associate with GCase, is transported to the lysosome by a pathway dependent on binding to mannose-6-phosphate receptors [48].

(b) GCase purified from fibroblasts labelled with [^{32}P]phosphate is not phosphorylated (J.M.F.G. Aerts, unpublished), as found by an immunoaffinity procedure. This entailed chromatographing the cell extracts on anti-GCase MAb's covalently coupled to Sepharose beads; enzyme labelled with [^{35}S]methionine could be eluted from the column in high recovery, using ethylene glycol (see **3.** above).

(c) Transport of GCase to the lysosome involves rendering the high-mannose oligosaccharides more complex. Growing U937 cells (a human monoblastoid cell line) in the presence of the trimming inhibitors

swainsonine or 1–deoxymannojirimycine results in the arrest of the
enzyme in a prelysosomal compartment [49]. These inhibitors suppress
mannosidases 2 and 1 respectively, two enzymes involved in cleavage
of mannosidic linkages in high–mannose oligosaccharide chains.

For trimming and for the initial step in *N*–glycosylation the
following inhibitors have been reported [details: 50, 51].-
- *Assembly of functional dolichol–linked oligosaccharides:*
tunicamycin; 2-deoxyglucose.
- *Oligosaccharide trimming:* deoxynojirimycin (glucosidases 1 & 2);
castanospermine (glucosidase 1); bromoconduritol (glucosidase 2);
1-deoxymannojirimycin (mannosidase 1 $^A/_B$); swainsonine (mannosidase 2).

In elucidating how proteins are transported through the cell,
evidently it is important to know the nature of glycoprotein oligosaccharide chains. Their characterization is aided by endoglycosidases, specific for the oligosaccharide type concerned.-
Cleavage between Asn *and* GlcNAc: Asn┼GlcNAc—GlcNAc—~~
- complex and high–mannose chains: glycopeptidase A, *N*–glycanase;
Cleavage between GlcNAc's: Asn—GlcNAc┼GlcNAc—~~
- high–mannose chains: endoglycosidase H;
- complex and high–mannose chains: endoglycosidase F;
- complex chains after removal of terminal sialic acid, GlcNAc and
galactose: endoglycosidase D.

It must be realized that deglycosylation of glycoproteins is
not facile. The oligosaccharide moiety should be properly exposed,
often achievable by boiling the glycoprotein in the presence of SDS
- which itself has an inactivating effect on endoglycosidases [52].
Moreover, proteolysis should be excluded. It is therefore wise and
essential to perform enzymic deglycosylation in the presence of
protease inhibitors and/or non–glycoproteins as competing protease
substrates and to limit the amount of SDS present in the final incubation. Inescapably, getting precise information on the nature of
the oligosaccharide moiety hinges on isolating it from, e.g., pronase-
treated glycoprotein and characterizing it by suitable analytical
procedures [53].

The above-mentioned observations provide substantial evidence
for an alternative route effective in intracellular transport of
GCase and possibly other lysosomal membrane constituents (see also
[49]).

6. GLUCOCEREBROSIDASE BIOSYNTHESIS IN FIBROBLASTS FROM GAUCHER PATIENTS

As pointed out in the opening sketch, there are three subtypes
of Gaucher disease [1]. By immunoblotting using monoclonal and polyclonal anti-GCase Ab's, detergent extracts of control fibroblasts
have shown three species having kM_r's of 59, 62 and 65-68; they
correspond to the main mol. wt. species observed during pulse-chase

experiments [41, 42].   In cells from type 1 Gaucher patients the
same three mol. wt. species are observed with similar intensity to
control fibroblasts when identical amounts of protein are used [54].
In extracts from types 2 and 3 cells, however, only minor amounts
of protein are observed with kM_r's between 62 and 67 [54]. Sometimes
also a trace of mature (M_r 59) GCase can be detected (S. van Weely,
unpublished). The conclusion should be noted that with type 1 cells
the amount of cross-reactive material in the cell is not reduced or
only moderately so, whereas the activity present in fibroblast homogenates is greatly reduced. This suggests a reduced s.a. per enzyme
molecule for the type-1 enzyme. Similar results were obtained with
enzyme purified from the spleen of type 1 Gaucher patients
(J.M.F.G. Aerts, unpublished).

Pulse-chase experiments using type 1 fibroblasts confirmed the
presence, as in normals, of the molecular forms occurring during
biosynthesis.   The initially synthesized protein is of kM_r 62, is
converted to a higher mol. wt. form and is subsequently degraded
to the mature species of kM_r 59 [42]. However, the latter had somewhat
reduced stability. When leupeptin, an inhibitor of thiol proteases,
was present during cell labelling, both type 1 and control fibroblasts
showed an increased signal, but more so in the type 1 cells [42].

Similar experiments with types 2 and 3 cells showed an apparently
normal synthesis of precursor GCase but a rapid degradation of the
newly synthesized protein.   In pulse-chase experiments no effect
of leupeptin was seen, but continuous pulsing of these fibroblasts
with [^{14}C]leucine for 2-5 days gave an increased signal due to the
inhibition of thiol proteases. An increase in enzyme activity, too,
could be observed when type 2 or 3 fibroblasts were grown for 14 days
in the presence of leupeptin. These results indicate the synthesis
of a highly unstable enzyme in fibroblasts from types 2 and 3 Gaucher
patients [42].

7. SUBCELLULAR LOCALIZATION OF ENZYME IN GAUCHER FIBROBLASTS

Besides having the kinetic capability of catalysing the lipid
substrate's hydrolysis, the enzyme activity must be present at the
substrate-accumulation site (the lysosome). To investigate the subcellular localization of GCase in cells from Gaucher patients, two
types of experiment have been performed, viz. immuno-electron microscopy and Percoll density-gradient centrifugation. The former involves
fibroblast fixation followed by ultra-thin sectioning of frozen cells
and incubation of the sections with specific Ab's, which are visualized
by a second incubation with protein A onto which electron-dense
particles are adsorbed.* With control fibroblasts [55] this technique
reveals, as expected, the presence of gold particles (and thus GCase)
in the lysosome, associated with the membrane.   In type 1 Gaucher

---

*See #A-12, *this vol.- Ed.*

cells the enzyme is likewise present in the lysosome, in apparently normal amounts. In type 2 and 3 cells, however, little if any label was observed in the lysosome, suggesting rapid degradation of the enzyme therein or before it reaches the lysosome.

The second technique exploits the difference in density between secondary lysosomes and other organelles (Golgi, e.r., etc.). Gentle homogenization of the cells under isotonic conditions, followed by a slow spin to remove nuclei and non-disrupted cells, leads to a so-called post-nuclear supernatant. This mixture of vesicles and organelles is layered onto a self-generating gradient (Percoll) and centrifuged for a set time. Density regions are harvested from the gradient, and assayed for marker enzymes, following the separation. Secondary lysosomes are observed in the high-density region, and pre-lysosomal compartments including Golgi and e.r. at lower densities [56].

With control fibroblasts GCase is found in light and heavy regions, especially the latter. Its distribution parallels that of the marker enzyme hexosaminidase. Type 1 cells give an identical distribution, in accord with the above immunocytochemical observations (S. van Weely, unpublished). With type 2 or 3 cells the situation is slightly different: in some experiments the enzyme is absent from or low in the heavy lysosomal fraction, whereas in other experiments an apparently normal distribution is seen (unpublished). Obviously these non-reproducible findings indicate that the condition of the cells and the experimental set-up determine whether GCase is observed in the lysosome. However, the results accord with the conclusion that the cells contain only minor amounts of the enzyme and that its instability in types 2 and 3 Gaucher cells precludes a significant lysosomal content.

8. OTHER APPROACHES TO SUBCELLULAR LOCALIZATION

As considered above, protein localization is achievable by cell fractionation techniques or by light- or electron-microscopic immuno-cytochemistry. Alternatively, intra-organelle pH can be exploited, as in the approach of Oude Elferink et al. [57]. They made use of the ability of glycyl-L-phenylalanine-β-naphthylamide (GPN) to accumulate in acidic organelles. It behaves as a weak base and can act as a substrate for the lysosomal enzyme cathepsin C [58]. There is accumulation of one of the products (Gly-Phe), causing osmotic shock to the acidic organelle and subsequent lysis and loss of latency of the enzyme of interest if present therein.

A second approach is to make use of the difference in cholesterol concentration existing between intracellular membranes, and hence the effect on membrane integrity of complexing with digitonin. Upon incubating intact fibroblasts with increasing concentrations of digitonin, first the plasma membrane is affected, leading to substrate accessibility (latency loss) of cytosolic (marker) enzymes. At higher concentrations the lysosomal and then the mitochondrial membranes are disrupted; but fibroblast peroxisomes are disruptable only with very high concentrations [59].

These two methods provide additional tools for exploring the localization of proteins in the cell. It should be noted that in both cases the disruption is detected as loss of enzyme latency, not sedimentability. Before loss of enzyme sedimentability occurs, latency diminishes because the membrane concerned becomes more permeable to substrate. Moreover, membrane-bound enzymes may not readily lose sedimentability but may be freely accessible to substrate.

Acknowledgements

Our studies are supported by grants from the National Gaucher Foundation (U.S.A.) and from the Netherlands Organization for the Advancement of Pure Research (ZWO) under the auspices of MEDIGON.

References

1. Stanbury, J.B., Wijngaarden, J.B., Frederickson, D.S., Goldstein, J.L. & Brown, M.S., eds. (1983) *The Metabolic Basis of Inherited Diseases*, McGraw-Hill, 1862 pp.
2. Von Figura, K. & Hasilik, A. (1984) *Trends Biochem. Sci. 9*, 29-31.
3. Tager, J.M. (1985) *Trends Biochem. Sci. 10*, 324-326.
4. Tager, J.M., Jonsson, L.M.V.. Aerts, J.M.F.G., Oude Elferink, R.P.J., Schram, A.W., Erikson, A.H. & Barranger, J.A. (1984) *Biochem. Soc. Trans. 12*, 902-905.
5. Kornfeld, S. (1986) *J. Clin. Invest. 77*, 1-6.
6. Brady, R.O. & Barranger, J.A. (1983) in *The Metabolic Basis of Inherited Diseases*, as for 1., pp.842-856.
7. Brady, R.O., Kanfer, J.N. & Shapiro, D. (1965) *J. Biol. Chem. 240*, 39-43.
8. Patrick, A.D. (1965) *Biochem. J. 97*, 17c-17d.
9. Strasberg, P.M. & Lowden, J.A. (1982) *Canad. J. Biochem. 60*, 1025-1030.
10. Grabowski, G.A. & Dagan, A. (1984) *Anal. Biochem. 141*, 267-279.
11. Murray, G.J., Youle, R.J., Gandy, S.E., Zirzow, G.C. & Barranger, J.A. (1985) *Anal. Biochem. 147*, 301-310.
12. Chaoy, F.Y.N. (1985) *Anal. Biochem. 156*, 515-520.
13. Furbish, F.S., Blair, H.E., Shiloach, J., Pentchev, P.G. & Brady, R.O. (1977) *Proc. Nat. Acad. Sci. 74*, 3560-3563.
14. Aerts, J.M.F.G., Donker-Koopman, W.E., Murray, G.J., Barranger, J.A., Tager, J.M. & Schram, A.W. (1986) *Anal. Biochem. 154*, 655-663.

15. Salvayre, R., Maret, A., Negre, A. & Douste-Blazy, L. (1982) in
    Gaucher Disease, a Century of Delineation and Research
    (Desnick, R.J., Gatt, S. & Grabowski, eds.), Liss, New York, pp. 443-452.
16. Tsuji, S., Choudary, P.V., Martin, B.M., Winfield, S.,
    Barranger, J.A. & Ginns, E.R. (1986) *J. Biol. Chem. 261*, 505-
    513.
17. Takasaki, S., Murray, G.J., Furbish, F.S., Brady, R.O.,
    Barranger, J.A. & Kobota, A. (1984) *J. Biol. Chem. 259*, 10112-
    10117.
18. Tager, J.M., Aerts, J.M.F.G., Jonsson, L.M.V., Murray, G.J.,
    von Weely, S., Styrland, A., Ginns, E.I., Reuser, A.J.J.,
    Schram, A.W. & Barranger, J.A. (1987) in *Enzymes of Lipid
    Metabolism* (Freysz, L., Dreyfus, H., Massarelli, R. & Gatt, S.,
    eds.), Plenum, New York, pp. 735-745.
19. Maret, A., Salvayre, R., Negre, A. & Douste-Blazy, L. (1981)
    Eur. J. Biochem. 115, 455-461.
20. Maret, A., Salvayre, R., Negre, A. & Douste-Blazy, L. (1983)
    Eur. J. Biochem. 133, 283-287.
21. Aerts, J.M.F.G., Donker-Koopman, W.E.. van der Vliet, M.K.,
    Jonsson, L.M.V., Ginns, E.I., Murray, G.J., Barranger, J.A.,
    Tager, J.M., Schram, A.W. (1985) *Eur. J. Biochem. 150*, 565-574.
22. Coyle, P.J., Chiao, Y.B., Glew, R.H. & Labow, R.E. (1981) *J.
    Biol. Chem. 256*, 13004-13113.
23. Peters, S.P., Coyle, P.J. & Glew, R.H. (1976) *Arch. Biochem.
    Biophys. 175*, 569-582.
24. Strasber, P.M. & Lowden, J.A. (1982) *Clin. Chim. Acta 118*,
    9-20.
25. Basu, A., Prence, E., Garrett, K., Glew, R.H. & Ellingson, J.S.
    (1985) *Arch. Biochem. Biophys. 243*, 28-34.
26. Ho, M.W. & O'Brien, J.S. (1971) *Proc. Nat. Acad. Sci. 68*,
    2813-2819.
27. Aerts, J.M.F.G., Donker-Koopman, W.E., van Laar, C., Brul, S.,
    Murray, G.J.,   Wenger,   D.A.,   Barranger, J.A.,   Tager, J.M. &
    Schram, A.W. (1987) *Eur. J. Biochem., 163*, 583-589.
28. Prence, E., Chakravorti, S., Basu, A., Clark, L.S., Glew, R.H.
    & Chambers, J.A. (1985) *Arch. Biochem. Biophys. 236*, 98-109.
29. Aerts, J.M.F.G., Donker-Koopman, W.E., Koot, M., Barranger, J.A.,
    Tager, J.M. & Schram, A.W. (1986) *Clin. Chim. Acta 158*, 155-
    164.
30. Oude Elferink, R.P.J., Brouwe-Kelder, E.M., Surya, I.,
    Strijland, A., Kroos, M., Reuser, J.J.A. & Tager, J.M. (1984)
    Eur. J. Biochem. 139, 489-495.
31. Aerts, J.M.F.G., Donker-Koopman, W.E., Koot, M., Murray, G.J.,
    Barranger, J.A., Tager, J.M. & Schram, A.W. (1986) *Biochim.
    Biophys. Acta 863*, 63-70.
32. Pentchev, P.G., Brady, R.O., Hibbert, S.A., Gal, A.E. &
    Shapiro, D. (1973) *J. Biol. Chem. 218*, 5256-5261.
33. Dale, V.L., Villacorte, D.G. & Beutler, E. (1976) *Biochem.
    Biophys. Res. Comm. 71*, 1048-1053.

34. Blonder, E., Klibansky, C. & de Vries, A. (1976) *Biochim. Biophys. Acta 431*, 45-53.

35. Dale, G.L. & Beutler, E. (1977) *Proc. Nat. Acad. Sci. 73*, 4672-4674.

36. Radin, N.S. (1984) in *Molecular Basis of Lysosomal Storage Disorders* (Barranger, J.A. & Brady, R.O., eds.), Academic Press, London, pp. 93-112.

37. Christomanou, H., Aignesberger, A. & Linke, R.P. (1986) *Biol. Chem. Hoppe-Seyler 367*, 879-890.

38. Hasilik, A. & Neufeld, E.G. (1980) *J. Biol. Chem. 255*, 4937-4945.

39. Kornfeld, R. & Kornfeld, S. (1985) *Annu. Rev. Biochem. 54*, 631-644.

40. Von Figura, K. & Hasilik, A. (1986) *Annu. Rev. Biochem. 55*, 167-193.

41. Erickson, A.H., Ginns, E.I. & Barranger, J.A. (1985) *J. Biol. Chem. 260*, 14319-14324.

42. Jonsson, L.M.V., Murray, G.J., Sorrell, S.H.., Strijland, A., Aerts, J.M.F.G., Ginns, E.I., Barranger, J.A., Tager, J.M. & Schram, A.W. (1987) *Eur. J. Biochem.*, in press.

43. Sly, W.S. & Fischer, H.D. (1982) *J. Cell Biochem. 18*, 67-85.

44. Gonzalez-Noriega, A., Grubb, H., Talkad, V. & Sly, W.S. (1980) *J. Cell Biol. 85*, 839-852.

45. Van Dongen, J.M., Willemsen, R., Ginns, E.I., Sips, H.J., Tager, J.M., Barranger, J.A. & Reuser, A.J.J. (1985) *Eur. J. Cell Biol. 39*, 179-189.

46. Leroy, J., Ho, M.W., MacBrinn, M.C., Zehlke, K., Jakob, J. & O'Brien, J.S. (1972) *Pediat. Res. 6*, 752-757.

47. Hickman, S. & Neufeld, E.F. (1972) *Biochem. Biophys. Res. Comm. 49*, 992-999.

48. Fujibayashi, Y. & Wenger, D.A. (1986) *J. Biol. Chem. 261*, 15339-15343.

49. Aerts, J.M.F.G., Brul, S., Donker-Koopman, W.E.. von Weely, S., Murray, G.J., Barranger, J.A., Tager, J.M. & Schram, A.W. (1986) *Biochem. Biophys. Res. Comm. 141*, 452-458.

50. Fuhrmann, U., Bause, E. & Ploegh, H. (1985) *Biochim. Biophys. Acta 825*, 95-110.

51. Schwarz, R.T. & Datema, R. (1984) *Trends Biochem. Sci. 4*, 32-34.

52. Trimble, R.B. & Maley, F. (1984) *Anal. Biochem. 141*, 515-522.

53. Overdijk, B., Beem, E.P., van Steijn, G.J., Trippelvitz, L.A.W., Lisman, J.J.W., Paz Perente, J., Cardon, P., Leroy, Y., Fournet, B., van Halbeek, H., Mutsaers, H.G.M. & Vliegenthart, J.F.G. (1985) *Biochem. J. 232*, 637-641.

54. Ginns, E.I., Brady, R.O., Pirucello, S., Moore, C., Sorrell, S., Furbish, F.S., Murray, G.J., Tager, J.M. & Barranger, J.A. (1982) *Proc. Nat. Acad. Sci. 79*, 5607-5610.

55. Willemsen, R., Van Dengen, J.M., Ginns, E.I. Sips, H.J., Schram, A.W., Tager, J.M., Barranger, J.A. & Reuser, A.J.J. (1987) *J. Neurol. 234*, 44-51.

56. Rome, L.H., Garvin, A.J., Allietta, M.M. & Neufeld, E.F. (1979) *Cell 17*, 143-153.

57. Oude Elferink, R.P.J., Van Doorn-van Wakeren, J., Strijland, A., Reuser, A.J.J. & Tager, J.M. (1985) *Eur. J. Biochem* *153*, 55-63.

58. Jadot, M., Colmant, C., Wattiaux-De Coninck, S. & Wattiaux, R. (1984) *Biochem. J. 219*, 965-970.

59. Wanders, R.J.A., Kos, M., Roest, B. Meijer, A.J., Schrakamp, G., Heymans, H.S.A., Tegegelaers, W.H.H., van den Bosch, H., Schutgens, R.B.H. & Tager, J.M. (1984) *Biochem. Biophys. Res. Comm. 123*, 1054-1061.

#A-12

CYTOCHEMICAL TECHNIQUES APPLIED TO WHOLE TISSUE AND TISSUE HOMOGENATES

J.T.R. Fitzsimons

Electron Microscopy Unit
Department of Neurophysiology
University of Southampton
Bassett Crescent East, Southampton SO9 3TU, U.K.

Biochemical studies can benefit from various approaches now outlined: enzyme cytochemistry applied to homogenates; negative staining of cell fragments, bacteria and viruses; autoradiography for fine-structural localization; tracer use of peroxidase whereby specific Ig's may be labelled; and gold probes. X-ray and computer aids to quantitation are indicated.

Cytochemistry, especially that relating to enzyme localization, has advanced dramatically since the pioneering work of Gomori [1] and Takamatu [2], who back in 1939 developed a method for alkaline phosphatase localization. Although many enzymes have been histo-chemically localized at light-microscopy level, only a small minority have been localized cytochemically at ultrastructural level. Many such cytochemical procedures are centered around the formation of a heavy-metal precipitate with one of the hydrolysis products of enzyme action. In addition to phosphatase demonstration other techniques are now available for localizing a host of other enzymes including those concerned with oxidation-reduction reactions.

The tissue incubation stage is usually a two-step reaction:

$$\text{substrate} \xrightarrow{\textit{enzyme}} \text{primary reaction product (PRP)} \xrightarrow[\textit{agent, e.g. } Pb^{2+}]{\textit{capturing}} \text{final reaction product (FRP)}$$

In the second step the PRP formed by substrate hydrolysis in the first step is 'captured' with an appropriate capturing agent, quite often a metal cation such as lead, copper or calcium, to form an insoluble electron-dense precipitate (FRP). However, false localizations can sometimes be encountered due to possible diffusion of PRP prior to

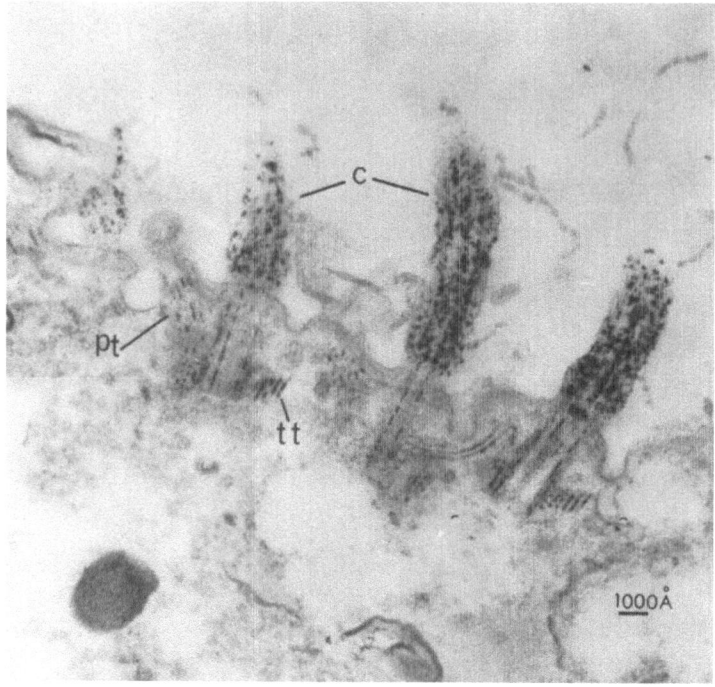

Fig. 1. Unstained section (e.m.) through the cortex of *Tetrahymena vorax*, fixed in 1% glutaraldehyde for 30 min and incubated (20°) in an ATPase medium for 1 h. The microtubules of the kinetosome and cilia (c) together with the post-ciliary (pt) and transverse (tt) micro-tubules all exhibit nucleoside triphosphatase activity (Pb deposits).

its capture. We overcame this in the demonstration of lysosomal acid phosphatase by a novel technique [3]:

$$\text{gold-AMP} \xrightarrow[\text{phosphatase}]{\text{acid}} \text{gold-adenosine + phosphate} \quad [\text{i.e. PRP} \equiv \text{FRP}]$$

The substrate is hydrolyzed to produce gold-adenosine which itself is an insoluble, electron-dense precipitate and does not require a capturing agent. Thus the possibility of diffusion artefact is reduced.

    Subcellular fractions.- Enzyme cytochemical studies on sub-cellular fractions have been little exploited in comparison with whole-tissue investigations. Difficulties arose with primary fixation in trying to maintain a balance between retention of reasonable enzyme activity and good morphological preservation. Satisfactory results were obtained [cf. *Editor's note* at end of refs. list] by using low concentrations of primary fixatives (<1% if glutaraldehyde, *vs.* the 1% level appropriate for intact cells as in Fig. 1), or even omitting

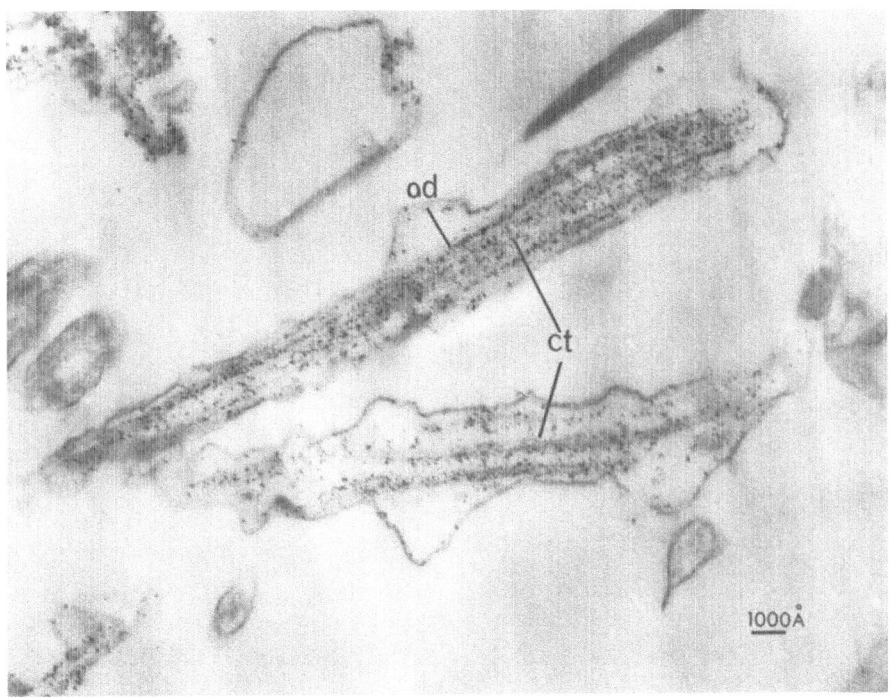

Fig. 2. Unstained cilia fraction from *Tetrahymena vorax* fixed in
0.5% glutaraldehyde for 30 min (at 20°) and incubated for 15 min
(at 20°)in an ATPase medium. Deposits of lead phosphate indicating
ATPase activity are found on both central (ct) and outer doublet
(od) microtubules.

them, and by employing short fixation times [4]. With such gentle
fixation procedures Sharp et al. [5] were able to demonstrate ATPase
localizations on ciliary and other microtubules of intact Tetrahymena
(Fig. 1). Subsequently [6] intact cilia were prepared, using dibucaine
in the deciliation step, and ATPase activity was demonstrated therein
(Fig. 2).   Enzyme cytochemistry therefore need not be confined to
sections of whole tissue but can also be a valuable tool to the biochemist in the examination of tissue homogenates and fractions.

VARIOUS AIDS TO THE BIOCHEMIST OR IMMUNOCHEMIST

Ultrastructural information on cell fragments, bacteria and
viruses can be obtained by the rapid technique of negative staining.
Fig. 3 shows a membranous fraction prepared from mitochondria and
negatively stained with sodium phosphotungstate. Electron transport
particles on   the inner mitochondrial membrane are readily recognizable. Autoradiography has made a successful transition from light
to electron microscopy for the fine-structural localization of radiolabelled compounds.

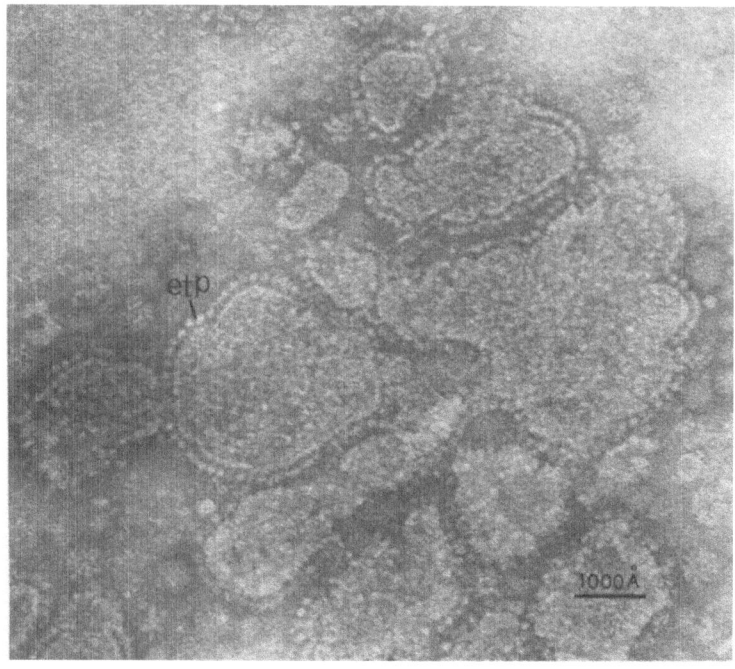

Fig. 3. Membranous fraction prepared from mitochondria and negatively stained with 2% Na phosphotungstate pH 6.8. Electron transport particles (etp) seen attached to membrane fragments.

In 1966 Graham & Karnovsky [7] introduced a method for the localization of peroxidase which became popular as a cytochemical tracer.
About the same time, immuno-peroxidase for demonstrating antigens
were described [8]. Modifications followed to improve sensitivity
and thus allow antigens to be detected in smaller amounts. Peroxidase-
anti-peroxidase methods [9] are perhaps the most widely employed
of these modifications.

Cytochemistry entered a new era with the introduction of colloidal
gold as a cytochemical marker when in 1971 Faulk & Taylor [10] adsorbed
rabbit anti-salmonella serum to colloidal gold and used it as a direct
immunocytochemical labelling probe for identifying surface antigens
on Salmonellae. This probe is excellent for such techniques at both
transmission and scanning e.m. levels. They can be produced as small
as 2 nm to give more efficient marking, and are marketed in different
sizes whereby multiple labelling is feasible.

Gold particles seen by e.m. are usually too small to be observed
at light-microscope level. However the immunogold silver staining
(IGSS) technique of Holgate et al. [11] allows the gold particles

Fig. 4. Approach for localizing
antigenic sites by light microscopy, entailing successive use
of a primary Ab, an Ab raised to
this Ab and labelled with gold,
and silver for visualization.

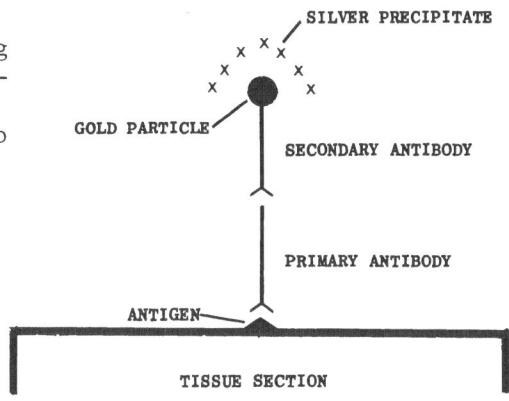

SILVER PRECIPITATE

GOLD PARTICLE

SECONDARY ANTIBODY

PRIMARY ANTIBODY

ANTIGEN

TISSUE SECTION

that are localized at antigenic sites to be revealed by light microscopy
by a silver precipitation reaction. This is shown schematically
in Fig. 4 where gold-labelled secondary antibodies (Ab's) are raised
to primary Ab's under investigation and these in turn reacted with
silver. This technique, applicable to conventional paraffin wax
sections, has become a very useful tool in pathology. Specimens
which gave negative results with peroxidase anti-peroxidase (PAP)
labelling techniques responded to the IGSS method [12].

Besides gold-labelled secondary Ab's as a means of tracing
antigen-bound primary Ab's, use can be made of Protein A, a cell-wall
component of *Staphylococcus aureus*, which binds to many species and
subclasses of Ab. Gold-labelled Protein A is widely used in the
ultrastructural localization of antigens in both resin-embedded
tissues [13] and ultrathin frozen sections [14]. The latter are
preferable because of penetration problems with conventionally fixed
plastic sections. With such techniues Geuze [14] was able by double
labelling to identify two proteins, sialoglycoprotein and albumin,
within separate compartments of the Golgi system.

In the ultrastructural localization of nucleic acids, Bendayan
[15] used enzyme-gold complexes. The specificity of the enzyme-subs-
trate interactions combined with the size and electron-density of
the gold particles and the good ultrastructural preservation of the
tissue resulted in a very specific labelling with high resolution.
Their results were impresssive and demonstrate the possibility of
detecting substrates by means of enzyme-gold complexes at the e.m.
level.

Cytochemical procedures which in the past were only qualitative
in nature can with the introduction of X-ray microanalysis [16] be
examined quantitatively, and computer analysis of electron micrographs
is providing a new wealth of information to biochemists, pathologists
and molecular biologists.

References

1. Gomori, G. (1939) *Proc. Soc. Exp. Biol. Med. 42*, 23-26.
2. Takamatu, H. (1939) *Acta Soc. Path. Japan 29*, 429 (& see [19]).
3. Fitzsimons, J.T.R., Gibson, D.E. & Barrnett, R.J. (1970) *J. Histochem. Cytochem. 18*, 673-674.
4. Fitzsimons, J.T.R. (1983) *Int. J. Biochem. 15*, 267-275.
5. Sharp, G.A., Fitzsimons, J.T.R. & Kerkut, G.A. (1979) *Comp. Biochem. Physiol. 63A*, 253-260.
6. Fitzsimons, J.T.R. & Kerkut, G.A. (1983) *Comp. Biochem. Physiol. 74A*,   739-743.
7. Graham, R.C. Jr. & Karnovsky, M.J. (1966) *J. Histochem. Cytochem. 14*, 291-302.
8. Nakane, P.K. & Pierce, G.M. (1966) *J. Histochem. Cytochem. 14*, 929-931.
9. Sternberger, L.A., Hardy, P.H., Cuculis, J.J. & Meyer, H.G. (1970) *J. Histochem. Cytochem.18*, 315-333.
10. Faulk, W.P. & Taylor, G.M. (1971) *Immunochemistry 8*, 1081-1083.
11. Holgate, C.S., Jackson, P., Cowen, P.N. & Bird, C.C. (1983) *J. Histochem. Cytochem. 31*, 938-944.
12. Hacker, G.W., Springall, D.R., Van Noorden, S., Bishop, A.E., Grimelius, L. & Polak, J.M. (1985) *Virchow's Arch. [Path. Anat.] 406*, 449-461.
13. Roth, J. (1982) in *Techniques in Immunocytochemistry*, Vol. 1 (Bullock, G.R. & Petrusz, P., eds.), Academic Press, London, pp. 107-133.
14. Geuze, H., Slot, J.W., Van der Ley, P., Schuffer, R. & Griffith, J. (1981) *J. Cell Biol. 89*, 653-665.
15. Bendayan, M. (1981) *J. Histochem. Cytochem. 29*, 531-541.
16. Chandler, J.A. (1977) in *Practical Methods in Electron Microscopy* (Glauert, A.M., ed.), North-Holland, Amsterdam, pp. 327-518.

Senior Editor's note.-   The following refs. (and, in this series, A.A. El-Aaser in Vols. 1 & 3) also deal with enzyme-localization techniques and difficulties, especially for phosphatases in isolated hepatic vesicles:

17. El-Aaser, A.A., Fitzsimons, J.T.R., Hinton, R.H., Norris, K.A. & Reid, E. (1973) *Histochem. J. 5*, 199-223.
18. El-Aaser, A.A. & Reid, E. (1975) in *Electron Microscopy of Enzymes*, Vol. 4 (Hayat, M.A., ed.), Van Nostrand Reinhold, New York, 177-201.

The pioneer study in ref. 2 is also described in:

19. Takamatu, H. (1940) *Gann 34*, 81-83.

#NC(A)

NOTE and COMMENTS related to

INVESTIGATIVE APPROACHES TO ORGANELLE DISTURBANCES

Comments on #A-1: R. Wattiaux et al. - ISCHAEMIC LIVER
　　　　and #A-2, A-3: T.J. Peters - BIOPSY SAMPLES; FATTY LIVER

Wattiaux, replying to C.A. Pasternak.- Chlorpromazine is not effective on lysosomes *in vitro*. **T.J. Peters replied to J.M. Graham** concerning Golgi-membrane density in vertical-rotor sucrose gradients: it is 1.12-1.14 in control liver and, surprisingly, the same with fatty liver notwithstanding the higher triglyceride content. **D. Allan remarked** that a liver membrane fraction rich in cholesterol will not necessarily have low membrane fluidity; in Golgi fractions similar to those isolated by Peters the free cholesterol was mostly in the Golgi contents rather than the membrane, i.e. the cholesterol is mainly membrane-bounded, not membrane-bound. The Golgi membrane itself resembles e.r. in its lipid composition (at least in rat liver) and thus is probably relatively fluid.

Comments on #A-5: R.K. Berge et al. - PEROXISOME PROLIFERATORS
　　　　and #A-7: G.G. Gibson & R. Sharma - P-450 ISOENZYMES

T. Berg asked Berge how quickly clofibrate yields the active form, clofibric acid, and how this related to the formation of mRNA for the peroxisomal enzymes that are induced. **Response:** we do not know as yet.[*] **C.A. Pasternak asked Gibson** why immunologically assessed differences between P-450 isoenzymes induced by different agents are so clearcut, i.e. whether turnover is very rapid. **Reply:** yes; exposure to inducers is long in comparison with the half-life of proteins. **W.H. Evans.-** How good is the evidence for your assumption that the microsomes used in your studies are derived from the parenchymal cells? Might the P-450 and associated enzymes be originating also from, e.g., sinusoidal or RES cells? **Gibson's reply.-** All the evidence points to the location of these activities being parenchymal, but exclusiveness is difficult to demonstrate. **J.K. McDonald.-** Iron-carbonyl complexes being (I recollect) photo-dissociable, will the Fe^{2+}-CO complex form only in the absence of light and so be difficult to work with? **Reply.-** This complex is notably stable even in the presence of light, and so presents absolutely no analytical problems.

---

[*]*Culled by Senior Ed. from Vol. 7:* clofibrate is "immediately and quantitatively hydrolyzed" *in vivo* [Gugler, R. & Jensen, C. (1976) *J. Chrom. 117*, 175-179].

Comments on #A-8: W. Lijinsky - MITOCHONDRIAL PROLIFERATION
        and #A-9: B.A. Fowler - METAL-INDUCED CELL INJURY

Lijinsky, answering M. Milton.- Analysis of the methapyrilene-
induced peroxisomal proliferation was solely morphometric, unaccom-
panied by biochemical analysis. Methapyrilene is unique among antihistaminics and their analogues in producing the proliferation.
Reply to H. Osmundsen.- Examination of subcellular fractions for
binding of the drug has shown two-thirds of radiolabelled drug is
associated with mitochondria.

Comments (to Fowler) by C.A. Pasternak.- Concerning the stimulation of mitochondrial membrane enzymes by arsenate, it is to be
noted that arsenate (and other inducers of cellular stress) elevates
the activity of a p.m. protein (glucose transport protein) indepen-
dently of synthesis of protein or mRNA, maybe by an effect on membrane
're-cycling' [Warren, A.P., James, M.H., Menzies, D.E., Widnell, C.C.,
Whitaker-Dowling, P.A. & Pasternak, C.A. (1986) *J. Cellul. Physiol.*
128, 383-388]. Similarly to lead, several toxins (e.g. melittin)
have a mitogenic effect (through an action on the p.m.) at sub-cytotoxic
doses (ref. [5] in #C-1). **Question by H. Sjöström.**- Are morphometric
effects on rough e.r. and smooth e.r. reflected in overall protein
synthesis rate? **Reply.**- Such experiments have been done, with labelled
leucine, and under certain conditions a correlation may be seen.

Comments on #A-10: M.J. O'Connell et al. - IRON OVERLOAD
        and #A-11: A.W. Schram et al. - GLUCOCEREBROSIDASE

Remark (to O'Connell) by R. Wattiaux.- One would like reassurance
that the hepatic iron-loaded lysosomes are of the same cell-type
origin as non-loaded lysosomes. **Questions to Schram.**- J.K. McDonald
asked whether the 'glucocerebrosidase' (GCase) is to be considered
identical with lysosomal 'β-glucosidase', or whether the latter is
unable to remove terminal β-glucosidic residues from a glucocerebro-
side. **Reply:** the two lysosomal enzymes are identical, and the lysosomal
GCase is the gene product deficient in Gaucher's disease; but there
also exist β-glucosidases in the cytosol and in an unknown membrane
location (see #A-11). **Reply to P. Bohley:** in normal cells GCase
has a half-life of 40 h. **Comment by Wattiaux:** the seemingly high
turnover of GCase is unusual for a lysosomal enzyme.

M. Mareel (to Schram): since you have shown that the complex
form of GCase is needed for routing to the lysosome, why does this
N-linked glycoprotein not go first to the p.m.? **Reply:** it does indeed
go first to the p.m., and then arrives in the lysosome. **Query by
T. Berg (reply:** "not known"): is GCase - a membrane-bound enzyme
in the lysosomes - bound to a receptor which, like the mannose-6-
phosphate receptor, transports the enzyme to the lysosomes?

[CONTINUED on p. 139

#NC(A)-1

A Note on

IMMUNOELECTRON MICROSCOPICAL LOCALIZATION OF LYSOSOMAL HYDROLASES IN NORMAL AND I-CELL FIBROBLASTS

R. Willemsen, A.T. Hoogeveen,
A.J.J. Reuser and J.M. van Dongen*

Department of Cell Biology and Genetics
Erasmus University, P.O. Box 1738
3000 DR Rotterdam, The Netherlands

In fibroblasts newly synthesized lysosomal enzymes are transported to the lysosomes by means of the mannose 6-phosphate marker-receptor system [1]. This transport mechanism is defective in fibroblasts from patients with Mucolipidosis II (I-cell disease), since most of the precursor forms of the lysosomal enzymes are missing the mannose-6-phosphate recognition site. The primary defect in I-cell disease is a deficiency of UDP-*N*-acetylglucosamine-1-phosphotransferase [2, 3]. This enzyme catalyzes the transfer of *N*-acetylglucosamine phosphate to some of the mannose residues of lysosomal precursors.

In cultured skin fibroblasts from the patients the intracellular activities of many lysosomal enzymes (notably acid α-glucosidase, *N*-acetyl-β-hexosaminidase) are low, whereas their extracellular activities are high [4, 5], indicating a secretion of non-phosphorylated precursor forms. However, normal activities for a few lysosomal enzymes (particularly glucocerebrosidase, acid phosphatase) have been reported. Accordingly, we performed immuno-e.m. on ultrathin frozen sections with gold probes to study the localization of acid α-glucosidase, *N*-acetyl-β-hexosaminidase, acid phosphatase (tartrate-inhibitable) and glucocerebrosidase in cultured skin fibroblasts from control subjects and patients with I-cell disease.

Fibroblasts were harvested and fixed for immunocytochemistry [6]. Ultracryotomy was carried out with an LKB NOVA ultrotome, equipped

* Collaborators for some of the work: Dr. J.M. Tager (Medical Enzymology & Metabolism Section, Academic Medical Centre, University of Amsterdam) and Dr. J.A. Barranger (Molecular & Medical Genetics Section, NIH, Bethesda, MD).

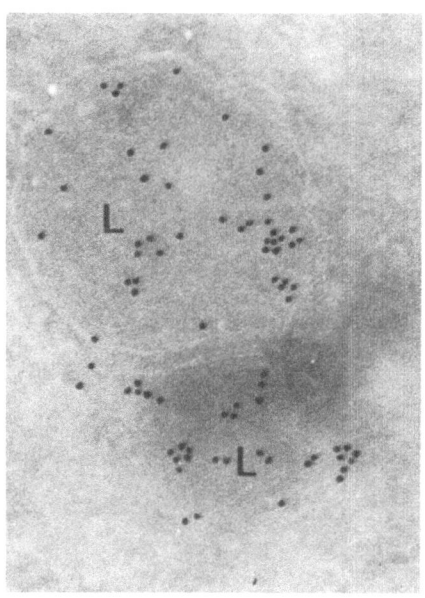 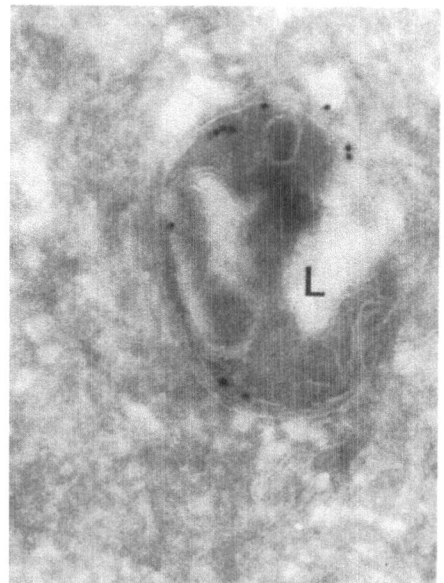

Fig. 1. Localization of acid α-glucosidase in normal human fibroblasts. (Fixation: in 1% acrolein/0.4% glutaraldehyde in 0.1 M phosphate buffer, pH 7.3.) Gold particles are randomly distributed within the lysosomes. × 95,000.

Fig. 2. Localization of gluco-cerebrosidase in normal human fibroblasts. Gold particles are located on the lysosomal membrane. × 95,000.

with the Cryo Nova, at -110°. The sections were collected on Formvar-coated grids and incubated with our specific rabbit antibodies. Antigen-antibody complexes were visualized using goat anti-rabbit immunoglobulins conjugated with 10 nm colloidal gold particles (GAR 10, Janssen Pharmaceutica, Belgium). Sections were stained with uranyl salts and subsequently embedded in 1.5% (w/v) methyl-cellulose (400 centipoises; Fluka) [7].

In control fibroblasts a random distribution of acid α-glucosidase and *N*-acetyl-β-hexosaminidase within the lysosomes was observed, whereas glucocerebrosidase and acid phosphatase were mainly found to be localized on or near the lysosomal membrane and in association with membraneous material present within the lysosomes. Our observations confirm the soluble character of the former pair of enzymes and the membrane-bound character of the latter, as already deduced from biochemical data. Representative results are shown in Figs. 1 & 2.

In I-cell fibroblasts an abnormal localization of the two soluble enzymes was found. Labelling in lysosomes was very weak, but instead, small 'presumptive' vesicles containing both enzymes were detected throughout the cytoplasm and close to the plasma membrane.

In contrast, a normal membrane-bound lysosomal localization was observed for glucocerebrosidase and acid phosphatase. It is concluded that the intracellular transport of these enzymes in the lysosomes can occur, even when the mannose 6-phosphate recognition system is defective.

References

1. Sly, W.S. & Fisher, H.D. (1982) *J. Cell Biochem. 18*, 67-85.
2. Hasilik, A., Waheed, A. & van Figura, K. (1981) *Biochem. Biophys. Res. Comm. 98*, 761-767.
3. Reitman, M.L., Varki, A. & Kornfeld, S. (1981) *J. Clin. Invest. 67*, 1574-1579.
4. Leroy, J.G., Ho, M.W., MacBrinn, M.C., Zielke, K., Jacob, J. & O'Brien, J.S. (1972) *Pediat. Res. 6*, 752-757.
5. Hickman. S. & Neufeld, E.F. (1972) *Biochem. Biophys. Res. Comm. 49*, 992-999.
6. van Dongen, J.M., Willemsen, R., Ginns, E.I., Sips, H., Tager, J.M., Barranger, J.A. & Reuser, A.J.J. (1985) *Eur. J. Cell Biol. 39*, 179-189.
7. Griffiths, G., Brands, R., Burke, S., Louvard, D. & Warren, G. (1982) *J. Cell Biol. 95*, 781-782.

Refs. to the specific antibodies (added at Editors' suggestion):

8. Reuser, A.J.J., Kroos, M., Oude Elferink, R.P.J. & Tager, J.M. (1985) *J. Biol. Chem. 260*, 8336-8342.
9. Ginns, E.I., Brady, R.O., Pirrucello, S., Moore, C., Sorrell, S., Furbish, F.S., Murray, G.J., Tager, J.M. & Barranger, J.A. (1982) *Proc. Nat. Acad. Sci. 79*, 5607-5610.

*Comments (continued from p. 134) on #***A-11***:* A.W. Schram

Schram, answering Wattiaux and Willemsen.- Besides the mannose-6-phosphate receptor, others exist that can route lysosomal enzymes to the lysosome; enzyme transport to the lysosome occurs in cells deficient in this receptor, and in fibroblasts where there is a second mannose-6-phosphate receptor. Sphingomyelinase, a membrane-associated enzyme, is deficient in MLII fibroblasts, which might indicate that membrane association is not the only factor involved in getting such enzymes to the lysosome. **Reply to query by McDonald** on attempted therapy of lysosomal storage diseases.- Many trials, e.g. enzyme replacement, have been published; but a long-term study and follow-up of these trials has not as yet been published.

Comments on #NC(A)-1: R. Willemsen et al. - Immuno-e.m. of lysosomal
                                                    hydrolases

   Remark (to Willemsen) by R. Wattiaux.- It may be pertinent that some lysosomal hydrolases (e.g. acid phosphatase) that are located in the lysosomal membrane do not depend on mannose-6-phosphate receptors.

Supplementary refs. contributed by Senior Editor

Disturbances involving mitochondria

   Mouse heart mitochondria were examined for membrane effects of 4'-epi-adriamycin, investigated as hopefully less cardiotoxic than the anti-cancer drug adriamycin: exposure *in vitro* but not *in vivo* caused inactivation of respiratory-chain complexes with membrane rigidification and enhanced lipid peroxidation.- Praet, M., Laghmiche, M., Pollakis, G., Goormaghtigh, E. & Ruysschaert, J.M. (1986) *Biochem. Pharmacol. 35*, 2923-2928.

   In patients with circulatory shock, skeletal muscle biopsies gave evidence of oxidative (free radical) damage to the electron transport chain.- Corbucci, G.G., et al. (1985) *Circ. Shock 15*, 15-26.

   The neurotoxin 1-methyl-4-phenyl-1,2,3,6-tetrahydropyridine (MPTP), as used in primates to simulate Parkinson's disease, appears to be activated by mitochondrial monoamine oxidase; the toxic product accumulates in mitochondria and impairs processes such as NADH-linked respiration.- Kindt, M.V., Nicholas, W.J., Sonsalla, P.K. & Heikkila, R.E. (1986) *Trends Pharmacol. Sci. 7*, 473-475.   .

   Methylglyoxal bis(guanylhydrazone) (MGBG), a polyamine-type drug, damages mitochondria and inhibits β-oxidation. With liver, heart and skeletal muscle mitochondria, MGBG *in vitro* (like other polyamines) caused aggregation. It increased membrane rigidity and blocked carnitine acyltransferases.- Brady, L.J., Brady, P.S. & Gandour, R.D. (1987) *Biochem. Pharmacol. 36*, 447-452.

   There is a classical case of a female patient with chronic hypermetabolism despite normal thyroid function: skeletal muscle mitochondria were remarkably abundant, and manifested loss of respiratory control.- Luft, R., Ikkos, D., Palmieri, G., Ernster, L. & Afzelius, B. (1962) *J. Clin. Invest. 41*, 1776-1804.

Another classical situation is autoimmunity to the mitochondrial inner membrane as seen in cases of primary biliary cirrhosis.- Berg, P.A., Roitt, I.M., Doniach, D. & Cooper, H.M. (1969) *Immunology 17*, 281-283. Defective mitochondrial fatty acid oxidation underlies various 'mitochondrial myopathies' (inborn errors).- Turnbull, D.M. & Sherratt, H.S.A. (1985) *Biochem. Soc. Trans. 13*, 645-647.

Toxin activation; free radicals

"Reactive metabolites as a cause of hepatotoxicity" (especially halogenated hydrocarbons and paracetamol).- Prescott, L.F. (1983) *Int. J. Clin. Pharmacol. Res. 3*, 437-441. Two surveys by Gillette, J.R., Lau, S. & Monks, T.J.: (1984) *Biochem. Soc. Trans. 12*, 4-7; (with L.R. Pohl) in *IUPHAR Proc. (9th Int. Congr. Pharmacol.)* (Paton, W.R., et al., eds.), Macmillan, Vol. 2, pp. 251-257 [in Section on free-radical injury].

Prevention of CCl_4 hepatotoxicity in mice (due to the $\cdot CCl_3$ free radical) by methoxsalen, a 'suicide substrate' for cytochrome P-450. - Labbe, G.,......& Pessayre, D. (1987) *Biochem. Pharmacol. 36*, 907-914. Microsomal binding/activation of etoposide.- Haim, N., et al., *ibid.* 527-536.

Epoxide hydrolases; peroxisome proliferators

Investigations on activities in particular fractions, cytosolic and particulate, and of peroxisomal proliferator influences have been reported recently in *Biochem. Pharmacol.*:- (1986) *35:* 1299-1308: Kaur, S. & Gill, S.S.; (1987) *36:* 345-351: Schladt, L., Hartmann, R., Timms, C., Strolin-Benedetti, M., Dostert, P., Worner, W. & Oesch, F.; 815-821: Lündgren, B., Meijer, J. & DePierre, J.W.

Factors pertinent to examining lipid peroxidation in liver microsomes include O_2 tension.- Reiter, R. & Burk, R.F. (1987) *ibid. 36*, 925-929.

Drug inhibition of microsomal lipid peroxidation by etoposide VP-16, an anti-cancer drug: this effect, and in general the cytotoxicity, was attributable to a quinone metabolite.- Sinha, B., Trush, M.A. & Kalyaraman, B. (1985) *Biochem. Pharmacol. 34*, 2036-2040.

Iron overload: an animal model, and hepatic ferritin iron response to chelators.- Longueville, A. & Crichton, R.R. (1986) *ibid. 35*, 3669-3674.

Organelle labilization by Hg^{2+} featured in Vol. 4, this series (Roels, H., et al.), studied *in vitro* with liver large-granule material.

Clostridial toxins have been reviewed: Simpson, L.L. (1986) *Annu. Rev. Pharmacol. Toxicol. 26*, 427-453.

Other organelle refs., including **nucleus** *and* **ribosomes** *(p. 240), are in later #NC Sections. Ion movements, notably Ca^{2+}, and other background refs.: see especially #NC(C)-1 (B.F. Trump).*

Section #B

APPROACHES DEPENDING ON SURFACE DISTINCTIONS

#B-1

FREE FLOW ELECTROPHORESIS: ITS APPLICATION TO
THE SEPARATION OF CELLS AND CELL MEMBRANES

J.M. Graham, R.B.J. Wilson and K. Patel

Department of Biochemistry
St. George's Hospital Medical School
Cranmer Terrace, London SW17 0RE, U.K.

FFE* enables preparative separation of cells, membranes and proteins, on the basis of their zeta potential, thus complementing centrifugation which separates on the basis of size or density. The sample is injected into a curtain of buffer which moves downwards between two glass plates (the separation chamber) and across which is imposed an electric field. The emerging buffer stream is divided into 90 fractions. Generally the buffer flow rate considerably exceeds the sedimentation rate of biological particles at $g = 1$; hence the size of the particle is not normally important.

FFE is of particular use in separating particles which tend to aggregate during centrifugation. Although the technique is primarily a preparative one it can be used in an analytical mode. Indeed, although there are certain restrictions on the composition of the separation buffer which do not apply to the more widely used technique of cytopherometry (microelectrophoresis), the ability of FFE to provide electrophoretic mobility data on millions of cells very rapidly is a distinct advantage over cytopherometry where measurements are made on single cells.

Detailed examination of the theoretical aspects of FFE will not be attempted in this article, which aims rather to describe the operation of a modern commercial machine with reference to the separation of cells and cell membranes, for which FFE is a useful preparative tool. Especial attention is given to the operational parameters which can influence the efficacy of such separations.

---

* FFE, free flow electrophoresis; p.m., plasma membrane.

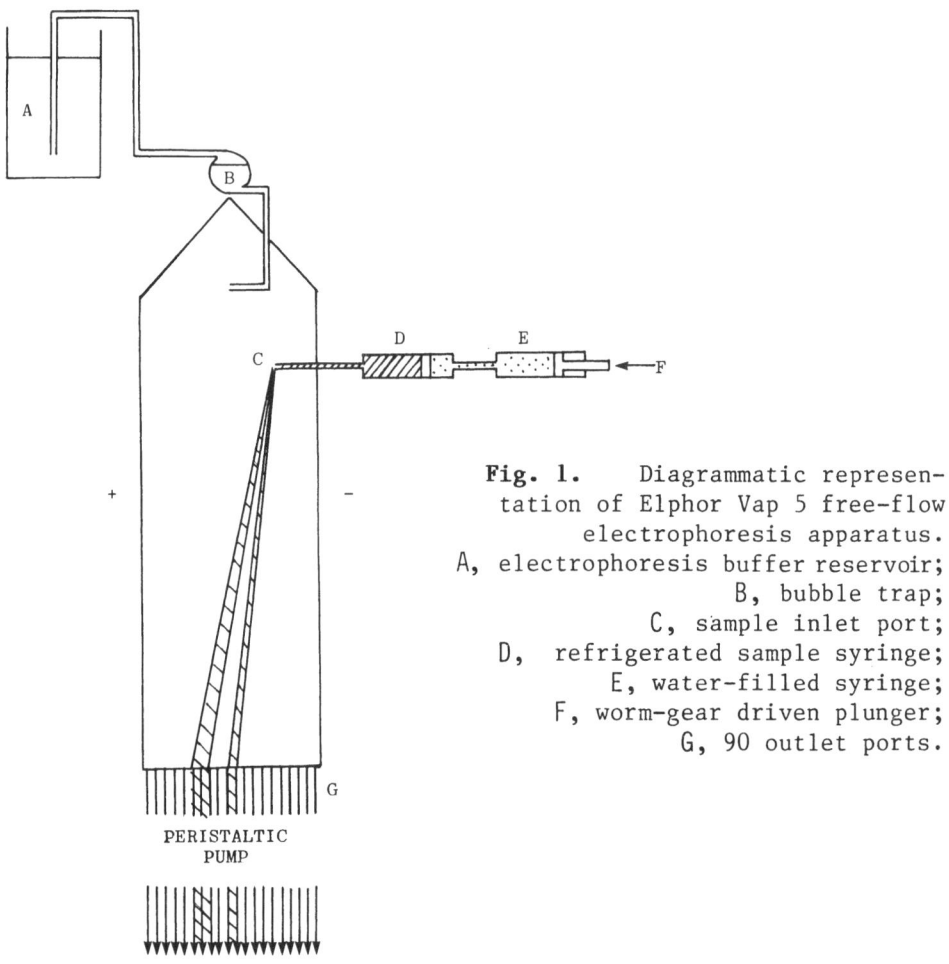

Fig. 1.   Diagrammatic represen-
tation of Elphor Vap 5 free-flow
electrophoresis apparatus.
A, electrophoresis buffer reservoir;
B, bubble trap;
C, sample inlet port;
D,   refrigerated sample syringe;
E, water-filled syringe;
F, worm-gear driven plunger;
G, 90 outlet ports.

THE APPARATUS

The most widely available machines are those produced by Bender
& Hobein (Munich; U.K. supplier: Biotech Insts., Luton). The machines
are distinguished by the name Elphor Vap, followed by a model number.
There are more recent, more sophisticated machines than the widely
available Elphor Vap 5 model; but they are similar in operation,
and it is the Elphor Vap 5 that will be described.

It consists of a vertical glass chamber (Fig. 1), ~100 mm wide,
600 mm high and 0.7 mm deep. The rear glass plate is silvered and
is in contact with a copper cooling plate. At the top of the chamber
an inlet port connects with a reservoir of electrophoresis buffer
via a bubble trap. At the bottom the buffer emerges in 90 fractions:
the rate of buffer flow is controlled by a peristaltic pump through
which the 90 outlet tubes pass.

On either side of the chamber run two ion-exchange strips (a cation- and an anion-exchange resin on the positive and negative sides respectively). Behind these strips are two Pt electrodes: electrical continuity between the strips and the electrodes is maintained by an electrode buffer which circulates over and between each electrode via a reservoir and water pump. This continuous exchange buffer between the electrodes together with the ion-exchange strips serves to minimize any changes in pH or ionic composition across the electrophoresis buffer.

The sample is contained within an oscillating syringe within a refrigerated chamber (say at 4-6°). A second syringe whose piston is advanced by a fine-pitched worm gear communicates with the sample syringe via a column of water and serves to inject the sample into the vertically flowing electrophoresis buffer through an inlet port (Fig.1) which is normally 2 cm from the right-hand edge of the chamber and ~8 cm from the top.

As the sample is moved down the chamber by the buffer flow (and as most cells and membranes have a net negative charge in buffers in the pH range 6-9), it will be deflected towards the anode to an extent which depends on the magnitude of the surface zeta potential, the applied electric field and the rate of buffer flow, according to the following equations:

$$\text{Apparent electrophoretic mobility} \quad \overline{U} = \frac{1 \times A \times K}{I \times t}$$

where 1 = migration distance (cm), A = cross-sectional area of chamber (cm^2), K = specific conductivity of medium (Siemens.cm^{-1}),

I = current (A), t = residence time in chamber (sec): $t = \frac{2}{3} \cdot \frac{V_k}{Q_k}$

where V_k = chamber vol. (cm^3),
Q_k = chamber buffer flow rate (cm^3.sec^{-1}).

In all instruments except the most recent models, the anode is on the left; accordingly, the lower the fraction number (1-90) in which a particular fraction elutes from the chamber, the closer it will be to the anode, hence the greater will be its zeta potential.

PRACTICAL CONSIDERATIONS

The buffer flow rate can be varied from ~50 to 750 ml/h, depending on the machine, although different ranges of flow rates can be obtained on individual machines by changes to the gear system driving the peristaltic pump. Although the routine operating temperature is 4-6°, temperatures up to 20° can be used. In earlier instruments (e.g. Elphor Vap 5) the maximum permissible buffer conductivity was 1500 μmho/cm, while in later models (e.g. Elphor Vap 11) this factor was increased to 7500 μmho/cm. This has enabled the maximum field strength of 120 V/cm for the earlier models to be increased to 145 V/cm in the later machines.

The sample application rate (up to 8 ml/h) depends to some extent on the electrophoresis buffer flow rate. For maximum resolution, the sample must be applied into the buffer at the mid-point between the glass plates: the broader and deeper the sample band, the lower the resolution.

Buffer composition

The upper limit set on the buffer conductivity of earlier models imposed severe retrictions on the ionic content of the buffer. To achieve electric fields sufficiently high to allow a significant electrophoretic migration of the particles, NaCl concentrations exceeding 10-15 mM were not permissible. For the separation of membranes this is not an important consideration, because generally the media which are best suited to the retention of functional activities have a low ionic strength. Indeed, the most common type of membrane suspension medium – 0.25 M sucrose buffered with low concentrations (5-10 mM) of an organic buffer such as Tris, Hepes or Tricine at pH 7.4 – is an ideal electrophoresis buffer.

Cells, however, are relatively intolerant to such low ionic-strength media; but with Vap 5 machines there is no alternative to providing most of the osmolarity of the separation buffer by an organic solute. Few cell types tolerate sucrose, and it is more common to use glycine, glycerol, glucose or mannitol, again with an organic buffer such as triethanolamine, Tris, Hepes or Tricine. Even the low concentrations of ions permitted in the Vap 5 machines are advantageous to recovering viable cells, and inclusion of 5-10 mM NaCl together with 1-2 mM $MgCl_2$ or $CaCl_2$ is quite a common practice. The retrictions imposed by the later models (Vap 11, Vap 22, etc.) are much less severe and at least half of the osmotic component can be provided by salt. The following are some of the more common separation buffers.-

- (a) 0.3 M Tris/borate.
- (b) 0.015 M triethanolamine (TEA)/acetic acid, 0.24 M glycine.
- (c) 0.03 M Hepes/NaOH, 0.25 M glycine.
- (d) 0.0058 M phosphate, 0.29 M glycine.
- (e) 0.01 M TEA/acetic acid.
- (f) = (e) but with 0.005 M glucose, 0.28 M sucrose, 5 µM $CaCl_2$ or $MgCl_2$, 5 mg bovine serum albumin (BSA).
- (g) 0.015 M TEA, 0.01 M glucose, 0.004 M potassium acetate, 0.24 M glycine.
- (h) 0.010 M TEA/acetic acid, 0.25 M glucose, 0.002 M NaCl.

It is difficult to recommend a particular buffer for cell electrophoresis. Buffers (b), (c), (g) or (h) supplemented with $CaCl_2$/$MgCl_2$ and BSA, as in (f), may be a good starting point.

APPLICATIONS TO CELLS *

Platelets

Background.- The use of FFE to purify platelets from human blood demonstrates one of the advantages of this technique (details: Wilson & Graham [1]) over the more widely used centrifugation methods. Even in the presence of chelating agents such as EDTA or EGTA, the sedimentation of platelets, to furnish either a pellet or a concentrated band on top of a density barrier, results often in their aggregation. Although such aggregated material is usually quite easily dispersed by gentle shearing forces (e.g. repeated uptake into and expulsion from a pipette or syringe), this can result in activation of the platelets. FFE on the other hand produces a slight dilution of the platelets by the electrophoresis separation buffer when the sample is injected into the chamber, and aggregation is rarely observed. Indeed, because of this, EDTA is not required in the separation buffer.

FFE conditions.- Electrophoresis separation buffer is (h) above (but NaCl 1 mM), pH 7.6; flow rate 600 ml/h. Electrode buffer contains 0.1 M triethanolamine/acetic acid, pH 7.6. Separations are at 6° with 1100 V applied; ~10 ml of a platelet-rich fraction from 150 ml blood can be fractionated in ~1 h.

Results.- Two major bands are recoverable from the chamber eluate. Cell recoveries are expressed relative to the platelet-rich starting material: fractions 47-52 *vs. 53-57:* : platelets 1-2%, *94-95%*; red cells: 97-99%, *0.1-0.2%*; white cells: 100%, *0%*.

Evidently the red and white cells elute in the former band and the much less electronegative platelets in the latter. Compared with any centrifugation technique, recoveries are as good if not better. Functionally, as measured by the retention of their nucleotides and their release upon aggregation, the platelets produced by FFE are superior to those produced by centrifugation [1].

Trypanosomes

Background.- In the laboratory *Trypanosoma brucei* is grown in the bloodstream of rats, and the most often used technique for their separation from host blood is by filtration through a bed of anion-exchange resin which retains the erythrocytes and white cells, while permitting the less negatively charged trypanosomes to elute with the plasma. The yield is ~60-80% and, the capacity of the resin being finite, ultimately host cells will emerge from the resin bed. It is the big difference in the magnitude of the surface charge between trypanosomes and erythrocytes that makes them ideal for separation by FFE (method details: Graham & Agbe [2]).

---

*Ed.'s citation of Vol. 8 arts.: Heidrich & Hannig, FFE theory & various cells; Lanham, trypanosomes etc.(not FFE); Crawford, platelets etc.(not FFE).

Table 1. Effect of pH and Mg^{2+} on the elution positions of trypanosomes and host rat erythrocytes: peak fractions, and cathodic shift
due to Mg^{2+}.

Cell type	pH 7.4/+Mg^{2+}	pH 7.4/-Mg^{2+}	Shift	pH 8.0/+Mg^{2+}	pH 8.0/-Mg^{2+}	Shift
Trypanosomes	62–64	44–46	18	48–50	32–34	16
Erythrocytes	48–50	36	12–14	38–40	34–36	4

FFE conditions.- Electrophoresis separation buffer [cf. (h)
above] contains 0.25 M glucose, 1 mM NaCl, 0.5 mM $MgSO_4$, 10 mM tri-
ethanolamine/acetic acid, pH 8.0; flow rate 450 ml/h. Electrode
buffer contains 100 mM triethanolamine/acetic acid, pH 8.0.
Separations are at 6° with 900 V applied.

Results.- Under the above optimized conditions, ~99% of the
erythrocytes elute in fractions 34–42, while 99% of the typanosomes
elute in fractions 44–54. Table 1 shows the differential effects
of pH and Mg^{2+} on the electrophoretic mobility of the two types of
cell. The cathodic shift in the trypanosome peak caused by inclusion
of Mg^{2+} is always greater than the erythrocyte shift. Whilst this
differential effect is noticeable at both pH 7.4 and pH 8.0, that
at 8.0 was much more marked; indeed, Mg^{2+} was virtually without effect
on the erythrocytes at this pH. The other obvious difference between
the cell types is that in the absence of Mg^{2+} the influence of pH
on erythrocyte mobility was nil; only with Mg^{2+} present was there
a significant cathodic shift when the pH was decreased from 8.0 to
7.4, whereas trypanosomes showed such a shift in the presence or
absence of Mg^{2+}.

This differential effect may reflect the marked difference in
the source of the cell-surface charge: no sialic acid has been detected
at the surface of *T. brucei* [3]. Whatever its cause, these observations
underline the importance of manipulating the ionic composition and
pH of the electrophoresis separation buffer. In most circumstances
the addition of cations and lowering the pH will tend to titrate
some of the negatively charged groups on the surface of the biological
particles. As a result the zeta potential will drop and the electrophoretic mobility of the particles be reduced. However, we have
shown that under certain conditions, particularly at pH <7.4, low
concentrations of NaCl (1–10 mM) may actually cause an increase in
mobility which is reversed at higher salt concentrations [4]. What
is more important is that modulation of these buffer parameters,
together with the buffer flow rates and the applied voltage seems
to affect different cell types to different extents.

Other cell types

Many other examples of the use of FFE in cell separations have been reported, including the following: kidney cells [5-7], T and B lymphocytes [8, 9], B lymphocyte subpopulations [10, 11], plasmodium-infected and non-infected erythrocytes and malarial parasites ([12] & Vol. 8, this series).

APPLICATIONS TO MEMBRANES

Guinea pig enterocyte membranes

Background.- As an example of the ability of FFE to separate not only different membrane types but also different domains of the same membrane, the fractionation of an enterocyte mitochondrial fraction will be presented. It demonstrates not only the resolving power of the technique but also its aptness for samples which show a pronounced tendency to aggregate (as considered above for platelets). The contamination of the enterocyte fraction with mucus often leads to aggregation during sedimentation of membranes into a pellet or banding in a sucrose gradient; hence a method in which the sample is diluted by the electrophoresis buffer is an advantage.

FFE conditions.- Electrophoresis separation buffer contains 0.25 M sucrose, 10 mM triethanolamine/acetic acid, pH 8.0; flow rate 400 ml/h. Electrode buffer contains 100 mM triethanolamine/acetic acid, pH 8.0. Separations are at 6° with 900 V applied; the rate of sample injection is 2 ml/h.

Results.- To obtain optimal resolution it was necessary to inject the sample slowly so as to give minimum band broadening in the chamber. Fig. 2 shows the resolution of a number of enzymes. The mitochondrial band (succinate-cytochrome **c** reductase) appears sharply in the middle of the eluate, while the basolateral p.m. (Na^+/K^+-ATPase) is predominantly in the more electronegative fractions and the brush-border p.m. (5'-nucleotidase, alkaline phosphatase and aminopeptidase) appears only in the less electronegative fractions. In this particular separation the acid phosphatase (lysosomes) overlapped the mitochondria and the brush border fractions.

To resolve these three membrane types more efficiently, a guinea pig enterocyte mitochondrial fraction was subjected to electrophoresis according to the conditions outlined above. The fractions containing Na^+/K^+-ATPase were discarded (fractions 1 & 2 - see Fig. 2) and then the total residual material in the eluate was bulked together, concentrated and run through the FFE system at a higher voltage (1100 V) and a lower electrophoresis buffer flow rate (300 ml/h). These conditions now permitted better resolution of the brush border, mitochondria and lysosomes (Fig. 3).

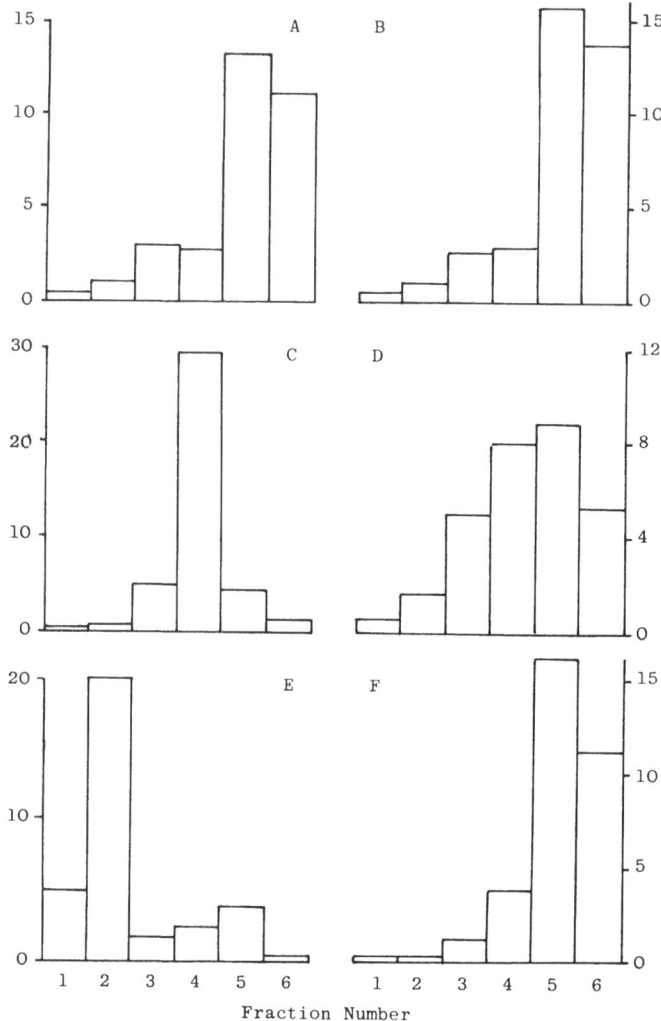

Fig. 2. Subfractionation of a guinea pig enterocyte mitochondrial fraction: isolation of basolateral and brush-border plasma membrane (p.m.). Electrophoresis conditions: see text. The product (in tubes 25-53) was consolidated into 6 fractions: 25-29, **1**; 30-32, **2**; 33-36, **3**; 37-43, **4**; 46 & 47, **5**; 48-53, **6**. **A**, 5'-nucleotidase; **B**, alk. phosphatase; **C**, succinate-cytochrome **c** reductase; **D**, acid phosphatase; **E**, Na^+/K^+-ATPase; **F**, aminopeptidase. All ordinates are relative specific activities, viz. fraction *vs.* homogenate.

Other membranes

Other types of membrane separated by FFE include: gastric membrane vesicles [13], normal and malignant colon p.m. [14], renal medulla membranes [15], rat duodenum basolateral membranes [16], human lysosomes [17], human platelet surface and intracellular membranes [18].

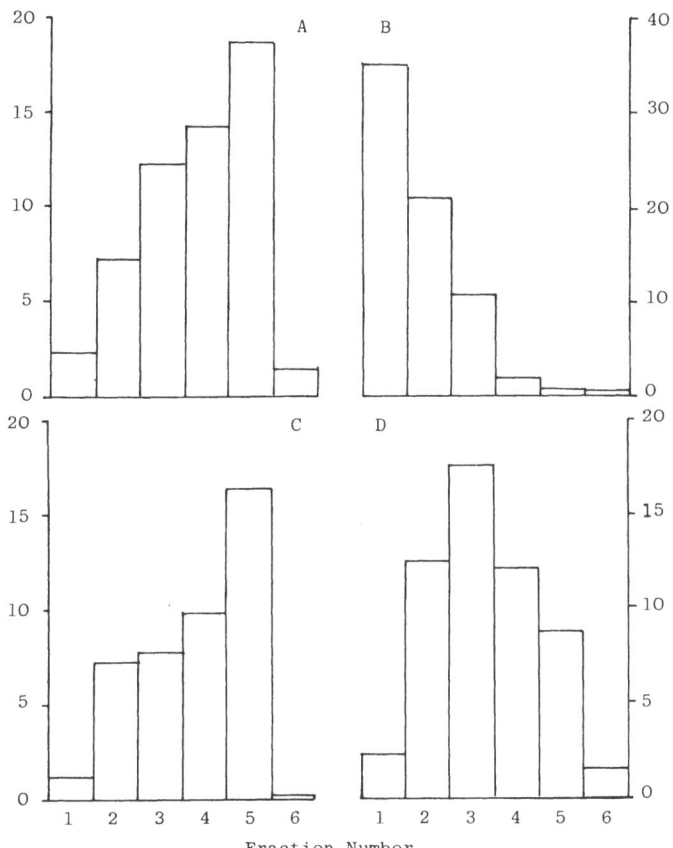

Fig. 3. Subfractionation of a guinea pig enterocyte mitochondrial fraction: isolation of lysosomes, mitochondria and brush border membrane. Electrophoresis (see text) performed on pooled fractions **3-6** from the previous separation (Fig. 2).
The products (in tubes 9-42) were consolidated into 6 fractions: 9-13, **1**; 14-18, **2**; 19-24, **3**; 25-28, **4**; 29-34, **5**; 34-42, **6**.
A, 5'-nucleotidase; **B**, succinate-cytochrome c reductase; **C**, alk. phosphatase; **D**, acid phosphatase. All ordinates are relative s.a.'s as in Fig. 2.

CONCLUSIONS

FFE permits the separation of cells and membranes under very mild conditions. The separation media are iso-osmotic and the viability of the collected material is normally very good. Modern machines permit the use of relatively high concentrations of salts (up to ~155 mOsM); earlier machines required the osmotic component of the separation medium to be mainly an organic solute. There is no upper limit to the amount of material that can be processed, and aggregation effects are minimal.

Acknowledgements

The authors thank the Cell Surface Research Fund, the British Heart Foundation, the Science & Engineering Research Council and Abbott Laboratories for generous financial support.

References

1. Wilson, R.B.J. & Graham, J.M. (1986) *Clin. Chim. Acta 159*, 211-217.
2. Graham, J.M. & Agbe, S.A.O. (1981) in *Cell Electrophoresis in Cancer and other Clinical Research* (Preece, A.W. & Light, P.A., eds.), North Holland, Amsterdam, pp. 309-312.
3. de-Lederkremer, R.M., Casal, O.L., Tanaka, C.T. & Colli, W. (1978) *Biochem. Biophys. Res. Comm. 85*, 1268-1274.
4. Graham, J.M., Pasternak, C.A., Wilson, R.B.J., Alder, G.M. & Bashford, C.L. (1986) in *Electrophoresis '86* (Dunn, M.J., ed.), VCH Verlagsgesellschaft mbH, Weinheim, FRG, pp. 77-85.
5. Kreisberg, J.I., Sachs, G., Pretlow, I.G. & McGuire, R.A. (1977) *J. Cell Physiol. 93*, 169-172.
6. Heidrich, H-G. & Dew, M. (1977) *J. Cell Biol. 74*, 780-788.
7. Vandewalle, A. & Heidrich, H-G. (1980) *Eur. J. Cell Biol. 22*, 595-601.
8. Shortman, K. (1977) *Meths. Cell Separation 1*, 229-249.
9. Vassor, E., Levy, M. & Brocks, D.E. (1976) *Cellular Immunology 21*, 257-271.
10. Zeiller, K., Rascher, G. & Hannig, K. (1976) *Immunology 31*, 863-880.
11. Shortman, K., Fidler, J.M., Schlegel, R.A., Nussel, G.J.V., Howard, M., Lipp, J. & van Boehmer, H. (1976) *Contemporary Topics in Immunology 5*, 1-45.
12. Heidrich, H-G., Russman, L., Bayer, B. & Jung, A. (1979) *Z. Parasitenkd. 58*, 151-159.
13. Saccomani, G., Stewart, H.B., Shaw, D., Levin, M. & Sachs, G. (1977) *Biochim. Biophys. Acta 465*, 311-330.
14. Jackson, R.J., Stewart, H.B. & Sachs, G. (1977) *Cancer 40*, 2487-2496.
15. Iyengar, R., Mallman, D.S. & Sachs, G. (1978) *Am. J. Physiol. 234*, F247-F254.
16. Mircheff, A.K., Sachs, G., Hanna, S.D., Sabiner, C.S., Rabon, E., Douglas, A.P., Walling, M.W. & Wright, E.M. (1979) *J. Membr. Biol. 50*, 343-363.
17. Harms, E., Kern, H. & Schneider, J.A. (1980) *Proc. Nat. Acad. Sci. 77*, 6139-6143.
18. Menashi, S., Weintraub, H. & Crawford, N. (1981) *J. Biol. Chem. 250*, 4095-4101.

#B-2

COUNTER-CURRENT CHROMATOGRAPHY AS A TOOL
IN SUBCELLULAR STUDIES

Ian A. Sutherland and Deborah Heywood-Waddington

National Institute for Medical Research
Mill Hill, London NW7 1AA, U.K.

In CCC* there is the advantage of no solid support: the eluting phase passes over a stationary liquid phase held by low-speed centrifugation. A new 'coil planet centrifuge' with toroidally wound coils is applicable to purification of molecules such as peptides and drugs using aqueous/organic phase systems. On the analytical scale it can cope with the viscosity of polymer phase systems, allowing cells and subcellular elements to be fractionated by use of two immiscible aqueous phases.

The application of CCC to the separation of viable biological material depends on the availability of suitable phase systems. These must be able to act as a gentle host medium for the cells and subcellular elements they contain and have the appropriate hydrodynamic properties for CCC operation.

Albertsson first introduced the concept of partition between immiscible aqueous polymer phases for separating cells in the mid-1950's [1]. But at least 20 years elapsed before the technique became widely used and accepted. The use of two-phase aqueous systems for separating cells, organelles and particles has been well reviewed ([2-5] and G. Johansson in Vol. 2, this series). The way aqueous phases can be manipulated to achieve separation will not be considered here, but it is an important prerequisite to applying the technology to CCC.

CCC can be regarded either as liquid-liquid chromatography without a solid support or as a continuous form of liquid-liquid extraction.

---

*Abbreviations include CCC = counter-current chromatography, PEG = polyethylene glycol.

Since it was first introduced in the 1970's, the process has been through a number of development stages which have improved its design and efficiency. Most schemes involve continuous coils of PTFE tubing rotating in some form of planetary motion without rotating seals. The process is now being used for the analytical/preparative separation and purification of a wide range of natural products and soluble biopolymers using aqueous/organic phase systems [6, 7] and for analytical fractionations of membranes and organelles with double aqueous phase systems [8, 9]. In other reviews [10, 11] various types of CCC apparatus have been outlined and guidelines given for use and application. This article reviews the latest CCC techniques used for separations of subcellular particles with aqueous/aqueous phase systems.

There are two essential requirements for successful CCC: the retention of one of the phases in the coil and the adequate mass transfer of sample constituents as the other (mobile) phase passes through and mixes with the retained phase. Successful retention is largely a function of the physical properties of the phase systems and interactions with the coil walls. Mass transfer depends on effective mixing which is determined by hydrodynamic factors. It is only now that the coil planet centrifuge [12-14] is emerging as a clear front runner giving increased retention of the stationary phase, shorter separation times, greater capacity and better reliability.

COUNTER-CURRENT CHROMATOGRAPHY WITH DOUBLE AQUEOUS PHASE SYSTEMS

When certain polymers are mixed with water, two immiscible aqueous phases are formed and, with suitable additives, can provide a gentle host medium for cells and organelles [2, 5]. Unfortunately these aqueous phase systems have increased viscosity, lower density difference and considerably reduced interfacial tension, when compared to aqueous/organic phase systems. This makes them unsuitable for use in the conventional high-resolution coils, such as the multi-layer coil planet centrifuge [12-14], due to their long mixing/settling cycle times. However, this does not preclude their use with CCC on an analytical scale. Special toroidal cells can be mounted perpendicularly to the force field, producing cascade mixing (Fig. 1) when one phase is flowing relative to the other [15].

Coils can be wound around the drum of an epicyclic coil planet centrifuge [16] or alternatively mounted circumferentially on a rotating disc [17]. The coil is initially filled with one of the phases and then the plate is rotated at 800 rpm while the other phase is pumped in. As the phases intermix, centrifugal force ensures that the lighter and heavier phases are respectively retained in the inner and outer halves of each coil unit. The pumped phase partially displaces the other phase from the coils until the cascade mixing scheme shown in Fig. 1 is established.

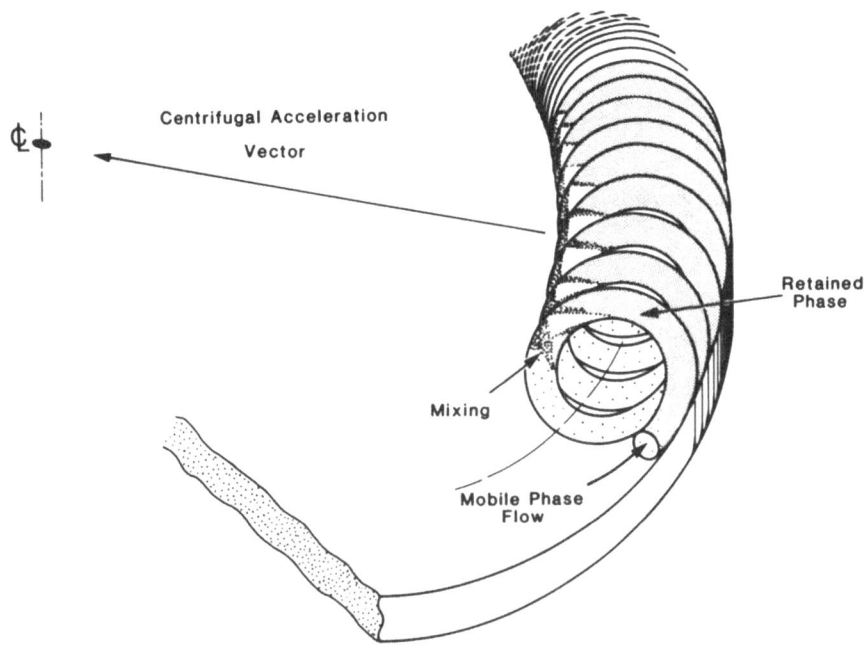

Centrifugal Acceleration
Vector

Retained
Phase

Mixing

Mobile Phase
Flow

Fig. 1. Cascade mixing produced by the flow of one phase relative to a retained phase in toroidal coils mounted circumferentially on a rotating disc.

A typical operating system is shown schematically in Fig. 2. The sample is injected with the mobile phase (using a conventional liquid chromatography sample loop) and undergoes a series of mixing and settling steps before it eventually elutes to the fraction collector. Sample components partitioning towards the mobile phase will elute early while those favouring the stationary phase or interface will be retained. As there is no solid support, either phase can be used as the mobile phase, or even a mixture of the two. Adding a small proportion of the stationary phase in the above example would accelerate the elution of all the retained components and clear the coil system for another sample loading.

APPLICATIONS

The resolution of spinach chloroplasts into three subpopulations by Albertsson & Baltescheffsky in 1963 [18] was one of the first fractionations of subcellular material using aqueous polymer phase systems. Later, Albertsson's thin layer counter-current distribution technique [19] was used for a number of subcellular fractionations including that of Golgi apparatus by Hino et al. [20]. Recently affinity partition methods have been proposed for industrial-scale enzyme and protein purification [5, 21].

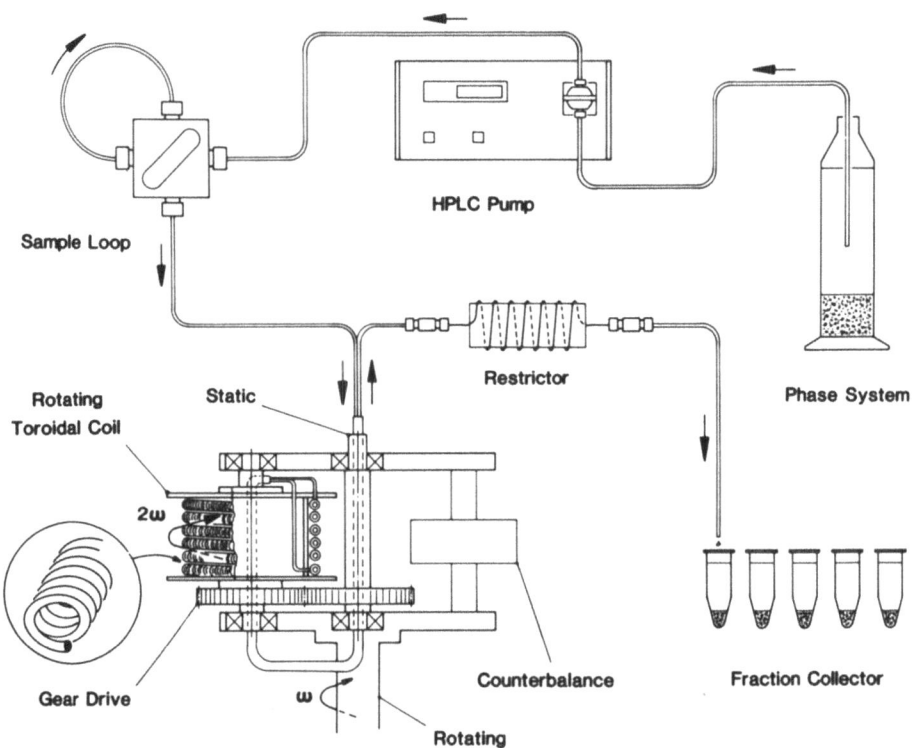

Fig. 2. Schematic layout of CCC operating system. Note the toroidal coils wound on a epicyclic coil planet centrifuge. A restrictor would be needed only when using aqueous/organic phase systems.

The use of CCC for separating viable biological material based on partition in aqueous two-phase poylmer systems has recently been reviewed [8, 9]. Special toroidally wound coils have been used on the epicyclic coil planet centrifuge to achieve successful fractionations of organelles, membranes and bacterial cells. However, large cells such as erythrocytes still require a more complicated rotor to avoid sedimentation effects.

Rat-liver homogenate has been successfully fractionated (Fig. 3) using a phase system containing 3.3% (w/w) dextran T-500, 5.4% PEG-6000, 10 mM Na phosphate/phosphoric acid buffer (pH 7.4), 0.26 M sucrose, 0.05 mM Na_2EDTA and 1 mM ethanol, in both the toroidal coil [17] and the epicyclic coil planet centrifuge [16]. Sample preparation and enzyme assay procedures have been described [22]. Fractionations in either machine are qualitatively similar with plasma membrane eluting early, lysosomes shortly after and endoplasmic reticulum spread over three possible fractions. Rat-liver homogenate has been

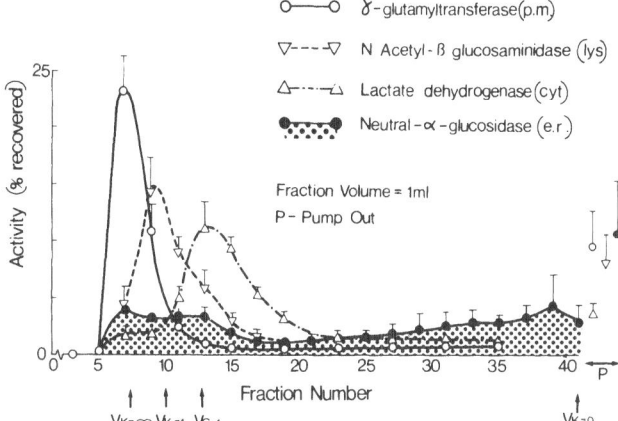

Fig. 3. Analytical subcellular fractionation of rat-liver homogenate using a continuous toroidal coil centrifuge (mean and S.D. from
7 expts.). (Enzyme assignments indicated parenthetically.) Activity
recoveries range from 70% to 90% [17].

used by us [22] as a standard fractionation for studying a number
of coil operating parameters, e.g. rotational speed, flow rate, coil
geometry and sample loading. The process was shown not to be critical,
and small changes in these parameters did not significantly affect
the order of elution or resolution of the process, provided that
certain boundary conditions were met.

Torpedo electropax membranes, enriched in nicotinic cholinergic
receptor sites, have been successfully purified (Fig. 4) by Flanagan
et al. [23] using affinity partitioning techniques with phase systems
operating near to the critical point. They have defined the procedures
for operating the toroidal coil near to the critical point, and examined
the effects of sample loading; in contrast to the phase systems used
for rat-liver fractionation, it was essential to use an emulsion
of upper and lower phases (~10:1 ratio) to elute the membranes from
the coil.   They also concluded that sample loading was limited by
the ability to obtain a sufficiently concentrated sample.

Enhanced-gravity counter-current distribution methods have been
developed by Åkerlund [24] to reduce separation times and thus minimize
the possibility of changes in partition with time. Chloroplast vesicles
[24] and membrane-bound opiate receptors [25] have been successfully
fractionated in this way.

CONCLUSIONS

CCC is easy to perform, has a wide range of applications and
is suited to automation. Its major asset is its low cost and high

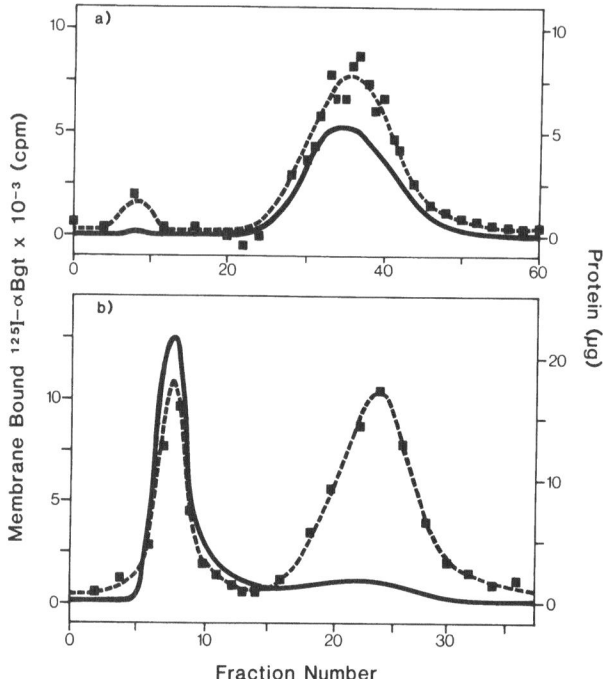

Fig. 4. Chromatograms of *Torpedo* membranes in the toroidal coil centrifuge with the phase system 4.6% (w/w) dextran T-500, 3.8% (w/w) PEG-8000, 5 mM Na phosphate pH 7.4, and 15 mM NaCl [23]. The distributions of [^{125}I]-α-bungarotoxin-labelled membrane (———) and protein (----) are shown, a) with no ligand PEG, and b) with hexaethonium-PEG replacing 0.05% of the PEG. *From [23], by permission of Marcel Dekker, Inc.*

sample recoveries. The lack of solid support minimizes adsorption problems and allows one set of coils to be used with a variety of different phase systems. However, it is still an emerging process and while polymer phase systems have been used for certain subcellular fractionations, the physical properties of these phase systems are not ideal for use with CCC. Consequently resolution is limited and application is restricted to the analytical scale.

The way ahead for the technology in the future is for new bio-compatible phase systems to be identified with physical properties more suited for use with CCC.

Acknowledgements

The authors thank David Stokes for his contribution towards the preparation of artwork for Fig. 2 and Kate Bunker for her help in preparing the manuscript.

References

1. Albertsson, P-Å. (1956) *Nature 177*, 771-774.
2. Albertsson, P-Å. (1986) *Partition of Cell Particles and Macromolecules*, 3rd edn., Wiley-Interscience, New York,
3. Albertsson, P-Å. (1977) *Endeavour 1*, 69-74.
4. Albertsson, P-Å., Andersson, B., Larsson, C. & Åkerlund, H-E. (1982) in *Methods of Biochemical Analysis* (Glick, D., ed.), Vol. 28, Wiley-Interscience, New York, pp. 115-150.
5. Walter, H., Brooks, D.E. & Fisher, D., eds. (1985) *Partitioning in Aqueous Two-phase Systems: Theory, Methods, Uses and Applications to Biotechnology*, Academic Press, London, 704 pp.
6. Conway, W.D. & Ito, Y. (1984) *LC Magazine 2 (5)*, 9 pp.
7. Martin, D.G., Biles, C. & Peltonen, R.E. (1986) *Am. Lab. (Oct.)*, 21-26.
8. Sutherland, I.A. (1985) *Chromatog. Internat. 7*, 11-15.
9. Sutherland, I.A. (1987) in *Countercurrent Chromatography: Methods and Applications* (Mandava, N.B. & Ito, Y., eds.), Dekker, New York, in [press.
10. Ito, Y. & Conway, W.D. (1984) *Anal. Chem. 56*, 534A-554A.
11. Ito, Y. (1984) in *Advances in Chromatography* (Giddings, J.C., Grushka, E., Cazes, J. & Brown, P.R., eds.), Vol. 24, Dekker, New York, pp. 181-226.
12. Ito, Y. & Bowman, R.L. (1977) *Anal. Biochem. 82*, 63-68.
13. Ito, Y. & Bowman, R.L. (1978) *J. Chromatog. 147*, 221-231.
14. Ito, Y. (1980) *J. Chromatog. 188*, 33-42.
15. Sutherland, I.A., Heywood-Waddington, D. & Ito, Y. (1987) *J. Chromatog. 384*, 197-207.
16. Sutherland, I.A., Heywood-Waddington, D. & Peters, T.J. (1985) *J. Liq. Chromatog. 8*, 2315-2336.
17. Sutherland, I.A., Heywood-Waddington, D. & Peters, T.J. (1984) *J. Liq. Chromatog. 7*, 363-384.
18. Albertsson, P-Å. & Baltescheffsky, H. (1963) *Biochem. Biophys. Res. Comm. 12*, 14-20.
19. Albertsson, P-Å. (1965) *Anal. Biochem. 11*, 121-125.
20. Hino, Y., Asano, A. & Sato, R. (1978) *J. Biochem. (Tokyo) 83*, 935-942.
21. Johansson, G., Joelsson, M. & Åkerlund, H-E. (1985) *J. Biotech. 2*, 225-237.
22. Heywood-Waddington, D., Sutherland, I.A., Morris, W.B. & Peters, T.J. (1984) *Biochem. J. 217*, 751-759.
23. Flanagan, S.D., Johansson, G., Yost, B., Ito, Y. & Sutherland, I.A. (1984) *J. Liq. Chromatog. 7*, 385-402.
24. Åkerlund, H-E. (1984) *J. Biochem. Biophys. Meths. 9*, 133-141.
25. Olde, B. & Johansson, G. (1985) *Neuroscience 15*, 1247-1253.

#B-3

FLOW CYTOMETRY: A TOOL IN STUDYING CELLULAR DERANGEMENTS

M.G. Ormerod

Institute of Cancer Research
The Haddow Laboratories
Clifton Avenue, Sutton, Surrey SM2 5PX, U.K.

The flow cytometer enables several parameters to be measured on large numbers of individual cells; it yields information about cell populations. Using appropriate fluorescent probes, changes in a variety of parameters may be measured. It is also possible to separate cells on the basis of the measured parameters.

The most widespread application is the measurement of the DNA content of cells using fluorescent dyes which bind to nucleic acids. Thereby, the distribution of cells in the cell cycle can be estimated and any interference, e.g. by a drug, can be followed. If cells are incubated with deoxybromouridine (BUDR), then those cells that are actively synthesizing DNA can be visualized using a monoclonal antibody (MAb) against BUDR. Combined with a label for DNA, this gives a very powerful method for following derangements in the cell cycle.

Another common application is the enumeration of cell sub-sets. Cells are distinguished by the amount of light scattered and by the expression of characteristic cell-surface epitopes, detected by using fluorescently labelled MAb's. Fluorescent probes may also be used to visualize, inter alia, *changes in membrane potential, intracytoplasmic pH and intracellular calcium.*

WHAT IS A FLOW CYTOMETER?

A flow cytometer measures certain properties of cells and other particles in a flow system. It is also possible to sort physically cells of a desired property and recover them for further study. All instruments measure one or more optical properties of the particles in the sample, viz. fluorescence at one or more wavelengths, scattered light and extinction (optical density). Some instruments can also record the Coulter volume.

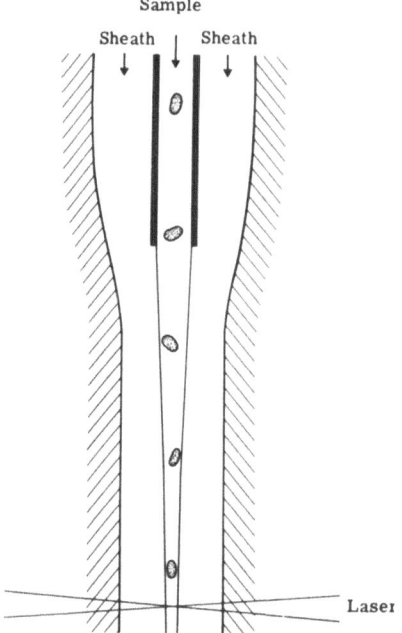

Sample

Sheath | Sheath

Laser

Fig. 1. Representation of a flow cell. The channel cut in the rectangular quartz cell is shown.

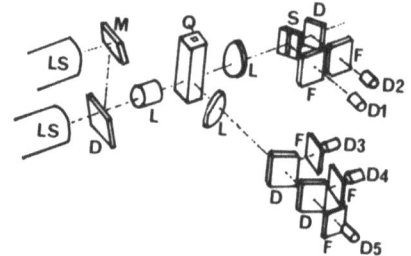

Fig. 2. The optical system of a typical flow cytometer.
LS – laser
M – mirror
D – dichroic mirror
L – lens
Q – quartz flow cell
S – reflective strip (reflects laser beam)
F – barrier filter.

Detectors:
D1 – direct light (for extinction measurement)
D2 – forward-scattered light
D3 – orthogonally scattered light
D4 – red fluorescent light
D5 – green fluorescent light.

The flow cell is at the heart of any flow cytometer (Fig. 1). It typically consists of a rectangular quartz cell, having a square channel through which sheath liquid, usually water or saline, flows under pressure. The particles under study (usually, but not necessarily, biological cells) are injected into the centre of the stream. The sample stream is focused hydrodynamically by the sheath liquid so that the particles are constrained to flow singly through the detection system.

Light, typically from a laser, is focused on the sample stream. Scattered and fluorescent light is collected by suitably placed lenses, separated into appropriate wavelengths by dichroic mirrors and barrier filters, and detected by photomultipliers (Fig. 2).

The use of an optical quartz cell is not universal. In some instruments the laser beam interrogates the flow stream after it has emerged from the flow cell ('stream in air'). Optically this has disadvantages as, inevitably, there is more scattered light in the system. It has some advantages if the cells are to be sorted (see below).

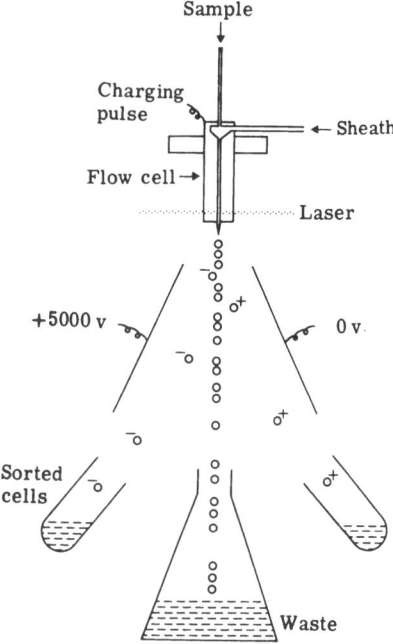

Fig. 3. Sorting cells.
For detailed explanation,
see text.

High quality commercial instruments can measure up to 4 parameters on each particle. Instruments which can measure as many as 8 have been built. These parameters may be measured on, say, 20,000 cells in 10 sec. This wealth of data can only be analyzed satisfactorily by a computer, an essential element of a modern flow cytometer.

SORTING CELLS *

One of the most powerful uses of flow cytometry is the sorting of cells (Fig. 3). To achieve this, the flow stream is forced from the flow cell through a narrow orifice (50-100 μm) to give a fine jet. The flow cell is vibrated by a transducer at a typical frequency of 30,000 Hz which causes the stream, after it emerges from the flow cell, to break into droplets. If the sheath liquid is conductive (e.g. saline), then applying an electric charge to the flow cell will charge the droplets, which can then be deflected by an electrostatic field.

When a particle passes through the laser beam, the instrument decides, according to parameters previously determined by the operator, whether it is to be sorted. If so, a charging pulse is applied to the flow cell just as the droplet containing the required particle is being formed. The charged droplet is then deflected either left

---

* In Vol. 8, this series, M.J. Owen & M.J. Crumpton describe B- and T-lymphocyte separation by Fluorescence-Activated Cell Sorter, FACS; likewise G. Blackledge in Vol. 11, with attention also to DNA and chromosome analysis.- *Ed.*

or right from the main stream by two plates carrying a fixed electrostatic charge.

There is a delay of ~100-1000 μsec between a particle passing
through the laser beam and the formation of its droplet. The instrument
has to be calibrated so that the charging pulse is issued after the
appropriate delay. Any fluctuation in the fluidics of the instrument
(e.g. in the sheath pressure) will affect the delay time and the
wrong droplet will be charged. It is therefore advantageous to keep
the distance between the laser beam and the break-off point in the
stream to a minimum. This is why some flow cytometers have a 'stream-
in-air' configuration.

APPLICATIONS

Most of these involve the labelling of cells with fluorescent
compounds. They fall broadly into three classes:

- the labelling of molecules (e.g. DNA) with fluorescent dyes;
- the detection of specific molecules using fluorescently labelled
Ab's;
- the use as probes of fluorescent compounds whose properties reflect
a particular property of the cell (e.g. pH-dependent dyes).

Examples of these applications are given later.

The flow cytometer makes multi-parametric measurements on large
numbers of particles at a fast rate (up to 5000/sec). It gives
statistically accurate information on populations of cells, enabling
different sub-sets of cells to be distinguished and counted. In
performing this task, it is faster and more accurate than a human
observer with a light microscope. However, apart from a few highly
specialized instruments, the flow cytometer gives less information
about detail within each cell, and for this type of analysis a different
approach is needed. For example, on a cytological smear, the human
observer will pick out a cell of interest and then analyze details
within, e.g. to determine whether the nucleus appears dyskaryotic.
The other feature of the flow cytometer is that cells must be in
suspension; hence with a tissue any information regarding spatial
relationships between cells is lost.

The analytical precision of the instrument can be used with
the sorter to purify sub-sets from a population of cells. Purities
of ~98% are not uncommon. With most instruments, the flow rate is
limited to ~5000 cells/sec. In practice, 2000 cells/sec is often
the maximum flow rate achieved and this limits the number of cells
that can be sorted. Sometimes it is necessary to perform a pre-
purification using another technique such as centrifugal elutriation,
or free-flow electrophoresis [see J.M. Graham, #B-1, this vol.- *Ed*.].

Fig. 4. DNA histogram from
a cell line derived from
human mammary cells. The
cells were harvested to give
a single-cell suspension and
fixed in alcohol. After re-
hydration they were treated
with RNase and propidium
iodide was added. (Sample
prepared by Mr. G. Lilley.)
D, diploid cells; T, tetra-
ploid cells having twice
the DNA content. The phases
of the cell cycle are
marked.

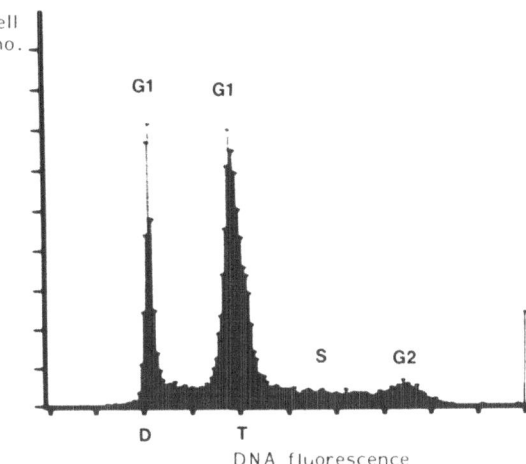

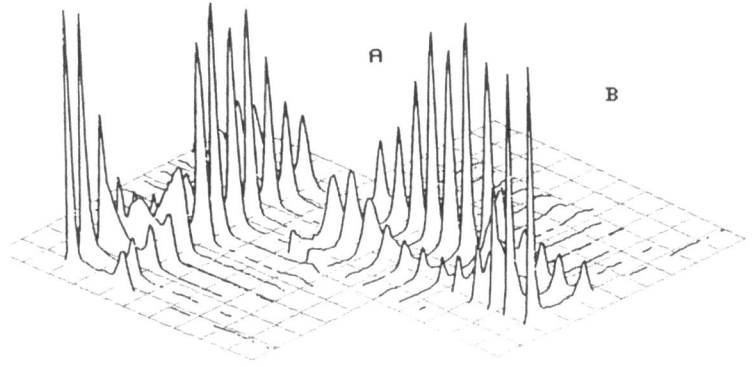

Fig. 5. DNA histograms from a drug-sensitive line of rat Walker
tumour cells. Sulphur mustard was added to the cells and samples
were taken then and at 3-h intervals thereafter. The histograms are
shown 'stacked' one behind another so that the changes can be easily
visualized; one view in A, and B shows the same picture from the
other side. (Samples prepared by Mr. F. Friedlos.)

QUANTITATION OF DNA

Various fluorescent dyes will bind to DNA stoichiometrically
so that the fluorescence from a cell can be used to quantitate its
DNA. Cell ploidy can thereby be determined (Fig. 4). Ploidy changes
in cultured cell lines are observed not infrequently. Since changes
can affect the response of cells to radiation or to drugs, monitoring
them is important for these studies.

The context of Fig. 5 is that the DNA content of a pre-mitotic
cell increases as it grows; hence the cell cycle can be visualized.

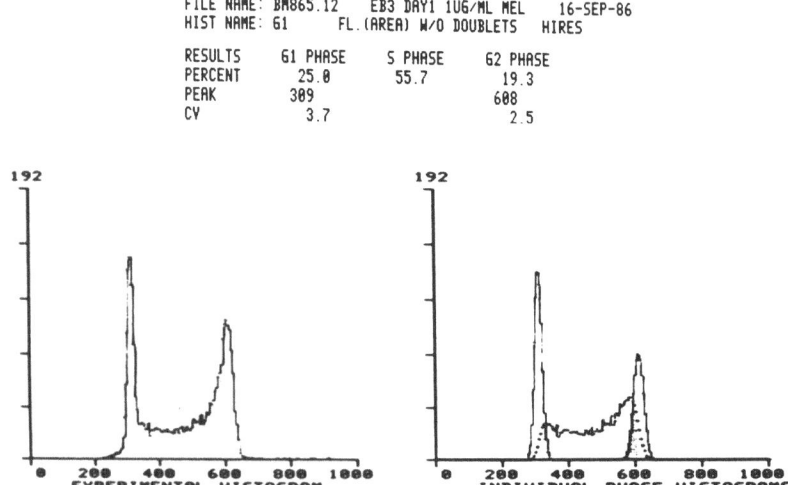

FILE NAME: BM865.12   EB3 DAY1 1UG/ML MEL   16-SEP-86
HIST NAME: G1   FL.(AREA) W/O DOUBLETS HIRES

RESULTS   G1 PHASE   S PHASE   G2 PHASE
PERCENT   25.0   55.7   19.3
PEAK   309   608
CV   3.7   2.5

Fig. 6. DNA histogram from cells of a human lymphoid line (EB3) which had been treated with melphalan 24 h previously. *Left*, the experimental histogram; *right*, the de-convolution by computer. The numerical results are also shown; the C.V. gives an estimate of the spread of the peaks. The Fig. is a photograph of the graphics screen of the Ortho 2150 computer system. [Program written by M.G. Ormerod, A.W.R. Payne & J.V. Watson (paper in preparation); cells prepared by Drs. J. Bell & B. Millar.]

This is a powerful technique for studying the mechanism of action of any treatment that interferes with cell proliferation.

Fig. 5 shows a set of histograms for cultured rat Walker-tumour cells after treatment with sulphur mustard. The first histogram shows the cells at the time of treatment. They were approaching the plateau phase of growth and few were in G0/G1 phase. Then the cells were diluted and started to grow: their movement into S phase could be observed. As the cells reached G2, further progress was blocked and the cells accumulated in this phase; G1 was almost completely depleted. Eventually some of the cells overcame the block and started to cycle again. However, cell division was abnormal, as evidenced by the broader distribution of the DNA in G1 at this time. Using other techniques, it could be shown that these cells did not survive [1].

Using a suitable computer program, the histogram of cell number against fluorescent intensity can be used to give an estimate of the % of cells in G0/G1, S and G2/M phases of the cell cycle (Fig. 6). Such a de-convolution gives, of course, only an estimate, and undue weight must not be placed on the detailed values.

Fig. 7. An isometric plot of a
cytogram obtained from Chinese
hamster V79 cells labelled in
culture (20 min) with BUDR. The
fixed cells have been labelled
with fluoresceinated anti-BUDR
and propidium iodide (PI).
Green fluorescence, BUDR; red,
DNA-PI. S-phase cells (showing
the former) are separated from
the cells in G1 and G2. (Sample
prepared by Mrs. K. Steele.)

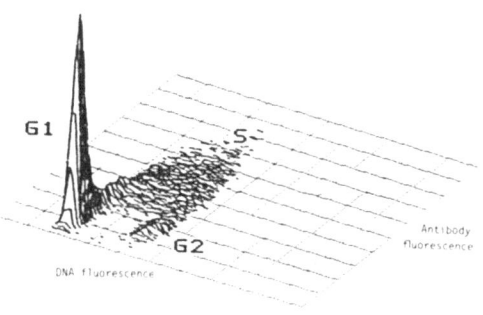

A more accurate estimation of the no. of cells in the different
compartments of the cell cycle can be obtained by pulse-labelling
cells with BUDR, which is incorporated into DNA in place of thymidine.
The S-phase cells can be visualized by fixing the cells and reacting
them with an anti-BUDR MAb labelled with fluorescein.   After RNase
treatment, added propidium iodide will label all the cellular DNA.
The laser is tuned to 488 nm and the fluorescence from fluorescein
measured at 520 nm (green) and that from propidium iodide at >600 nm
(red).   Display of red *vs*. green fluorescence gives separation of
the individual compartments of the cell cycle (Fig. 7 and ref. [2]).
Individual chromosomes can also be analyzed and sorted [3].   This   is
not a trivial application.

ENUMERATION OF CELLULAR SUB-SETS

This is most commonly applied to immunological problems in which
sub-sets of lymphocytes are enumerated.   The cells are labelled with
a MAb which defines a particular sub-set; in the analysis, light
scatter serves to define the cells of interest.   The scattering of
light by cells is a complex phenomenon.   Forward light-scatter is
approximately related to cell size, while orthogonal scatter reflects
parameters such as granularity and differences in refractive index
between the cell and the surrounding medium.

For   sheep   lymphocytes   collected   from   a   duct   draining   the
intestine,   Fig. 8 shows light scattered in a narrow forward angle
vs. that scattered orthogonally.   This defines two populations: the
numerically larger consists of small lymphocytes, the lesser comprises
larger lymphocytes.   A computer was used to define regions outlining
these populations, to display their Ab fluorescence and to calculate
the number of positive and negative cells for each region.   This
exemplifies well the use of a computer to enhance a flow-cytometric
analysis.

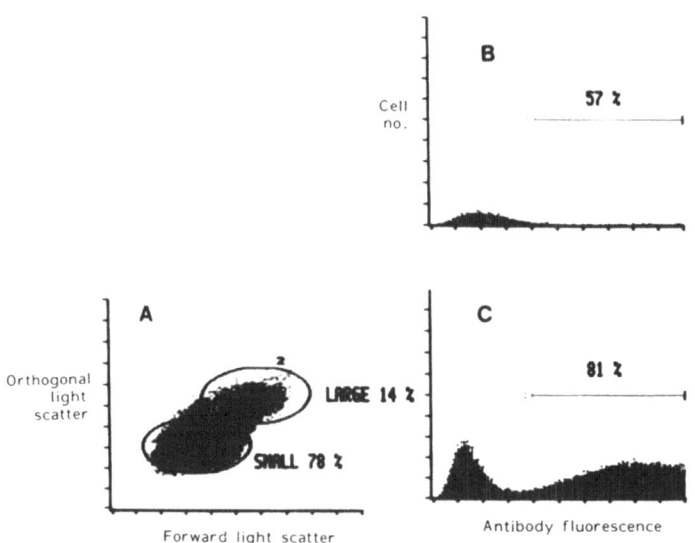

Fig. 8. Sheep intestinal lymphocytes labelled with rat anti-sheep T-cells followed by fluoresceinated rabbit anti-rat Ig. A cytogram of orthogonal *vs.* forward light scatter was displayed. Two regions were defined by computer. The histograms of each region were then displayed and the % of positive cells calculated. (Sample prepared by Ms. M. Sohatha.)

FLUORESCENT PROBES

Some of the more commonly used probes [ref. given] are as follows:
- hexamethyloxacarbocyanine, for membrane potential [4];
- quin-2, for intracellular calcium [5];
- 1,4-diacetoxy-2,3-dicyanobenzol, for intracellular pH [6];
- various, for enzyme activities [7].

The following example relates to use of electroporation at 0° to punch holes in the plasma membrane (p.m.) of mammalian cells. The holes may be re-sealed by a brief incubation at 37°. This method is used to introduce DNA into cells. The process can be monitored and optimum conditions established by flow cytometry.

Fluorescein diacetate (FDA) is taken up by cells. In the cytoplasm, esterases remove the acetate group to produce fluorescein which fluoresces green and which cannot cross the p.m. FDA can be used to distinguish cells whose p.m. is intact. Propidium iodide (PI), a red-fluorescing dye which binds to nucleic acids (see above), is normally excluded by the p.m. of live cells.

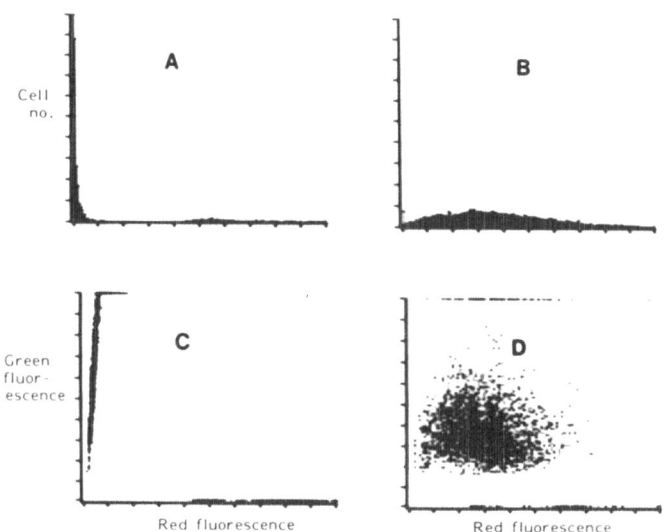

Fig. 9. Electroporation of a cell line derived from rat mammary cells. PI was added before electroporation and FDA afterwards. Light scatter was used to select the cells from any other particles that might be present, and red *vs.* green fluorescence was displayed. For amplification, see text. (Samples prepared by Dr. M.J. O'Hare and Mr. G. Lilley.)

Cells were electroporated in the presence of PI; those cells which were successfully porated took up the dye. They were then incubated briefly at 37°. FDA was added and the cells incubated at room temperature for 5 min. The laser was tuned to 488 nm, forward and orthogonal light scatter was used to select the cells (as distinct from any other particles in the sample), and green *vs.* red fluorescence was displayed (Fig. 9). Cells which had had their membrane punctured fluoresced red; if the membrane had then re-sealed they also fluoresced green. Only red fluorescence was shown by dead cells, and only green by cells which were not porated.

CONCLUDING REMARKS

This article has outlined some of the main applications of flow cytometry and shown a few examples which it is hoped might stimulate the reader's interest. More detailed treatments of the subject will be found in two recently published books [8, 9].

Acknowledgements

This work was supported by grants from the Cancer Research Campaign and the Medical Research Council. I thank Mr. P.R. Imrie for his technical assistance with the flow cytometry.

References

1. Roberts, J.J., Friedlos, F., Scott, D., Ormerod, M.G. & Rawlings, C.J. (1986) *Mutation Res. 166*, 169-181.
2. Gray, J.W. & Mayall, B.H., eds. (1985) *Monoclonal Antibodies Against Bromouridine, Cytometry 6*, no. 6.
3. Carrano, A.V., Gray, J.W., Moore, D.H., Minkler, H.J., Mayall, B.H. & Van Dilla, M.A. (1979) *Proc. Nat. Acad. Sci. 76*, 1382-1384.
4. Shapiro, H.M. (1980) *Cytometry 1*, 301-312.
5. Tsien, R.Y., Pozzan, T. & Rink, T.J. (1982) *Nature 295*, 68-71.
6. Valet, G., Raffael, A., Moroder, L., Wunsch, E. & Ruhensroth-Bauer, G. (1981) *Naturwiss. 68*, 265-266.
7. Watson, J.V. (1980) *Cytometry 1*, 143-148.
8. Van Dilla, M.A., Dean, P.N., Laerum, O.D. & Melamed, M.R., eds. (1985) *Flow Cytometry: Instrumentation and Data Analysis*, Academic Press, London, 288 pp.
9. Shapiro, H.M. (1985) *Practical Flow Cytometry*, Alan R. Liss, New York, 294 pp.

#B-4

INTERACTION OF PAF-ACETHER WITH THE HUMAN PLATELET: RECEPTOR BINDING AND PHARMACOLOGY

*D.P. Tuffin and   P.J. Wade

*ICI plc, Pharmaceuticals Division
Mereside, Alderley Park
Macclesfield, Cheshire SK10 4TG, U.K.

Interest in the ether-linked phospholipid inflammatory agonist 'platelet-activating factor' has grown rapidly in recent years. Different cell populations possess specific, high-affinity binding sites for this agent. However, since the measurement of binding sites does not necessarily reflect the existence of true, physiological receptor sites, a critical analysis of binding kinetics, receptor occupancy and response coupling is essential. Here we describe progress in the development of a relatively simple human platelet binding assay which can be used to study the kinetics of platelet-activating factor binding to washed human platelets and for routine evaluation of putative antagonists.

Platelet-activating factor (Paf-acether, 1-*O*-alkyl-2-acetyl-sn-glycero-3-phosphocholine) is an extremely potent platelet agonist capable of inducing shape change, aggregation and a secretory response in guinea pig, rabbit and human platelets at nanomolar concentrations. However, the platelet aggregatory response to Paf-acether appears to be independent of either arachidonate cyclo-oxygenase pathway activation or secretion of ADP and 5-HT, since Paf-acether-induced aggregation *in vitro* and *in vivo* remains unchanged in the presence of specific inhibitors [1, 2].

In a broader context the pharmacology and pathology of the ether lipid mediators is currently of great interest. Apart from its potent stimulatory effect on platelets, Paf-acether has now been shown to cause aggregation and secretion in polymorphonuclear leucocytes [3], leucocyte chemotaxis [4], increased vascular permeability [5], and contraction of ileal [6] or bronchial [7] smooth muscle. Consequently, Paf-acether has been implicated as a potential mediator in a number of clinical disorders including thrombosis, inflammatory conditions, psoriasis and bronchial asthma.

Most cell types are now thought to possess specific, high-affinity receptor populations for Paf-acether, and it is widely believed that its pharmacological actions are a direct consequence of receptor occupancy. Specific high-affinity binding sites have been demonstrated by ourselves and others in platelets [8-11], neutrophils [12] and human lung tissue [13]. Moreover, Paf-acether receptor subtypes may exist amongst different tissues and species [14]. Clearly receptor binding studies are of great importance in understanding the pharmacology of this powerful endogenous mediator and, moreover, are a valuable tool in the search for selective and potent Paf-acether receptor antagonists.

BINDING ASSAY METHODOLOGY

The binding assay was carried out using twice washed human platelets re-suspended in pH 7.4 buffer: 0.15 M NaCl/0.15 M Tris-HCl (9:1 by vol.). Briefly, blood samples were obtained by antecubital venepuncture from male volunteers who had taken no aspirin-containing or other medications for at least 14 days. The butterfly cannula and syringe contained 0.1 vol. (relative to the blood) of 3.2% Na_3 citrate as anti-coagulant. The blood was gently mixed by inversion, transferred to 50 ml polycarbonate tubes, and centrifuged (200 **g**, 20 min) to obtain a platelet-rich plasma (PRP) supernatant. In some experiments the prostaglandin PGI_2 was added (final concn. 300 ng/ml) to assist in preventing platelet activation; it had no effect on binding parameters obtained.

The PRP was further centrifuged at 400 **g** for 20 min to obtain an initial platelet pellet which, to obtain the working preparation, was twice washed with pH 7.4 buffer (0.15 M NaCl, 0.15 M Tris-HCl, 77 mM EDTA; 45:4:1 by vol.). If the suspension of platelets in this medium were visibly contaminated with erythrocytes, a further brief centrifugation at only 100 **g** was performed. The final suspensions, with 3.0 × 10^8 cells/ml, had <1% contamination with leucocytes by light microscopy.

Some 3 years ago, when we commenced assay development, tritiated Paf-acether had recently become commercially available, but consisted of a crude mixture of C-16, C-18 and C-20 ether-linked derivatives with low overall specific activity, which hindered interpretation of results. Consequently, a sample of R-1-O-[^{3}H]octadecyl-2-acetyl-sn-glycero-3-phosphoryl choline, [^{3}H]-C-18-Paf-acether, was prepared [8] and used as ligand throughout the studies described here. It appeared >96.5% pure by TLC on silica gel plates developed in $CHCl_3$/

CH$_3$OH/NH$_4$OH (70:35:7 by vol.) and had high specific activity, ~100 Ci/mmol; it was stored at -28°. The decision to synthesize a quantity of pure ligand *de novo* was vindicated by subsequent experiments, in which the binding affinity of closely related Paf-acether derivatives was shown to differ appreciably.

The binding assay procedure was performed throughout in capped 1.5 ml Eppendorf tubes. Paf-acether (labelled and unlabelled) was dissolved for use in a standard Tris assay buffer (above) containing 0.25% bovine serum albumin (BSA; without this carrier, Paf-acether deteriorates rapidly in aqueous solution). Bioassay experiments on human platelets in the aggregometer demonstrated that the ligand remained stable in this medium for at least 4-6 h. The assay procedure is as follows.-

Incubation mixture: (1) 300 µl assay buffer; (2) 50 µl buffer containing 0.25% BSA for total binding, **or** 50 µl unlabelled C-18-Paf-acether (363 nM) for non-specific binding; (3) 100 µl washed platelet suspension (3×10^7 platelets). **Procedure.-**
- Equilibrate (2 min, 37°); add [^{3}H]-C-18-Paf-acether ligand (a 50 µl aliquot containing 2-200 nCi; 0.04-4.0 nM final concn.).
- Vortex-mix, 2 sec; incubate at 37°, 15 min, and centrifuge (10,000 **g**, 2 min).
- Discard *supernatant.* and wash *platelet pellet* with 250 µl BSA-buffer; re-centrifuge (10,000 **g**, 2 min).
- Discard *supernatant,* dissolve *washed pellet* in 50 µl 10% SDS.
- To each tube add 1.5 ml Bioflour scintillant. Thoroughly vortex-mix. Count the bound [^{3}H]-Paf-acether.

Total binding and non-specific binding (excess unlabelled *R*-C-18-Paf-acether present, 363 nM) were determined in parallel incubates; the difference represented specific binding. Non-specific trapping of the ligand in the pellet was found to be consistently low (0.52 ±0.02%; n = 4) when measured by estimating the pellet's retention of [^{14}C]sucrose (Amersham International).

THE NONSPECIFIC BINDING COMPONENT

Paf-acether is a highly lipophilic molecule. Hence the level of non-specific binding observed (i.e. Paf-acether non-specifically adsorbed into the lipid bilayers of the platelet membrane) is likely to be high. Fig. 1 shows data on the characteristics of [^{3}H]-C-18-Paf-acether binding to washed human platelets at equilibrium. Evidently the non-specific binding component measured is high, comprising 5-10% of the [^{3}H]-C-18-Paf-acether added, and ranging from 40% to 80% of the total [^{3}H]-C-18-Paf-acether bound depending on ligand concentration (% non-specific binding being necessarily greater at the higher ligand concentrations due to saturation of the specific binding sites).

Fig. 1. Equilibrium binding
of [^{3}H]-1-O-octadecyl Paf-acether
to washed human platelets at 37°:
specific (●), non-specific (■)
and total (□) binding. Each
data point is the mean of
determinations in triplicate,
and the results are typical
of curves achieved in >20
experiments.

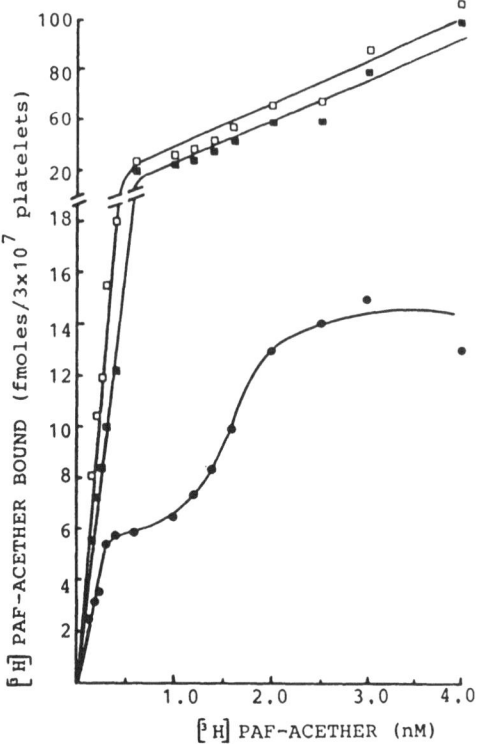

Different techniques were explored in an attempt to lower the
level of non-specific binding evident in the assay. These included
washing the platelet pellet by re-suspending in a small volume (250 µl)
of BSA-buffer, spinning the platelets through sucrose (2% w/v) or
silicone oil. Interestingly, the data demonstrated that merely washing
the pellet, whilst not altering the specific binding component,
achieved the best reduction of non-specific binding; nevertheless
this remained high in most assays, and hence each treatment was usually
carried out in triplicate.

In the final assay protocol (above) the incubation was terminated
by rapid centrifugation at 10,000 **g**, such that the platelets were
pelleted within 10 sec. Filtration is commonly used to separate
bound ligand from free; but we disfavoured this alternative, for
two reasons. We anticipated significant adsorption – possibly
enhancing error in estimating specific binding – of [^{3}H-Paf-acether
onto commonly used types of filter [15] (in our method, adsorption
onto the assay tube surface or into the pellet was <1% of total binding).
The second contra-indication was operational impracticability:
reproducibility in the additional manipulation would have called
for routine handling of a very large number of assay tubes.

CHARACTERIZATION OF [³H]-PAF-ACETHER BINDING AT EQUILIBRIUM

Having established the above methodology, we attempted to characterize the nature of the human platelet [³H]-Paf-acether binding observed. In Vol. 13, this series [16], Laduron outlined 8 criteria which should be satisfied before a binding site may be considered a true receptor site. In summary these comprise the following.-

(i)   Displacement of ligand by drug.
(ii)   Correlation between drug affinity *in vitro* and pharmacological potency *in vitro* and *in vivo*.
(iii) A regional distribution of binding sites.
(iv)   A subcellular distribution of binding sites.
(v)   Stereospecificity of ligand binding.
(vi)   Saturability of specific ligand binding.
(vii) Reversible nature of ligand binding.
(viii) Evidence of specific high-affinity binding.

A true receptor site will, on binding with specific ligand, trigger a measurable physiological response under the appropriate experimental conditions, i.e. the response is dependent on, and proportional to, receptor occupancy by ligand. Since binding sites in other systems have been recognized which do not produce a physiological response, termed 'acceptor sites', despite demonstrating high affinity, reversibility, saturability and stereospecificity, it is particularly important to satisfy criteria (i) and (ii) if the binding sites studied are to be classified as true receptor sites.

We have previously shown that in terms of the physiological responses in the platelet comprising shape change and aggregation, threshold responses to Paf-acether correlate closely with measured occupancy of binding sites [8]. However, although several rather diverse drug types affect [³H]-Paf-acether binding [17-19], it is only relatively recently that specific antagonists have been described [20, 21]. Consequently, we continue at present to refer to binding sites rather than receptor sites. The characteristics of [³H]-Paf-acether binding to washed human platelets determined thus far are described below.

(a) Existence of high and low affinity binding sites

As Fig. 1 illustrates, the binding of [³H]-Paf-acether to washed human platelets proved saturable in our hands at ligand concentrations of ~2.0 nM. However, the curve is clearly biphasic, with an initial binding plateau between 0.5 and 1.0 nM, indicating that there may be two binding sites.

Scatchard analysis of the binding curve data reflects two apparent binding sites (Fig. 2): a high-affinity site with a K_D value of 0.259 ±0.033 nM (245 ±30 sites per platelet) and a second lower affinity site with K_D 9.22 ±1.17 nM (1616 ±165 sites per platelet) [8]. Note

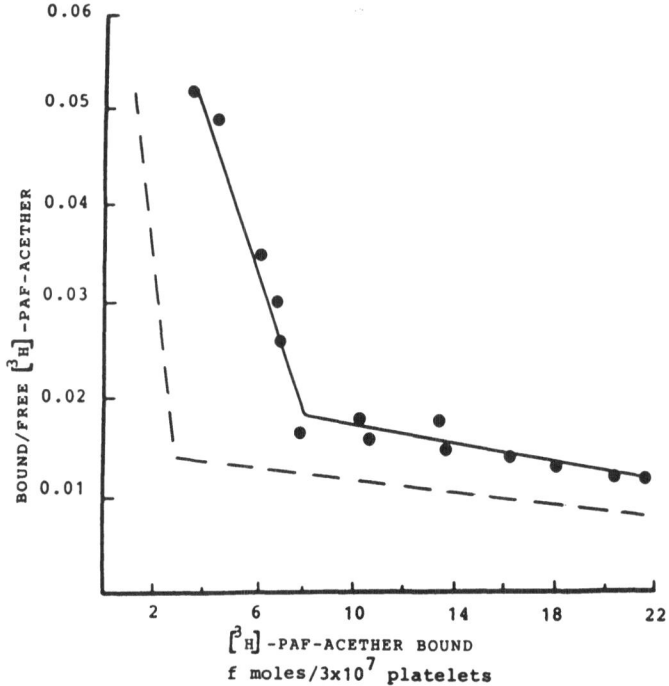

Fig. 2. Scatchard analysis (data from a single representative experiment) of [^{3}H]-Paf-acether binding to washed human platelets at 37°. Each data point is the mean of triplicate determinations. Broken line, $- - -$: the data transformed by computer (see text).

that Fig. 2 shows, besides raw data, computer-analyzed curves (NIH Scatpack MED 41 Program [22]) where the analysis takes account of a 2-phase binding system such that binding observed at each data point was corrected for either a low or a high affinity component: i.e. even at the lower ligand concentrations, a small proportion of the ligand is bound to the low-affinity sites. Consequently, the resultant transformed Scatchard plot does not run parallel to the raw data plot, since at those data points closest to the change from high-affinity to low-affinity binding, there is a progressively greater correction factor [cf. 23].

In Table 1 our results are summarized and compared with those from other laboratories. The data derived for high-affinity binding compare favourably with data reported elsewhere [11, 18], as does the presence of a two-component [^{3}H]-Paf-acether binding curve with human platelets. Interestingly, in three instances [10, 12, 24] the kinetic data differed, such that the K_D value and number of specific 'high' affinity binding sites per platelet more closely resembles our values for low affinity binding at 37° [cf. 25].

Table 1. Binding of [^{3}H]-Paf-acether to whole human platelet preparations. Separation methods (SEP.): cent, centrifugation; filt, filtration; **n** signifies no. of binding sites. The data for K_D and No. of sites (per platelet) are means ± S.E.M.

Material; assay temp.	SEP.	n	K_D, nM	No. of sites	Ref.
Human washed platelets; 37°	Cent.	2	0.259±0.033 9.22 ±1.17	245 ±30 1616 ±165	Tuffin et al. [8]
ditto; 4°	Cent.	1	1.20 ±0.05	300 ±10	see text (c)
Human gel-filtered platelets; 22°	Filt.	1	18.86±4.82[#]	242 ±64	Kloprogge & Akkerman [11]
ditto; 25°	Cent.	2	0.10 ±0.02 0.55 ±0.07	320 ±38 733 ±49	Chesney & Pifer [18]
Human washed platelets; 20°	Cent.	2	37.0 ±5.3 + ∞	1399 ±188 binding	Valone et al. [10]
ditto; 22°	Filt.	2	20.1 ±3.6 + ∞	1577 ±266 binding	Bussolino et al. [12]
ditto; 37°	Cent.	2	1.58 ±0.21 + ∞	1983 ±226 binding	Inarrea et al. [24]

∞ signifies infinite (non-receptor?) binding.   [#] K_a ($\times 10^9 M^{-1}$)

(b) Association and dissociation of binding

In physiological terms Paf-acether is a rapidly acting agonist on most cells and tissues. If, then, the binding sites observed are true receptors, association with the binding sites should be equally rapid. This was indeed found for association of Paf-acether (0.4 M) to the high-affinity site at 37°, half-maximal binding being achieved within 30-45 sec and maximum binding by 2 min (Fig. 3).

Dissociation of binding was determined by incubating the platelets with [^{3}H]-Paf-acether for 20 min at 37°, removing 2 aliquots (to estimate total binding), and adding excess unlabelled Paf-acether (363 mM). Further aliquots were removed at 10 sec intervals and the decrease in binding determined. For both high- and low-affinity binding (Fig. 4), an initially rapid dissociation was observed (t½ <10 sec) which was followed by a second phase of slower dissociation (t½ 71±11 and 70±10 sec for the respective binding sites). The data imply that the initially rapid dissociation component was from the low-affinity binding site and the subsequent slower dissociation rate from the high-affinity site. These findings are, however, in

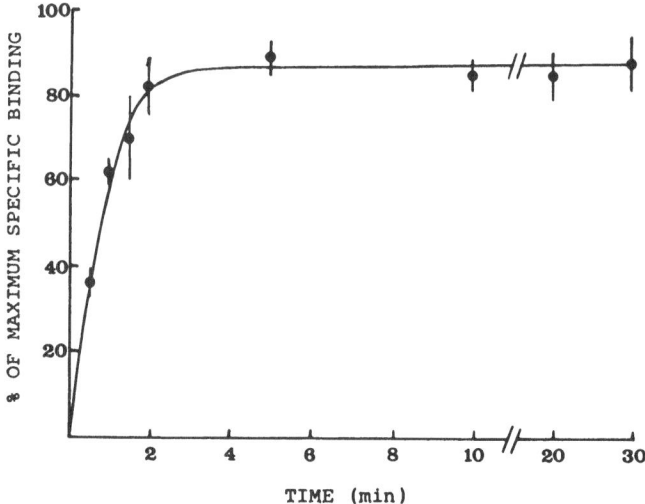

Fig. 3. Association of [³H]-Paf-acether (0.4 nM) to washed human platelets at 37°. Each data point is mean ±S.E.M. from 6 experiments.

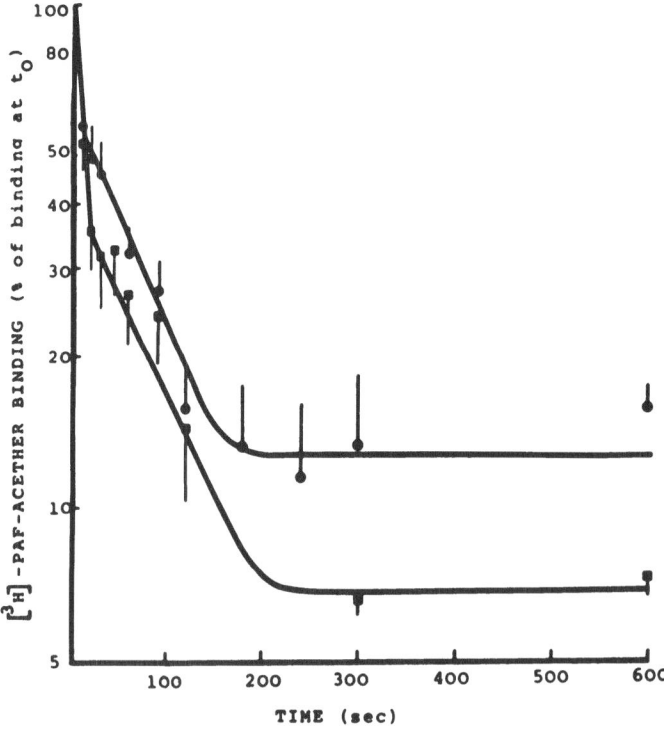

Fig. 4. As for Fig. 3, but dissociation; ligand concentrations corresponded to high (0.4 nM, ●) and low (2.0 nM, ■) affinity binding respectively. Each data point is mean ±S.E.M. from 7 experiments.

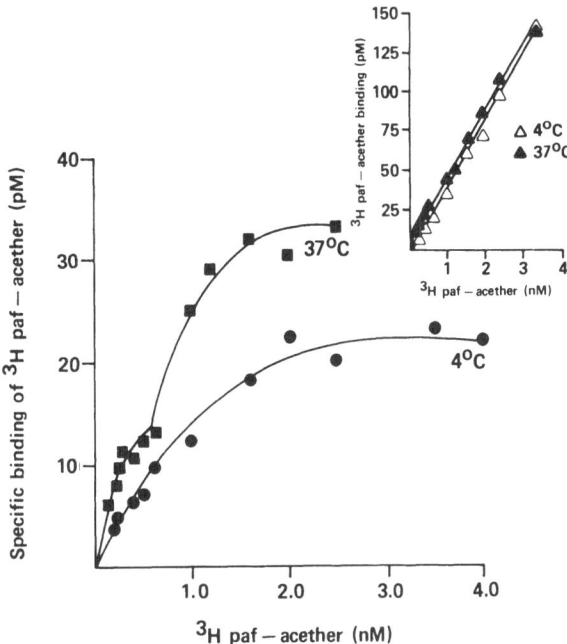

Fig. 5. Comparison of specific and non-specific binding of Paf-acether to washed human platelets at 4° and 37° (results represen-tative of 6 separate expts.). Each data point is the mean of triplicate determinations at each ligand concentration.

direct contrast to those of Kloprogge & Akkerman [11] who observed a significantly slower rate of dissociation, with only 20% of bound ligand dissociated after 5 min. These authors have suggested that at higher [^{3}H]-Paf-acether ligand concentrations ($\geq$1.0 nM) irrever-sible ligand-receptor complexes may appear; but in our system we havé seen no evidence of this.

(c) Kinetics of [^{3}H]-Paf-acether binding at 4°

Assay with incubation at 4° instead of 37° was next investigated. Specific binding was found to be similar to that at 37°, in that it was clearly dose-dependent and saturable (~2.5 nM). Maximum binding was lower than at 37°, but the non-specific binding component was markedly reduced (Fig. 5). There appeared to be no inflection in the ligand binding curve at 4°, and Scatchard analysis revealed only a single binding site whose kinetics were similar to the high-affinity site at 37° (K_D = 1.20 ±0.05 nM with 300 ±10 binding sites per platelet, Table 1).

Furthermore, at 4° the rate of association of ligand with the binding site was rather slower; not till 40 min was maximum binding

Fig. 6. Dissociation of [³H]-
Paf-acether from washed human
platelet binding sites at 4°.
Each data point is the mean
±S.E.M. of data from 6 separate
experiments with different
donor platelets.

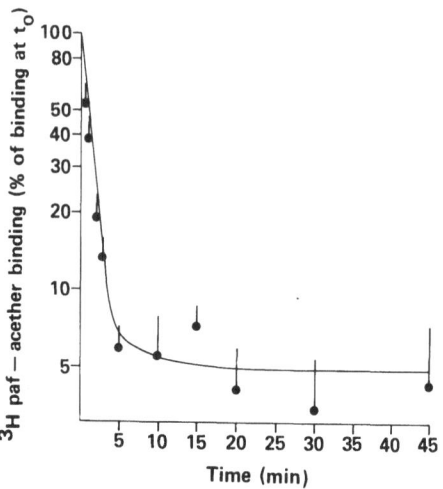

attained.   Moreover, only one dissociation rate was evident at 4°
(Fig. 6), though the rate observed (t½ 53.0 ±5.3 sec) did not differ
significantly from that of the high-affinity site at 37° (t½ 70.0
±10.0 sec).   Although caution is advisable in interpreting these
data, since even association to the high-affinity site did not attain
equilibrium until at least 40 min had elapsed at 4°, it does appear
that the lower-affinity binding observed at 37° is an energy-dependent
phenomenon.

(d) Physiological role of the low-affinity site

    For Paf-acether binding to its platelet receptor site, little
is currently known about the nature of receptor processing, and the
question of whether binding of this ligand causes a change of some
kind in receptor conformation (itself an implicit requirement for
stimulus-response coupling) remains unclear.   As we have shown, the
incubation of [³H]-Paf-acether with washed whole human platelets
at 37° exhibits complex binding kinetics which, on Scatchard analysis,
are indicative of two binding-site populations of high and low affinity
respectively.   Preliminary studies suggest that a change in cell
shape - the initial step in the Paf-acether-induced platelet aggregatory response - correlates with occupancy of the high-affinity binding
sites [8].   Thus, for the low-affinity site observed only at 37°
in our studies a possible explanation (1) is that it is merely a
site of non-specific membrane uptake and metabolism.   Alternatively,
(2) it may reflect true binding-site heterogeneity, (3) negative
site-site co-operativity within a single receptor class, or (4) may
even be a desensitized, lower-energy form of the high-affinity binding
site resulting from a change in receptor conformation [26].

    Clearly our data suggest that (1) is unlikely for the low-affinity
binding at ligand concentrations between 0.4 nM and 2.0 nM, since

Table 2. Inhibition of high-affinity ligand binding, and comparative aggregatory potency. K_D values were derived according to Cheng & Prusoff [27]. For the washed human platelet suspensions used (see text), shape change and aggregation were measured by standard turbidometric methods at 37° with stirring at 1000 rpm, after adding $CaCl_2$ (to 1.0 mM) and human fibrinogen (to 1.0 mg/ml). The means (±S.E.M.) are from at least 4 determinations; n.d. = not done.

Test (values are nM)	R–C–18–Paf	R–C–16–Paf	(R–S)–C–18–Paf	R–C–18–lyso–Paf
Binding assay, IC_{50}	0.48±0.03	0.08 ±0.01	1.06 ±0.19	>300
Binding assay, K_D	0.19 ±0.01	0.03 ±0.003	0.42 ±0.07	>118
Shape change, EC_{50}	0.5–1.0	0.1–0.3	n.d.	>1000
Aggregation, EC_{50}	10.0 ±2.7	2.7 ±0.7	n.d.	>1000

it is saturable and since dissociation of bound ligand from the low-affinity sites remains evident even following incubation for 20 min at equilibrium. Accordingly, the favoured explanation is either (2), (3) or (4), and additional detailed kinetic analyses are needed to address this issue. It may be relevant that in similar test systems, investigation of the binding of other lipophilic platelet ligands such as thrombin [28], prostaglandin I_2 [29] and the thromboxane mimic $[^3H]$-15(S)-9,11-epoxymethano-PGH_2 [30] to human platelets has also resulted in similar multi-component binding isotherms.

(e) Stereospecificity of the high–affinity binding site

Binding of (R)-$[^3H]$-C-18-Paf-acether (0.4 M) to this site was inhibited in a competitive, concentration-dependent fashion by unlabelled (R)-C-18-Paf-acether on an equimolar basis (Table 2). (R)-C-16-Paf-acether was 6-fold more potent, in agreement with aggregatory potency in our laboratory. As expected, since (S)-C-18-Paf-acether is pharmacologically inactive [31], the racemate (R-S)-C-18-Paf-acether exhibited half the potency of the pure R-isomer, whilst (R)-C-18-lyso-Paf-acether was inactive at the concentrations tested.

Thus, the binding measured in the assay appeared to be stereo-specific, and the inhibitory activity of the analogues correlated closely with aggregatory potency on washed human platelets *in vitro* (Table 2) and *in vivo* [2]. Previously we have shown that the pharmacological potency (platelet shape change and aggregation) of 1-O-octadecyl-Paf-acether correlates closely with occupancy of the platelet high-affinity binding sites. Others [11] have reported that 0.5 nM Paf-acether (i.e. sufficient to saturate platelet high-affinity binding sites) causes aggregation, exposure of the platelet fibrinogen receptor, and alterations in turnover of polyphosphatidylinositides.

CONCLUSIONS

[^{3}H]-Paf-acether binds to washed human platelet preparations in a dose-dependent and saturable manner. Scatchard analysis of binding at 37° revealed a 2-component system, manifesting high-affinity (K_D 0.259 ±0.033 nM) and low-affinity (9.22 ±1.17 nM) sites, whereas at 4° the data revealed only a single binding site similar in kinetics to the high-affinity binding observed at 37°. Binding to the high-affinity site was independent of buffer calcium concentration [17], inhibited on an equimolar basis by unlabelled R-1-O-octadecyl-Paf-acether, but remained unchanged in the presence of lyso-Paf-acether. Association to the high-affinity site at 37° was rapid, maximal at 2 min, and remained constant for at least 20 min. Association to the single site observed at 4° was rather slower, maximal binding being achieved only after incubation for 40 min. At 37° an initially rapid dissociation from both high- and low-affinity sites was observed ($t\frac{1}{2}$ = 10 sec) which was followed in both cases by a second phase of slower dissociation ($t\frac{1}{2}$ = 70 sec). In contrast, at 40° only one rate of dissociation was evident, but was not significantly different ($t\frac{1}{2}$ = 53 sec) from that of the high-affinity site at 37°.

Studies to date suggest that the high-affinity Paf-acether binding site is a true physiological receptor on the washed platelet; but evaluation of potent and selective antagonists is necessary before the criteria outlined by Laduron [16] are fully satisfied. Further studies are also required to clarify the nature and role of the low-affinity binding site apparent at 37° and to reconcile the differences in binding parameters observed between laboratories.

Acknowledgements

The authors would like to thank G. Healey for advice on statistical analysis and K. Tuckman for typing the manuscript.

References

1. Chignard, M., Le Couedic, J.P., Tence, M., Vargaftig, B.B. & Beneviste, J. (1979) *Nature 279*, 799-800.
2. Honey, A.C., Lad, N. & Tuffin, D.P. (1986) *Thromb. Haemostas. 56*, 80-85.
3. Henson, P.M. (1970) *J. Exp. Med. 131*, 287-304.
4. Shaw, J.O., Pinckard, R.N., Ferrigni, K.S., McManus, L.M. & Hanahan, D.J. (1981) *J. Immunol. 127*, 1250-1255.
5. Humphrey, D.M., McManus, L.M., Satouchi, K., Hanahan, D.J. & Pinckard, R.N. (1982) *Lab. Invest. 46*, 422-427.
6. Findley, S.R., Lichtenstein, L.M., Hanahan, D.J. & Pinckard, R.N. (1982) *Am. J. Physiol. 241*, C130-133.
7. Stimler, N.P., Bloor, C.M., Hugli, T.E., Wykle, R.L., McCall, C.E. & O'Flaherty, J.T. (1981) *Am. J. Path. 105*, 64-69.

8. Tuffin, D.P., Davey, P., Dyer, R.L., Lunt, D.O. & Wade, P.J.
    (1985) in *Mechanisms of Stimulus-Coupling in Platelets*
    (Westwick, J., Scully, M.F., MacIntyre, D.E. & Kakkar, V.V.,
    eds.), Plenum, New York, pp. 83-96.
9. Shaw, J.O. & Henson, P.M. (1980) *Am. J. Path. 98*, 791-810.
10. Valone, F.H., Coles, E., Reinhold, V.R. & Goetzl, E.J. (1982)
    J. Immunol. 129, 1637-1641.
11. Kloprogge, E. & Akkerman, J.W.B. (1984) *Biochem. J. 223*, 901-909.
12. Bussolino, F., Tetta, C. & Camussi, G. (1984) *Agents & Actions
    15*, 15-17.
13. Hwang, S.B., Lan, M.H. & Shen, T.Y. (1985) *Biochem. Biophys.
    Res. Comm. 128*, 972-979.
14. Lambrecht, G. & Parnham, M.J. (1986) *Br. J. Pharmac. 87*, 287-
    289.
15. Birdsall, N.J.M. & Hulme, E.C. (1984) in *Investigation of
    Membrane-Located Receptors* [Vol. 13, this series] (Reid, E.,
    Cook, G.M.W. & Morré, D.J., eds.), Plenum, New York, pp. 7-15.
16. Laduron, P.M. (1984) as for 15., pp. 31-46.
17. Wade, P.J., Lunt, D.O., Lad, N., Tuffin, D.P. & McCullagh, K.G.
    (1986) *Thromb. Res. 41*, 251-262.
18. Chesney, C.M. & Pifer, D.D. (1985) *Thromb. Haemostas. 54*, 177.
19. Godfroid, J.J., Robin, J.P. & Braquet, P. (1985) *Prostaglandins
    30*, 686.
20. Shen, T.Y., Hwang, S.B., Chang, M.N., Doebber, T.W., Lam, M.T.,
    Wu, M.S., Wang, X., Han, G.Q. & Li, R.Z. (1985) *Proc. Nat.
    Acad. Sci. 82*, 672-676.
21. Braquet, P. (1985) *Prostaglandins 30*, 687.
22. Creede, C., Jaffe, M.L., Munson, P.J. & Rodbard, D. (1980)
    Scatpack Computer Program (Med 41), U.S. Biomedical Computing
    Technology Information Center, Washington DC.
23. Valone, F.H. (1983) *J. Pharmacologie 14 (Suppl. 1)*, 19.
24. Inarrea, P., Gomez-Cambronero, J., Nieto, M. & Crespo, M.S.
    (1984) *Eur. J. Pharmac. 105*, 309-315.
25. Hwang, S.B., Lee, C., Cheah, M.J. & Shen, T.Y. (1983)
    Biochemistry 22, 4756-4763.
26. Wiland, G.A. & Molinoff, P.B. (1981) *Life Sci. 29*, 313-330.
27. Cheng, Y-C. & Prusoff, W.H. (1973) *Biochem. Pharmacol. 22*,
    3099-3108.
28. Tollefsen, D.M., Feagler, J.R. & Majerus, P.W. (1974)
    J. Biol. Chem. 249, 2646-2651.
29. Siegl, A.M., Smith, J.B., Silver, M.J., Nicolau, K.C. & Ahern, D.
    (1979) *J. Clin. Invest. 63*, 215-220.
30. Armstrong, R.A., Jones, R.L. & Wilson, N.H. (1983) *Br. J.
    Pharmac. 79*, 953-964.
31. Wykle, R.L., Miller, C.H., Lewis, J.C., Schmitt, J.D.,
    Smith, J.A., Surles, J.R., Piantadosi, C. & O'Flaherty, J.T.
    (1981) *Biochem. Biophys. Res. Comm. 100*, 1651-1658.

#NC(B)

COMMENTS related to

APPROACHES DEPENDING ON SURFACE DISTINCTIONS

Comments on #B-1: J.M. Graham et al. - FREE-FLOW ELECTROPHORESIS

Remarks by J.K. McDonald.- The use of FFE to purify proteins is facilitated by introduction into the chamber at a pH that is identical with the pI of the protein of interest, which will then move directly through the chamber with a minimum of diffusion while impurities will migrate away. However, I should warn that cacodylate buffers should not be used, because the electrode decomposition product of the dimethylarsinate ion is arsene gas, which is extremely poisonous. **Replies to M.G. Ormerod.**- Separation of ~10^9 cells/h can be managed by FFE; cells labelled with an Ab having a charged group have been successfully separated.

Comments invited by Senior Editor on -

USEFULNESS OF FFE - W.H. Evans (NIMR, London)

Supplementing the examples in #B-1, the technique has also been applied in subcellular fractionation studies to analyze the complexity of endosome and Golgi fractions. Thus, Debanne et al. [1] and Evans & Flint [2] showed that rat-liver endosome vesicles containing various radio-iodinated ligands recovered at low density (1.11-1.13) by fractionation in sucrose and Nycodenz gradients (cf. #E-1) could be resolved further in the Bender & Hobein FFE apparatus (Elphor, VaP5). The technique also shows promise in separating rat-liver Golgi apparatus fractions introduced after treatment with amylases followed by mild physical disruption [3].

1. Debanne, M., Evans, W.H., Flint, N. & Regoeczi, E. (1982) *Nature* *298*, 398-400.
2. Evans, W.H. & Flint, N. (1985) *Biochem. J. 232*, 25-32.
3. Morré, D.J., Morré, D.M. & Heidrich, H.S. (1983) *J. Cell Biol.* *31*, 263-274.

Comments on #B-3: M.G. Ormerod - FLOW CYTOMETRY

Remarks by N. Crawford.- It would be useful, e.g. to distinguish cells at rest and at different levels of activation, to be able to set different thresholds for the same fluorochrome or one threshold and sort cells above and below; e.g. Quin-2 might be used to sort out cells differing in cytosol [Ca^{2+}], say 50-100, 200-500 and >500 μM. Thereby one would have cells at rest and at different levels of activation. **Ormerod** felt that a feasibility study was indeed warranted.

Comment on #**B-4**: D.P. Tuffin & P. Wade – PLATELET BINDING STUDIES

Question to Tuffin by W.H. Evans.– Your binding studies were mostly carried out at 4°. Are platelets active in endocytosis? The different binding results at 37° might be caused, in part, by endocytic uptake of Paf-acether. **Reply.**– We have no evidence, but it seems unlikely that the binding data are affected by endocytosis.

Supplementary 'alertings' by Senior Editor

Techniques described in arts. #B-1 and #B-3 have also featured earlier in this series (Vols. 2-4, 6, 8 & 11; list facing title p.), and the whole of Vol. 13 is devoted to receptors (cf. #B-4). For rat-liver organelles, counter-current partition has been compared with gradient centrifugation by Morris, W.B., Smith, G.D. & Peters, T.J. (1986) *Biochem. Pharmacol. 35*, 2187-2191.

Section #C

PLASMA MEMBRANE AND CYTOSKELETON

#C-1

MEMBRANE-MEDIATED CYTOTOXICITY: MEASUREMENT OF CHANGES AND THEIR PREVENTION BY DIVALENT CATIONS

C.A. Pasternak

Department of Biochemistry
St. George's Hospital Medical School
Cranmer Terrace, London SW17 0RE, U.K.

Many haemolytic agents damage non-red cells without causing lysis. This is because such cells are able (a) to withstand colloid osmotic swelling better than red cells and (b) possess mechanisms for membrane repair absent in red cells. We have measured membrane damage by monitoring (i) transmembrane potential, and leakage of (ii) monovalent cations, (iii) low mol. wt. phosphorylated metabolites, and (iv) proteins such as lactate dehydrogenase (LDH). Under conditions of low agent:cell ratio, little (iv) leakage of proteins occurs.

Cytotoxic agents that are haemolytic include certain paramyxo-viruses and other enveloped viruses at low pH, animal toxins such as the bee venom protein melittin, bacterial toxins such as the α and δ toxins of S. aureus, the membrane attack complex of complement (C5b-8; 9n) and other endogenous proteins, besides detergents and other synthetic compounds. When membrane damage induced by each of these agents is measured as above, a common pattern of action emerges: characteristically the leakage is inhibited by the presence of divalent cations, with relative potency $Zn^{2+} > Ca^{2+} > Mg^{2+}$ in each case. Thus agents quite dissimilar in structure induce similar types of membrane damage in susceptible cells. Because Ca^{2+} and Zn^{2+} inhibit membrane changes at concentrations of <0.1 and <1 mM respectively, clearly these two cations can protect cells against certain cytotoxic agents in vivo.

Membrane damage is the cause of cytotoxicity in many pathological situations. Specific examples are given elsewhere in this book, e.g. by P. Druet et al. (#F-7; derangements in toxin-induced auto-immune nephropathies), R. Wattiaux et al. (#A-1; ischaemic effects on subcellular structures) and B.A. Fowler (#A-9; metal-induced cell injury). In this article, cytotoxic agents that damage the plasma membrane (p.m.) by making it more leaky, and thus lead to certain cell surface diseases' [1], will be discussed.

The agents under discussion are all haemolytic, i.e. they lyse erythrocytes. However, other cells are merely damaged, not lysed, as shown below. Because the pathophysiological consequence of attack by a haemolytic agent is generally *not* the lysis of erythrocytes, the membrane-mediated damage here described has considerable clinical relevance. The types of agent include the following:
- viruses, notably those of the paramyxovirus family, e.g. Sendai virus or the related Newcastle disease virus of veterinary importance, and other enveloped viruses at low pH;
- bacterial toxins such as those (e.g. α) secreted by pathogenic strains of *Staphylococcus aureus*;
- animal toxins such as melittin (the active ingredient of bee stings);
- endogenous agents such as the membrane attack complex (C5b-9) of complement;
- synthetic agents such as polylysine or detergents (e.g. Triton X-100, Lubrol, Nonidet).
Each of the agents listed induces membrane lesions that are sensitive to inhibition by Ca^{2+} or Zn^{2+} [2-5].

MEASUREMENT OF PERMEABILITY CHANGES

In order to assess the degree to which the p.m. has become leaky, it is important to use appropriate measurements. The smallest type of lesion is best detected by measurement of transmembrane potential; the largest by measurement of protein leakage. Between these two extremes are lesions of diam. ~1-2 nm that may be detected by leakage of sucrose or low mol. wt. phosphorylated metabolic intermediates such as sugar phosphates, phosphorylcholine or inositol phosphates, mononucleotides, and so forth. It must be stressed that the size of the lesion is an arbitrary estimate only.- Many small lesions may be operationally equivalent to a few large lesions; likewise, because it is likely that lesions 'flicker' between an 'open' and a 'closed' state, a small lesion that is open for a long time may be operationally equivalent to a large lesion that is open briefly; moreover, most agents induce lesions that are heterogeneous in size. Table 1 summarizes our methodology.

The rationale behind these methods is as follows. Membrane potential is measured by the use of readily permeant ionic dyes which equilibrate in accordance with the Nernst equation: if the potential is negative inside, cations will accumulate, but anions will remain outside. If there is a change in potential, e.g. depolarization, cations will be expelled and anions will enter. Because the fluorescence (and absorbance) of the ions depends on their milieu (the cellular milieu being more hydrophobic than that of the surrounding medium), assay of the fluorescence (and absorbance) of a cell suspension can be calibrated to give a good measure of the membrane potential [6].

Table 1. Methodology for measuring permeability changes in cells affected by haemolytic and other cytotoxic agents.

Estimated lesion size[*] (pore diam.)	Most relevant parameter to measure	Most suitable type of measurement[†]	Ref.
< ~0.7 nm	Membrane potential	Absorbance changes with oxonol-V	6-9
~0.7-1 nm	Leakage of K^+ and Na^+	Atomic absorption of cell extracts	9, 10
~1-2 nm	Uptake of α-methyl-glucoside or sucrose	^3H or ^{14}C in cells after exposure to the labelled mol.	11-13
	Leakage of phosphorylated metabolites (dGlc[⊗], choline)	^3H or ^{14}C in cells and medium after pre-incubation with the labelled mol.	9,10, 14
> ~2 nm	Leakage of small proteins	LDH in cells and medium	9, 10

[*] See text for proviso regarding size estimates. [⊗] 2-deoxyglucose.
[†] The types are those routinely used in the author's laboratory and are easy to perform; but many other methods have been described.

Generally it is the cationic dyes, e.g. the cyanines [15], that are employed. Because the potential across the mitochondrial membrane normally exceeds that across the p.m. (e.g. -180 mV compared with -60 mV), such dyes accumulate within mitochondria and do not properly register p.m. potential unless mitochondrial inhibitors are present. The same is true of radioactively labelled cations such as $[^3H]$tetraphenylphosphonium, which in addition have a reduced permeability across membranes, and hence their response time is slower.

Anions, on the other hand, can be used without recourse to mitochondrial inhibitors, and we favour one such dye, namely oxonol-V. This has the additional advantage of a substantially altered absorption spectrum in a hydrophobic milieu, so that the differential absorption at two wavelengths – $A_{630-590}$ – can be used instead of fluorescence for measurement of membrane potential; this obviates many complicating factors such as cell density, etc. [8]. Although the use of dyes has been criticized on some grounds, e.g. not discriminating properly between surface potential (zeta potential) and transmembrane potential, we have not found this to be so. In our hands oxonol-V has proved to be an effective indicator of transmembrane potential in monolayer cultures [8] as well as in cell suspensions [7, 16], and yields essentially the same results as electrophysiological recordings (unpublished experiments with N.G. Bryne and W.A. Large).

The measurement of cellular cations by atomic absorption is relatively straightforward, and yields more direct data than, e.g., the uptake or leakage of tracers such as $^{86}Rb^+$ or $^{36}Cl^-$. Cells in suspension are best separated from surrounding medium by spinning through oil, cells in monolayer by rapid repeated washes. Because many cells have very active electroneutral exchange mechanisms for ions such as Na^+ or K^+, there is a tendency to lose Na^+ when washing or merely diluting in, e.g., choline-containing medium [17], and hence we now favour adding a cell suspension directly to oil and spinning immediately [9].

Leakage of phosphorylated metabolic intermediates is best assessed by use of isotopically labelled compounds such as 2-deoxy-glucose (dGlc) or choline at μM concentrations (Fig. 1). Both compounds enter cells by specific uptake mechanisms and are then rapidly phosphorylated by active kinases, such that the intracellular concentration of free dGlc or choline remains low [18]. Further metabolism of dGlc-6-P is very limited, while that of phosphorylcholine is slow (because of a relatively large pool of unlabelled phosphorylcholine, as a result of which radioactivity from labelled choline is effectively trapped as labelled phosphorylcholine [17]). Hence the radioactivity of cells exposed to ^3H- or ^{14}C-labelled dGlc or choline is an excellent indicator of the low mol. wt. 'acid-soluble' pool, which leaks out of cells only when the p.m. is damaged.

Other methods, in which adenine nucleotides [17], cyclic AMP, cyclic GMP, inositol phosphates, and so forth, are measured by radio-active or enzymic means could also be employed. In each case, measurement of medium as well as of cellular content improves the accuracy of assessing leakage.

For measurement of protein leakage, we have favoured enzymic assay of an abundant cytoplasmic enzyme such as LDH [9, 10]. Again other methods are available, including leakage of radiolabelled proteins from cells following pre-incubation with [^{35}S]methionine or other amino acid [10], as well as the uptake of toxic proteins from the medium [19]. The latter method can be very sensitive, as a single molecule of, e.g., diphtheria toxin is sufficient to kill a cell [20].

INCREASED PERMEABILITY WITHOUT LYSIS (Criteria: Table 2)

Contrary to popular belief, haemolytic agents do not necessarily lyse non-erythroid cells. In fact at low concentrations of agent, the normal outcome is cell damage, not cell lysis. The reason is that most haemolytic agents do not induce pores > ~2 nm across the p.m. of affected cells, and hence proteins such as haemoglobin do not leak out. Erythrocytes lyse because the entry of Na^+ into cells, which is accompanied by Cl^- and H_2O, causes osmotic swelling, which results in rupture of the membrane and loss of haemoglobin. Erythrocytes treated with haemolytic agent in the presence of

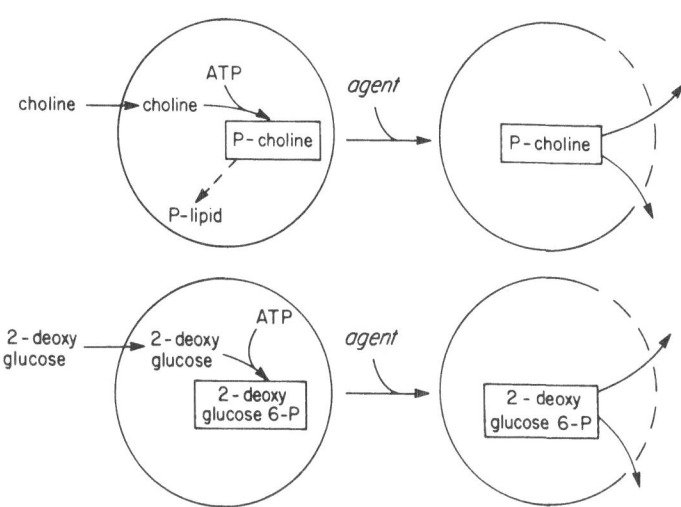

Fig. 1. Use of labelled choline or dGlc to assess leakage of phosphorylated metabolites induced by cytotoxic agents.

Table 2. Criteria for establishing lack of lysis in cells affected by haemolytic agents. The refs. are largely to work from the author's laboratory; similar techniques in respect of other agents have been widely used elsewhere, and the listings are merely illustrative.

Technique	Agent	Ref.
Demonstration of pore size < ~2 nm		
Lack of trypan blue uptake	Sendai virus	24
Lack of ^{51}Cr leakage from pre-labelled cells	Sendai virus	24
Lack of LDH leakage	Sendai virus, *S. aureus* α-toxin, melittin, complement, polylysine, Triton X-100	5, 9, 10
Prevention of haemolysis by osmotically active cpds., e.g. albumin	Sendai virus, complement	25, 26
Maintenance, or stimulation, of normal cell function		
0$_2$ uptake	Sendai virus, melittin, complement	5
Hormone secretion	Melittin	27–31
Prostaglandin release	Complement	32
Prostacyclin release	*S. aureus* α-toxin	33
Recovery of cell function		
Membrane potential, excitability, heart beat, etc.	Sendai virus	34
Membrane potential	Complement	35

osmotically active non-permeant compounds such as albumin or sucrose
do not swell or lyse, and retain their haemoglobin.   Non-erythroid
cells do not lyse even in the absence of albumin or sucrose, though
such compounds do afford protection at high concentrations of haemolytic agent.   The reason for the lack of lysis in non-erythroid cells
is that these possess mechanisms for volume expansion [21] and membrane
repair [22, 23], absent in erythrocytes.

The above argument applies to those agents that haemolyse by
forming pores or other physical distortion of p.m. structure, but
not to those agents, typified by phospholipases present in snake
venom or by detergents at high concentrations, that literally dissolve
the membrane away.   Such agents lyse non-erythroid cells as well
as erythrocytes.   Hence it is important to assess the extent of lysis,
as distinct from an increased permeability, caused by a particular
haemolytic or otherwise cytotoxic agent.   Table 2 lists some of the
criteria that have been employed.

PROTECTION BY DIVALENT CATIONS

Some of the agents listed at the start of this article have
been examined in detail, and - despite marked differences in their
structure or in the exact steps leading up to the generation of membrane
damage - have been found to cause lesions that have many properties
in common [2-5].   (1) The induction of leakage appears to be sequential,
in that pores of ~0.7 nm open before pores of ~1.0 nm, and so on.
(2) The induction shows positive cooperativity with respect to amount
of agent, and when two agents are present together they exhibit synergy.
(3) Leakage is depressed in media of low ionic strength, and it can
be prevented entirely by the presence of divalent cations such as
Ca^{2+} or Zn^{2+} (Fig. 2).   Observation (3) is of potential clinical
interest, especially as concentrations of Ca^{2+} in the physiological
range are effective [36].

Because of current interest in situations where the intracellular
concentration of Ca^{2+} becomes elevated, and because it is known that
a rise in cytoplasmic Ca^{2+} blocks leakage across intercellular
junctions (pore size ~1-2 nm) [37], it is important to determine
whether Ca^{2+} and Zn^{2+} act intracellularly or extracellularly.   Table 3
shows some of the methods that have been applied.   In fact no one
method gives a clear-cut result.   One reason for this is that a pore
of > ~0.7 nm itself allows entry of added protective divalent cation
such as Ca^{2+} [38-41] or Zn^{2+}.

Measurement of radioactive cation level in cells exposed to
different concentrations of extracellular radioactive divalent cation
[39] has obvious difficulties of interpretation; measurement of
divalent cation level in cells by dyes such as quin 2 under conditions
of increasing extracellular divalent cation has not yet been carried
out (Method 1 in Table 3).   Addition of A23187 to cells exposed to

Ca^{2+} plus agent (method 2) indicates an extracellular site, in that agent-induced leakage from A23187-treated cells is increased, not decreased. Experiments with planar lipid membranes (Method 3) are difficult to interpret, in that divalent cations act *cis* or *trans* to agent (though in different manners). Experiments with membrane vesicles exposed to divalent cation inside or outside (Method 4), on the other hand, are more straightforward to interpret. Correlation with altered surface charge (Method 5) indicates an extracellular site, but this is circumstantial evidence only, since the measurement of surface charge has so far been carried out only in the absence of haemolytic agent. The same is true of Method 6; thus, a concentration of Ca^{2+} that inhibits Na^+/K^+-ATPase in cell extracts has no effect on it in intact cells. On balance, it would appear from Table 3 that divalent cations act extracellularly, but this conclusion must remain tentative for the time being.

CONCLUSION

The methodologies that have been described in this article have been employed to study the types of lesion formed across the p.m. of susceptible, non-erythroid, cells by certain cytotoxic agents. All agents so far tested (p. 190) induce relatively small lesions (at low concentrations of agent) that are in every case sensitive to inhibition by certain divalent cations. They are therefore of use as biological tools for the permeabilization of cells and, like electric shock [45,46], have been used to introduce ions or molecules such as Ca^{2+} and Ca^{2+} chelates [47-51], GTP analogues [52, 53], antibiotics [54] and proteins [19] into cells for specific studies.

Table 3. Criteria for establishing the site at which divalent cations block cell leakage. See text for further details. Only examples are given. Agents shown ( ); S.v. = Sendai virus, *S.a.* = *S. aureus*.

Technique (& agent)	Conclusion	Ref.
1. Measurement of intracellular divalent cation level (S.v.)	Inconclusive	39
2. Manipulation of intracellular cation level with ionophores (S.v., melittin, *S.a.* α-toxin, polylysine	Extracellular	4
3. Efficacy of divalent cation *cis* or *trans* to agent in planar lipid membranes (*S.a.* α-toxin)	Both sides	42
4. Efficacy of divalent cation inside or outside membrane vesicles	Extracellular	43
5. Correlation with altered surface charge	Extracellular	44
6. Comparison of effect of divalent cation on p.m. enzymes in cell extracts *vs.* intact cells	Extracellular	⊗

⊗ C.L. Bashford, J.M. Graham & C.A. Pasternak, unpublished work.

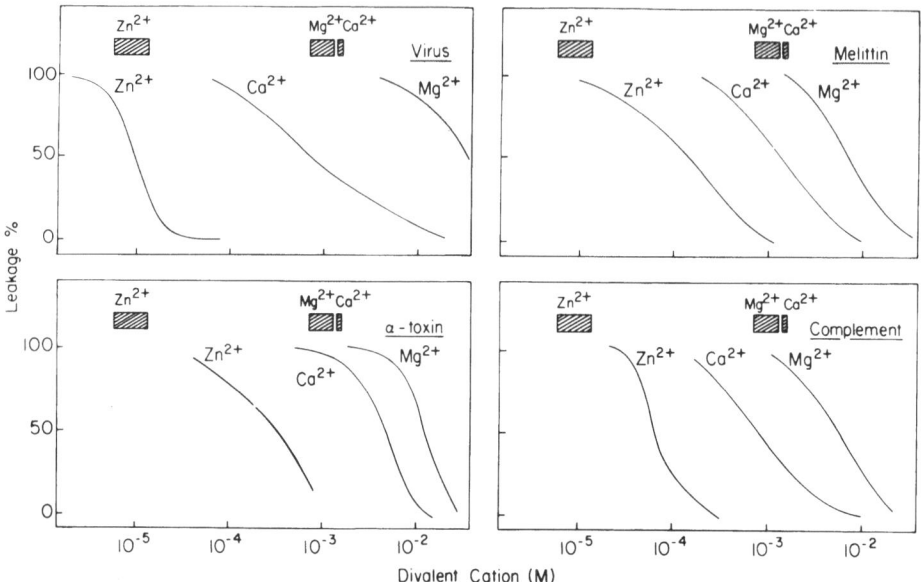

Fig. 2. Inhibition of leakage by divalent cations. The effect of Zn^{2+}, Ca^{2+} and Mg^{2+} on leakage of phosphorylated metabolites from Lettré cells induced by Sendai virus, melittin, *S. aureus* α-toxin or activated complement is indicated. In each case the cells (pre-labelled with [³H]choline; Fig. 1) were exposed to an amount of agent such as to give ~50% of maximum leakage in 10-60 min. The results have been pooled from a number of experiments performed during 1984-86. Shaded bars indicate the approx. levels of the free cations in human plasma. *From ref. [36], by permission.*

The methodologies may now be used to investigate lesions formed by other cytotoxic, pore-forming agents such as the cytolytic granular proteins from cytotoxic lymphocytes [55] or eosinophils [56], the cytolytic protein 'amoebapore' from *Entamoeba histolytica* [57], high concentrations of cations such as Hg^{2+}, Cu^{2+}, Cd^{2+} or Zn^{2+} that themselves induce leakage [36], and many other agents.

Acknowledgements

The results presented above owe much to fruitful collaboration with C.L. Bashford, J.M. Graham, K.J. Micklem and other colleagues over the past 15 years. They were made possible by generous support from the Medical Research Council, the Cancer Research Campaign, the Wellcome Trust and the Cell Surface Research Fund.

References

1. Griffin, G. & Pasternak, C.A. (1982) *Clin. Sci. 63*, 1-9.
2. Bashford, C.L., Alder, G.M., Patel, K. & Pasternak, C.A. (1984) *Biosci. Rep. 4*, 797-805.

3. Pasternak, C.A., Bashford, C.L. & Micklem, K.J. (1985) *J. Biosci. (Suppl.) 8*, 273-291.

4. Pasternak, C.A., Alder, G.M., Bashford, C.L., Buckley, C.L., Micklem, K.J. & Patel, K. (1985) in *The Molecular Basis of Movement through Membranes* (Pasternak, C.A. & Quinn, P.J., eds.), *Biochem. Soc. Symp. 50*, 247-264.

5. Bashford, C.L., Alder, G.M., Menestrina, G., Micklem, K.J., Murphy, J.J. & Pasternak, C.A. (1986) *J. Biol. Chem. 261*, 9300-9308.

6. Bashford, C.L. (1981) *Biosci. Rep. 1*, 183-196.

7. Bashford, C.L. & Pasternak, C.A. (1984) *J. Membr. Biol. 79*, 275-284.

8. Bashford, C.L., Alder, G.M., Gray, M.A., Micklem, K.J., Taylor, C.C., Turek, P.J. & Pasternak, C.A. (1985) *J. Cellular Physiol. 123*, 326-336.

9. Bashford, C.L., Micklem, K.J. & Pasternak, C.A. (1985) *Biochim. Biophys. Acta 814*, 247-255.

10. Poste, G. & Pasternak, C.A. (1978) in *Cell Surface Reviews 5* (Poste, G. & Nicolson, G.L., eds.), North-Holland, Amsterdam, pp. 306-349.

11. Pasternak, C.A. & Micklem, K.J. (1974) *Biochem. J. 144*, 593-595.

12. Sims, P.J. & Lauf, P.K. (1980) *J. Immunol. 125*, 2117-2125.

13. Sims, P.J. (1981) *Proc. Nat. Acad. Sci. 78*, 1838-1842.

14. Pasternak, C.A. & Micklem, K.J. (1973) *J. Membr. Biol. 14*, 293-303.

15. Sims, P.J., Waggoner, A.S., Wang, C.H. & Hoffman, J.F. (1974) *Biochemistry 13*, 3315-3330.

16. Bashford, C.L. & Pasternak, C.A. (1985) *Biochim. Biophys. Acta 817*, 174-180.

17. Impraim, C.C., Foster, K.A., Micklem, K.J. & Pasternak, C.A. (1980) *Biochem. J. 186*, 847-860.

18. Wohlheuter, R.M. & Plagemann, P.G.W. (1980) *Int. Rev. Cytol. 64*, 171-240.

19. Yamaizumi, M., Uchida, T. & Okada, Y. (1979) *Virol. 95*, 218-221.

20. Yamaizumi, M., Mekada, E., Uchida, T. & Okada, Y. (1978) *Cell 15*, 245-250.

21. Knutton, S., Jackson, D., Graham, J.M., Micklem, K.J. & Pasternak, C.A. (1976) *Nature 262*, 52-54.

22. Ramm, L.E., Whitlow, M.B., Koski, C.L., Shin, M.L. & Mayer, M.M. (1983) *J. Immunol. 131*, 1411-1415.

23. Campbell, A.K. & Morgan, B.P. (1985) *Nature 317*, 164-166.

24. Pasternak, C.A. & Micklem, K.J. (1973) *J. Membr. Biol. 14*, 293-303.

25. Green, H., Barrow, P. & Goldberg, B. (1959) *J. Exp. Med. 110*, 699-713.

26. Knutton, S. & Pasternak, C.A. (1979) *Trends Biochem. Sci. 4*, 220-223.

27. Heisler, S., Reisine, T.D., Hood, U.Y.H. & Axelrod, J. (1982) *Proc. Nat. Acad. Sci. 79*, 6502-6506.

28. Dunlop, M., Christianthou, A., Fletcher, A., Veroni, M., Woodman, P. & Larkins, R. (1984) *Biochim. Biophys. Acta 801*, 10-15.

29. Grandison, L. (1984) *Endocrinol. 114*, 1-7.

30. Martin, T.F.J. & Kowalchyk, J.A. (1984) *Endocrinol.115*, 1517-1526.

31. Morgan, N.G. & Montague, W. (1984) *Biosci. Rep. 4*, 665-671.

32. Hansch, G.M., Seitz, M., Martinotti, G., Betz, M., Rauterberg, E.W. & Gemser, D. (1984) *J. Immunol. 133*, 2145-2150.

33. Suttorp, N., Seeger,W., Dewein, E., Bhakdhi,S.& Roka, L. (1985) *Am. J. Physiol. 248*, C127-C134.

34. Forda, S.R., Gillies, G., Kelly, J.S., Micklem, K.J. & Pasternak, C.A. (1982) *Neurosci. Lett. 29*, 237-242.

35. Wiedmer, J. & Sims, P.J. (1985) *J. Biol. Chem. 260*, 8014-8019.

36. Pasternak, C.A. (1987) *BioEssays 6*, 14-18.

37. Rose, B. & Lowenstein, W.R. (1975) *Nature 254*, 250-252.

38. Volsky, D.J. & Loyter, A. (1978) *J. Cell Biol. 78*, 465-479.

39. Impraim, C.C., Micklem, K.J. & Pasternak, C.A. (1979) *Biochem. Pharmacol. 28*, 1963-1969.

40. Campbell, A.K., Daw, R.A., Hallett, M.B. & Luzio, J.P. (1981) *Biochem. J. 194*, 551-560.

41. Hallet, M.B., Fuchs, P. & Campbell, A.K. (1982) *Biochem. J. 206*, 671-674.

42. Menestrina, G. (1986) *J. Membr.Biol. 90*, 177-190.

43. Schulz, I. & Heil, K. (1979) *J. Membr. Biol. 46*, 41-70.

44. Graham, J.M., Pasternak, C.A., Wilson, R.B.J., Alder, G.M. & Bashford, C.L. (1986) in *Electrophoresis* (Dunn, M.J., ed.), VCH Verlagsgesellschaft, Wernheim), pp. 77-85.

45. Knight, D.E. & Scrutton, M.C. (1986) *Biochem. J. 234*, 497-506.

46. Baker, P.F. & Knight, D.E. (1978) *Nature 276*, 620-622.

47. Miller, M.R., Castellot, J.J. & Pardee, A.B. (1978) *Biochemistry 17*, 1073-1080.

48. Dunn, L.A. & Holz, R.W. (1983) *J. Biol. Chem. 258*, 4989-4993.

49. Wilson, S.P. & Kirshner, N. (1983) *J. Biol. Chem. 258*, 4994-5000.

50. Gomperts, B.D., Baldwin, J.M. & Micklem, K.J. (1983) *Biochem. J. 210*, 737-745.

51. Ahnert-Hillger, G., Bhaki, S. & Gratzl, M. (1985) *J. Biol. Chem. 260*, 12730-12734.

52. Gray, M.A., Austin, S.A., Clemens, M.J., Rodrigues, L. & Pasternak, C.A. (1983) *J. Gen. Virol. 64*, 2631-2640.

53. Barrowman, M.M., Cockroft, S. & Gomperts, B.D. (1986) *Nature 319*, 504-507.

54. Cameron, J.M., Clemens, M.J., Gray, M.A., Menzies, D.E., Mills, B.J., Warren, A.P. & Pasternak, C.A. (1986) *Virology 155*, 534-544.

55. Henkart, P.A. (1985) *Annu. Rev. Immunol. 3*, 31-58.

56. Gleich, G.J., Loegering, K.G., Mann, K.G., & Macdonald, J.E. (1976) *J. Clin. Invest 57*, 633-640.

57. Gitler, C., Calef, E. & Rosenberg, I. (1984) *Phil. Trans. Roy. Soc. B.. 307*, 73-85.

#C-2

COMPLEMENT-MEDIATED PLASMA-MEMBRANE DAMAGE:
EFFECTS ON CELL PHYSIOLOGY, PROSPECTS OF CELL RECOVERY
AND IMPLICATIONS FOR DISEASE MECHANISMS

[1]J. Paul Luzio, [1]Arefaine Abraha, [1]Peter J. Richardson,
[⊗][1]Richard A. Daw, [2]Caroline A. Sewry, [3]B. Paul Morgan
and [3]Anthony K. Campbell

[1]Department of Clinical Biochemistry
University of Cambridge, Addenbrooke's Hospital
Hills Road, Cambridge CB2 2QR, U.K.

[2]Jerry Lewis Muscle   [3]Department of Medical
   Research Centre   Biochemistry, University of
Hammersmith Hospital   Wales College of Medicine
Ducane Road, London W12 OHS   Heath Park, Cardiff, CF4 4XN

The formation of the MAC on appropriate target-cell membranes
provides the principal mechanism of complement-mediated cell killing
and plays a role in defence against bacterial infection. However,
lesions may also be caused in the p.m. of host cells and can give
rise to sublytic cell damage which may be important in some autoimmune
disorders. The molecular events associated with MAC insertion into
target membranes have been investigated using MAb's to the terminal
complement components C8 and C9. It is proposed that sublytic complement attack on mammalian cells can cause metabolic damage as a result
of a rise in cytosolic Ca^{2+} concentration, which can trigger membrane
shedding from the cell surface which is required for cell recovery.*

The insertion of the terminal complement component C9 into
appropriate target membranes provides the principal mechanism of
complement-mediated cell killing [1] and may also be responsible
for sublytic cell damage important in some autoimmune disorders [2-4].

[⊗]now at: Regional Blood Transfusion Centre, Vincent Drive,
Edgbaston, Birmingham B15 2SG, U.K.

[*]Abbreviations: MAC, membrane attack complex; MAb, monoclonal antibody; p.m., plasma membrane; DTT, dithiothreitol.

There is increasing evidence that understanding the mechanism of complement-mediated immune attack will also provide insights into cell-mediated attack, since killer lymphocytes produce a protein responsible for cell killing that is closely related to C9 with common antigenic determinants [5, 6]. Although insertion of C9 into the MAC and resulting cell damage have been widely described, there remains controversy about the molecular events associated with C9-mediated membrane damage, the consequences of this damage, and the possibility of cell recovery from sublytic damage.

MONOCLONAL ANTIBODIES (MAb's) AS REAGENTS TO STUDY THE MAC

The MAC is made up of complement components C5b, C6, C7, C8 and C9. The study of C9 incorporation into the complex has been facilitated by the preparation of MAb's to different epitopes on C9 [7]. These MAb's were used as probes to select cDNA coding for C9 in an expression cloning system, thereby allowing the primary sequence of the C9 polypeptide chain and the MAb antigenic sites to be determined [8, 9]. The MAb's were also used to show that C9 becomes a trans-membrane protein when inserted into a target membrane [10] and were instrumental in developing a two-step molecular model for the incorporation of C9 into the MAC [11]. It is not yet clear whether the subsequent polymerization of 12-16 molecules of C9 in the membrane to form poly-C9 [12, 13] is essential to complement-mediated cell damage or whether monomeric C9 is sufficient [14].

Despite the absolute requirement for C8 in the membrane-bound C5b-8 complex to allow C9 binding, direct interaction in the membrane between C8 and C9 has not been demonstrated [15]. The C5b-8 complex is thought to catalyze the polymerization of C9 [13], with C8 participating directly in this process. Human C8 is a serum glycoprotein with an apparent M_r of 151,000 composed of 3 non-identical polypeptide chains - α (M_r 64,000), β (M_r 64,000 and γ (M_r 22,000). These subunits exist as a covalently linked $\alpha\gamma$ dimer non-covalently associated with the β subunit [16]. Solution binding studies using purified complement components have suggested that the β subunit mediates the binding of C8 to C5b-7, with the α subunit being capable of binding to C9 [17]. Recently we have prepared MAb's to C8 and used them to interfere with formation of the MAC.

MAb's to C8 were prepared by standard methods [18] as previously described for C9 [7] using purified C8 donated by Dr. A. Esser as antigen [19]. Of the 10 MAb's that were prepared, 6 were shown by immunoblotting to react with the α chain, 2 with the β chain and 2 with the γ chain (Fig. 1). MAb's to γ showed strong cross-reactivity with α, and MAb's to β weak cross-reactivity with α. No cross-reaction with β or γ was shown by MAb's to α. None of these MAb's interacted with C9, though in other studies Ab cross-reactions with C9, C8 and other components of the MAC have been observed [5, 20]. Epitope mapping of the MAb's was conducted using each Ab in solution phase to interfere

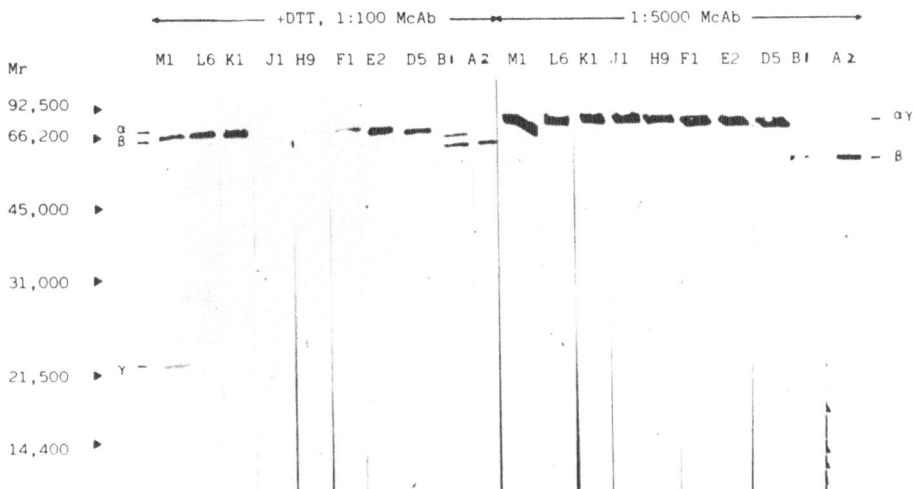

Fig. 1. Immunoblotting of C8 with MAb's. SDS-PAGE of C8 (1 µg/track) with or without reduction by DTT, and transfer to nitrocellulose, was followed by incubation with ascites fluid at concentrations indicated and detection with immunoperoxidase-labelled second Ab.

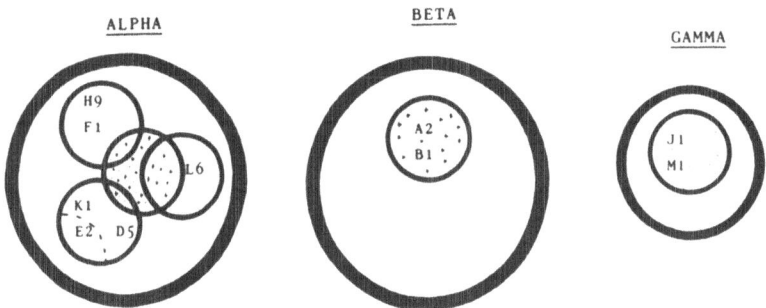

Fig. 2. Epitope map summarizing the binding of 10 mouse MAb's to human C8. The central ring in the α subunit representation indicates cross-reactions of MAb's to β and γ.

with the binding of ^{125}I-labelled C8 to each Ab on a solid phase [21]. These studies suggested that whereas the MAb's to γ and β bound to a single epitope on each subunit, the MAb's to α were directed to 3 sites (Fig. 2). When cross-reacting with α the MAb's to β interacted with all 3 sites and the MAb's to γ with two sites.

Two of the MAb's to C8, coded H9 (anti-α) and A2 (anti-β), were used to interfere with MAC formation. Pigeon erythrocyte ghosts containing the Ca^{2+}-sensitive photoprotein obelin were used as target cells, since a rapid C9-dependent influx of Ca^{2+} has been clearly demonstrated in these cells after formation of a MAC. The ghosts

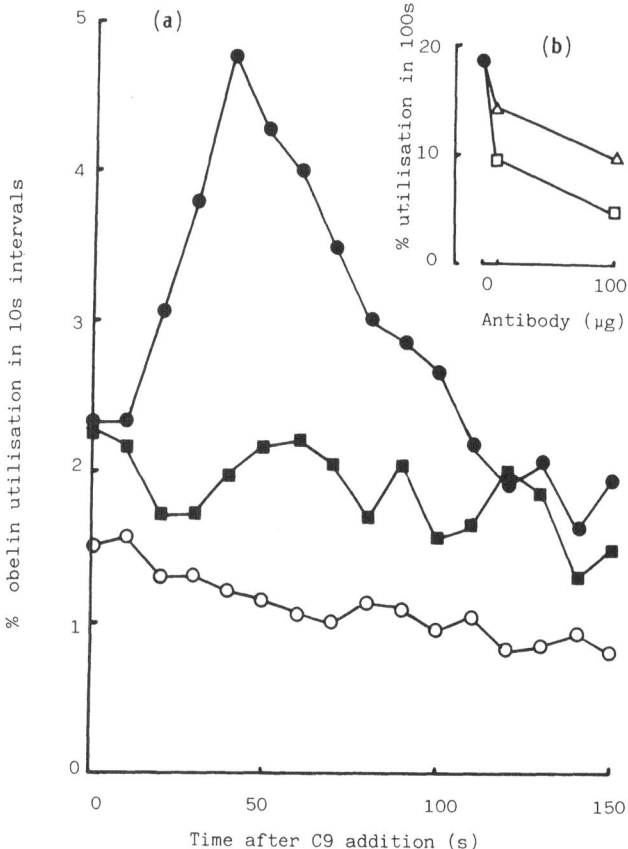

Fig. 3. Effects of anti-C8 MAb's on MAC formation. Pigeon erythro-cyte ghosts (10^8 in 100 μl) loaded with obelin as in [22] were incub-ated with guinea pig anti-erythrocyte serum (1:100; 2 min, 37°) then with C9-depleted human serum (prepared as in [7]; 1:20; 2 min, 37°), and washed after each incubation. MAb's to C8, or control IgG, were then added; after 2 min at 37°, C9 (50 ng, 50 μl) was added. Obelin luminescence in response to Ca^{2+} influx was measured at 10 sec inter-vals. (a) o, no C9 added; ●, control mouse IgG (100 μg X63); ■, anti-C8 IgG (100 μg A2). (b) ●, as in (a); □, A2 Ig; Δ, H9 Ig. % Obelin utilization in 100 sec recorded after subtracting no-C9 basal value.

were incubated with anti-cell Ab and C9-depleted serum [7] to form cell-surface C5b-8. MAb's to C8 were then added followed by sufficient C9 to cause a Ca^{2+} flux resulting in measurable but sub-maximal obelin consumption (Fig. 3a). Addition of both Ab's A2 and H9 before addition of C9 prevented the stimulation of obelin luminescence in a dose-dependent manner (Fig. 3b), suggesting that both α and β subunits of C8 play a role in the functional insertion of C9 into the MAC.

Ca^2 INFLUX AND METABOLIC DAMAGE

Experiments with pigeon erythrocyte ghosts containing the photo-protein obelin showed that the earliest measurable ionic event following formation of MAC's at the cell surface was a C9-dependent Ca^{2+} influx [22]. Comparison of obelin luminescence in erythrocyte ghosts with luminescence in standard Ca^{2+} solutions showed that in response to MAC formation cytosolic Ca^{2+} concentration rose to between 1 and 10 µM and plateaued at this level for at least 1-2 min. Such a cytosolic Ca^{2+} concentration is above the normal physiological range yet cannot be due to lysis since the extracellular Ca^{2+} concentration is 1 mM. It was, then, proposed that a rise in cytosolic Ca^{2+} concentration in the µM range would allow sublytic complement concentrations to cause pathological cell damage. Tbe observation that complement concentrations causing <20% haemolysis of pigeon erythrocytes inhibited adrenaline-sensitive adenylate cyclase by >80% [23] lent support to this proposal.

Further evidence for the role of cytosolic free Ca^{2+} in mediating C9-dependent complement-mediated sublytic cell damage came from experiments with polymorphonuclear leucocytes. These cells have a natural chemiluminescence due to the production of peroxide moieties, and the luminescence signal may be enhanced by adding luminol. A variety of agents can stimulate the signal including MAC formation [24]. The MAC effect was shown to be C9-dependent and occurred in the absence of cell lysis and in response to a rise in cytosolic Ca^{2+} concentration [25].

Evidence that MAC formation at the cell surface can result in sublytic damage playing a role in autoimmune disease has been obtained in immunohistochemical studies of muscle fibres from patients with myositis [3, 26]. Besides necrotic fibres showing intense staining with MAb's to C9, fibres were seen which although apparently normal by other light-microscopic histochemical criteria had punctate localizations of C9 and C8 on the cell surface (Fig. 4). Whilst MAC formation and membrane insertion of C9 have been suggested to be involved in the pathology of other autoimmune diseases [27-32] it is not yet known whether localization of C9 on the surface of morphologically undamaged cells is a widespread phenomenon in such a disease.

RECOVERY OF CELLS FROM SUBLYTIC COMPLEMENT DAMAGE

Although complement attack may lead to cell death as the result of a threshold event [33], the existence of sublytic cell damage implies that cells have protective mechanisms for the inactivation and/or removal of MAC's from their surface. A clue as to the nature of such mechanisms came from experiments with rat adipocytes. When these were attacked by Ab and complement, specific p.m. vesicles

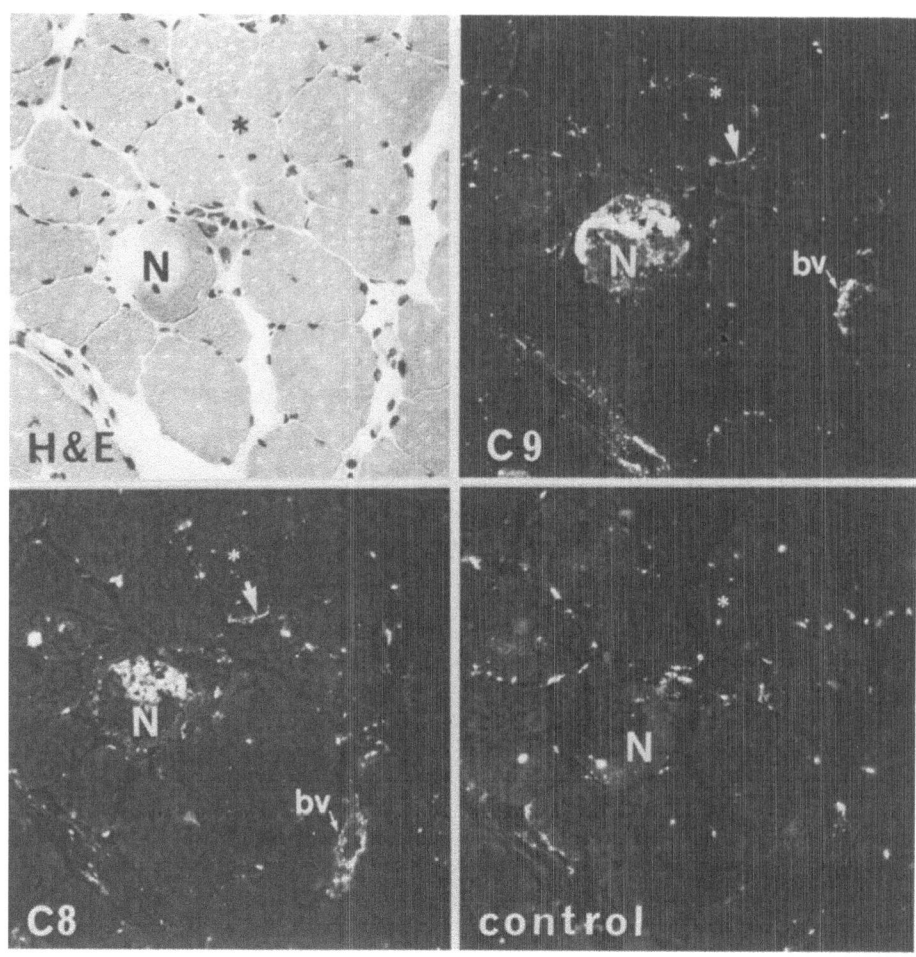

Fig. 4.   Immunolocalization of C8 and C9 on muscle fibres in a biopsy
from a patient with polymyositis.   Serial sections stained with H & E
and with MAb's to C9 (C9-47) and C8 (E2) using an immunofluorescent
biotin-streptavidin Texas Red technique.   A necrotic fibre (N) shows
high levels of C9 and C8 internally.   A non-necrotic fibre (*) shows
a peripheral area labelled with anti-C9 and -C8 *(arrows)*.   A blood
vessel (bv) also contains C9 and C8.   Only lipofuscin autofluores-
cence is seen in the control section stained without primary Ab. ×185.

were formed which could not be obtained in response to homogenization
or osmotic shock [34].   The vesicles were identified from the distribution of 5'-nucleotidase (a p.m. marker) in sucrose density gradients
(Fig. 5) and the presence of visible bands coincident with the two
densest peaks of 5'-nucleotidase activity.

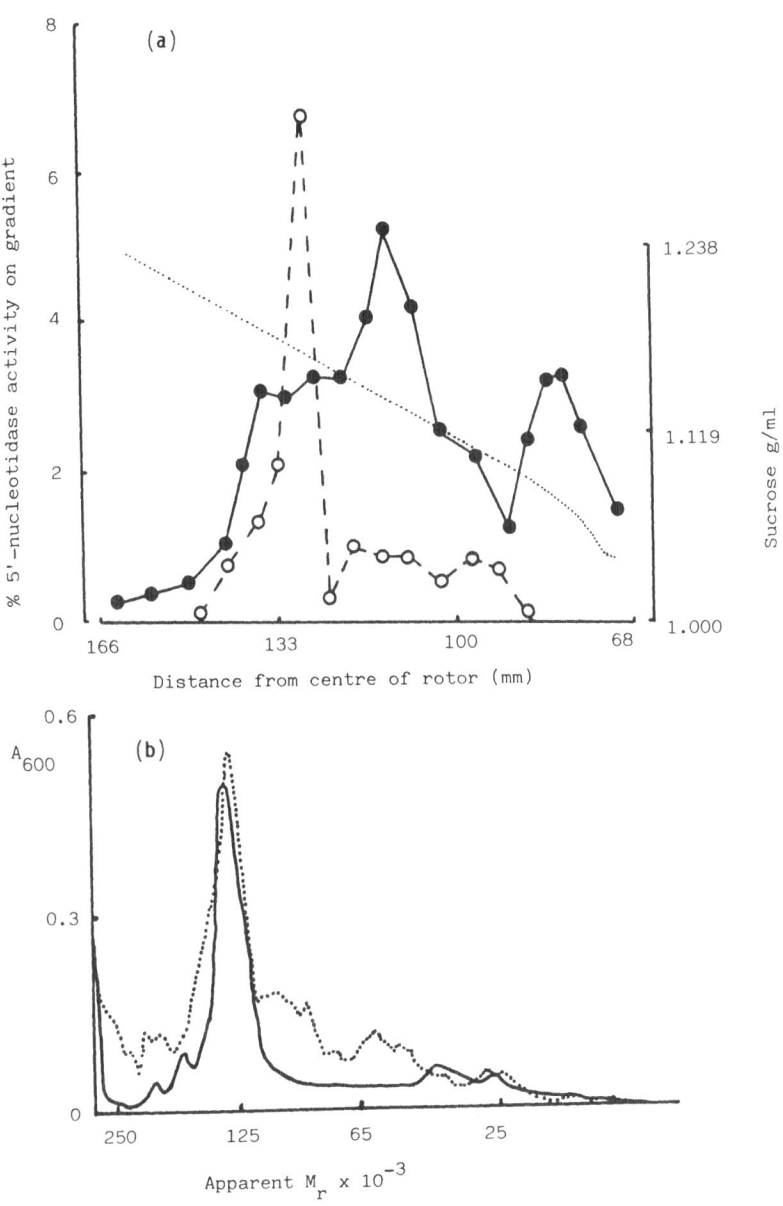

Fig. 5. Formation of p.m. vesicles from rat adipocytes.
(a) Distribution of 5'-nucleotidase on a sucrose gradient after
lysis of the cells in Krebs-Ringer/Hepes buffer, pH 7.4 (KRH) with
Ab and guinea pig complement (●) as in [34], or treatment of the
same number of adipocytes with 20 μM A23187 at 37° for 2 h (○).
(b) SDS-PAGE of the p.m. vesicles produced by A23187 (·········) or
complement (———) and isolated as in (a). Membrane protein loaded
60 μg/track. Positions of mol. wt. markers shown on the ordinate.

Interestingly, Ca^{2+} ionophore treatment of adipocytes produced a membrane vesicle band equivalent to the denser one occurring in response to complement attack [35]. Membranes from the complement- and ionophore-derived vesicle bands had a similar protein profile on SDS-PAGE and a restricted protein composition relative to adipocyte p.m. [34, 35] (Fig. 5) In these experiments no evidence was obtained for the vesicles being formed during sublytic complement attack on adipocytes, nor for the vesicles containing components of the MAC. However, similarities to membrane vesicles shed from the erythrocyte cell surface in response to the Ca^{2+} ionophore A23187 [36] were noted [34]. In view of these experiments and of the observation of a C9-dependent rise in cytosolic Ca^{2+} concentration during complement attack [22] it was suggested that membrane shedding from the cell surface might be a protection mechanism against complement attack in nucleated cells [2, 35].

The first evidence that membrane shedding could remove MAb's from the surface of nucleated cells during sublytic complement attack was obtained from experiments with polymorphonuclear leucocytes. Using a MAb to detect cell-surface C9 it was shown that MAC's were rapidly lost from the surface of these cells when incubated at 37° [37]. Further experiments [38, 39] showed that ~63% of surface MAC's were shed on small membrane vesicles, the remainder being endocytosed and degraded. Recovery of the cells from sublytic complement attack was mediated by a rise in cytosolic Ca^{2+} concentration as a result both of Ca^{2+} influx from outside the cells and of mobilization of intracellular Ca^{2+} stores [40]. Evidence for Ca^{2+}-dependent membrane shedding playing a role in removing MAC's from the cell surface and allowing recovery from complement attack has also been obtained with blood platelets [41] and Ehrlich ascites tumour cells [42].

Although, as might be expected, protective mechanisms against complement attack appear to be better developed in nucleated mammalian cells than in enucleate erythrocytes, it is possible to demonstrate complement-mediated membrane shedding in such erythrocytes. Both Ca^{2+} ionophore- and complement-mediated shedding may be followed by the release of cell surface enzymes (Fig. 6a), and shed vesicles may be isolated by density-gradient centrifugation (Fig. 6b). Shed vesicles produced in response to Ca^{2+}-ionophore treatment are known to have a restricted membrane protein composition relative to total p.m. [43]. The protein compositions of erythrocyte membrane vesicles shed in response to Ca^{2+}-ionophore and complement are similar (Fig. 6c). Whilst the production of membrane vesicles in response to complement attack can be demonstrated in erythrocytes, such vesicles are difficult to produce experimentally in these cells. Nonetheless the experiments indicate a common mechanism in mammalian cells, triggered by a rise in cytosolic Ca^{2+} concentration that results in membrane shedding.

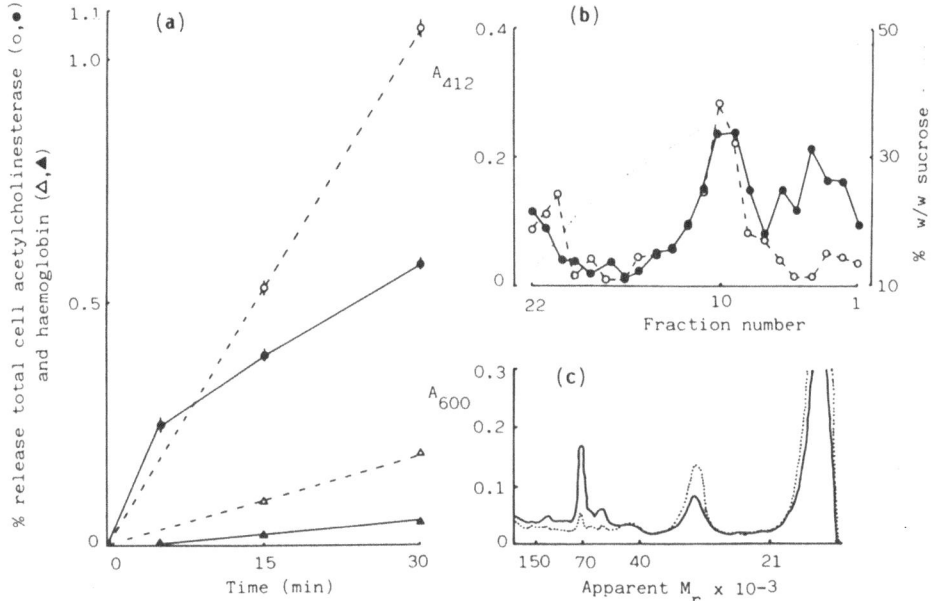

Fig. 6. Shedding of membrane vesicles from human erythrocytes.
(a) Differential release of erythrocyte acetylcholinesterase and Hb.
Erythrocytes (100 µl packed cells) were incubated in KRH (1 ml)
with 5 µM A23187 (o, Δ) at 37° for the times indicated. Alternatively, cells were incubated with 100 µl rabbit anti-erythrocyte
serum at 20° for 20 min, then washed, resuspended and incubated
with 10 µl fresh human serum at 37° for the times indicated (●,▲).
After centrifugation (15 sec, 8000 **g**), supernatants were assayed
for Hb (A_{540}; *triangles*) and acetylcholinesterase (*circles*) in the
presence of tetraisopropylpyrophosphoramide to inhibit serum cholin-
esterase [44].
(b) Identification of shed membrane vesicles produced as in (a) from
10 ml incubations, concentrated by centrifugation (30 min, 20,000 **g**),
resuspended in 2 ml KRH and loaded on 20 ml 23-50% w/w sucrose
gradients and centrifuged (18 h, 20,000 **g**): o, ionophore; ●, complement.
(c) SDS-PAGE of erythrocyte membrane vesicles produced by A23187
(⋯⋯⋯⋯) or complement (————) as in (a) and isolated as in (b).
The Hb peak in the ionophore vesicle traces reaches an A_{600} of 0.61
and in the complement vesicle trace an A_{600} of 0.41. The positions
of mol. wt. markers are shown on the ordinate.

CONCLUSIONS

The formation of the complement MAC on the mammalian cell surface
may lead to lysis or non-lethal cell damage. Many details of the
formation of the MAC remain to be established, and it seems likely
that specific MAb's will continue to be important reagents in resolving

these. Subsequent to formation of the MAC and insertion of C9 into the complex a rise in cytosolic Ca^{2+} concentration is responsible for metabolic damage. It can also trigger membrane shedding to remove cell-surface MAC and allow recovery from sublytic complement attack. Such attack may play an important role in tissue damage occurring in many autoimmune diseases.

Acknowledgements

The Medical Research Council, the Arthritis and Rheumatism Council, and Muscular Dystrophy Group of Great Britain and Northern Ireland, and the World Health Organization provided support.

References

1. Podack, E.R. & Tschopp, J. (1984) *Mol. Immunol. 21*, 589-603.
2. Campbell, A.K. & Luzio, J.P. (1981) *Experientia 37*,   1110-1112.
3. Morgan, B.P., Sewry, C.A., Siddle, K., Luzio, J.P. & Campbell, A.K. (1984) *Immunology 52*, 181-188.
4. Morgan, B.P., Campbell, A.K. & Compston, A. (1984) *Lancet ii*, 251-255.
5. Tschopp, J., Masson, D. & Stanley, K.K. (1986) *Nature 322*, 831-834.
6. Young, J.D-E., Cohn, Z.A. & Podack, E.R. (1986) *Science 233*, 184-190.
7. Morgan, B.P., Daw, R.A., Siddle, K., Luzio, J.P. & Campbell, A.K. (1983) *J. Immunol. Meths. 64*, 269-281.
8. Stanley, K.K., Kocher, H-P., Luzio, J.P., Jackson, P. & Tschopp, J. (1985) *EMBO J. 4*, 375-382.
9. Luzio, J.P., Stanley, K.K., Jackson, P., Siddle, K., Morgan, B.P. (1986) *this series 15* [Reid, E., et al., eds.], 247-253.
10. Morgan, B.P., Luzio, J.P. & Campbell, A.K. (1984) *Biochem. Biophys. Res. Comm. 118*, 616-622.
11. Stanley, K.K., Page, M., Campbell, A.K. & Luzio, J.P. (1986) *Mol. Immunol. 23*, 451-458.
12. Podack, E.R., Tschopp, J. & Muller-Eberhard, H.J. (1982) *J. Exp. Med. 156*, 268-282.
13. Tschopp, J. Podack, E.R. & Muller-Eberhard, H.J. (1985) *J. Immunol. 134*, 495-499.
14. Dankert, J.R. & Esser, A.F. (1985) *Proc. Nat. Acad. Sci. 82*, 2128-2132.
15. Monahan, J.B., Stewart, J.L. & Sodetz, J.M. (1983) *J. Biol. Chem. 258*, 5056-5062.
16. Steckel, E.W., York, R.G., Monahan, J.B. & Sodetz, J.M. (1980) *J. Biol. Chem. 255*, 11997-12005.
17. Stewart, J.L. & Sodetz, J.M. (1985) *Biochemistry 24*, 4598-4602.
18. Galfre, G. & Milstein, C. (1981) *Meths. Enzymol. 73*, 3-46.

19. Abraha, A., Bailyes, E.M., Richardson, P.J., Campbell, A.K. & Luzio, J.P. (1986) *Biochem. Soc. Trans. 14*, 779.
20. Tschopp, J. & Mollnes, T-E. (1986) *Proc. Nat. Acad. Sci. 83*, 4223-4227.
21. Soos, M. & Siddle, K. (1982) *J. Immunol. Meths. 51*, 57-68.
22. Campbell, A.K., Daw, R.A., Hallett, M.B. & Luzio, J.P. (1981) *Biochem. J. 194*, 551-560.
23. Campbell, A.K., Daw, R.A. & Luzio, J.P. (1979) *FEBS Lett. 107*, 55-60.
24. Hallett, M.B., Luzio, J.P. & Campbell, A.K. (1981) *Immunology 44*, 569-576.
25. Hallett, M.B. & Campbell, A.K. (1982) *Nature 295*, 155-158.
26. Abraha, A., Sewry, C.A., Campbell, A.K. & Luzio, J.P. (1987) *Biochem. Soc. Trans. 15*, in press.
27. Sahashi, K., Engel, A.G., Lambert, E.H. & Howard, F.M. (1980) *J. Neuropath. Exp. Neurol. 39*, 160-172.
28. Biesecker, G. (1983) *Lab. Invest. 49*, 237-249.
29. Schafer, H-J., Mathey, D., Hugo, F. & Bhakdi, S. (1986) *J. Immunol. 137*, 1945-1949.
30. Falk, R.J., Dalmasso, A.P., Kim, Y., Lam, S. & Michael, A. (1985) *New Engl. J. Med. 312*, 1594-1599.
31. Mollnes, T.E., Lea, T., Mellbye, O.J., Pahle, J., Grand, O. & Harboe, M. (1986) *Arthritis & Rheumatism 29*, 715-721.
32. Rus, H.G., Niculescu, F., Constantinescu, E., Cristea, A. & Vlaicu, R. (1986) *Atherosclerosis 61*, 35-42.
33. Edwards, S.W., Morgan, B.P., Hoy, T.G., Luzio, J.P. & Campbell, A.K. (1983) *Biochem. J. 216,*195-202.
34. Richardson, P.J. & Luzio, J.P. (1980) *Biochem J. 186*, 897-906.
35. Richardson, P.J. (1979) *PhD. Thesis*, University of Cambridge, pp. 85-103.
36. Allan, D., Billah, M.M., Finean, J.B. & Michell, R.H. (1976) *Nature 261*, 58-60.
37. Morgan, B.P., Campbell, A.K., Luzio, J.P. & Hallett, M.B. (1984) *Biochem. Soc. Trans. 12*, 779-780.
38. Campbell, A.K. & Morgan, B.P. (1985) *Nature 317*, 164-166.
39. Morgan, B.P., Dankert, J.R. & Esser, A.F. (1987) *J. Immunol. 138*, 246-253.
40. Morgan, B.P. & Campbell, A.K. (1985) *Biochem. J. 231*, 205-208.
41. Sims, P.J. & Wiedmer, T. (1986) *Blood 68*, 556-561.
42. Carney, D.F., Hammer, C.H. & Shin, M.L. (1986) *J. Immunol. 137*, 263-270.
43. Shukla, S.D., Berriman, J., Coleman, R., Finean, J.B. & Michell, R.H. (1978) *FEBS Lett. 90*, 289-292.
44. Ellman, G.L., Courtney, K.D., Andres, V. & Featherstone, R.M. (1961) *Biochem. Pharmacol. 7*, 88-95.

#C-3

ERYTHROCYTE MEMBRANE DERANGEMENTS ASSOCIATED WITH SICKLE CELL DISEASE

David Allan and Priti J. Raval

Department of Experimental Pathology
School of Medicine, University College London
University Street, London WC1E 6JJ, U.K.

During the accelerated ageing process which converts young sickle cells into irreversibly sickled cells (ISC's) a variety of changes occur in the plasma membrane (p.m.). These include breakdown and aggregation of polypeptides (particularly those associated with the membrane skeleton), loss of lipids by vesiculation of the bilayer portion of the membrane, and breakdown of polyphosphoinositides by an endogenous phosphodiesterase. Each of these changes, which may contribute to the abnormal rheological and morphological characteristics of ISC's, can be mimicked in normal erythrocytes by artifically increasing the intracellular concentration of Ca^{2+}, suggesting that the alterations in sickle cells result from Ca^{2+} accumulation. Recent evidence suggests that this may be due to an enhanced permeability to Ca^{2+} of the p.m. of these cells when exposed to sickle-unsickle cycling.

In order to analyze the biochemical changes which accompany the process of normal red-cell ageing or the analogous process which converts sickle cells into ISC's, one needs some reliable means of separating the cells into different age cohorts. This has often been achieved by making use of the observation that there is a relative increase in cell density with age, so that the cells can generally be separated by investigatation on a suitable density gradient. Normally one should avoid gradient media which themselves cause cell-density changes (e.g. sucrose or other osmotically active small molecules) or which would cause cell aggregation (e.g. dextran). The materials must also furnish a high enough density to resolve the densest red-cell fractions, whilst retaining a low enough viscosity to allow reasonable centrifugation times.

Several different density separations have been employed to fractionate red cells, each having certain advantages and disadvantages.

The earliest attempts utilized bovine serum albumin which worked very well but was rather expensive [1, 2]. Use is now more often made of Stractan, an arabinogalactan derived from larch wood [3], or Percoll/Renograffin [4] or Percoll/Hypaque [5] mixtures which give self-forming gradients. Red-cell separation into density classes has also been achieved merely by high-speed centrifugation of packed cells at 30° in an angle rotor in the absence of any gradient-forming extracellular compound [6].

The increased density of older cells seems to result from progressive dehydration, for which the most obvious cause is the so-called Gardos Effect [7, 8] whereby a small increase in intracellular Ca^{2+} gives rise to a huge increase in the membrane permeability to K^+ which escapes from the cells together with Cl^- ions and water (to maintain electrical and osmotic balance respectively). There is clear evidence that aged normal cells and sickle cells (particularly ISC's) have markedly elevated levels of intracellular Ca^{2+} which would activate the Gardos Effect if present in the cytosol [9, 10]. An increase in cytosolic Ca^{2+} also causes various other biochemical changes in normal human erythrocytes including lipid loss [11], proteolysis and aggregation of membrane proteins [12, 13], enhanced susceptibility of phospholipids to oxidation [14] and activation of polyphosphoinositide phosphodiesterase [12].

Attention is here drawn to the biochemical changes seen in the membranes of sickle cells which appear to be analogous to those alterations that have been observed in normal cells in which the cytosolic Ca^{2+} concentration has been increased. The data strongly support the idea that Ca^{2+} has an important role in the production of ISC's.

METHODS

Blood samples from patients with sickle cell disease were obtained from King's College Hospital, London. After three centrifugal washings (3000 rpm) in 0.15 M NaCl at 20°, cells were fractionated centrifugally either by Murphy's method [6] or on a discontinuous Stractan gradient. Murphy's procedure depends on a high-speed centrifugation in a relatively wide tube in an angle rotor which allows cells to circulate and thus sort themselves out according to density. It has the disadvantage that the densest cell fractions pack so firmly that it may be difficult to remove them from the tube; but because of the large number of cells that can be used, it does facilitate biochemical measurements. The fractions (1 ml) were removed from the top, using an automatic pipette with the end of the tip cut off, following centrifugation of the packed-cell sample (8 ml) in a 7.5 × 1.5 cm plastic tube for 1 h at 30° with the 8 × 10 ml angle rotor of the RC1B Sorvall centrifuge (15,000 rpm, 23000 g_{av}).

The Stractan gradient was constructed manually in 10 ml tubes of the Beckman SW21 rotor using a 1.5 ml Stractan cushion of d = 1.145, overlaid successively with 2 ml each of d = 1.14, 1.135, 1.13, 1.12 and finally with 2 ml of cells at 50% haematocrit. All solutions contained 0.15 M NaCl, 20 mM MOPS-NaOH buffer pH 7.1, 1 mM phosphate at pH 7.1, 1mM $MgCl_2$, 0.1 mM EGTA and 10 mM glucose. Centrifugaton was for 2 h at 35,000 rpm in a Beckman L2-65B ultracentrifuge maintained at 20°. Visible bands at the different density interfaces were removed using a Pasteur pipette with bent tip.

For each gradient, cell numbers in each fraction were counted in a haemocytometer, and % ISC's assessed on samples diluted in 0.15 M NaCl. Cells were classified as ISC's if under oxygenated conditions they retained a pointed projection or their length exceeded twice the width [15].

For determinations of mean cell volume (MCV) cells were counted in the light microscope, and packed cell volume was measured in a microhaematocrit centrifuge. Ghosts were prepared from normal and sickle cell fractions by lysis at 4° in 40 vol. of 20 mM tris-HCl pH 7.4 containing 1 mM EDTA and 0.1 mM phenylmethylsulphonyl fluoride to inhibit proteases, followed by three centrifugal washes (30,000 g_{av}) for 10 min in the same medium. Analysis of membrane polypeptides was carried out by SDS-polyacrylamide gel electrophoresis (PAGE) on 6% (w/w) gels [16]. Polypeptides were quantified by scanning the Coomassie Blue stained gels using a Bio-Rad scanning densitometer (model 1650) linked to a Shimadzu C-RIB integrator.

For phospholipid analysis [17], ghosts were extracted first with chloroform/methanol (1:2 by vol.) to remove all phospholipids except the polyphosphoinisotides, and then with chloroform/methanol/ conc. HCl (1:2:0.01) to extract di- and triphosphoinositides. The lipids from the first extract were separated on EDTA-impregnated TLC plates with chloroform/methanol/acetic acid/water (75:45:3:1) as solvent. Those from the second extract were separated on oxalate-impregnated TLC plates run in chloroform/methanol/acetone/acetic acid/water (40:13:15:12:7). Iodine-staining spots were analyzed for phosphorus by the procedure of Bartlett [18].

RESULTS AND DISCUSSION

Using Murphy's procedure (Fig. 1) it was clear that the cells sedimented according to their density, i.e. those with the lowest mean corpuscular volume went to tbe bottom of the centrifuge tube. The denser fractions were relatively depleted in phospholipid and accounted for most of the ISC's in this sample which was derived from a patient with an unusually high level of ISC's (39%). The marked changes seen in this experiment were observed only in an attenuated form in more typical samples having <10% ISC's. However, this emphasized the correlation between ISC characteristics, high density (i.e. dehydration) and loss of phospholipid.

Fig. 1. Fractionation of sickle cells by Murphy's method.   The cells were derived from a patient with an unusually high proportion (39%) of ISC's.   Packed cells were sedimented and fractionated as described under Methods.
•, % ISC's; △, MCV; □, total phospholipid in each fraction.

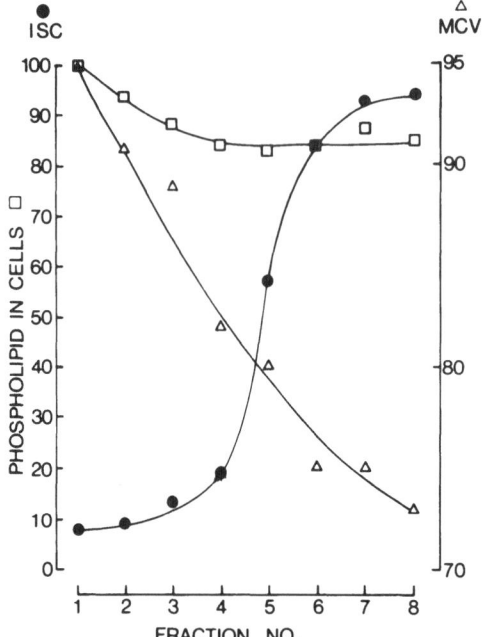

    Separation of sickle cells on a discontinuous gradient gave a similar enrichment of ISC's in the denser fractions (Fig. 2,S) where very few cells sedimented from samples of normal erythrocytes (Fig. 2,N).   It is noteworthy that the distribution of cells was bimodal with a well-defined trough at the 1.13/1.135 density interface. If the proportion of cells in each region is a reflection of the time which they spend in each phase of the transition from young to old (dense) cells, then this trough suggests a relatively rapid transition from the normal density to that which is characteristic of ISC's (fractions 4 & 5 in Fig. 2,S).

    Analysis of polypeptides from the different fractions showed significant differences both in normal and in sickle cells ([19] & Fig. 3a).   The most well-defined changes were reductions in the amount of polypeptide 4.1 and in a component of apparent M_r ~110,000, together with a rise in the amount of a polypeptide of M_r ~180,000 which is probably a breakdown product of ankyrin.   Some of the polypeptide which disappears could be present in polymeric material which does not enter the gel and which is visible in the denser fractions of normal and sickle cells.   It is noteworthy that the polypeptides which change in amount are phosphoproteins and probably constitute part of the membrane skeleton [19].

    The denser sickle fractions, which are rich in ISC's, show a relative loss of phosphoinositides compared with lighter sickle frac-

Fig. 2. Fractionation of erythrocytes, normal (N; 5 donors) and sickle (S; 6 donors) on Stractan density gradients. Results are expressed as means ±S.E. Successive fractions in order of increasing density are denoted 1-5. Block heights represent % of cells in each fraction; o, % ISC's in each sickle fraction.

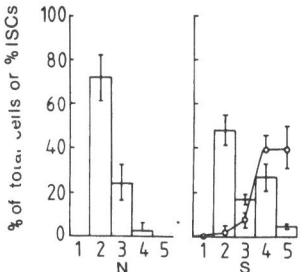

Figs. 2 & 3: courtesy of Elsevier, from ref. [19].

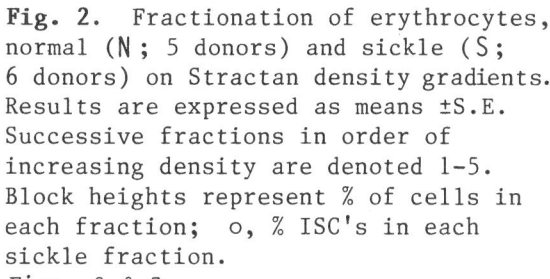

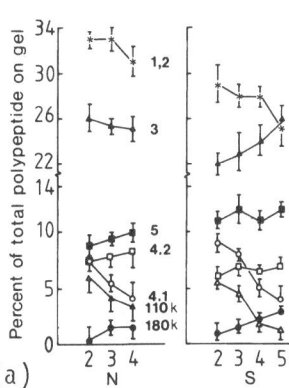

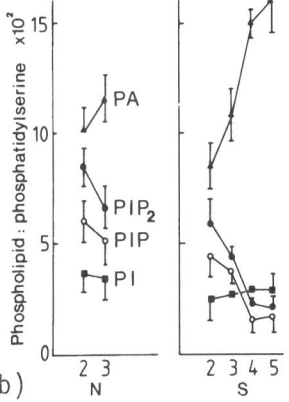

(a)   (b)

Fig. 3. Changes in (a) membrane polypeptide composition, and (b) membrane phospholipids in normal and sickle cells fractionated on Stractan density gradients. Cell samples were as in Fig. 2. Polypeptides and phospholipids were quantified as described under Methods. Polypeptides were numbered according to Steck's nomenclature [24]. ▲, phosphatidate; ■, phosphatidylinositol, and ●, its 4,5-bisphosphate; o, its 4-phosphate.

tions and normal cells (Fig. 3b). There is also a reciprocal increase in phosphatidate in the dense sickle fractions, and this change is precisely what would be expected if phosphoinositide diesterase had been activated to give 1,2-diacylglycerol which was then phosphorylated by diacylglycerol kinase [12]. Recently it has been suggested that polyphosphoinositides may be involved in mediating interactions between protein components of the membrane skeleton [20] so that breakdown of these lipids may influence cell shape [21].

It is important to note that all the above changes seen in ISC's can be reproduced in normal cells by treating them with Ca^{2+} and ionophore A23187. This treatment causes a rise in intracellular Ca^{2+} which (1) activates the Gardos Effect, causing efflux of KCl and water and resulting in acute cell shrinkage, (2) precipitates the release of up to 20% of cell lipid as microvesicles which are free of proteins of the membrane skeleton, (3) activates protease(s) which degrade some of the membrane skeletal proteins, including ankyrin and band 4.1, and (4) promotes the breakdown of endogenous polyphospho-inositide to 1,2-diacylglycerol which is then phosphorylated to give phosphatidate. Ca^{2+} also increases the amount of cross-linked membrane protein either by activating an endogenous transamidase [13] or by causing enhanced oxidation of protein and lipids [14].

Thus many of the properties of ISC's could be due to an accumula-tion of Ca^{2+} in the cells in terms of the biochemical changes undergone, especially since there is good evidence that the Ca^{2+} content of ISC's is very much higher than in normal cells [9, 10]. Recent work [22] has suggested that much of this Ca^{2+} is in a metabolically silent form (possibly sequestered in internal vesicles); but in view of the parallels between ISC's and Ca^{2+}-loaded normal cells, it seems that at least some of the Ca^{2+} must be present free in the cytosol. Interestingly, the Ca^{2+}-permeability of sickle cells seems to increase very markedly during sickle-unsickle cycling [23]; in this situation the cells accumulate unusual amounts of Ca^{2+} which may then cause the irreversible changes in the cell membrane that are referred to above. Some of these changes, particularly those which appear to destabilize the membrane skeleton - e.g. polypeptide changes and polyphosphoinositide breakdown - may be important factors in the release of bilayer lipids from the cells and in the development of the membrane rigidity characteristic of ISC's. However, the membrane changes and the dehydration process which accompany ISC formation all seem to be related to cellular accumulation of Ca^{2+}.

Acknowledgement

We thank the Medical Research Council for financial support.

References

1. Leif, R.C. & Vinograd, J. (1964) *Proc. Nat. Acad. Sci. 54*, 520-528.
2. Bishop, C. & Prentice, T.C. (1966) *J. Cell Physiol. 67*, 197-207.
3. Corash, L.M., Piomelli, S., Chen, H.C., Seaman, C. & Gross, E. (1974) *J. Lab. Clin. Med. 84*, 147-151.
4. Vettore, L., De Matteis, M.C. & Zampini, P. (1980) *Am. J. Hematol. 8*, 281-297.
5. Snyder, L.M., Leb, L., Piotrowski, J., Sauberman, N., Liu, S.C. & Fortier, J.R. (1983) *Br. J. Haematol. 53*, 379-384.
6. Murphy, J.R. (1973) *J. Lab. Clin. Med. 82*, 334-341.
7. Gardos, G. (1958) *Biochim. Biophys. Acta 30*, 653-654.
8. Glader, B.E. & Nathan, D.G. (1978) *Blood 51*, 983-989.
9. Eaton, J.W., Skelton, T.D., Swofford, H.S., Koplin, C.E. & Jacob, H.S. (1973) *Nature 246*, 105-106.

10. Palek, J. (1977) *Br. J. Haematol. 35*, 1-9.
11. Allan, D., Billah, M.M., Finean, J.B. & Michell, R.H. (1976) *Nature 261*, 58-60.
12. Allan, D. & Thomas, P. (1981) *Biochem. J. 191*, 433-440.
13. Lorand, L., Siefring, G.E. Jr. & Lowe-Krentz, L. (1979) *Seminars in Haematology 16*, 65-74.
14. Jain, S.K. & Shohet, S.B. (1984) *Blood 63*, 362-367.
15. Westerman, M.P. & Allan, D. (1983) *Br. J. Haematol. 53*, 399-409.
16. Laemmli, U.K. (1970) *Nature 227*, 680-685.
17. Raval, P.J. & Allan, D. (1985) *Biochem. J. 231*, 179-183.
18. Bartlett, G. (1959) *J. Biol. Chem. 234*, 466-468.
19. Raval, P.J. & Allan, D. (1986) *Biochim. Biophys. Acta 856*, 595-601.
20. Anderson, R.A. & Marchesi, V.T. (1985) *Nature 318*, 295-298.
21. Ferrell, J.E. & Huestis, W.H. (1984) *J. Cell Biol. 98*, 1992-1998.
22. Lew, V.L., Hockaday, A., Sepulveda, M-I., Somlyo, A.P., Somlyo, A.V., Ortiz, O.E. & Bookchin, R.M. (1985) *Nature 315*, 586-589.
23. Ohnishi, S.T., Horiuchi, Y.K. & Horiuchi, K. (1986) *Biochim. Biophys. Acta 886*, 119-126.
24. Steck, T.L. (1974) *J. Cell Biol. 62*, 1-19.

#C-4

THE MEMBRANE-CYTOSKELETAL AXIS IN PHAGOCYTOSING PMN-LEUCOCYTES AND VIRALLY TRANSFORMED FIBROBLASTS

N. Crawford, T.C. Holme and S. Kellie

Department of Biochemistry, Hunterian Institute
Royal College of Surgeons of England
Lincoln's Inn Fields, London WC2A 3PN,   U.K.

Actin-rich filamentous arrays comprising a submembraneous network are known to be especially evident in cells whose functions include adhesion to surfaces or to other cells, locomotion, chemotaxis, endo- and exocytosis, formation of pseudopodia, and other membrane-mediated phenomena. Many of these motile activities require a considerable topographical reorganization of membrane constituents within the lipid bilayer and also changes in the macromolecular character and disposition of cell actin and the associated proteins involved in regulating its assembly from monomer actin subunits to the cytosolic pool.

To investigate some of these changes and to explore the inter- dependence of cell surface membrane constituents and cytoskeletal components, we have studied two model systems - phagocytosing poly- morphonuclear leucocytes (PMNL's) and virally (RSV) transformed fibroblasts - whose actin status changes profoundly in response to stimuli. With the former we have followed the formation of micro- domains in the surface membrane with selective segregation of a major glycoprotein out of the forming phagocytic vesicle into the uninvolved membrane; this lateral migration appears to be linked in some way to changes in cytoskeletal integrity. In the fibroblast there are actin changes: the monomer (G)$\rightleftharpoons$ polymer (F) equilibrium shifts in the cytosol, and F-actin in the membrane-associated cytoskel- eton rises; there are concomitant changes in the distribution of vinculin and α-actinin in membrane adhesion plaques and the production of pp60^{vsrc}-associated protein kinase activity in response to src-gene expression. Mention is made of the techniques (amplification in [1-3]) for purifying cell-surface membranes and phagocytic vesicles using continuous-flow electrophoresis (CFE), for quantifying G- and F-actin by the DNase-I inhibition assay and for morphological monitoring using interference reflectance microscopy and immunofluor- escence.

The Singer-Nicolson fluid mosaic model [4] for cell membranes, in which integral proteins are depicted embedded in the lipid bilayer, is now widely accepted and better fits our concepts of the reorganization that can take place during cell stimulation (receptor revelation and clustering, formation of functional micro-domains, adhesion to surfaces, etc.). Later studies, however, have revealed that the freedom for membrane proteins to diffuse laterally through the fluid phase is not only subject to thermodynamic limitations but also to constraints due to interaction of integral membrane constituents with peripheral membrane proteins bound to either surface of the lipid bilayer.

Many ultrastructural studies have revealed that microfilament assemblies of varying complexity lie sub-adjacent to the cytoplasmic face of cell surface membranes and are often attached, if not directly, through some linkage protein to integral membrane constituents. The major constituent of these filamentous networks is the contractile protein actin. Whereas the structural disposition of actin is well organized in, e.g., heart and skeletal muscle sarcomeres, cellular actin is in a more dynamic state with fully formed filaments in the membrane-associated cytoskeleton and both monomer (G) and polymer (F) actin present in the cytoplasmic matrix. The nature of the monomer-polymer equilibrium state, as also the presence of other proteins which can cap, fragment, cross-link or bundle the actin, determines the sol/gel characteristics of the cytoplasm and the force-generating processes that occur in motile phenomena.

We have used two cell model systems to investigate some of the features of the interdependence of cell surface membrane and the cytoskeleton, viz. the phagocytosing polymorph and the virally transformed fibroblast. Both cells show profound changes in the character and disposition of cellular actin in response to the phagocytic or transforming stimulus. These changes will be compared and contrasted with respect to the operational role of the cytoskeletal/membrane axis in the expression of cell motile properties.

THE PHAGOCYTOSING POLYMORPHONUCLEAR LEUCOCYTE

The ingestion of particles by phagocytosis is an important property of specialized cells which participate in the generalized host-defensive mechanisms of the body, removing dead cells, bacteria, fungi and other harmful particulate substances from the circulation. The kinetic aspects, divalent cation dependence and energy demands of the phagocytic process have been studied in some detail, as also have the mechanisms involved in bactericidal and cytotoxic events and lysosomal digestion which take place in the internalized vesicles [5-9]. However, much less is known about the detailed molecular changes that occur at the surface membrane level in the initial adhesion phase and the particle-engulfment processes.

For detailed studies of the membrane/cytoskeletal axis and the changes occurring during phagocytosis, a highly purified surface-membrane fraction is required which is free from contaminating intra-cellular components and of course essentially unaltered by the isolation procedure in respect of any associated cytoskeletal structures. Most conventional membrane isolation procedures produce a heterogeneous fraction of membrane vesicles containing elements of both surface and intracellular origin, making interpretation of changes most difficult. We have combined a density-gradient procedure with continuous-flow electrophoresis (CFE) to produce highly purified surface membrane fractions from resting and phagocytosing neutrophils. Additionally, by using a phagocytic stimulus consisting of opsonized paraffin oil droplets into which has been incorporated [^{3}H]glycerol, our approaches have gained three main advantages.-

(a) The phagocytic process can be standardized and reproduced by quantitative determination of the radiolabelled oil droplets taken up by the cell.
(b) It is essentially a 'frustrated phagocytic event' in that it does not proceed to lysosome fusion. Thus analytical and enzymatic studies can be made on the phagocytic vesicle (phagosome) without the complexities inherent in dealing with lysosome components in the membrane of a phagolysosome.
(c) During density-gradient sedimentation of the leucocyte homogenate the phagocytic vesicles enclosing the oil float to a position close to the meniscus of the gradient and can be harvested cleanly from surface membrane vesicles arising from domains uninvolved in the internalization process.

With these procedures, direct comparisons can now be made between the surface membrane of the resting leucocyte and the surface membrane and phagocytic vesicle membrane of the phagocytosing leucocyte. Aspects of membrane reorganization during the phagocytic event can then be studied.

The full details of the conditions for phagocytosis, the preparation of the homogenates, the gradients and the various membrane fractions have been given elsewhere [1, 10]. A flow chart (Scheme 1) shows the essential features of the procedure and the appearance of the density gradients. For most of the studies, PMN's were harvested from the peritoneal cavity of rabbits after elicitation with sterile glycogen [2]. Actin in both monomeric and polymeric forms was determined in the whole cells and in the membrane and cytosol fractions using the DNase-I inhibition assay of Blikstad et al. [11]. Since the assay determines G-actin, filamentous (F) actin was depolymerized with 3 M guanidinium chloride. The inhibition assay details have likewise been published [1, 3], as have the surface membrane labelling procedures with ^{125}I-*Lens culinaris* lectin and the assay of the sub-cellular fractions for marker enzymes [7].

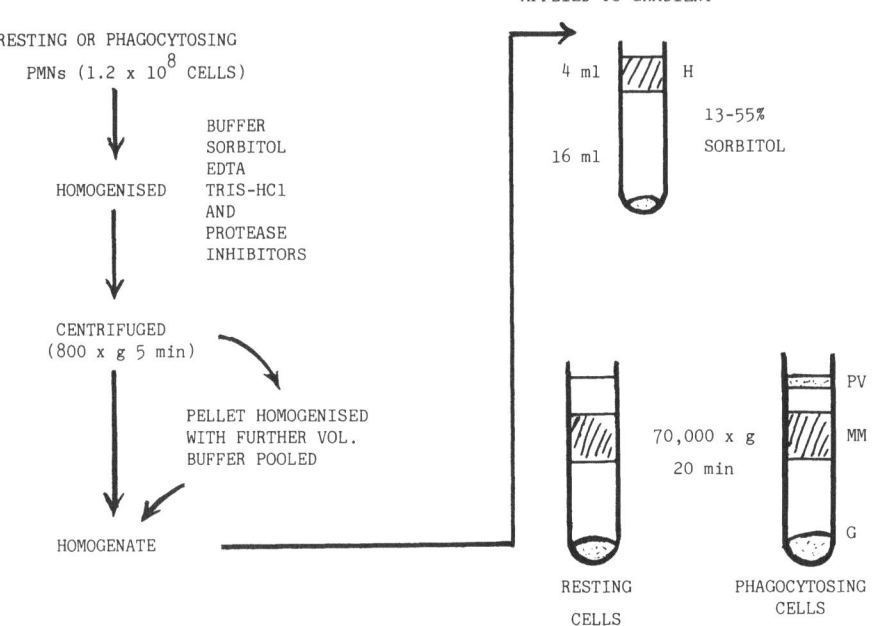

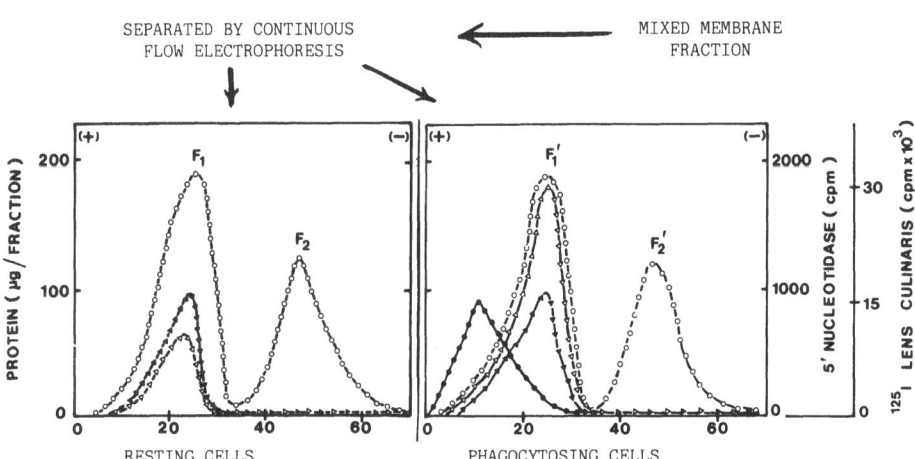

Scheme 1. Flow diagram for preparing purified plasma membrane and phagocytic vesicle membranes (phagosomes, PV) from resting and phagocytosing cells. Cells are treated exactly the same but PV are usually harvested separately from the sorbitol gradient (exclusive location of [³H]glycerol).   The CFE profiles are for mixed membrane fractions, for resting cells *(left)* and for phagocytosing cells *(right)* whose phagosomes separated from uninvolved plasma membrane. Protein, o; ¹²⁵I-*Lens culinaris*, ▲; 5'-nucleotidase, Δ or, for the separately run PV (phagosomes), ● (electrophoretic profiles superimposed).

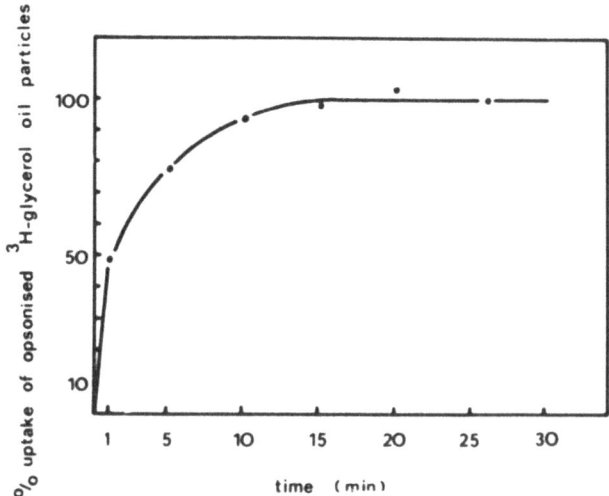

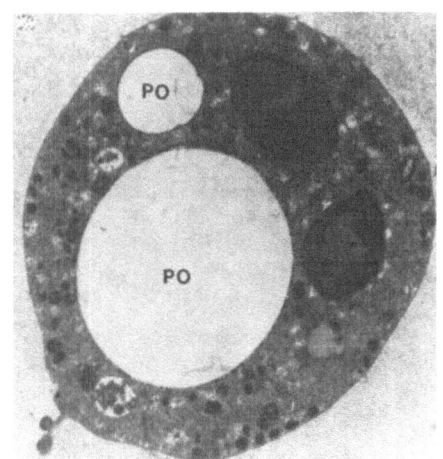

Fig. 1a *(above).*   Uptake of
opsonized [³H]glycerol/paraffin
oil particles by PMN-leucocytes.

Fig. 1b *(right).*   A typical PMN-
leucocyte with ingested paraffin
oil (PO) droplets.

    Fig. 1a shows a time curve for the uptake of opsonized paraffin
oil droplets.   Since with this phagocytic model the cell is unable
to process the inert oil and re-cycle surface membrane, the extent
of internalization is limited by the amount of surface membrane
depletion possible without loss of cell integrity.   For comparative
studies the phagocytosis was stopped at 20 min, well into the plateau
of the uptake curve.   Fig. 1b shows a typical polymorph with ingested
oil droplets.   Initially a study was made of the distribution of
total actin (G and F) in the cytosol and particulate fractions prepared
from resting and phagocytosing cells by high-speed centrifugation
(9×10^6 min) of homogenates.   Table 1 shows the data for total actin
distribution for 6 different preparations of resting cells and 5
preparations of phagocytosing cells.

Table 1. Total actin content of cytosol and particulate fractions from resting (n = 6) and phagocytosing (n = 5) polymorphs (mean ±S.D.).

Actin, µg:	total	cytosol	particulate
10^6 resting cells	56 ±3.6 (100%)	17 ±1.8 (31%)	39 ±2 (69%)
10^6 phagocytosing cells	57 ±5.0 (100%)	9 ±0.7 (16%)	48 ±5 (84%)

Table 2. Surface membrane-associated actin in resting and phagocytosing polymorphs (prepared by density gradient sedimentation and electrophoresis) as determined on the purified membranes (6 expts.). The values are µg actin/100 µg protein.

	1	2	3	4	5	6	Mean ±S.D.
Resting cells	10.9	5.4	8.2	5.7	4.1	8.5	7.1 ±2.5
Phagocytosing cells	15.7	8.0	12.3	7.9	5.4	12.3	10.2 ±3.8
% increase	*+44*	*+48*	*+50*	*+38*	*+28*	*+45*	*+43 ±6.5*

It can be seen that after phagocytosis for 20 min there is a significant redistribution of the actin pools. In the resting cells, 69% of the total actin was sedimentable whereas after phagocytosis 84% sedimented. In another study, the surface membrane from resting cells was isolated and purified as in Scheme 1 and the amount of actin associated with it quantified by the DNAase-I assay. The surface membrane uninvolved in the phagocytic events was also isolated and purified from phagocytosing polymorphs for comparison of the actin content. Table 2 shows the data from 6 preparations of resting cells and the corresponding cell preparations after phagocytosis. These data show that during phagocytosis there is a 43% increase in actin associated with the uninvolved surface membrane. Further studies showed that all of the actin associated with these membranes was filamentous. In some preparations, these purified surface membrane were treated with Triton X-100 and a pellet and supernatant prepared by high-speed centrifugation (100,000 **g**, 20 min).

This detergent dissection removed ~10% of the filamentous actin associated with the resting cell surface membranes and a significantly higher proportion (25-30%) from the membranes of the phagocytosing cells. In quantitative terms this actin solubilized by Triton X-100 accounted for almost all the newly assembled F-actin that became associated with the membrane during the phagocytic event. Moreover, similar Triton X-100 treatment of the isolated phagocytic vesicles revealed that the detergent liberated all the actin, suggesting that although associated with the vesicle membrane, sufficiently tightly to survive gradient sedimentation and external washing in the CFE

Table 3. Plasma membrane (p.m.) and intracellular membrane in membrane fractions from resting and pnagocytosing polymorphs: enrichment values, *vs.* homogenate = 1.0. The first two markers are for p.m., and the others are for intracellular membrane and nuclei; ± following a mean is the S.E.M. (n = 6).

Marker	Resting cell p.m.	Phagocytosing cell p.m.
5'-Nucleotidase	18.0 ±0.6	38.0 ±3.2
[125]I-*Lens culinaris*	11.0 ±0.7	10.0 ±1.1
Acid phosphatase	1.2	1.3
Glucose-6-phosphatase	1.3	1.3
DNA	<0.1	<0.1

chamber it was not stabilized in the same way as the Triton-resistant filamentous network intrinsic to the resting cell surface membrane or to the uninvolved membrane domains of the phagocytosing cells.

A membrane feature which may relate to this change in the distribution of actin emerged from a study of the cell surface enzyme 5'-nucleotidase and the surface marker [125]I-*Lens culinaris* to monitor enrichment of the membranes with respect to homogenate. The lectin was enriched to almost the same degree in the surface membrane fractions derived from both the resting and the phagocytosing cells (Table 3) as in the phagocytic vesicle membrane. However, the 5'-nucleotidase enrichment *vs.* homogenate was 2-3 times greater (38-fold) in the uninvolved surface membrane of the phagocytosing cells than in the surface membrane of the resting cells (18-fold; Table 3), and there was little enrichment in the phagocytic vesicles

The data in Table 3 suggest that whereas receptors for the labelled lectin remained uniformly distributed after phagocytosis there is a selective segregation of 5'-nucleotidase out of the forming vesicle into the uninvolved surface domain during the phagocytic event (Fig. 2). This movement out of the vesicle of one of the neutrophil's major sialoglycoproteins has now been confirmed histocytochemically by Cremer and colleagues [12] using latex beads as a phagocytic stimulus. Carraway et al. [13] using ascites tumour cell microvilli as a model membrane/cytoskeleton system have shown co-sedimentation of phalloidin-stabilized microfilaments with 5'-nucleotidase when Triton-insoluble pellets are applied to sucrose gradients. These and the present studies of the segregatory events in the phagocytosing polymorph membrane are, we believe, the first demonstration of a redistribution of a major cell-surface glycoprotein during a motile event *without* the use of a multivalent ligand, as in patch and cap studies with surface Ig movements in lymphocytes and in other cells.

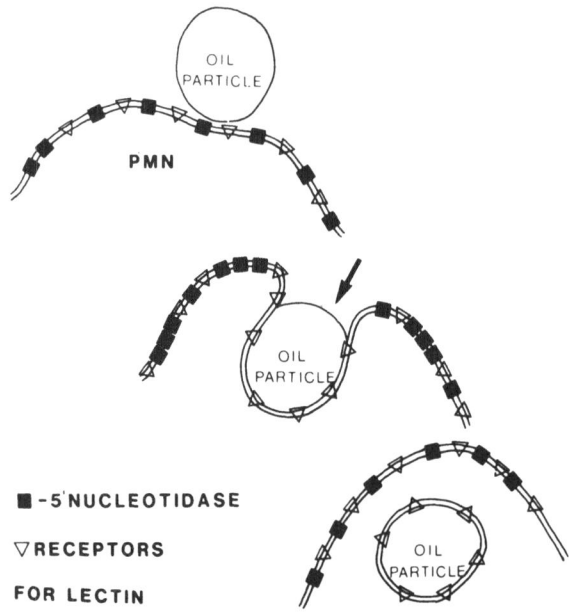

Fig. 2. Selective segregation of 5'-nucleotidase out of the forming phagocytic vesicle membrane into the uninvolved p.m. domain. Receptors for ^{125}I-*Lens culinaris* remain uniformly distributed.

The concomitant changes in the disposition and character of F-actin, although suggestive, do not as yet constitute sufficient evidence to indict a direct linkage between the migrating intrinsic membrane glycoprotein and cytoskeletal elements on the cytoplasmic face. There are a number of potential candidates for such linkages, notably α-actinin, actin-binding protein and a structurally similar protein, filamin, acumentin, and a range of filament end-blocking proteins of which some are known to promote actin nucleation and filament assembly [14]. However, few of these have been characterized in the granulocyte, and since there is increasing evidence that proteins not normally regarded as cytoskeleton proteins can bind to actin [14], the search may have to be widened if we are to understand the inter-relationship between cell membranes and the underlying features involved in regulating membrane reorganization and function. More recently, membrane-anchoring domains involving hydrophobic inter-actions of proteins with phospholipids, and particularly the phospho-inositides, are attracting some interest, and these too may have important structural and functional implications [15].

VIRAL TRANSFORMATION AND THE MEMBRANE-CYTOSKELETAL AXIS

Many viruses transform cells by the production of a single viral gene product [16, 17]. Rous sarcoma virus (RSV) is capable of transforming a variety of cell types and is associated with the production of a phosphoprotein of kM_r 60 which has tyrosine kinase activity, pp60[vsrc]. Since many transformed cells show a marked reduction in actin-containing microfilament bundles in the cytoplasm, it has been postulated that one of the cellular targets for the pp60-kinase activity may be a component of the cytoskeleton [18, 19]. The pp60 kinase appears to be closely associated with the p.m. of transformed cells and after viral transformation the microfilament actin present in the large multifilament bundles appears to reorganize into a looser submembranous network [20, 21]. Studies of the dynamic changes in actin may provide some information about the morphological consequences of neoplastic transformation by RNA tumour viruses, particularly with respect to membrane-mediated properties such as motility and adhesiveness.

For the present studies, Rat-1 fibroblasts transformed by infection with B77 strain RSV have been used for comparison with untransformed fibroblasts. The characteristics of this established and stable transformed cell line have been documented earlier [3]. Since, unlike the polymorph study, the fibroblasts have grown adherent to culture flasks the cells were either treated briefly with trypsin to effect their release or lysed *in situ* with a buffer which, besides having glycerol (15% v/v) and ATP (0.2 mM) to stabilize actin filaments, and prevent disassembly during the lysis procedure, contained 5 mM phosphate pH 7.6, 150 mM NaCl, 2 mM $MgCl_2$, 2 mM EGTA, 0.5% (v/v) Triton X-100, 0.1 mM dithioerythritol and 0.01 mM phenylmethyl-sulphonyl fluoride. From this lysate a cytosol fraction and a cyto-skeletal core (Triton-insoluble) were prepared by centrifugation pro-cedures.

In the study of the trypsin-released cells the enzyme activity was 'stopped' with soya bean trypsin inhibitor, and the cell suspension was treated with lysis buffer and centrifuged to provide cytosol and cytoskeletal core fractions as for the adherent cells. Actin in monomeric and polymeric forms was assayed as described above for the polymorph study. Table 4 shows the distribution of actin in the transformed and control fibroblasts grown in monolayer culture and released from their substrates by the lysis-buffer procedure.

Evidently the cytosol monomer-to-filamentous actin equilibrium state was shifted in the direction of filament assembly following transformation, with a decrease in F-actin from 38% to 20% of the total cell actin.

Accompanying this decrease in cytosol F-actin the cytosol core actin (F-actin) increases from 12.6 to 24.0% of the total cell actin content. Both these changes are highly significant. The data for the trypsinized cells support those from cells lysed *in situ*: these show an essentially similar shift in equilibrium between G and F and an increase in cytosol core actin from 22% to 36% of the total

Table 4. Distribution of G- and F-actin in control and RSV-transformed fibroblasts in cells lysed *in situ*. Results expressed as % of total cell actin, mean ± S.D. (n = 7).

	Controls	Transformed	Significance
Cytosol monomer actin (G)	49.3 ±5.5	55.6 ±8.7	NS
Cytosol filament actin (F)	38.3 ±5.1	20.6 ±6.4	$P < 0.001$
Cytoskeletal core actin	12.6 ±1.5	24.0 ±6.4	$P < 0.001$

cell content. In order to complement these findings morphologically, fluorescence staining of actin with NDB-phallacidin (8-nitrobenz-2-oxa-1,3-diazole phallacidin) was used on whole-cell preparations. This highly fluorescent derivative of the mushroom toxin phalloidin binds specifically to small and large F-actin filaments (Kd 2×10^{-8} M) and stabilizes them.

Figs. 3a and 3b show respectively the appearance of the control Rat-1 and RSV-transformed fibroblasts by phase-contrast microscopy, and Figs. 3c and 3d show the cells stained with NBD-phallacidin. The control fibroblasts (3a & 3c) are typically large flattened elongated cells showing large numbers of actin-containing microfilament bundles. By comparison the transformed cells were more refractile, contained only a few bundles but showed brightly stained aggregates of actin scattered throughout the cytoplasm.

CONCLUDING COMMENTS

Using two different cell model systems, the phagocytosing polymorph and the transformed fibroblast, it has been shown that the distribution of actin – the major cytoskeletal protein of both these cells – alters significantly during stimulated events. Both phagocytosis and viral infection involve much surface membrane reorganization and, concomitant with this, significant changes take place in the disposition and level of organization of F-actin associated with the submembranous cytoskeleton. The changes are complex, however, and may be due in part to changes in the ionic environment in the cytosol and in part to the action of regulatory proteins affecting the dynamics of the polymeric assemblies of actin and the subunit monomer pool from which these polymer forms are recruited.

The membrane-cytoskeletal axis is of unquestionable importance in normal cellular functions such as locomotion, intracellular motile events, exocytosis and endocytosis, receptor topography, cell-cell and cell-substrate interactions, embryogenesis, tumour growth and metastases, etc There has been an explosive growth over the last 4 years in the use of antibodies and other specific probes to study

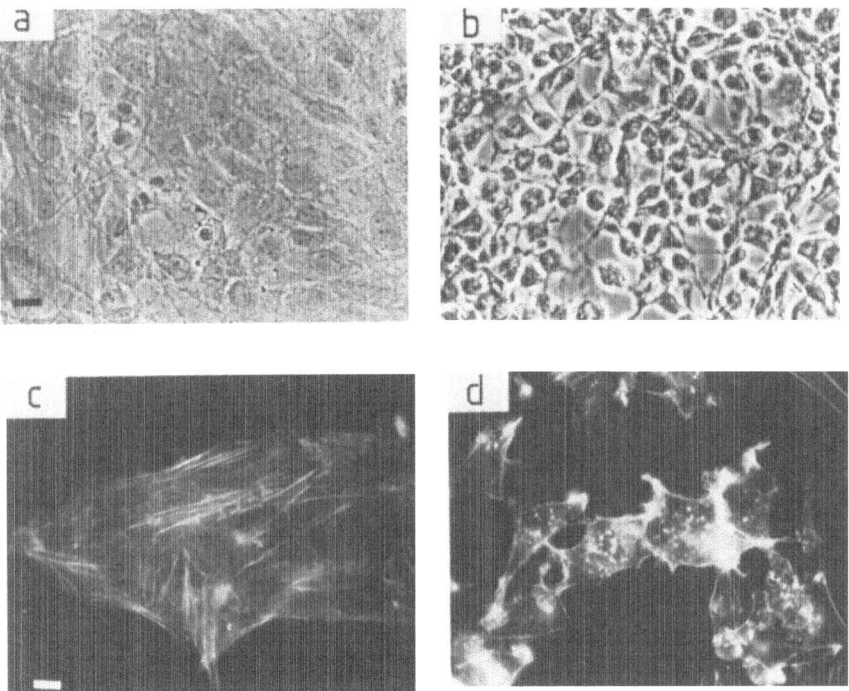

Fig. 3. (a) Rat-1 fibroblasts and (b) virally transformed cells: phase-contrast microscopy (bar = 50 μm). (c) and (d): Corresponding actin staining using NBD-phallicidin; note parallel actin bundles with cross-linking in the Rat-1 cells (c) and the dense submembranous assemblies in the transformed cells (d).

the disposition and level of macromolecular organization of cytoskeletal elements and their membrane associations. Such studies do not, however, furnish information about the dynamic nature of the cytoskeletal framework of cells. If we are to more fully understand cellular pathological manifestations related to changes in the membrane-cytoskeletal axis, new approaches are required whereby we can study the relationships between contractile and cytoskeletal filaments associated with the cytoplasmic face of cell membranes, the subunit protein pools in the cytosol compartments and the factors which regulate polymerization and disassembly. The work in this article has been largely focused upon only one cytoskeletal component, the microfilament protein actin and its organizational state. Clearly the opportunity to widen this approach should be seized and any cellular model system (normal or pathological) exploited to shed further light on this fascinating and important field.

Acknowledgements

We thank Mr. S. Gschmeissner of the Anatomy Department of this Institute for his help with the electron microscopy and Miss H. Watson for preparation and typing of the manuscript. We also express our appreciation to Dr. G. Poste (Vice-President, Research & Development) of Smith Kline & French Inc., Philadelphia, for generous support for some of our research costs.

References

1. Stewart, D.I.H. & Crawford, N. (1985) *Biochem. J. 225*, 807-814.
2. Stewart, D.I.H. & Crawford, N. (1981) *FEBS Lett. 126*, 175-179.
3. Holme, T.C. Kellie, S., Wyke, J.A. & Crawford, N. (1986) *Br. J. Cancer 53*, 465-476.
4. Singer, S.J. & Nicolson, G.L. (1972) *Science 175*, 720-721.
5. Sbarra, A.J. & Karnovsky, M.L. (1979) *J. Biol. Chem. 234*, 1355-1362.
6. Michell, R.H., Pancake, S.J., Noseworthy, J. & Karnovsky, M.L. (1969) *J. Cell Biol. 40*, 216-224.
7. Silverstein, S.C., Steinman, R.M. & Cohn, Z.A. (1977) *Ann. Rev. Biochem. 46*, 669-722.
8. Hallgren, R., Sjöström, P. & Bill, A. (1978) *Immunology 34*, 347-351.
9. Besterman, J.M. & Low, R.B. (1983) *Biochem. J. 210*, 1-13.
10. Crawford, N. (1985) in *Cell Electrophoresis* (Schutt, W. & Klukman, H., eds.), Walter de Gruyter, Berlin, pp. 225-246.
11. Blikstad, I., Markey, F., Carlsson, L., Persson, T. & Lindberg, U. (1978) *Cell 15*, 935-941.
12. Cremer, E., Bainton, D.F. & Werb, Z. (1982) *Proc. 20th Annual Mtg., Am. Soc. Cell Biol.*, Abs. CS 756.
13. Carraway, C.A., Sindler, C. & Weiss, M. (1986) *Biochim. Biophys. Acta 885*, 68-73.
14. Stossel, T.P., Chaponnier, C., Ezzell, R.M., Hartwig, J.H., Janmey, P.A., Kwiatkowski, D.J., Lind, S.E., Smith, D.B., Southwick, F.S., Yin, H.I. & Zaner, K.S. (1985) *Ann. Rev. Cell Biol. 1*, 353-402.
15. Low, M.G., Ferguson, M.A.J., Futerman, A.H. & Silman, I. (1986) *Trends Biochem. Sci. 9*, 212-215.
16. Erikson, R.L., Purchio, A.F., Erikson, E., Collett, M.S. & Brugge, J.S. (1980) *J. Cell Biol. 87*, 319-323.
17. Weiss R., Teich, N., Varmus, H. & Coffin, J., eds. (1982) *RNA Tumor Viruses*, Cold Spring Harbor Lab., New York.
18. Burr, J.G., Dreyfuss, G., Penman, S. & Buchanan, J.M. (1980) *Proc. Nat. Acad. Sci. 77*, 3484-3488.
19. Rohrschneider, L.R. & Rosok, M.J. (1983) *Mol. Cell Biol. 3*, 731-746.
20. Stoker, A.W., Kellie, S. & Wyke, J.A. (1986) *J. Virol. 58*, 876-883.
21. Kellie, S., Patel, B., Wigglesworth, N.M., Critchley, D.R. & Wyke, J.A. (1986) *Exp. Cell Res. 165*, 216-228.

#NC(C)

NOTES and COMMENTS related to

PLASMA MEMBRANE AND CYTOSKELETON

Comments on #**C-1**: C.A. Pasternak - MEMBRANE-MEDIATED CYTOTOXICITY
and #**C-2**: J.P. Luzio et al. - COMPLEMENT-MEDIATED DAMAGE

Question by S. Zucker.- How do you explain the requirement for calcium in natural lymphocyte killing of target cells in view of your evidence that external calcium actually is inhibitory to the haemolytic process? **Pasternak's reply**.- Calcium is required for the activity of the perforin-like molecules, but an excess of calcium or zinc can prevent the leakage. **Comment by R. Wattiaux** on protective effect of Ca^{2+} and especially of Zn^{2+}: both are lysosomal membrane stabilizers.

D. Allan asked Pasternak whether the membrane of enveloped viruses itself has the permeability properties which are conferred on cells by fusion with these viruses. **Reply**.- No! Although e.m. studies of haemolytic *vs*. non-haemolytic Sendai virus by negative stain suggest the presence of 'pores' across the envelope of haemolytic (but not of non-haemolytic) virus [Shimizu, Y.K., Ishida, N. & Homma, M. (1976) *Virology 71*, 48-60], it is difficult to carry out proper permeability studies with purified envelope vesicles that are sufficiently free of the detergent used to prepare them.

Remarks (to Luzio) by C.A. Pasternak.- There is no incompatibility between Ca^{2+} entry due to complement (along with other agents) and prevention of leakage (including Ca^{2+} entry) by high concentrations of Ca^{2+} (see ref. [39] in #D-1). Increasing the recovery of cells by vesiculation removes not only the damaging agent (i.e. complement) but perhaps also the mediating agent - internal Ca_i^{2+} - since vesicles extruded from cells are very rich in Ca^{2+} [see B.F. Trump, #NC(C)-1].

Concerning the inhibition of adrenaline-sensitive adenylate cyclase caused by Ca^{2+} influx into pigeon red cells, **D. Allan asked Luzio** whether this might be a Ca^{2+}-specific activation of cAMP phosphodiesterase rather than cyclase inhibition. **Reply**.- The ionophore A23187 in the presence of extracellular Ca^{2+} lowered adrenaline-stimulated cAMP degradation in pigeon red cells. This suggests an effect on the cyclase rather than on a phosphodiesterase [Campbell, A.K. & Siddle, K. (1976) *Biochem. J. 158*, 211-221].

Comments on #C-3: D. Allan - ERYTHROCYTE SICKLING
    and #C-4: N. Crawford et al. - MEMBRANE-CYTOSKELETAL AXIS

C.A. Pasternak asked Allan whether leakage of Ca^{2+} into sickled cells is specific, or whether leakage occurs with other ions, e.g. Rb$^+$ in the absence of Ca^{2+} so that the Gardos Effect is inoperative. **Reply.-** Yes, Rb$^+$ flux is indeed increased, but due to the Gardos Effect since this has been done in the presence of Ca^{2+}; I am unsure whether this has been done in the absence of Ca^{2+} also. **H. Hilderson** asked whether experiments in the converse direction can be performed *in situ* on reversibly sickled cells, e.g. by removing Ca^{2+} or by adding P1P2 (by means of transfer proteins) and thus enhancing membrane fluidity. **Reply.-** This would be difficult to do, since Hb comprises 90% of the dry wt. of the blood cells.

Crawford, answering J.M. Graham.- It is involvement of groups other than sialic acid, notably -SH, that causes the phagosome to be apparently more electronegative than the bulk of the surface membrane; also ruffling of the surface may link some negative charges, and therefore they are not expressed electrophoretically. **Remark by C.A. Pasternak.-** Since RSV-transformation has a low efficiency, one wonders about the feasibility of ascertaining whether the altered actin pattern of the transformed cells result from the expression of vital functions or from the selection of a minor pre-existing subpopulation of cells. **Response.-** Your comment is a valid one, but the co-expression of pp60^{src} in the same cells as have an altered actin pattern would seem to suggest that selectivity is always hard to exclude.
K. Donaldson sought Crawford's comments on the relationship of his findings with 5'-nucleotidase (5'N) in neutrophils and the well documented finding in macrophages that a dramatic fall accompanies macrophage activation. **Answer.-** I think one can carry the neutrophil/ macrophage analogies too far and it would be unwise to compare directly macrophage activation with our polymorph phagocytic process which is a frustrated phagocytosis since it does not proceed to lysosomal fusion! It was employed to allow harvesting of phagocytic vesicle membrane domains without lysosomal membrane inclusion. In our PMN model the net amount of 5'N does not change; it merely redistributes out of the vesicle domain into the uninvolved membrane regions. This selective migration takes place without change in the distribution of lectin receptors (Lens culinaris). Cremer et al. in California have confirmed histochemically the exclusion of 5'N from the phagocytic vesicle in PMN phagocytosis of opsonized latex particles. It therefore seems to be a real event concomitant with phagocytosis. Perhaps in the macrophage selective migration doesn't take place as in the PMN and surface-membrane 5'N is internalized and actually recycled!

Supplementary ref. from Senior Ed. [others on p. 239].- Erythrocyte and platelet membrane responses to amantadine, rimantadine and tetracaine suggest altered endosomal or lysosomal pH underlying modified virus-membrane interactions.- Donath, E., et al. (1987) *Biochem. Pharmacol. 36,* 481-487.

#NC(C)-1

A Note on

STUDIES ON CELL INJURY: EFFECTS OF ION REGULATION
ON CELL MEMBRANES AND THE CYTOSKELETON

Benjamin F. Trump and Irene K. Berezesky

University of Maryland School of Medicine
Department of Pathology
10 South Pine Street
Baltimore, MD 21201, U.S.A.

A major role of cellular ion regulation in cell injury and cell death has been proposed for many years. Recently, however, new technology has permitted significant improvement in the characterization of such ion movements and redistribution. The following outline summarizes these developments, concentrating on work from our laboratory.

The types of injury under study include inhibition of ATP synthesis (KCN or anoxia, uncouplers, indole acetic acid); direct damage to membrane integrity ($HgCl_2$ or PCMBS); and calcium ionophores (A23187 and ionomycin). The systems studied include Ehrlich ascites tumour cells (EATC), rat, human and rabbit proximal tubule primary cultures, and rat or human hepatocytes. The methods, in addition to well established light- and electron-microscopic techniques, include: fluorescent probes for pH, calcium and membrane peroxidation; immunofluorescence of the cytoskeleton; video microscopy, including digital image processing and analysis of structures as well as distribution and quantitation of fluorescent probes.

In these studies, typical early, reversible changes, as seen by light- and electron-microsocpy, include membrane blebbing at the cell surface, dilatation of the endoplasmic reticulum (e.r.), and mitochondrial condensation. The blebs, which can be readily visualized by phase- or Nomarski-microscopy, are associated with altered staining patterns of actin, tubulin and calmodulin, changes in intracellular calcium distribution and an increase of cytosolic sodium. These blebs occur much more rapidly after $HgCl_2$ treatment than with inhibition of ATP synthesis. Increases of mitochondrial calcium, in the form of hydroxyapatite deposits, are late changes, probably occurring

in most cases in irreversibly altered cells and only if the type
of injury does not initially inhibit mitochondrial function. The
early rise in cytosolic calcium precedes the onset of cell death
as measured by trypan blue or propidium iodide. In some cases, the
early rise in cytosolic calcium may represent redistribution within
the cell (e.g. release from e.r. or mitochondria) rather than entry
from the extracellular space.

We have hypothesized for some time that there are important
roles for ion deregulation in the pathogenesis of cell injury. The
application of new technology to the study of cell injury permits
the correlation of analytical information, e.g. local ion concentrations, with images of living or fixed cells which in turn can be
analyzed by digital image processing. All of this information can,
additionally, be correlated with the results of more classical biochemical and morphological approaches. Current experimentation has
already indicated that changes in cellular ion concentrations occur
at very early time periods following injury and that such ion deregulation plays an important role in initiating the later changes in
cellular structure and function.

============

Some citations (book articles) added by Senior Editor
- supplementing the foregoing *in absentia* Forum contribution

Trump, B.F., Berezesky, I.K., Phelps, P.C. & Jones, R.T. (1983) in
Cellular Pathobiology of Human Disease (Trump, B.F., Jones, R.T. &
Laufer, A., eds.), Gustav Fischer, New York, pp. 3-48. - *'An over-
view of the role of membranes in human disease'*.

Trump, B.F., Berezesky, I.K., Phelps, P.C. & Saladino, A.J. (1983)
in *Human Carcinogenesis* (Harris, C.C. & Autrup, H.N., eds.), Academic
Press, New York, pp. 35-84. - *'Ion regulation and the cytoskeleton
in preneoplastic and neoplastic cells'*.

Hirsimaki, P., Arstila, A.U., Trump, B.F. & Marzella, L. (1983) in
Pathobiology of Cell Membranes, Vol. 3 (Trump, B.F. & Arstila, A.U.,
eds.), Raven Press, New York, pp. 123-135. - *'Autophagocytosis'*.
[*Vol. 1* (1975) *includes mitochondrial damage and lysosome topics.*]

Trump, B.F. & Berezesky, I.K. (1984) in *Drug Metabolism and Drug
Toxicity* (Mitchell, J.R. & Horning, M.G., eds.), Raven Press, New
York, pp. 261-300. - *'The role of sodium and calcium regulation
in toxic cell injury'*.

Trump, B.F., Phelps, P.C. & Shamsuddin, A.M. (1984) in *Progress in
Cancer Research and Therapy*, Vol. 29 (Wolman, S.R. & Mastromarino, A.J.,
eds.), Raven Press, New York, pp. 23-49. - *'Cellular pathobiology of
human large intestine'*.

Trump, B.F. & Berezesky, I.K. (1985) *Curr. Topics Membs. Transport 25*, 279-
319.- *'Cellular Ion Regulation and Disease: A Hypothesis'*.

#NC(C)-2

A Note on

THE GLYCOPROTEINS AND FLUIDITY OF MULTINODULAR GOITER EUTHYROID PLASMA MEMBRANES [⊗]

[1]H. Depauw, [1]C. Peeters, [1]H.J. Hilderson, [1]M. De Wolf,
[2]G. Van Dessel, [1]A. Lagrou and [1,2]W. Dierick

[1]RUCA-Laboratory for Human   [2]UIA-Laboratory for
  Biochemistry   Pathological Biochemistry
                University of Antwerp
                Groenenborgerlaan 171, B-2020 Antwerp, Belgium

Changes in membrane composition and fluidity may reflect derangements in cellular functions. In plasma membranes (p.m.'s) they can modulate adenylate cyclase activity [1, 2]. Differences in both composition and physicochemical state could be demonstrated when comparing thyroid p.m.'s obtained (method as in [3]) from patients with normal thyroid and patients with multinodular goiter euthyroid gland. Normal thyroid membranes displayed higher protein-to-phospholipid ratios and a lower 5'-nucleotidase specific activity (expts. in triplicate on 1 goiter and 2 normal thyroid p.m. preparations; ±S.E.M.'s):

- protein:phospholipid (w/w): normal, 8.59 ±0.41; goiter, 4.26 ±0.15;
- cholesterol:phospholipid (molar ratio basis):   normal, 0.69 ±0.08; goiter, 0.38 ±0.03;
- 5'-nucleotidase (nmol/min/mg protein): normal, 8.3; goiter, 18.4.

Normal and goiter p.m.'s gave similar values for basal and TSH- or forskolin-stimulated adenylate cyclase [pmol cAMP/mg protein/30 min; mean of 3 expts. (one preparation for the goiter p.m.'s)]:

- *no* agonist: normal, 0.34; goiter, 0.35;
- TSH (10 µM): normal, 0.95; goiter, 0.98;
- forskolin (10 µM): normal, 1.47; goiter, 1.52;
- NaF (10 mM): normal, 6.46; goiter, 9.80;
- cholera toxin, activated with 20 mM DTT for 10 min at 30°: normal, 3.26; goiter, 4.80.

Whilst adenylate cyclase activity was less stimulated by NaF or cholera toxin in normal thyroid membranes, the similar s.a.'s with TSH or forskolin present suggest that all preparations used were contaminated to a similar degree with non-p.m. elements.

[⊗] An **Addendum** outlines lectin specificities (at Editors' suggestion).

GLYCOPROTEINS

After detergent solubilization of proteins (1.5% DOC, pH 7.5), SDS-gel electrophoresis, nitrocellulose transfer (25 mM Tris-192 mM glycine, 20% methanol, pH 8.3) and FITC-lectin 'staining', more proteins were manifest as an intense peak in the high mol. wt. region than with Coomassie blue. The densitograms of membrane proteins of normal thyroid, using different fluorochrome-labelled lectins, were distinct from those of membrane proteins of multinodular goiter euthyroid gland. Table 1 summarizes the apparent mol. wts. derived from these scans. Three bands of kM_r 185, 145 and 58 respectively were always manifest. They could originate from proteins that are detected by each of the lectins used, although it must be stressed that co-migration in SDS gels does not always mean 'identity'

Table 1 further shows that with the two FITC-agglutinins more glycoproteins were revealed in the multinodular goiter euthyroid membrane electroblots than in those obtained from normal thyroid membranes. RCA chiefly recognizes the two N-acetyl-lactosaminyl groups of the oligosaccharide chain whereby the Gal residues have to be minimally substituted (see **Addendum**). The FITC-RCA electroblot could in this respect suggest that in the former membranes more proteins are present with fewer substitutions on the Gal residues of the N-acetyl-lactosaminyl groups on the trimannosyl core. This is confirmed by the FITC-WGA electroblots. WGA chiefly recognizes the chitobiosyl-asparagine group although peripheral sialic acid residues and to a less extent N-acetylglucosaminyl and N-acetylgalactosaminyl groups enhance the lectin affinity. This could explain why the two kinds of electroblots do not coincide completely. The results match the observation that in abnormally proliferating cells the pattern of cell surface determinants (mostly glycoconjugated) is altered.

Table 1. Mol. wts. (kM_r, approx.) of human p.m. glycoproteins after SDS-PAGE (gels 7.5%, w/v), nitrocellulose transfer and visualization with FITC-lectins. '^{6}>200' signifies 6 bands of kM_r>200; 'some' signifies some minor bands. Each upper row = **normal**, lower = **goiter**.

FITC-Concanavalin (-Con A)

normal	230	^{6}>200		185		145	120		100	86	72-75	67	63	56	54	50		45
goiter	225	some	190		150	145	120	110	105	82	72				54		47	

FITC-Wheat germ agglutinin (-WGA)

normal			^{5}>200	185	145			80			58	51	some
goiter	265	240	215	185	140	115	105		73	65	56	49	

FITC-*Ricinus communis* agglutinin (-RCA)

normal			^{5}>200	185			145				87			58	some
goiter	260	240	225		180	160	140	130	110	100	88	77	67	58	

Table 2. Fluidity characteristics of human normal and multinodular goiter euthyroid p.m.'s at 37°. Parameters: $r_{S_{DPH}}$, $r_{S_{12-AS}}$ and $r_{S_{TMA-DPH}}$ = steady-state fluorescence anisotropy of, respectively, 1,6-diphenyl-1,3,5-hexatriene (DPH), 12-(9-anthroyl)stearic acid (12-AS) and 1-(4-trimethylammoniumphenyl)-6-phenyl-1,3,5-hexatriene (TMA-DPH); D_{diff} = lateral diffusion coefficient of pyrene, cm^2/sec; ± values represent S.E.M.'s.

	$r_{S_{DPH}}$	$r_{S_{12-AS}}$	$r_{S_{TMA-DPH}}$	D_{diff} $(\times 10^8)$
Normal	0.228 ±0.004	0.165 ±0.003	0.258 ±0.004	0.82 ±0.12
Goiter	0.151 ±0.003	0.132 ±0.002	0.169 ±0.003	1.22 ±0.18

MEMBRANE FLUIDITY

Evaluation of the steady-state fluorescence anisotropy (r_S) of different probes such as those named in the heading to Table 2 (DPH, 12-AS, TMA-DPH) combined with the measurement of the lateral diffusion coefficient of pyrene (D_{diff}) allowed us to monitor changes in membrane fluidity at different levels of the lipid bilayer. From Table 2 it is obvious that membranes derived from the multinodular goiter euthyroid gland display a higher fluidity at any depth of the bilayer than those prepared from normal tissue.

In thyroid a higher value of r_S can be correlated with more abundant neutral lipids. This was previously demonstrated using reconstituted vesicles derived from highly purified bovine thyroid p.m.'s [4]. From the same studies it also became apparent that D_{diff} could be lowered by the presence of more p.m. proteins. Hence it is not surprising that human multinodular goiter euthyroid membranes, diplaying lower protein:phospholipid and cholesterol:phospholipid ratios show lower r_S's and a higher D_{diff}. One cannot exclude the possibility that the observed increase in NaF and cholera-toxin stimulation in the goiter membranes is related to the higher fluidity in the former membranes.

COMMENT

These data support the view that the coupling between the α_S subunit (ADP-ribosylated when cholera toxin activates) of the stimulatory GTP-binding regulatory protein (N_S) with the catalyst of the adenylate cyclase system depends on membrane fluidity, whereas the rate-limiting coupling between the occupied stimulatory receptors and N_S is not as sensitive to the physico-chemical state of the membrane. The latter point also accords with the slight temperature-dependency of the activation of adenylate cyclase activity by TSH [5].

References

1. McOsker, C.C., Welland, G.A. & Zilversmit, D.B. (1983) *J. Biol. Chem. 258*, 13017-13026.
2. Salesse, R., Garnier, J., Leterrier, F., Daveloose, D. & Viret, J. (1982) *Biochemistry 21*, 1581-1586.
3. Orgiazzi, J., Williams, D.E., Chopra, I.J. & Solomon, D.J. (1976) *J. Clin. Endocrinol. Metab. 42*, 341-354.
4. Depauw, H., De Wolf, M., Van Dessel, G., Hilderson, H.J., Lagrou, A. & Dierick, W. (1985) *Biochim. Biophys. Acta 814,* 57-67.
5. Van Sande, J., Pochet, R. & Dumont, J.E. (1977) *Biochim.Biophys. Acta 585*, 282-292.
6. Debray, H., Decout, D., Strecker, G., Spik, G. & Montreuil, J. (1981) *Eur. J. Biochem. 117*, 41-55   *[cited in* **Addendum***]*.

Addendum: ASPECTS OF LECTIN SPECIFICITY *[Presentation revised by Ed.]*

$$[(\text{SA}-\text{Gal}-\text{HNAc})_{2\,1}-\text{Man}]_2-\text{Man}-\text{HNAc}-\text{HNAc}-\text{AsN}$$

| 1 | 2 | 3 | 4 | 5 | 6 | 7 | 8 |

[The **1-3** *chains,* ⩾2 *in all, may differ]*

Non-standard abbreviations: SA, sialic acid; H, hexose.
Lectins: see Table 1.

Extrapolation of known lectin specificities for oligosaccharides [6] to glycopeptides is not without risk, since inhibition properties may differ. Lectins recognize specific sugar residues, but in an oligosaccharide chain more than one residue may be recognized; it matters that some sugar residues are substituted and others not. Different lectins can recognize different sugar sequences in the same glycan structure, but do not overlap unduly and hence differences in fine structure as well as in environment (protein structure) are important. Referring to the notional structure above, main affinities are now summarized.

- **Con A.** The tri-Man core **4-5**, either substituted with 2 GlcNAc's, **3** (not other HNAc's) or with extra Man residues at **3**. ConA has a pronounced specificity for α-anomers of mannose (or glucose).
- **RCA.** Oligosaccharides having the Gal's at **2** minimally substituted. The AsN (**8**) is inhibitory, but less so in a peptide.
- **WGA.** Chitotriose or, as found in glycopeptides, a chitobiosyl group between **8** and the tri-Man core **4-5**. SA's (**1**) and to a less extent **3** where H = Glc or Gal increase the affinity. Hence very complex interactions occur and interpretation is difficult.

In the above thyroid studies, lectins were used with these considerations in mind. To summarize, "Lectins are able to bind sugars specifically, but it is well known that the monosaccharide residue in a terminal non-reducing position on a glycan is not the only carbohydrate moiety recognized" [6].

Supplementary refs. contributed by Senior Editor (ctd. from p. 232)

Membranes/ions/cell damage

Impaired Cl^- passage through apical membranes from airway epithelial cells studied by patch-clamping is observed in cystic fibrosis, but the defect seems to be in channel regulation rather than capacity.- Welsh, M.J.W. & Liedtke, C.M.L. (1986) *Nature 322*, 467-470.

In cultured hepatocytes exposed to 10 different toxins that damage the p.m. in various ways, the damage allowed entry of extracellular Ca^{2+} as the start of a final common pathway for cell death.- Schanne, F.A.X., Kane, A.B., Young, E.E. & Farber, J.L. (1979) *Science 206*, 700-702.

However, "it appears that influx of extracellular Ca^{2+} is not an obligatory step in the development of irreversible toxic injury in hepatocytes".- Orrenius, S. & Bellomo, G. (1986) in *Cell Calcium* Vol. 6 (Cheung, W.Y., ed.), Academic Press, New York, pp. 185-208.

In ischaemic brain damage, uncontrolled influx of Ca^{2+} into certain neurones (e.g. through p.m. damage or energy deprivation) can cause mitochondrial Ca^{2+} overload or extensive degradation of cell structure through proteolysis or lipolysis.- Siesjö, B.K. (1986) *Mayo Clin. Proc. 61*, 299-302.

In liver exposed *in vivo* or by perfusion to sporidodesmin, which causes obstructive jaundice, bile canalicular microvilli rapidly disappear.- Bullock, G.M., Eakins, M.N., Sawyer, B.C. & Slater, T.F. (1974) *Proc. Roy. Soc. 186B*, 333-356.

Loss of hepatocyte microvilli also results from ionophore action, notably dinitrophenol (which raises cytosolic Ca^{2+}); evidence of cell injury includes mitochondrial swelling and loss of contents, and a fall in the cell's ATP.- George, M., Chenery, R.J. & Krishna, G. (1982) *Toxicol. Appl. Pharmacol. 66*, 349-360.

When ischaemic myocardium is re-oxygenated or re-perfused, Ca^{2+} influx and mitochondrial deposition occurs, with damage for which different causes including p.m. disruption and radical formation have been suggested.- Poole-Wilson, P.A., Harding, D.P., Bourdillon, P.D. & Tones, M.A. (1984) *J. Mol. Cell. Cardiol. 16*, 175-187.

Studies with liposomes or adipocytes exposed to gossypol, a highly lipophilic drug, indicated non-specific perturbation of membrane lipid regions that may relate to its myriad disruptive effects, especially in cells particularly sensitive to variations in glucose availability.- De Peyster, A., Hyslop, P.A., Kuhn, C.E. & Sauerheber, R.D. (1986) *Biochem. Pharmacol. 35*, 3293-3300.

Nutrient uptake also appeared to be involved in effects of acute or chronic ethanol ingestion on liver p.m. enzymes.- Gonzalez-Calvin, J.L., ... & Williams, R. (1985) *ibid. 34*, 2685-2289.

Cellular aspects of diabetes mellitus have been considered.- Goldstein, S. (1984) *Path. Biol. 32*, 99-106.

Endothelial injury, including possible p.m. damage (as postulated for bleomycin), has been surveyed in respect of anti-neoplastic agents.- Lazo, J.S. (1986) *Biochem. Pharmacol. 35*, 1919-1923.

Forum contribution by C.R. Birkett (Univ. of Kent, Canterbury) on
THE BENZIMIDAZOLE CARBAMATE ANTI-MICROTUBULE DRUGS

Amongst anti-microtubule drug families, structurally unrelated, the benzimidazole carbamates have received especial attention because activity depends on the side-group. Particular drugs exhibit greater specificity for the tubulin (the target protein) of lower eukaryotes than of mammals, as tested with microtubule protein purified from parasitic nematodes. With live nematodes drug exposure causes intra-cellular derangements associated with loss of microtubules, especially in the highly active secretory cells of the gut.

Comment (Birkett agreed) by M. Mareel.- In mammalian cells micro-tubules do not seem to have a cytoskeletal function [De Brabander, M., et al. (1977) *Cell Biol. Internat. Repts. 1*, 177-183]. *Editor's note.-* In *The Cytoskeleton: A Target for Toxic Agents* (1986; Clarkson, T.W., et al., eds.; Plenum), the Canterbury work is presented by E.H. Byard & K. Gull (pp. 83-96); it includes benzimidazole carbamate inhibition of protein and TG secretion by hepatocytes [Birkett, C.R., et al. (1981) *Biochem. Pharmacol. 30*, 1629-1633]. They also cite Mareel et al. (cf. #E-8) for microtubule inhibitors as anti-invasive agents. T.L.M. Syversen et al. (pp. 23-34) survey toxic agents as tools for cytoskeletal studies, especially on tubulin, besides (e.g.) neuropathy due to vinca alkaloids. The book gives nil attention to the e.r.

Other supplementary refs. contributed by Senior Editor

Microtubules and microfilaments

With liver subcellular fractions from rats given colchicine or vinblastine, evidence was obtained for decreased incorporation of certain membrane-forming phospholipids and retarded membrane flow from one compartment to another.- Azhar, S., Hwang, S-F. & Reaven, E. (1985) *Biochem. Pharmacol. 34*, 3153-3159.

Perfused liver exposed to cytochalasin B, a microfilament poison, from rats whose lysosomes had been loaded with Triton WR-1339, showed no increased loss of the detergent or of lysosomal enzymes, although LDH loss was reduced.- Michelakakis, H. & Danpure, C.J. (1986) *ibid. 35*, 933-938. In Vol. 15 of this series there are surveys, notably by C.R. Birkett & K. Gull and by W.W. Franke & R.A. Quinlan, of ways to study microtubules and intermediate filaments (thick microfila-ments). The latter have been studied (M.G. Caldwell & co-workers, mid-1970's) in relation to chronic ethanol ingestion.

Nucleus, nucleic acids and ribosomes

These cell components feature in B.F. Trump's publications (cf. p. 234) but not in the present book. In an early sketch, E. Farber [(1971) *Biochem. Pharmacol. 20*, 1023-26] noted two patterns of response to inhibition of RNA synthesis. Agents that react primarily with DNA (e.g. actinomycin) cause dissociation of nucleolar components. Ribosome refs.: ischaemia, Bernelli-Zazzera, A., et al. (1971) *Exp. Mol. Path. 17*, 121-131; carcinogens *in vitro* (or centrifugation!): B.R. Rabin & co-authors (1978) *Biochem. J. 176*, 9-14.

Section #D

EPITHELIAL MEMBRANES IN DIGESTIVE TISSUES

#D-1

GASTRODUODENAL EPITHELIAL CELLS: THE ROLE OF THE APICAL MEMBRANE IN MUCOSAL PROTECTION

J.M. Wilkes, H.J. Ballard, J.A.E. Latham and B.H. Hirst

Department of Physiological Sciences
University of Newcastle upon Tyne, Medical School
Newcastle upon Tyne, NE2 4HH,   U.K.

The aim was to investigate apical membranes of the epithelial cells in the upper GI* tract as a possible barrier to luminal acid and other insults, forming part of the 'intrinsic protective mechanism' of the mucosa.   To study the integrity and transport capabilities of membrane vesicles, p.m.'s were isolated from duodenal enterocytes (furnishing BBMV's) and histamine-stimulated parietal cells (furnishing SAV's) and, for comparison, microsomes from resting parietal cells. The effect of trypsin on membrane vesicle integrity and function was determined. Both BBMV's and SAV's manifested a major population of vesicles resistant to trypsin.

These membrane types showed reproducible differences in lipid bilayer fluidity, monitored by the fluorescence polarization of diphenyl hexatriene (DPH). An increase in fluidity resulted from exposure of the membranes to ethanol or benzyl alcohol, and systematically led to increased salt-permeability of BBMV's. An assay for determining H^+ permeability was developed. The H^+-permeation rank order in the vesicle types was SAV's $\ll$ BBMV's $<$ gastric microsomes. These results correlate closely with the situation in vivo, *and indicate a function of apical membranes in acting as a barrier to luminal H^+. While no correlation is evident between native membrane fluidity and H^+ permeation, fluidizing agents increase H^+ permeation systematically.*

The mucosa of the upper GI tract is frequently exposed to aggressive stimuli, which can compromise its integrity.   The agents may be ingested (e.g. ethanol, aspirin) or may be produced by the gut itself (e.g. proteases, bile salt, acid).   The effect of aggressive

---

Abbreviations:* GI, gastrointestinal;   BBM(V), brush border membrane (vesicle);   SAV, stimulation-associated vesicle;   **iso, inside-out; **rso**, right-side out;   p.m., plasma membrane; *for* DPH, pNPP, MSEP, *etc.*, *see text.*

stimuli seems to be exacerbated by the presence of luminal acid [1]; indeed, acid is central to the pathogenesis of peptic ulceration, to the extent that current ulcer therapy is predominantly based on the maxim 'no acid, no ulcer' [2]. The gastroduodenal mucosa of a healthy individual possesses properties which enable it to resist these luminal insults, and peptic ulceration is currently thought to be due to an imbalance between the erosive properties of gastric juice and resistance factors of the mucosa.

Mechanisms of mucosal protection have been broadly divided into two categories: extrinsic and intrinsic protection [3]. Extrinsic protection is concerned with the ability of the mucosa as a whole to prevent entry of H^+ into the tissue or the accumulation of toxic levels of H^+ within the tissue. This is achieved by the specialized secretions of the mucosa (e.g. mucus, bicarbonate), by the high blood flow rate and by the tissue's ability to maintain an acid/base balance.

Intrinsic protection involves the properties of individual cells of the epithelial lining of the mucosa, and their ability to withstand influxing acid or, if damage is widespread, to rapidly restore the integrity of the epithelium. The apical membranes of cells forming the gastroduodenal epithelium will be exposed to luminal acid and other insults, and have inherent resistance to the permeation and/or accumulation of H^+ into the cells and to luminal proteases. These resistance capabilities, together with the epithelial tight junctions, constitute the 'gastric mucosal barrier' [1]. Noxious agents of the class known as 'barrier breakers', e.g. ethanol, detergents and aspirin, increase the permeability of the membrane to H^+ [4-6].

Gastric mucosal cells have the capacity to extrude H^+ by the operation of a Na^+/H^+ exchange mechanism [7], while the apical membrane of the parietal cell maintains proton gradients of 6 orders of magnitude, generated by the operation of a H^+/K^+-ATPase. The approach pursued in these studies is to isolate membrane vesicles from defined cell types of the gastroduodenal mucosa, and to use these as models of membrane resistance.

The following membrane types were selected for study.-
(a) Apical membranes of histamine-stimulated parietal cells; these SAV's possess H^+/K^+-ATPase, K^+ and Cl^- conductances and in vivo are exposed to H^+ gradients exceeding 10^6 Such membranes may be regarded as highly resistant to acid.
(b) Tubulovesicular membranes of resting   (H_2-receptor antagonized) parietal cells (gastric microsomes); these possess H^+/K^+-ATPase but no K^+ conductances. Their intracellular localization in vivo indicates that they will not encounter extremes of pH, or other insults.
(c) Microvillus (brush border) membranes of duodenal enterocytes; these BBM's possess enzymes and transport elements involved in digestion and absorption of nutrients, e.g. saccharide transport, and will be exposed to episodic pH changes and luminal barrier breakers.

PREPARATION AND CHARACTERIZATION OF MUCOSAL CELL MEMBRANES

Treatment of animals and preparation of tissue

New Zealand White rabbits (~3 kg) of both sexes were fasted over-night with free access to water. If stimulated gastric mucosa was required, free access to food was allowed for 20 min. Animals were injected s.c. with 1.0 ml of a sedative mixture (ketamine hydro-chloride, 60 mg/ml; xylazine, 60 mg/ml; acepromazine maleate, 1.2 mg/ml), 0.3 ml of chlorpheniramine maleate (15 mg/ml) and - repeated 15 min later - 0.3 ml of histamine acid phosphate (100 mM). After a further 10 min the animals were anaesthetized with i.v. Na pentobarbital. For the preparation of resting mucosa the animals were not fed, and chlorpheniramine and histamine injections were replaced by 2×0.3 ml of cimetidine (150 mg/ml). The remaining procedures were identical for the two secretory states.

Following anaesthesia the stomach was removed, divided longi-tudinally, washed free of luminal contents and immersed in ice-cold saline. The antrum was discarded; the fundus was blotted to remove mucus and scraped with glass microscope slides. The mucosa was trans-ferred to 20 ml of ice-cold 'MSEP' buffer (125 mM mannitol, 40 mM sucrose, 1 mM EDTA, 5 mM Pipes-Tris, pH 6.7). The tissue was minced with surgical scissors, diluted with 180 ml of MSEP, and homogenized by 16-18 passes of a Potter-Elvehjem homogenizer, with a loose-fitting teflon pestle operating at 200 rev/min (Braun Potter S).

To prepare BBMV's, animals were killed by cervical dislocation or i.v. pentobarbital, and 40 cm of proximal duodenum removed. It was flushed with ice-cold saline, opened longitudinally and blotted to remove adherent mucus. The mucosa was scraped with microscope slides and transferred to 20 vol. of homogenizing medium (50 mM mannitol, 2 mM Tris-Cl, pH 7.4). The tissue was homogenized for 2 min at 0° in a Serval Omnimixer operating at 3000 rev/min.

Preparation of membrane fractions

Gastric mucosa was fractionated by differential centrifugation followed by density-gradient centrifugation based on the method of Wolosin & Forte [8]. BBMV's were isolated from duodenum by Mg^{2+}-induced precipitation of subcellular organelles, followed by differential centrifugation, based on the method of Kessler et al. [9].

Preparation of SAV's. - The homogenate was centrifuged at 800 **g** for 10 min, and the pellet discarded. The supernatant was adjusted to pH 7.4 with 1 M Tris and centrifuged at 7000 **g** for 12 min. The pellet was re-suspended in 12.5 ml sucrose-Tris (300 mM sucrose, 5 mM Tris-HCl pH 7.4) and diluted with 1 vol. of 20% (w/v) Ficoll 400 in sucrose-Tris; 12 ml was layered onto 12 ml of 16% Ficoll 400 in the same medium, and sucrose-Tris layered on top. The gradient was

centrifuged at 135,000 **g** for 16 h. The material from each interface
('10%' and '16%') was diluted into 25-50 vol. of sucrose-Tris and
harvested by centrifugation at 140,000 **g** for 60 min. Pellets were
re-suspended in sucrose-Tris, and either used on the same day or
diluted with 0.8 vol. of 2 M sucrose, frozen in liquid nitrogen and
stored at -70° for up to 1 month. (SAV product: the '16%' material.)

   Preparation of gastric microsomes.- The homogenate was centrifuged
at 800 **g** for 10 min (the pellet being discarded in this and the next
two centrifugations). The supernatant was adjusted to pH 7.4, centrifuged at 2000 **g** for 10 min, and the supernatant centrifuged at 12,000 **g**
for 12 min. The supernatant was centrifuged at 140,000 **g** for 60 min
to yield the microsomal pellet, which was re-suspended in sucrose-Tris,
layered onto a discontinuous gradient made of 5, 10 and 15% (w/v)
Ficoll 400 in sucrose-Tris, and centrifuged for 16 h at 135,000 **g**.
The material accumulating at the interfaces was collected, and diluted,
harvested and further treated as for SAV's (microsomes at 5% interface).

   Preparation of duodenal BBMV's.- Duodenal mucosal homogenate was
stored at 0° and 0.01 vol. 1 M $MgCl_2$ added. Centrifugation, 15 min
later, was at 3000 **g** for 20 min. The supernatant was centrifuged
at 27,000 **g** for 40 min, and the pellet re-suspended in 300 mM mannitol/
5 mM Tris–HCl at a final protein concentration of 15-30 mg/ml; it
was either used immediately or frozen in liquid nitrogen and stored
at -70°.

Characterization of membrane fractions

   Protein was determined by Bradford's method [10] using γ–globulin
(Cohn fraction V) as standard. Zn^{2+}-resistant α-glucosidase activity
was determined fluorometrically by the method of Peters [11]. Samples
were incubated with 1 vol. of 20 mM $ZnSO_4$ for 20 min at 0°, and then
at 37° (in 0.25 ml) in the presence of 0.21 mM 4-methylumbelliferyl-
α-D-glucopyranoside and 0.1 M phosphate buffer, pH 8.0. The reaction
was stopped by adding 2 ml pH 10.4 50 mM glycine buffer. Liberated
4-methylumbelliferone was measured at 460 nm (excitation at 365 nm;
Perkin Elmer LS-5 spectrometer). The latency of this activity was
investigated by assaying in the presence of 300mM mannitol with or
without 0.1% (w/v) Triton X-100 present.

   p-Nitrophenyl phosphatase (pNPPase) activities were determined
in 1 ml of medium containing 10-50 µg of protein, 20 mM Tris–HCl
(pH 7.5), 4 mM $MgSO_4$, 0.2 mM EDTA, 1 mM ouabain and 10 mM NaCl or
KCl, initiated by addition of Na pNPP to 5 mM. K^+-pNPPase, an expression of the H^+/K^+-ATPase, was defined as the activity in the presence
of 10 mM K^+, after subtraction of that in the presence of 10 mM Na^+.

   Latent K^+-pNPPase activities were determined by incubation of
vesicles with the detergent n-octyl glucoside [12], 6-18 mM in 100 µl
sucrose-Tris, with a membrane sample containing 50 µg of protein;

Table 1. Marker enzyme characterization and transport properties of gastroduodenal membrane vesicles (abbreviations: footnote, title p.). Enrichment is expressed as relative specific activity (fraction *vs.* homogenate, on protein basis).

Type	Marker	r.s.a.	% latent	Transport property assayed
Gastric microsome	K^+-pNPPase	10-12	0	ATP-dependent H^+-accumulation (expressed only by vesicles with **iso** orientation)
SAV	K^+-pNPPase	6-9	50-70	
BBMV	Zn^{2+}-resistant α-glucosidase	15-18	0	Na^+-dependent glucose accumulation

after 10 min at room temperature, 9 vol. of ice-cold sucrose-Tris was added and 200 µl aliquots were assayed for K^+-pNPPase in the presence of 250 mM sucrose as osmotic protectant (i.e. total osmotic pressure ~300 mOsm).

Table 1 summarizes the characterization of the membrane preparations by enzymic analysis.

Transport properties of vesicles

H^+-uptake. - Vesicular acidification was followed by quenching of acridine orange fluorescence [12]. An aliquot of vesicles was diluted into a medium containing 140 mM KCl, 20 mM sucrose, 2.5 mM Tris-Tes (pH 7.2), 50 mM EDTA and 2 µM Mg-ATP, and fluorescence was monitored at 530 nm (excitation at 493 nm).

Fig. 1 summarizes the uptake of H^+ by SAV's and gastric microsomes. Both types of vesicle demonstrate ATP-dependent H^+ accumulation due to the action of the H^+/K^+-ATPase. The process in gastric microsomes is dependent on valinomycin, indicating impermeability to K^+, whereas SAV's possess a conductance for this ion. Due to the enzyme's orientation, only sealed vesicles with an **iso** orientation may be expected to demonstrate H^+ accumulation. Both preparations thus contain **iso** vesicles, exclusively so for gastric microsomal vesicles whereas the presence of latent K^+-pNPPase (and hence H^+/K^+-ATPase) in SAV's indicates a mixed population of **iso** and **rso** vesicles. Evidence from freeze-fracture electron micrographs of SAV's indicates that the vesicles are predominantly **rso** [13].

Na^+-dependent glucose uptake. - BBMV's were assayed for their ability to support Na^+-dependent glucose accumulation [14]. Vesicles (15-30 mg protein/ml) were added to 1 vol. of 0.15 M NaCl/0.2 mM Hepes-HCl (pH 7.4)/0.2 mM glucose containing [U-^{14}C]glucose, at room temperature. At intervals aliquots were removed and filtered through nitrocellulose filters (mesh size 0.45 µm) in a vacuum filtration manifold (Amicon VFM1). Filters were washed with 2 × 5 ml aliquots

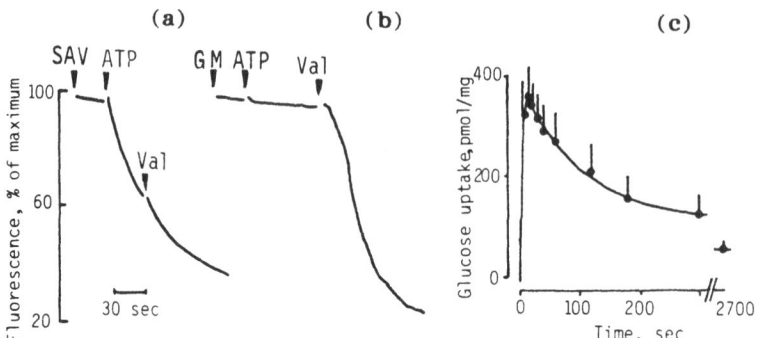

Fig. 1. Transport properties of membrane vesicles isolated from (a, b) gastric or (c) duodenal mucosa. Val = 10 μM valinomycin. (a): ATP-dependent H$^+$-accumulation by SAV's (method as in text). (b): ATP-dependent H$^+$-accumulation by gastric microsomes. (c): Na$^+$-dependent glucose accumulation by BBMV's. Points represent mean ±S.E.M. (n = 6).

of ice-cold 0.5 mM phloridzin in 0.15 M NaCl, allowed to dry, left overnight in scintillant (5 ml; LKB Optiphase Safe), and counted.

The accumulation of glucose by BBMV's in response to an imposed Na$^+$ gradient demonstrates the presence of a population of sealed vesicles. The lack of latent Zn^{2+}-resistant α-glucosidase activity (an ecto enzyme *in vivo*) indicates that the vesicles are exclusively **rso**. The glucose retained at equilibrium indicates an intravesicular volume of 0.5-1 μl/mg protein.

Resistance of luminal membrane vesicles to trypsin

Treatment with trypsin (bovine, type I) was performed at 30° in sucrose-Tris (gastric microsomes and SAV's) or 300 mM mannitol/5 mM Tris-HCl, pH 7.4 (BBMV's), the wt. ratio of trypsin to membrane protein being varied from 1:100 to 1:5. At intervals aliquots were added to soya bean trypsin inhibitor (preceding trypsin in the controls) in a 10-fold wt. excess over trypsin.

Trypsin effects on the gastric vesicles.- Fig. 2 shows the effect on latent and non-latent K$^+$-pNPPase activity and on ATP-dependent H$^+$-accumulation. K$^+$-pNPPase activity associated with gastric microsomes was almost totally destroyed by 10 min incubation with trypsin at a wt. ratio of 1:10, and SAV non-latent activity was reduced to 10% while latent activity was unaffected. This result indicates that apical membranes of parietal cells are intrinsically resistant to trypsin. Treatment with trypsin also produced total inhibition of H$^+$ accumulation, not directly linked to the reduction in K$^+$-pNPPase activity. At a 1:100 wt. ratio, K$^+$-pNPPase activity in gastric

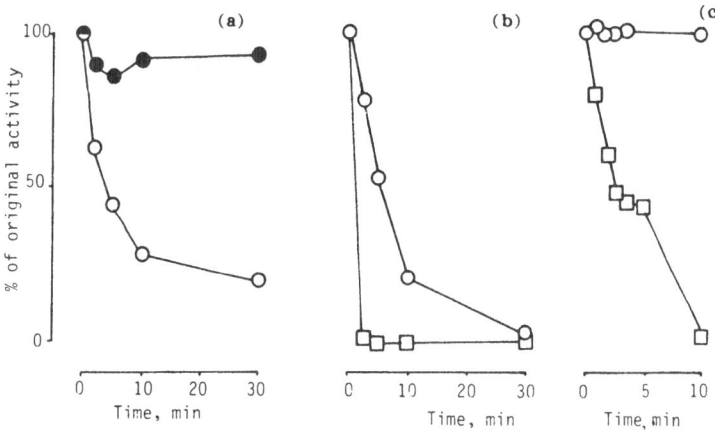

Fig. 2. K$^+$-pNPPase activity and ATP-dependent H$^+$ accumulation as affected by trypsin, in wt. ratio of 1:10 or (c) 1:100 to vesicle protein. Activities expressed at % of the zero-time value.
(a) SAV's: effect on latent (●) and non-latent (o) pNPPase.
(b) and [with less trypsin] (c): gastric microsomes - effects on total pNPPase (o) and H$^+$ accumulation (□).

microsomes was unaffected, but H$^+$ accumulation was progressively reduced. One interpretation of this result is that trypsin has permeabilized the vesicles, and subsequently allowed transported protons to leak back into the medium.

Trypsin effects on the duodenal vesicles. - For BBMV's, Fig. 3 summarizes the effect of trypsin (a) on the Na$^+$/glucose accumulation, and (b) on the average enclosed volume. Na$^+$-driven glucose uptake was demonstrable in both control and trypsin-treated vesicles, reaching after 20 sec a peak of 464 ±90 (control) or 326 ±68 (treated) pmol/mg protein (means ±S.E.M.'s; n = 4). The glucose retained at equilibrium was reduced from 68.3 ±17.1 (n = 8) to 33.8 ±5.5 (n=8) pmol/mg. The ratio of peak to equilibrium uptake (control: 8.2 ±0.7) was unaffected by trypsin (9.8 ±1.4), indicating that the functional properties of the Na$^+$/glucose transporter are unaffected by trypsin.

Fig. 3 (b) shows the time course of the trypsin-induced reduction in vesicular volume. It fell from 0.53 ±0.08 (n = 14) to 0.32 ±0.06 (n = 7) μl/mg protein after 30 min and to 0.26 ±0.07 (n = 8) after 80 min of incubation with trypsin at a 1:10 wt. ratio. With 1:5 there were similar reductions, while with 1:50 the parameters measured showed no significant changes. These results indicate the presence of two populations of sealed membrane vesicles, both capable of Na$^+$-dependent glucose transport. The first is sensitive to trypsin which disrupts it rapidly. The second, identical with the first with respect to glucose transport, is resistant. The developmental relationship between these two populations may be of interest.

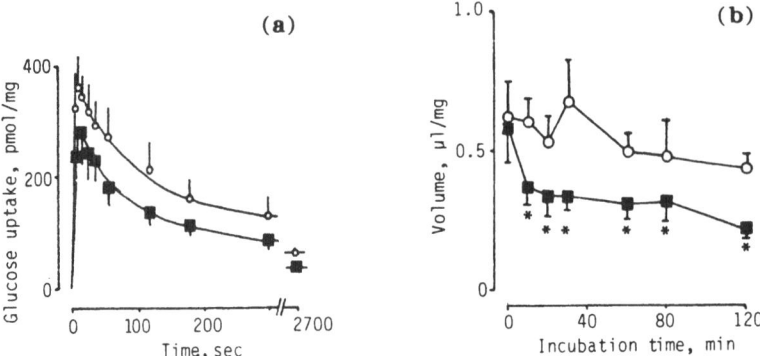

Fig. 3. Na$^+$-dependent glucose accumulation (**a**) and average enclosed volume (**b**) of BBMV's as affected by incubation with trypsin, in wt. ratio of 1:10. Values are means ±S.E.M.; o, control; ■, trypsin. (**a**): 30 min incubation; n = 6. (**b**): n = 8-10; * $p < 0.05$.

Membrane fluidity: effects on membrane permeability and integrity

The fluidity (inverse of the microviscosity) of the hydrophobic region of membrane vesicles was monitored from the degree of fluorescence polarization of 1,6-diphenyl-1,3,5-hexatriene (DPH). Vesicles were suspended in sucrose-Tris in the presence of 2 μM DPH (added as a 2 mM stock solution in tetrahydrofuran). Fluorescence anisotropy (**P**) was measured in a Perkin Elmer LS-5 luminescence spectrometer fitted with an automatically controlled polarizing accessory, and the temperature was maintained by a thermostatted cuvette holder attached to a circulating water bath. P was related to the fluorescence intensities (360 nm excitation, 430 nm emission) when the emission polarizer was oriented parallel ($I_{//}$) and perpendicular ($I_\perp$) to the vertically polarized excitation beam:

$$P = \frac{I_{//} - g \cdot I_\perp}{I_{//} + g \cdot I_\perp} \qquad \text{where} \quad g = \frac{I'_{//}}{I'_\perp}$$

g being a correction factor for the different intensities of parallel and perpendicularly polarized light; primed intensities are emission intensities measured with horizontally polarized excitation beams. The degree of fluorescence polarization was related to apparent microviscosity ($\bar{\eta}$) by the following equation [15]:

$$\bar{\eta} = \frac{2P}{0.46 - P}$$

Fluidity is the inverse of this value, and has units of poise^{-1}. Agents were added to the membrane suspension and the system allowed to equilibrate for 2 min before readings were taken.

Fig. 4a shows the effect of temperature on the degree of fluorescence polarization, measured in the presence of the three membrane

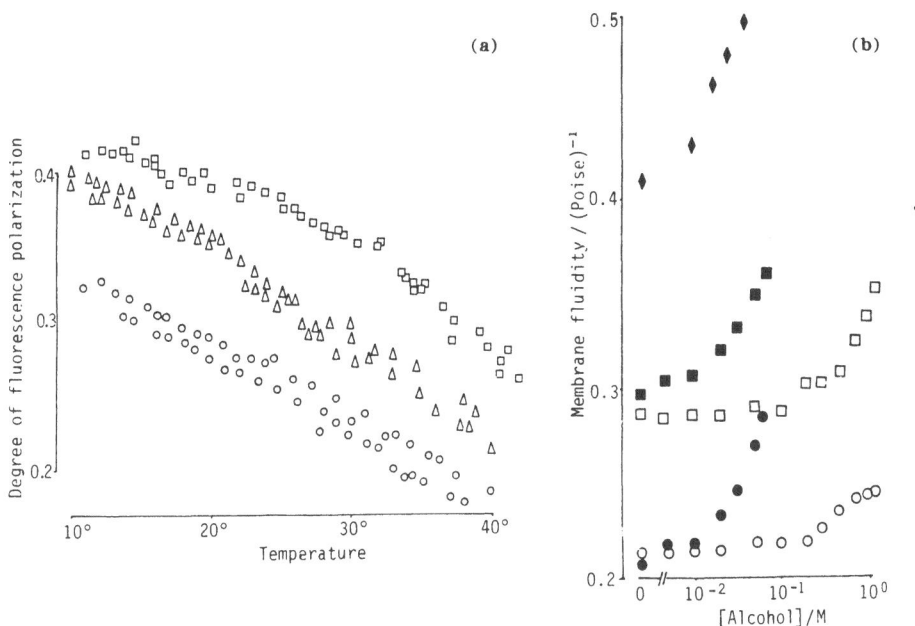

Fig. 4. Fluidity of membranes, assessed by fluorescence anisotropy of DPH. (a) Effect of temperature on the anisotropy (2 expts.) of: □, BBMV's; △, SAV's; o, gastric microsome membranes. (b) Effect of benzyl alcohol (●, ■, ◆) and ethanol (o, □) on apparent membrane fluidity of BBMV's (●,o), SAV's (■, □) and gastric microsomes (◆).

preparations. These have characteristic fluorescence polarization values, and fluidity is increased (i.e. fluorescence polarization decreased) by increasing temperatures. The order of fluidity is BBMV < SAV < gastric microsomes. The effect of increasing concentrations of benzyl alcohol and ethanol on membrane fluidity is shown in Fig. 4b. Both agents increase fluidity in all membrane types studied. Aspirin and bile salts at concentrations below their critical micellar concentrations produced no effect on membrane fluidity as assessed by DPH fluorescence polarization.

Effect of membrane fluidization on permeability

The permeability of BBMV's to NaCl was determined from the rate of re-swell of a suspension of vesicles following exposure to a hypertonic pulse of salt. The rate of re-swell, and hence the time course of scattered-light intensity as used to monitor it, has been shown to follow first-order kinetics [16].

BBMV's (0.1 mg protein/ml) in sucrose-Tris and 50 mM benzyl alcohol were equilibrated at 30° (instrumentation as above; 450 nm excitation, and emission grating removed); 0.1 vol. of 2.5 M NaCl

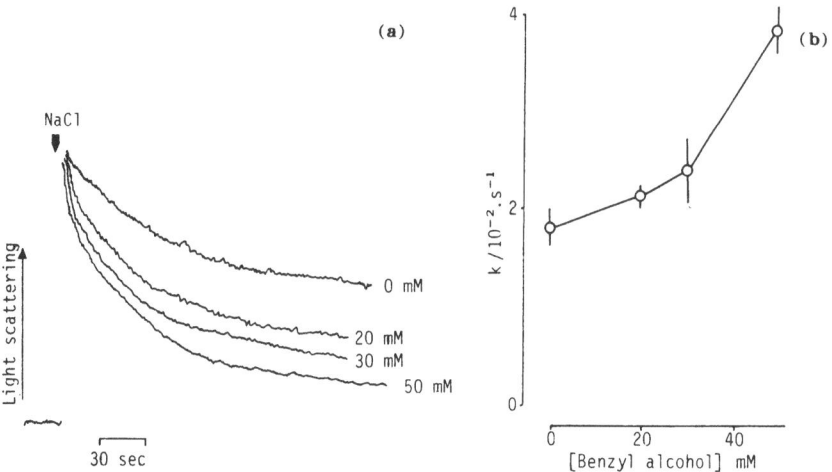

Fig. 5. Effect of benzyl alcohol (0–50 mM) on NaCl permeability of BBMV's. (a) Time course of re-swell (monitored by light-scattering decrease) following exposure to a hypertonic pulse of NaCl (0.25 M). (b) First-order rate constants determined from the re-swell time course; mean ±S.D. (n = 3).

in sucrose-Tris was added, the contents mixed by rapid aspiration into a pipette, and light-scattering monitored. Time courses were ascertained, and first-order rate constants calculated after the method of Rabon et al. [16].

The effect of increasing concentrations of benzyl alcohol on the time course of re-swell, and the rate constant for re-swell, are shown in Fig. 5. Increasing concentrations increased vesicle permeability, as demonstrated by the magnitude of the rate constant for re-swell.

Proton permeability of membranes

The permeability of membrane vesicles to H^+ was measured from the rate of collapse of pre-formed pH gradients, monitored by acridine orange fluorescence quenching [17]. The vesicles were equilibrated overnight in a medium containing 300 mM sucrose/10 mM Hepes-Tris, pH 6.5. H^+ permeation was measured by diluting vesicles to a final concentration of 0.1–0.2 mg/ml into 6 µM acridine orange in the same medium but of pH 8.0 and with 0–50 mM benzyl alcohol present. Addition of vesicles resulted in a 50–70% decrease in the acridine orange fluorescent signal, whose subsequent time course was followed (494 nm excitation, 530 nm emission).

The recovery of fluorescence followed a first-order process (Figs. 6 & 7). The observed order of proton permeabilities was SAV's

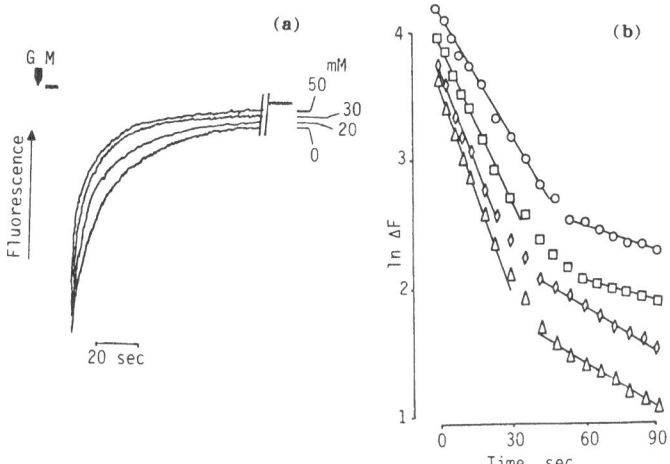

Fig. 6. H^+ permeability of gastric microsomes: acridine orange fluorescence quenching *vs.* time. (**a**) Relation to concentration of benzyl alcohol (added to alter fluidity). **G M** denotes gastric microsomes. (**b**) Curves as described by a bi-exponential function:

$$\Delta F_t = \Delta F_o (A \cdot e^{k't} + B \cdot e^{k''t})$$

A, B are proportionality constants (A + B = 1); k', k'' are first-order rate constants; $\Delta F_o, \Delta F_t$ = fluorescence quenching, initial and at time t. Curves analyzed are for 0 (o), 10 (□), 25 (◇) & 50 (△) mM benzyl alcohol.

Fig. 7. H^+ permeability of vesicles: acridine orange fluorescence quenching *vs.* time, with different benzyl alcohol concentrations. Analysis of the time courses gave rate constants: k' (o) and k'' (●) for gastric microsomes; k' (⊔) for BBMV's; k' (△) for SAV's.

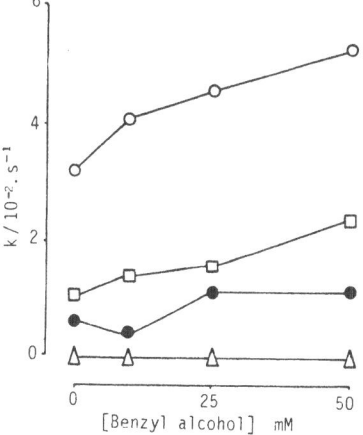

(nil) < BBMV's < gastric microsomes, correlating well with their physiological situations. The parietal cell membrane has to maintain large pH gradients, and any proton leak would short-circuit the K+/H+-ATPase responsible for H^+ secretion. BBM's will be periodically exposed to low pH, while gastric microsomes will be exposed only to intracellular pH, which is maintained between narrow limits by active processes. There is no correlation between membrane fluidity (BBMV's < SAV's < gastric microsomes) and intrinsic proton permeability (SAV's < BBMV's < gastric microsomes); but the presence of the fluidizing agent benzyl alcohol increases proton permeability (Fig. 6b) in both gastric microsomes and BBMV's. The experiments were not sensitive enough to demonstrate any change in the permeability of SAV's.

Acknowledgements

Support came from an MRC grant G8418056CA, a University of Newcastle Research Grant 563022 and a grant from the SmithKline Foundation. J.A.E.L. was supported by a grant from the Physiological Society (Dale Fund).

References

1. Davenport, H.W. (1975) *Mayo Clin. Proc. 50*, 507-514.
2. Schwartz, K. (1910) *Beitrage Klin. Chirurg. 67*, 96-128.
3. Powell, D.W. (1984) in *Mechanisms of Mucosal Protection in the Upper Gastrointestinal Tract* (Allen, A., Flemström, G., Gorner, A., Silen, W. & Turnberg, L.A., eds.), Raven Press, New York, pp. 1-6.
4. Guteknecht, J. & Tosteson, D.C. (1970) *J. Gen. Physiol. 55*, 359-374.
5. Helenius, A. & Simons, K. (1975) *Biochim. Biophys. Acta 415*, 29-79.
6. McLaughlin, S.G. & Dilger, J.P. (1980) *Physiol. Rev. 60*, 825-863.
7. Olender, E.J., Fromm, D., Furukawa, T. & Kolis, M. (1984) *Gastroenterology 86*, 698-705.
8. Wolosin, M. & Forte, J.G. (1981) *J. Biol. Chem. 256*, 3149-3152.
9. Kessler, M., Acuto, O., Storelli, C., Murer, H., Müller, M. & Semenza, G. (1978) *Biochim. Biophys. Acta 506*, 136-154.
10. Bradford, M. (1976) *Anal. Biochem. 72*, 248-254.
11. Peters, T.J. (1976) *Clin. Sci. 51*, 557-574.
12. Hirst, B.H. & Forte, J.G. (1985) *Biochem. J. 231*, 641-649.
13. Black, J.A., Forte, J.G. & Hirst, B.H. (1985) *J. Physiol. 371*, 136P.
14. Hopfer, U., Nelson, K., Perrotto, J. & Isselbacher, K.J. (1973) *J. Biol. Chem. 248*, 25-32.
15. Shinitzky, M. & Barenholz, Y. (1978) *Biochim. Biophys. Acta 515*, 367-394.
16. Rabon, E., Takeguchi, N. & Sachs, G. (1980) *J. Memb. Biol. 53*, 109-117.
17. Ives, H.E. & Verkman, A.S. (1985) *Am. J. Physiol. 249*, F933-F940.

#D-2

STRUCTURAL AND FUNCTIONAL ASPECTS OF THE
INTESTINAL BRUSH BORDER

G.M. Cowell, E.M. Danielsen, S.U. Friis, S.U. Gorr,
G.H. Hansen, O. Norén, H. Sjöström and H. Skovbjerg

Department of Biochemistry C
The Panum Institute, University of Copenhagen
Blegdamsvej 3C, DK-2200 Copenhagen N, Denmark

The small intestinal brush border membrane contains several peptidases and glycosidases of importance for the final digestion of proteins and carbohydrates. These enzymes typically have a main portion in the intestinal lumen and a small N-terminal portion anchoring them to the membrane. Certain maldigestion diseases can be explained by deficient expression of these enzymes. The enzymes are synthesized in the rough e.r.* and transported to the brush border membrane via the Golgi apparatus, where they are modified. Disturbances in these processes may cause depressed enzyme activities; thus sucrase-isomaltase is not fully active if incompletely glycosylated.

In coeliac disease the microvillar structure is deranged because of a reaction to wheat gliadin proteins, which also give rise to circulating Ab's. The disease was formerly attributed to malfunction of some digestive enzyme, but conceivably the gliadin peptides have a direct toxic effect on the membrane structure.

The brush border membrane is continuously renewed by the insertion of exocytotic vesicles, which in some part of their route may gain a protein coat. To help elucidate the mechanism for p.m. renewal, methods have been developed for purifying coated vesicles from the small intestine of the pig.

The small intestinal mucosa is typically folded into villi which are interrupted by regenerative crypts and are supported by a connective tissue layer containing blood and lymph vessels, nerve fibres, smooth muscle cells and fibroblasts. The monoepithelial mucosal layer comprises mainly enterocytes but also goblet and endocrine cells.

---

*Ed.'s abbreviations: c.v., coated vesicle; e.r., endoplasmic reticulum; p.m., plasma membrane; Ab, antibody.

The enterocyte is a highly polarized columnar epithelial cell with mainly digestive and absorptive functions. For these purposes the area of the apical membrane is greatly enhanced by microvilli that form the brush border. The basolateral membrane has its own distinct morphology and function, and consequently differs in biochemical composition.

Cell and membrane isolation *

It is easy to isolate enterocytes, either as free cells or as sheets, since these cells are only loosely attached to the connective tissue layer. For many biochemical purposes large amounts of mucosa may be obtained by scraping the washed intestine with a microscope slide. For preparative purification purposes, it is common to subject intestinal pieces to vibration [1]; this procedure is easy and rapid although giving somewhat lower yield.

Intact brush borders can be isolated by careful homogenization of an enterocyte suspension in the presence of EDTA [2]. Thereby the p.m. is disrupted at the tight junctions and in the basolateral membrane, but because of a well-developed cytoskeleton the apical membrane is kept intact. Contaminating organelles are removed by differential centrifugation. By changing the conditions the brush borders can be disrupted, giving small microvillar vesicles [3].

For preparative biochemical purposes the introduction of the divalent ion precipitation technique was of great importance [4]. In this technique, enterocytes are homogenized in a Waring-type blender and the intracellular and basolateral membrane fragments removed by precipitation with Ca^{2+} or Mg^{2+}. Small amounts of contaminants can be removed by the use of immunadsorbent chromatography or adsorption chromatography on agarose beads [5, 6].

Microvillar membranes prepared by any of these methods are mainly right-side-out vesicles. Their interior is filled with a filamentous material originating from the microvillar cytoskeleton. A pure membrane preparation may be obtained by removal of this material by treatment with high concentrations of LiSCN [7] or by alkaline pH [8].

MICROVILLAR MEMBRANE PROTEINS

Aminopeptidase N, sucrase-isomaltase and maltase-glucoamylase are major proteins of the microvillar membrane [9]. They are integrated into the microvillar membrane by a short N-terminal peptide segment, the anchor. This property makes it possible to release them in an enzymatically active form by proteolytic treatment, leaving the anchor

---

* See also B.A. Lewis et al. in Vol. 6, this series, and the index entry 'Sidedness' in Vols. 6, 13 & 15.- *Ed.*

Table 1. Immunoadsorbent purification of intestinal microvillar enzymes by hypotonic elution, with 0.1% Triton X-100 in the buffer, viz. 2 mM Tris-HCl pH 8.0 or 1 mM K phosphate pH 7.4. *From ref. [10].*

Enzyme	Elution buffer	Recovery	Purific'n factor	Ref.
Aminopeptidase N	Tris-HCl	70%	5.6	11
Dipeptidyl peptidase IV	Tris-HCl	56%	48	12
Sucrase-isomaltase	Tris-HCl	40%	8.0	13
Lactase-phlorizin	K phos.	32%	21	14
hydrolase	K phos.	30%	32	15
Maltase-glucoamylase	K phos.	32%	8.3	16

in the membrane. To solubilize the intact molecule, however, detergents are required. [For detergent approaches (non-intestinal) see Vol. 13.—*Ed.*]

The purification of detergent-released proteins causes methodological problems, as the binding of detergent obviously gives the different molecules very similar properties, resulting in poor resolution with conventional chromatography methods. As this does not apply to proteolytically released enzymes, these were purified and specific Ab's raised. By using a new principle for the elution of proteins from immunoadsorbents, hypotonic elution [10], three glycosidases and two peptidases as listed in Table 1 were purified, all from the pig (also human in the case of lactase-phlorizin hydrolase). In this method the immunoadsorbent is saturated with the protein to be purified (seemingly essential) and is then, after washing, eluted by a buffer of very low ionic strength or with water. The method has the obvious advantage that it does not destroy the biological activity of the proteins to be purified and allows the use of the immunoadsorbent several times. Thus the same column was used up to 20 times without significant decrease in either capacity or specificity.

STUDIES ON BIOSYNTHESIS OF THE ENZYMES

The characterization of the biosynthesis, the intracellular transport and processing of these enzymes might be expected to suggest new mechanisms behind depressed expression of these enzymes in the microvillar membrane. Such an approach has successfully unravelled several different mechanisms behind the low levels of the LDL receptor as a cause of familial hypercholesterolaemia.

For biosyntnesis studies organ culture of small intestinal mucosa sheets (porcine) was used. When mucosa and submucosa was dissected from the rest of the intestine and placed on a grid in tissue culture medium, the tissue stayed alive for up to 24 h [17]. [^{35}S]Methionine

was added to the medium for defined periods of time, chased in some
experiments by cold methionine. The tissue was homogenized, microvillar membrane vesicles separated from the other membrane vesicles
by the divalent cation precipitation technique, and the enzymes
solubilized with detergent.

The different enzymes were purified by a modification of line
immunoelectrophoresis. This method [18] has the advantage that it
allows the purification of up to 4 different enzymes in the same
experiment. In short, the antigen was applied in a 1 x 1 cm square
and different specific Ab's cast into 1 x 1.5 cm squares anodically
to the antigen sample. After electrophoresis, the agarose gel was
pressed, re-swollen in 0.5 M NaCl, pressed again, re-swollen in deionized water, and finally pressed before excision of the immunoprecipitates by means of a scalpel. If the immunoprecipitates were not
readily detectable, they were visualized by staining with Coomassie
Brilliant Blue and the gel re-swollen by immersion in deionized water
before excising the precipitates.

Our results have been summarized in a recent review [19]. The
enzymes labelled with a 10 min pulse were all exclusively located
in the Ca^{2+}-precipitated membrane fraction rather than occurring in
a soluble form. Thus, membranes other than the microvillar membrane
must be translocationally active for insertion of these enzymes.
By cell-free translation of intestinal mRNA, it was revealed that
aminopeptidase N was integrated into dog microsomal membranes, arguing
in favour of a co-translational membrane insertion of the microvillar
enzymes into the rough e.r. During this process they became high-
mannose-glycosylated as evidenced by their sensitivity to endoglycosidase H.

The intracellular transport and processing of the enzymes was
studied by pulse-chase labelling in combination with subcellular
fractionation (Fig. 1). At 40-60 min after a 10-min pulse a new
molecular form was detected in the Ca^{2+}-precipitated membrane fraction
that was insensitive to endoglycosidase H, suggesting that it was
further subjected to so-called complex glycosylation. After 60-90 min
of chase these latter forms appeared in the microvillar membrane.
These experiments argue for a common, membrane-bound route of transport
for the transfer of microvillar enzymes from their site of synthesis
to their final destination.

The use of different inhibitors with defined effects gave further
information on the intracellular transport and processing. When
pig intestinal explants were labelled in the presence of monensin
(which impairs the function of the Golgi complex, the endosomes and
the lysosomes without affecting protein synthesis), the intracellular
traffic of newly synthesized enzymes to the microvillar membrane
and their processing were arrested. Colchicine (known to interfere
with the function of microtubules) was also found to inhibit the
transport of pig intestinal enzymes to the microvillar membrane.

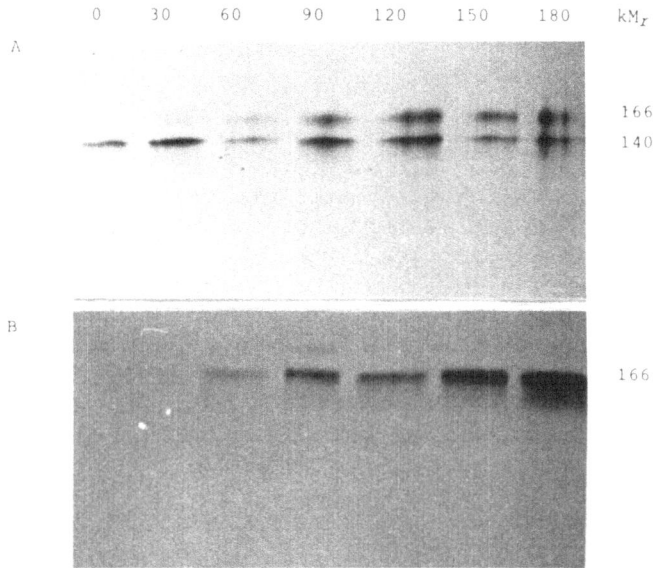

Fig. 1. Pulse-chase labelling of aminopeptidase N from the solubilized Ca^{2+}-precipitated membrane fraction (**A**) and microvillar membrane fraction (**B**) from pig mucosa.   *From [20], by permission.*

Use of endoglycosidase H showed that the enzymes are N-glycosyl-
ated. The considerable difference in mol. wt. between the high-mannose
and the complex glycosylated forms suggests, however, that the enzymes
might also be O-glycosylated, a hypothesis supported by use of the
inhibitor swainsonine in labelling experiments and by exposure of
the enzymes to alkaline pH or endoglycosidase F.

For the elucidation of enzyme deficiency states it was of
considerable interest to know whether the carbohydrate processing
is of an importance for the enzymatic activity. By treatment of
a Ca^{2+}-precipitated membrane fraction with papain prior to solubilization with detergent and subsequent purification by immunoadsorbent
chromatography, essentially pure forms of high-mannose-glycosylated
aminopeptidase N, maltase-glucoamylase and sucrase-isomaltase were
obtained [21]. The explanation for this is that probably the papain
treatment releases the complex glycosylated forms present in contaminating microvillar vesicles, or in intracellular membranes that have
not vesiculated in a way that protects the membranes from the action
of the proteinase. Comparison of specific enzyme activities (measured
as enzymatic activity/immunological reactivity) of high-mannose and
complex glycosylated forms showed no difference for aminopeptidase N
or maltase-glucoamylase, but for sucrase-isomaltase the high-mannose-
glycosylated form had a markedly lower activity.

DEPRESSED ENZYME EXPRESSION

Some chronic diarrhoeas may be explained by absent or mal-functioning digestive enzymes. Adult-type hypolactasia is the most frequent condition caused by disappearance of lactase-phlorizin hydrolase in childhood. By using crossed immunoelectrophoresis in combination with specific enzymatic staining of the precipitates we demonstrated that the low enzymatic activity in adults with hypo-lactasia is caused by a low amount of the lactase protein and not by a malfunctioning enzyme [22]. As the immunological technique also measures possible inactive, intracellularly localized lactase molecules, our results also show that no significant amount of inactive lactase is present within the cell.

Another group with depressed enzyme expression are the rarer inborn deficiencies of sucrase-isomaltase, maltase-glucoamylase or lactase-phlorizin hydrolase. In a study of 9 sucrose-intolerant patients carried out in our laboratory [23], 3 had residual isomaltase activity and a corresponding isomaltase precipitate in crossed immuno-electrophoresis. By polyacrylamide gel electrophoresis (PAGE) in the presence of SDS followed by immunoblotting, the residual isomaltase was shown to be connected to a single polypeptide chain of $kM_r \sim 145$. By quantitative crossed immunoelectrophoresis it was also demonstrated that one patient had an almost total deficiency of maltase-gluco-amylase, whereas 5 patients had depressed amounts of this enzyme. Defective intracellular processing followed by proteolysis may be an explanation of depressed levels whereas absence is more probably correlated to deficient synthesis.

The first demonstration of a disturbed intracellular transport of a brush border membrane hydrolase was recently reported [24]. An intestinal biopsy from a 5-year old girl lacked sucrase but possessed low residual isomaltase activity. By immunoelectron microscopy with monoclonal Ab's against sucrase-isomaltase the enzyme was found almost exclusively in trans-Golgi cisternae and in associated vesicular structures, while no specific labelling was associated with the microvillar membrane. Immunoprecipitation experiments with iodinated mucosal homogenates suggested that the synthesized enzyme was high-mannose glycosylated.

STUDIES IN COELIAC DISEASE

The function of the enterocyte is greatly disturbed in coeliac disease, a condition caused by intolerance to wheat gluten. More specifically, the enterocyte is damaged by gliadin, the ethanol-extractable protein fraction of gluten. The incidence in Northern European countries is 1/800, the main symptom is chronic diarrhoea and the treatment is life-long diet avoiding all food containing gluten.

The mucosa of gliadin-exposed intestine of coeliacs shows lympho-cyte infiltration and loss of normal villous architecture. The entero-cyte is damaged with short distorted microvilli and numerous vesicles of different sizes in the cytoplasm.

The molecular mechanisms behind the toxicity of gliadin in coeliac disease are unsettled.   It was long believed that the reason was the absence of a digestive enzyme necessary for the digestion of one or several gliadin components.   This enzyme should function as a detoxifier of a toxic gliadin fraction.   The demonstration [25] that a brush border sample from patients with coeliac disease can totally digest a gliadin sample made this hypothesis less plausible.

It must therefore be suggested that the toxic peptide(s) interact with the brush border membrane before being digested and consequently that there is a difference in this interaction between normal and sensitive individuals.   It may be that gliadin is taken up in both normal and predisposed individuals but is abnormally handled in the diseased patient.   Alternatively, gliadin may be taken up only in the sensitive patient.

Gliadin consists of >40 different polypeptides, which can be separated into 4 groups by electrophoresis at acidic pH: the α-, β-, γ- and ω- fractions.   Even though many attempts have been made to identify the toxic component(s) of gliadin both *in vivo* and *in vitro*, this problem has not been unequivocally solved.

Immunological approaches

Untreated coeliac patients have circulating IgA and IgG Ab's to gliadin.   In order to investigate whether the patients have a characteristic antigen-Ab pattern, we have analyzed the reactivity of the circulating Ab's to the 4 isolated gliadin fractions using an immunoblotting technique [26].   Gliadin was prepared from wheat flour and the 4 peptide fractions purified by ion-exchange chromato-graphy on an SP-Sepharose C-50 column.   The fractions were run by SDS-PAGE (10% gel), transferred to nitrocellulose sheets, incubated with patient sera in appropriate dilutions and visualized by a peroxidase-conjugated second Ab.

Untreated patients show a characteristic reaction pattern (Fig. 2).   All but one of the 15 tested reacted with two γ-fraction bands of kM_r 35 and 45.   In active coeliac disease, especially with pronounced intestinal symptoms, reactivity was also observed in α-, β- and ω-fractions. Sera from other patient groups, including coeliac patients in remission, usually displayed some reactivity against α- and β-gliadin.

By immunizing rabbits with a gliadin sample we could show that the peptides of kM_r 35 and 45 in the γ-fraction were not more immunogenic

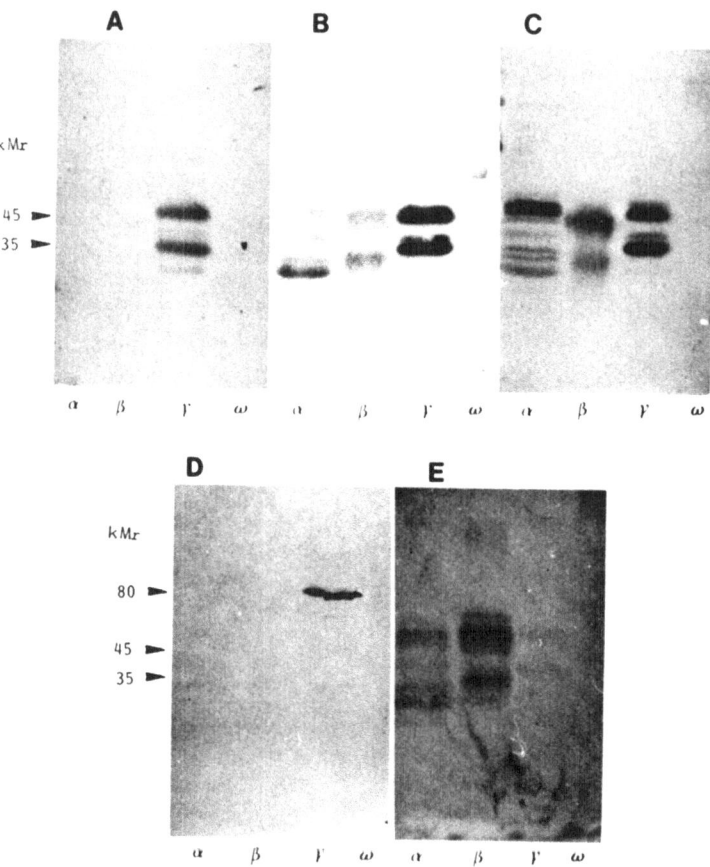

Fig. 2. Immunoblotting of the polypeptides in the α-, β-, γ- and ω-fractions. A, B, C, D: untreated coeliac disease; E, ulcerative colitis. *From ref. [26], by permission.*

than most of tne other peptides. It is therefore striking that patients with active coeliac disease, unlike other patients with disturbed intestinal function, characteristically produce Ab's against certain polypeptides in the γ-fraction. This might be due to a specific uptake and presentation to the immunocompetent cells of certain peptides and it argues against the presence of these gliadin Ab's as a secondary immunological phenomenon.

If ingestion of gliadin is continued, this might lead to damage to the enterocytes by activation of the complement system. Ab's against other gliadin fractions, as observed in coeliac disease patients with considerable intestinal symptoms, may be a consequence of immunization due to secondary barrier damage. Such an argument is strengthened by the Ab response to polypeptides in the α- and β-fractions seen in sera of other groups of patients, where there is damage to the intestinal mucosa, e.g. by bacteria, causing leakage and hence a more general immunization with gliadin.

PURIFICATION OF COATED VESICLES (c.v.'s)

The brush border membrane is continually renewed by the fusion of vesicles containing newly synthesized material. In a state of damage to the cell, as in coeliac disease where apical p.m. area is reduced, there is a disproportion between removal of p.m. (by budding or endocytosis) and insertion of new material by exocytosis.

The purification and characterization of exocytic vesicles would be expected to illuminate the mechanisms for membrane renewal and for the generation of polarity in a polarized cell. This seems, however, to be an extraordinarily difficult task, probably because of the paucity of vesicles present at a certain instant, even in (e.g.) virus-infected cells synthesizing membrane glycoproteins, and also of the absence of known markers.

Vesicles have, however, been isolated from different tissues on the basis of their characteristic coat. These vesicles are best characterized from cells active in receptor-mediated endocytosis [27], but have also been suggested to be important in exocytic pathways [28]. In electron micrographs of the intestine, c.v.'s have been observed most often in the region of the Golgi apparatus, but also near the p.m. (Fig. 3), especially in samples from adult, cultured or foetal intestine.

Two approaches were used for the preparative purification of such c.v.'s (Scheme 1): (I) destruction of contaminating vesicles by solubilization with Triton X-100, or (II) removal of microvillus vesicles by precipitation with Ab's and of smooth vesicles by aggreg-ation with Mg^{2+}

Method I is based on the finding that despite solubilization of the membrane inside the coat, the coat is able to keep the material inside. There are 4 major steps (Scheme 1): a differential centri-fugation procedure to get a crude c.v. preparation, an initial Triton X-100 treatment of the crude c.v.'s, gel filtration on Sephacryl S-100 in the presence of Triton X-100 and finally centrifugation of the fraction to separate soluble from vesicular material. This purification method gives c.v.'s containing ~1% contaminating vesicles (400 vesicles counted/preparation) and small amounts of amorphous material. The intestinal c.v.'s exhibit the characteristic lattice structure found in c.v.'s from other sources (Fig. 4A).

The protein pattern of the purified vesicles (Fig. 5) was dominated by clathrin of kM_r 180. There were other prominent bands of kM_r ~100, 80, 70 and 50. In addition, several minor components were found including a distinct double band having M_r's ~33. This

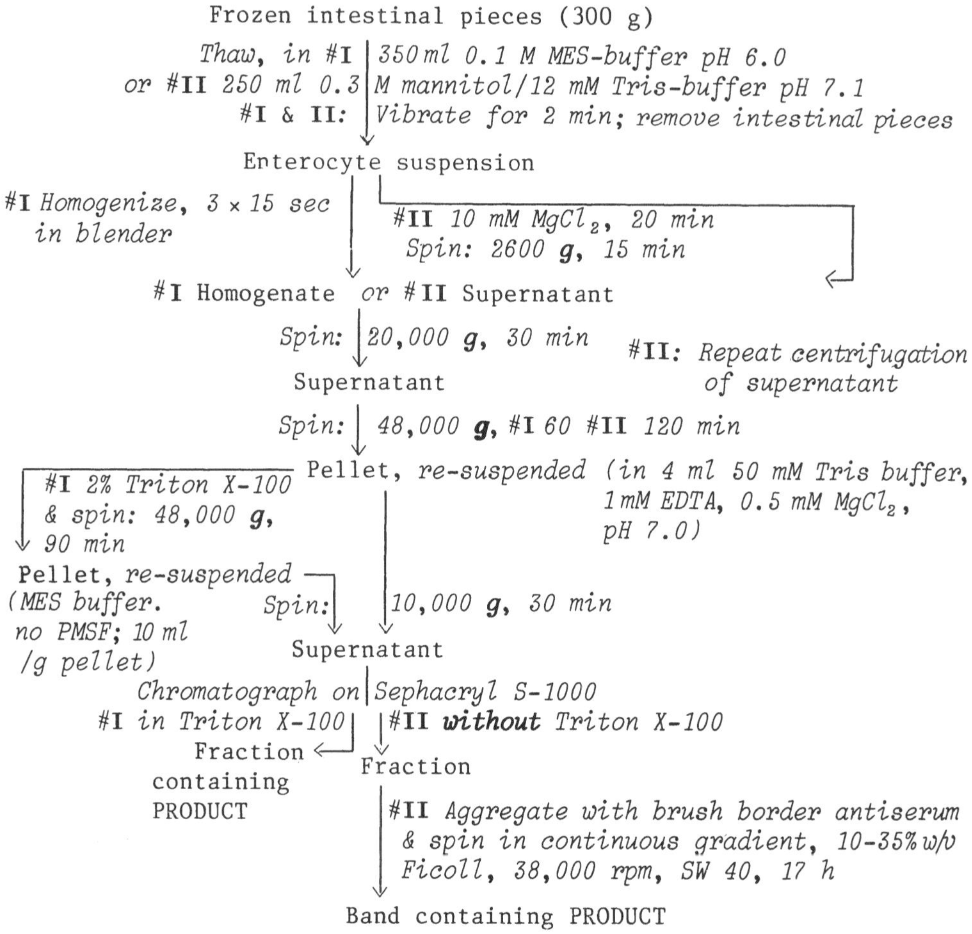

Frozen intestinal pieces (300 g)

Thaw, in #I | 350 ml 0.1 M MES-buffer pH 6.0
or #II 250 ml 0.3 | M mannitol/12 mM Tris-buffer pH 7.1
#I & II: | Vibrate for 2 min; remove intestinal pieces

Enterocyte suspension

#I Homogenize, 3 × 15 sec   *#II 10 mM MgCl$_2$, 20 min*
in blender   *Spin: 2600 g, 15 min*

#I Homogenate or #II Supernatant

Spin: | 20,000 g, 30 min   *#II: Repeat centrifugation*
                                            of supernatant
Supernatant

Spin: | 48,000 g, #I 60 #II 120 min

#I 2% Triton X-100 Pellet, *re-suspended (in 4 ml 50 mM Tris buffer,*
& spin: 48,000 g,   *1 mM EDTA, 0.5 mM MgCl$_2$,*
90 min   *pH 7.0)*

Pellet, *re-suspended* —
(MES buffer.   Spin: | 10,000 g, 30 min
no PMSF; 10 ml
/g pellet)   Supernatant

Chromatograph on | Sephacryl S-1000
*#I in Triton X-100 | | #II **without** Triton X-100*
Fraction ← Fraction
containing
PRODUCT
                          #II Aggregate with brush border antiserum
                          & spin in continuous gradient, 10-35% w/v
                          Ficoll, 38,000 rpm, SW 40, 17 h

Band containing PRODUCT

Scheme 1. Purification of pig small intestinal c.v.'s, with variant procedures, #I and #II. MES buffer: 0.1 M 2-[**N**-morpholino]-ethansulphonic acid, 1 mM MgCl$_2$, 0.5 mM EDTA, 0.02% NaN$_3$, 0.1 mM phenyl-methane sulphonyl chloride (PMSF).

pattern is more heterogeneous than those reported for purified c.v.'s from other organs, perhaps because the Triton-solubilized proteins non-specifically adhere to the outside of the coat and thereby co-purify.

Method II was therefore developed, not using detergent. The supernatant after precipitation of basolateral and intracellular membranes with Mg^{2+} and sedimentation of the bulk of the microvillar membranes by centrifugation was shown to contain c.v.'s, together with microvillar and smooth vesicles. Vesicles larger than the c.v.'s were removed by chromatography on Sephacryl S-1000. Addition of brush border antiserum preferentially aggregates remaining microvillar

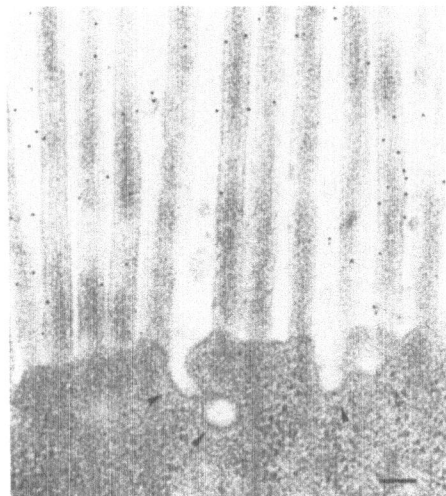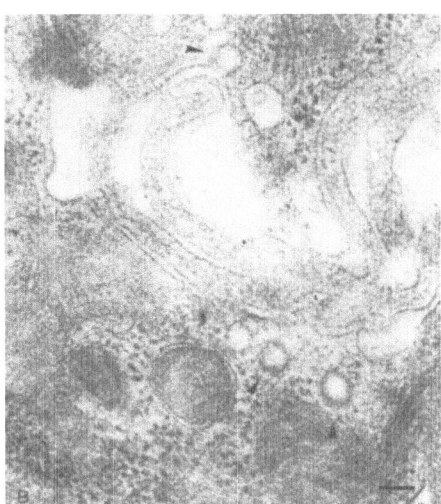

Fig. 3.   Coated pits and coated vesicles *(arrows)* in sections of
cultured small intestine labelled for aminopeptidase N by an immunogold technique. *Left,* brush border region; *right,* Golgi region.
Bar = 100 nm.

vesicles of the same size as the c.v.'s. These aggregates and remaining
smooth vesicles of the same size as the c.v.'s were separated by
isopycnic centrifugation in a Ficoll gradient.   The purity of this
preparation was ~90% (Fig. 4B).

Interestingly,   such a preparation displays a heterogeneous
protein pattern similar to that previously demonstrated for c.v.'s
purified by the detergent-extraction method.   This special pattern
of protein in intestinal c.v.'s may be explained by different functions
of c.v.'s in various organs.   Most of the hitherto characterized
coated vesicle preparations originate from organs where endocytosis
is a main feature.   However, in normal intestine endocytosis is not
prominent.   It may therefore be that the majority of c.v.'s from
the intestine represent a population of such vesicles taking part
mainly in the exocytosis of newly synthesized membrane material in
the intestine.

Acknowledgements

The results in the authors' laboratory could not have been obtained
without the work carried out by the skilful technicians. The authors
are indebted to the Danish Medical Research Council, the Novo Research
Foundation,   the Danish Cancer Society and the Carl P. Pedersen's
Foundation for financial support.

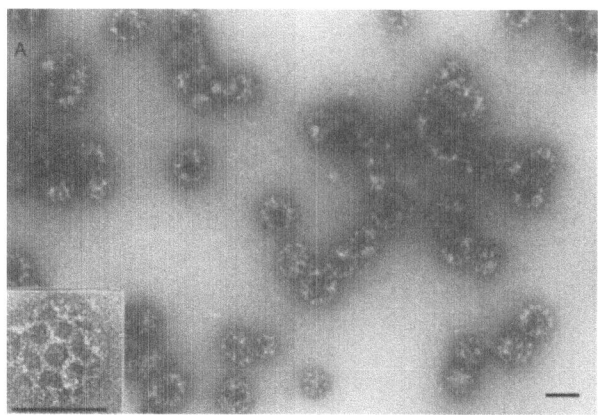

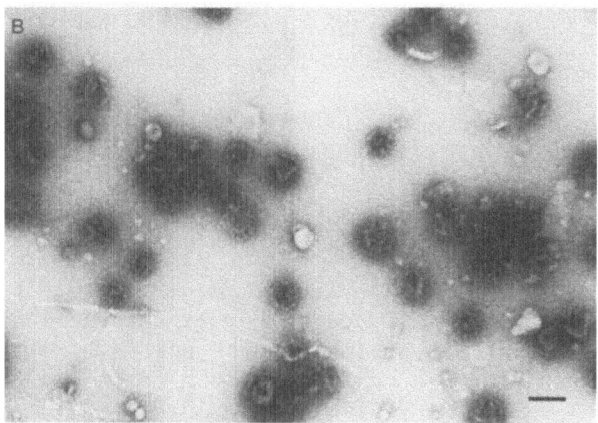

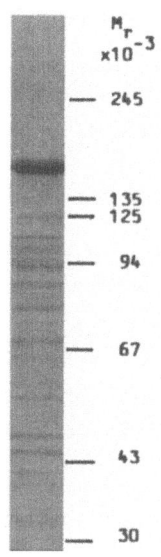

Fig. 4. Electron micrograph of c.v.'s purified by procedure **I** (A) and procedure **II** (B); negative-stained with 1% uranyl acetate. Bars = 100 nm [mag. for INSERT 2.7 times that of main (A) e.m.]

Fig. 5. SDS-PAGE of purified c.v.'s (45 µg protein). Coomassie Brilliant Blue staining.

References

1. Kessler, M., Acuto, O., Storelli, C., Murer, H., Müller, M. & Semenza, G. (1978) *Biochim. Biophys. Acta 506*, 136–154.
2. Miller, D. & Crane, R.K. (1961) *Anal. Biochem. 2*, 284–286.
3. Forstner, G.G., Sabesin, S.M. & Isselbacher, K. (1968) *Biochem. J. 106*, 381–390.
4. Schmitz, J., Preiser, H., Maestracci, D., Gosh, B.K., Cerda, J.J. & Crane, R.K. (1973) *Biochim. Biophys. Acta 323*, 98–112.
5. Carlsen, J., Christiansen, K. & Bro, B. (1982) *Biochim. Biophys. Acta 689*, 12–20.

6. Carlsen, J., Christiansen, K. & Bro, B. (1983) *Biochim. Biophys. Acta 727*, 412-415.
7. Hopfer, U., Crowe, T.D. & Tandler, B. (1983) *Anal. Biochem. 131*, 447-452.
8. Cowell, G.M. & Danielsen, E.M. (1984) *FEBS Lett. 172*, 309-314.
9. Norén, O., Sjöström, H., Danielsen, E.M., Cowell, G.M. & Skovbjerg, H. (1986) in *Molecular and Cellular Basis of Digestion* (Desnuelle, P., Sjöström, H. & Norén, O., eds.), Elsevier, Amsterdam, pp. 335-365.
10. Danielsen, E.M., Sjöström, H. & Norén, O. (1982) *J. Immunol. Meth. 52*, 223-232.
11. Sjöström, H., Norén, O., Jeppesen, L., Staun, M., Svensson, B. & Christiansen, L. (1978) *Eur. J. Biochem. 88*,503-511.
12. Svensson, B., Danielsen, E.M., Staun, M., Jeppesen, L., Norén, O. & Sjöström, H. (1978) *Eur. J. Biochem. 90*, 489-498.
13. Sjöström, H., Norén, O., Christiansen, L., Wacker, H. & Semenza, G. (1980) *J. Biol. Chem. 255*, 11332-11338.
14. Skovjberg, H., Sjöström, H. & Norén, O. (1981) *Eur. J. Biochem. 114*, 653-661.
15. Skovjberg, H., Norén, O., Sjöström, H., Danielsen, E.M. & Enevoldsen, B.S. (1982) *Biochim. Biophys. Acta 707*, 89-97.
16. Sørensen, S.H., Norén, O., Sjöström, H. & Danielsen, E.M. (1982) *Eur. J. Biochem. 126*, 559-568.
17. Danielsen, E.M., Sjöström, H., Norén, O., Bro, B. & Dabelsteen, E. (1982) *Biochem. J. 202*, 647-654.
18. Danielsen, E.M. & Cowell, G.M. (1983) *J. Biochem. Biophys. Meth. 8*, 41-47.
19. Danielsen, E.M., Cowell, G.M., Norén, O. & Sjöström, H. (1984) *Biochem. J. 221*, 1-14.
20. Danielsen, E.M. (1982) *Biochem. J. 204*, 639-645.
21. Sjöström, H., Norén, O. & Danielsen, E.M. (1985) *J. Ped. Gastroent. Nutr. 4*, 980-983.
22. Skovbjerg, H., Gudmand-Høyer, E. & Fenger, H.J. (1980) *Gut 21*, 360-364.
23. Skovbjerg, H. & Krasilnikoff, P.A. (1986) *J. Ped. Gastroent. Nutr. 5*, 365-371.
24. Hauri, H-P., Roth, J., Sterchi, E.E. & Lentze, M.J. (1985) *Proc. Nat. Acad. Sci. 82*, 4423-4427.
25. Bruce, G., Woodley, J.F. & Swan, C.H.J. (1984) *Gut 25*, 919-924.
26. Friis, S.U., Norén, O., Sjöström, H. & Gudmand-Høyer, E. (1986) *Clin. Chim. Acta 155*, 133-142.
27. Steinman, R.M., Mellman,I.S., Muller, W.A. & Cohn, Z.A. (1983) *J. Cell Biol. 96*, 1-27.
28. Rothman,J.E. & Fine, R.E. (1980) *Proc. Nat. Acad. Sci. 77*, 780-784.

#D-3

STUDIES OF THE MECHANISMS OF MUCOSAL IRON UPTAKE

Robert J. Simpson and Timothy J. Peters

Division of Clinical Cell Biology
MRC Clinical Research Centre
Watford Road, Harrow, Middlesex HA1 3UJ, U.K.

Brush border membrane can be isolated as closed vesicles, and transport components of $^{59}Fe^{2+}$ and $^{59}Fe^{3+}$ uptake by brush border membrane vesicles have been identified. This brush border membrane transport is subject to adaptive regulation in response to varying whole-body iron requirements and accounts quantitatively and qualitatively for in vivo *mucosal $^{59}Fe^{2+}$ uptake but not $^{59}Fe^{3+}$ uptake. Investigation of the mechanism of $^{59}Fe^{2+}$ transport by rabbit brush border membrane vesicles has led to the isolation of a membrane Fe^{2+}-binding constituent, identified as non-esterified fatty acid (NEFA).*

The study of mechanisms of intestinal iron absorption requires investigation of Fe uptake by isolated brush border membrane vesicles. Studies of Fe transport are complicated by the instability of Fe^{2+} and Fe^{3+} solutions at neutral pH and by the high affinity of Fe^{3+} for various biological molecules. These factors signify that Fe does not behave as a simple solute, hence demonstration that Fe uptake by brush border membrane vesicles is actually a membrane transport process is difficult. Fe^{2+} and Fe^{3+} uptake by brush border membrane vesicles represents accumulation of Fe with respect to the Fe concentration in the medium (Fig. 1). Thus some binding or hydrolysis of Fe is occurring, either within or outside the vesicles.

The conventional approach to demonstrating that an uptake process is in fact transport is to investigate the dependence of uptake on the osmotically active space within the vesicles. This is done by adding high concentrations of solutes (e.g. sugars, salts) to the outside of the vesicles, leading to shrinkage. Fig. 2 shows that Fe^{2+} uptake by mouse duodenal brush border membrane vesicles manifests a variety of responses, depending on the nature of the solute used to shrink the vesicles. For such experiments to be valid, retention of vesicles by the filters used in the uptake determination must

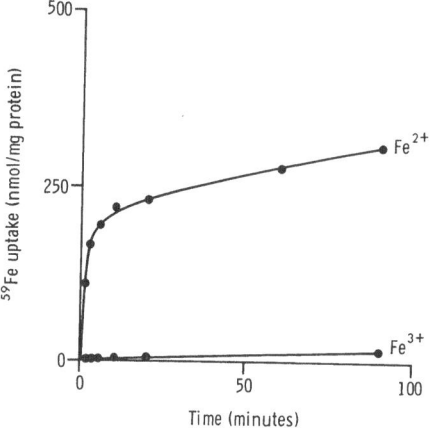

Fig. 1. Uptake of Fe^{2+} and Fe^{3+} by mouse duodenal brush border membrane vesicles. The vesicles, isolated as in [1, 2], were incubated at 37° in ^{59}Fe-containing media, and uptake determined by rapid millipore filtration [2, 3]. Incubation media contained 90 μM Fe, either 180 μM nitriloacetic acid, NTA (Fe^{3+}) or 1.9 mM Na ascorbate (Fe^{2+}), 0.1 M NaCl, 0.1 M mannitol and 20 mM Hepes (final pH 7.25). Equilibration of Fe across the vesicle membrane would be represented by an uptake of 0.09 nmol/mg protein in both cases.

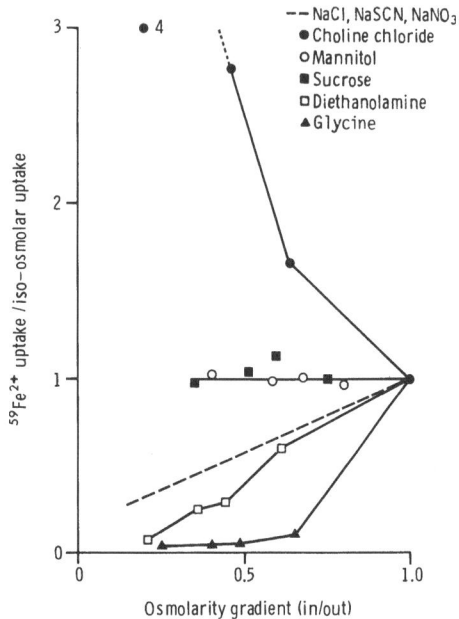

Fig. 2. Uptake of Fe^{2+} in the presence of high concentrations of various solutes, generating high extravesicular osmolarities. Vesicles and uptake (incubation for 1 min) as in Fig. 1. Uptake plotted against ratio of intra- to extra-vesicular osmolarity (calculated assuming ideal solute behaviour).

be unaffected by the shrinking agent. Retention of fresh mouse duodenal brush border membrane vesicles by Millipore GSWP filters is >99% (determined by assays of brush border marker α-glucosidase), even when they are maximally shrunk by NaCl [3]. Clearly, no simple osmotic experiment can determine whether Fe^{2+} uptake by brush border membrane vesicles is transport, as opposed to binding to the outside of the vesicles. Na salts weakly inhibit Fe uptake ($K_{I\ app}$ >10 times the Fe concentration) and thus give a result predicted for a classic osmotic shrinkage experiment.

A better way to investigate whether uptake is in fact transport is to use inhibitors of the uptake. Addition of NaCl stops the uptake, but does not displace $^{59}Fe^{2+}$ taken up prior to addition of the salt. This suggests that the Fe^{2+} uptake is not a simple binding process which is inhibitable by NaCl. Protons are also potent inhibitors of Fe^{2+} uptake by brush border membrane vesicles.

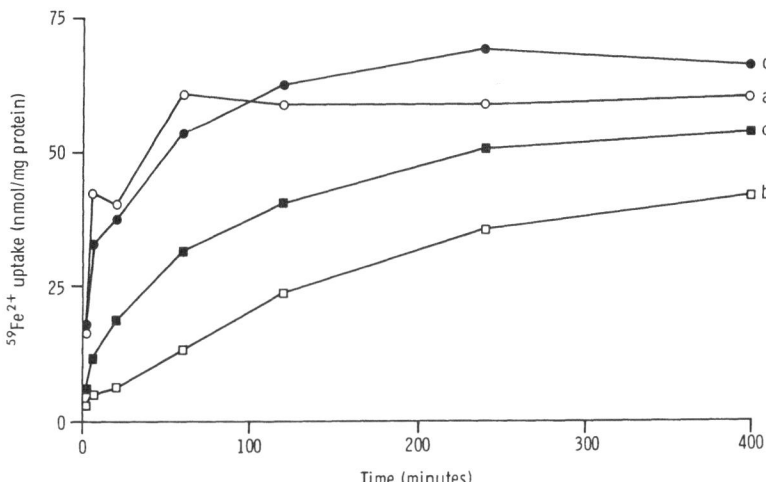

Fig. 3. Effect of lowering the pH and addition of 8-hydroxyquino-
line on Fe²⁺ uptake by brush border membrane vesicles. (a) pH 7.2
(& conditions as in Fig. 1); (b) pH 6.5 – as also in (c) & (d) where
1.8 and 9 μM, respectively (final concn.), 8-hydroxyquinoline was
present in the incubation medium, added as hemi-sulphate salt.

Fig. 3 shows the effect of lowering the pH of the medium on
Fe^{2+} uptake by vesicles. Evidently the Fe^{2+} uptake rate, but not
the end-point, is potently inhibited by lowering the medium's pH.
This is consistent with uptake representing a pH-sensitive step with
a subsequent pH-insensitive step. Addition of lipid-soluble Fe
ligands, e.g. 8-hydroxyquinoline, restores the original uptake rate
without greatly affecting the end-point (Fig. 3). These experiments,
together with studies with cholate [4, 5], suggest that uptake is
predominantly a two-step process, namely membrane transport followed
by intravesicular binding. Several other studies also suggest that
Fe^{2+} uptake by mouse duodenal brush border membrane vesicles is
primarily a transport process [2, 6].

Studies of membrane Fe^{2+}-binding components revealed a chloro-
form/methanol- and cholate-extractable component of rabbit duodenal
brush border membrane vesicles implicated in membrane transport [5];
it was identified as free fatty acid.

Fig. 4 shows that incorporation of oleic acid into egg phosphati-
dylcholine/cholesterol liposomes (containing ferrozine, a Fe^{2+}-sensitive
dye) rendered them permeable to Fe^{2+}. Pre-incubation studies showed
that no leakage of ferrozine occurred, suggesting that the transport
is specific for the metal and is not a detergent effect. The concentration of fatty acid used (7.5% of the total lipid) is within the range
of NEFA levels in mouse duodenal brush border membrane vesicles.
Fe^{2+} transport by several fatty acids (not illustrated) shows pH

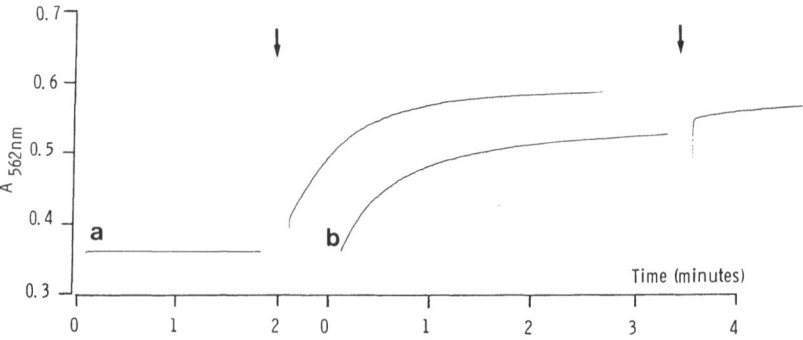

Fig. 4. Effect of oleic acid on Fe^{2+} transport by liposomes made from 10 mg phosphatidylcholine and 2.5 mg cholesterol (a) or these lipids plus 1 mg oleic acid (b). Pre-incubation (37°): 10 min for liposomes, 5 min for freshly prepared solution as added at zero time: 200 μM Fe^{2+}, 4 mM Na ascorbate, 0.1 M NaCl, 0.1 mM mannitol, 20 mM Hepes (final pH 7.2); 400 μl. Ionophore A23187 (to 5 μM) was added *(arrows)* to mediate Fe^{2+} uptake by control liposomes [7] in order to demonstrate the uptake endpoint and to confirm the integrity of the liposomes. These endpoints for both (a) and (b) correspond to ~9 nmol total uptake of Fe^{2+} by the liposomes.

optima near 7.0 (similar to brush border membrane vesicles). Fe^{2+} transport by oleic acid is weakly inhibited by NaCl ($K_{I\ app}$ 1 M). The NaCl liposome inhibition experiments were conducted at zero osmolarity gradient across the liposome membrane; thus the NaCl effect is solely due to inhibition.

These studies suggest that Fe^{2+} uptake by mouse and rabbit proximal intestine brush border membrane vesicles is a two-step process, namely transport followed by intravesicular binding. The membrane transport step can be explained by transport mediated by free fatty acids [cf. 8].

Acknowledgements

We thank A. Evans, K. Raja and H. Grindley for technical assistance and P. McAree for typing this manuscript.

References: 'BBA' signifies *Biochim. Biophys. Acta*

1. Kessler, M., Acuto, O., Storelli, C., Murer, H., Müller, M. & Semenza, G. (1978) *BBA 506*, 136-154.
2. Simpson, R.J. & Peters, T.J. (1984) *BBA 772*, 220-226.
3. Simpson, R.J. & Peters, T.J. (1985) *BBA 814*, 381-388.
4. Simpson, R.J. & Peters, T.J. (1986) *BBA 856*, 109-114.
5. Simpson, R J. & Peters, T J. (1986) *BBA 859*, 227-236.
6. Simpson, R.J., Weir, M.P., Gibson, J.F. & Simpson, R.J. (1984) *Biochem. Soc. Trans. 12*, 859.
7. Young, S.P. & Gomperts, B.D. (1977) *BBA 469*, 281-291.
8. Hauser, H., Howell, K., Dawson, R.M.G. & Bowyer, D.E. (1980) *BBA 647*, 567-577.

#D-4

PANCREATIC ACINAR CELL PLASMA-MEMBRANE ENZYMES
INVOLVED IN STIMULUS-SECRETION COUPLING

R.L. Dormer and A.R. Al-Mutairy

Department of Medical Biochemistry
University of Wales Collegeof Medicine
Heath Park, Cardiff CF4 4XN, U.K.

The aim was to purify the plasma membrane (p.m.) from pancreatic acinar cells (rat) in order to investigate its role in intracellular Ca^{2+} homeostasis and in activation of the cell by stimulators of enzyme secretion. One purification approach investigated was attachment of isolated pancreatic acini to concanavalin A-Sepharose 4B followed by hypotonic lysis of the cells and elution of the membranes with α-methyl mannoside; however, this gave p.m. contaminated with intracellular organelles which bound to the Con A-Sepharose after cell lysis.

The other approach entailed homogenization of isolated acini or whole pancreas followed by differential centrifugation and p.m. separation by Percoll density-gradient centrifugation. The p.m. thus isolated had <0.1% of the total cell endoplasmic reticulum (e.r.) and mitochondria, and could be vesiculated in the presence of a Ca^{2+}-activated photoprotein, allowing measurement of control of free Ca^{2+} in isolated membrane vesicles. In addition, the purity of these membranes has allowed us to demonstrate a Ca^{2+}-activated Mg-ATPase, distinct from that of mitochondria or rough e.r. Its properties suggest that the p.m. ATPase is cardinal to control of cytoplasmic free Ca^{2+} in the stimulated cell. Investigations on p.m. from exocrine acinar cells could aid pathophysiological studies, e.g. on cystic fibrosis.

In the pancreatic acinar cell, as in many cell types, a resting cytoplasmic ionized calcium concentration of the order of 10^{-7} has been measured [1-3]. This is maintained against the extracellular Ca^{2+} concentration, ~10^{-3} M, by the relative impermeability of the p.m. to Ca^{2+} and by intracellular Ca^{2+} being bound or sequestered in organelles [4, 5]. In order to keep total cell Ca^{2+} constant, there must be mechanisms by which Ca^{2+} can leave the cell. In general these can be energy-linked extrusion of Ca^{2+} across the p.m. [6]

and via the secretory pathway by exocytosis [7]. There is considerable evidence that stimulation of enzyme secretion is triggered by
a rise in cytoplasmic Ca^{2+} [1-3, 8-10], and that this Ca^{2+} originates
from an intracellular store [11, 12]. However, increased Ca^{2+} entry
is also necessary to maintain the stimulated rate of enzyme release
[11].

We have previously used subcellular fractionation techniques
to investigate these processes in two ways [5, 13, 14].- (a) By
isolating specific membrane types and studying their Ca^{2+}-transloc-
ating properties, we sought to assess their role in intracellular
Ca^{2+} homeostasis. Thus, we have purified rough e.r. membranes and
demonstrated both ATP-dependent $^{45}Ca^{2+}$ uptake and Ca^{2+}-activated
Mg-ATPase activity [13, 14] which, being sensitive to free Ca^{2+} (half-
max. activity at 0.16 µM), might be important in maintaining the
resting cytoplasmic Ca^{2+} concentration; moreover, active sequestration
of Ca^{2+} by the e.r. could provide access to the secretory pathway and
hence exit from the cell. (b) The second approach has been to stimulate
the intact cell and then isolate different organelle-enriched
fractions, as rapidly as possible, in order to determine whether
their Ca^{2+}-handling properties are altered in a manner consistent
with a role in the stimulation of secretion. These experiments have
shown that the rough e.r. loses Ca^{2+} following cholinergic stimulation
of the acinar cell and have provided evidence that this is the site
of Ca^{2+} release.

The discovery of the inositol phosphate pathways [15] provided
evidence for a second messenger linking receptor occupancy at the
p.m. with Ca^{2+} release from the intracellular store. Thus, stimulators
of enzyme secretion cause rapid (within 5 sec) formation of inositol-
1,4,5-triphosphate [16] which has a direct action, releasing Ca^{2+}
from rough e.r. membranes [16, 17].

As outlined in part elsewhere [18], we have sought to develop
methods for isolating p.m. from the pancreatic acinar cell, in order
to study three main functions which are involved in stimulus-secretion
coupling:
(1) to determine whether there is a Ca^{2+}-activated ATPase which could
be an important mechanism by which Ca^{2+} can leave the cell either
at rest or during stimulation of secretion; distinction from the
rough e.r. ATPase which we have already characterized needs to be
made;
(2) to investigate the mechanism for Ca^{2+} entry into the cell and
how it is linked to receptor occupancy;
(3) to study the phospholipase responsible for inositol phosphate
formation from phospholipid precursors in the membrane and whether
this has any interaction with the Ca^{2+} entry or exit pathways.

A major problem in purifying p.m. from the acinar cell is the
large area of rough e.r. Bolender [19] showed that 60% of the total

membrane area is rough e.r. and only 5% is p.m. The two schemes
described below for isolating p.m. were designed to satisfy two aims.
One aim was to isolate large membrane fragments which could be
vesiculated in the presence of an indicator of free Ca^{2+}, e.g. a
Ca^{2+}-activated photoprotein, in order to measure Ca^{2+} transport
directly, as previously done for erythrocyte membrane 'ghosts' ([20,
21]; cf. J.P. Luzio, #C-2, this vol.). Secondly, the procedures
should be as rapid as possible because the activities we wish to
study may be labile and we need a 'snapshot' of pre-isolation
stimulation effects, e.g. on translocation activity and on membrane
Ca^{2+} and phospholipid levels.

(A) ISOLATION OF PLASMA MEMBRANE FOLLOWING HYPOTONIC LYSIS

The erythrocyte p.m. is probably unique in its resistance to
fragmentation during hypotonic lysis. However, it has been reported
that various agents can stabilize the membranes of mammalian cells
in hypotonic media [22, 23]. We have investigated one of these,
borate, as described by Bauer et al. [24]. We used isolated pancreatic
acini [25] which are as responsive as tissue fragments to physiological
secretagogues, suggesting that proteolytic damage to the p.m. during
the dissociation process is minimal.

A suspension of isolated rat pancreatic acini was added dropwise
to 20 vol. of 20 mM sodium borate, pH 8.0, stirring at 10°. After
10 min, <10% of whole cells were demonstrable in the suspension.
The particulate material was removed centrifugally (6000 g_{av}, 10
min), and the supernatant assayed for the cytosolic enzyme lactate
dehydrogenase (LDH). In 4 experiments the % of the total LDH recovered
in the supernatant was 95.9 ±4.2 (mean ±S.E.M.), indicating a high
degree of cell lysis. However, the membrane fragments obtained in the
pellets were aggregated and difficult to re-suspend. To try to reduce
this aggregation and obtain rapid separation of p.m., the following
scheme in which isolated acini are attached to a solid support before
hypotonic lysis was investigated (all centrifugations at room temp.).-

Scheme **A**
1. Centrifuge isolated acini for 3 min at 50 **g**.
2. Re-suspend in 0.9% (w/v) NaCl at ~6 x 10^4 acini/ml.
3. Add (beads/acini ratio ~2:1) to a pellet of Con A-Sepharose 4B
(washed x5 in 0.9% NaCl), from ~1 ml of gel for each 5 ml of acini.
4. Mix for 10 min at room temp. (plastic tube; invert every 5 sec.).
5. Centrifuge for 3 min at 200 **g**.
6. Add pellet to 20 vol. 20 mM borate, pH 8.0, with stirring for
10 min at 10°.
7. Centrifuge for 3 min at 200 **g**.
8. Wash pellet twice with 10 vol. 20 mM borate, pH 8.0.
9. Stir pellet for 30 min at 10° with 5 vol. of 10 mM 'Tes' buffer,
pH 7.0, containing α-methyl mannoside at a chosen concentration.

Table 1. Binding of organelles to Con A-Sepharose 4B beads.
Isolated acini were bound to the beads and hypotonically lysed in
borate medium (text, Scheme **A**). The values are % bound, viz. the
activity of marker enzymes as % of total activity in a homogenate
of the original acini. Each value represents a separate experiment.
N.D. signifies activity not determined. Cyt. = cytochrome **c**.

Addition to lysis medium	Na,K- ATPase	Cyt. oxidase	Amylase	NADPH/Cyt. reductase	LDH
None	95, 57.3	94.5, 94.8	17.4, 19.6	11.2, 27.4	N.D., 6.9
50 mM α-Me mannoside	93.9	95.7	N.D.	N.D.	50.0
250 mM α-Me mannoside	92.9, 80.6	92.0, 88.2	N.D., 15.	N.D., 31.9	N.D., 24.0

10. Centrifuge for 3 min at 200 **g**.
11. Wash with 2.5 vol. of buffer containing α-methyl mannoside as
in 9. above.
12. Combine supernatants from α-methyl mannoside elutions, and
centrifuge for 20 min at 30,000 **g**.

Following step 4, >95% of the isolated acini were observed by
phase-contrast microscopy to be attached to the Con A-Sepharose beads.
After hypotonic lysis in borate buffer (step 6) the supernatants
from steps 7 and 8 were combined and assayed for the following marker
enzymes: (Na$^+$+K$^+$)-activated ATPase for the p.m. [13]; cyanide-
sensitive NADPH-dependent cytochrome **c** reductase for the e.r. [26];
cytochrome **c** oxidase for mitochondria [27]; and amylase for zymogen
granules [28]. These activities were expressed relative to homogenate activity to give the % binding of the organelles (Table 1).

Evidently >50% of the p.m.'s remained attached to the beads
following hypotonic lysis under the different conditions tried, but
~95% of the mitochondria also became attached. Following lysis in
borate buffer alone, >90% of the LDH was recovered in the supernatants,
showing that only 7% of the bead-attached organelles could be accounted
for by whole cells. The organelles other than the p.m. must become
attached after cell lysis. In attempts to prevent this binding without
p.m. loss, different concentrations of α-methyl mannoside were added
to the lysis buffer. Whilst with 50 or 250 mM (Table 1) 80-94% of
the p.m. remained bound after lysis, 88-96% of the mitochondria were
also still bound. The LDH values show that lysis was only 50-76%
complete and that the presence of whole cells on the beads might
therefore account for the remaining zymogen granules and e.r.

In two experiments the % recovery of organelles in step 12,
after elution with 250 or 500 mM α-methyl mannoside, was 6.8-12.2%

for p.m., 38.2-45.7% for mitochondria, 2.8-10.4% for zymogen granules
and 2.3-17.0% for e.r. These recoveries were not improved by adding
KCl (to 140 mM) to the elution buffer or replacing 10 mM Tes, pH 7.0,
by 0.1 M borate, pH 6.5.

This method having failed to give pure p.m., other agarose-type
supports and other lectins will be tested, aiming to reduce the post-
lysis binding of other organelles and increase p.m. recovery. However,
the method described yields a preparation containing up to 80% of
the acinar cell p.m. attached to a solid support, possibly with the
cytoplasmic surface exposed. Since the beads are readily separable
from isolated zymogen granules, this preparation will be investigated
as a means of studying the interaction of p.m. and zymogen granules
in vitro, as a model for exocytosis.

(B) ISOLATION OF PLASMA MEMBRANES USING PERCOLL DENSITY GRADIENTS

This approach was initially aimed at recovering the larger p.m.
fragments from homogenized, isolated acini, by low-speed centrifugation ($<10,000$ g) followed by Percoll density-gradient fractionation
to obtain rapid separation. Several parameters were investigated.-

(a) The homogenization medium used initially was 0.3 M sucrose
containing 2 mM $MgCl_2$ and 1 mM benzamidine. Since the p.m. fractions
furnished by Percoll were aggregated, homogenization in sucrose without
$MgCl_2$ and containing 0.1 mM EDTA with or without 100 mM KCl was
tried. Aggregation was undiminished.

(b) The number of passes of the Potter homogenizer, the protein
concentration of the homogenate and the centrifugation speed to
obtain the low-speed pellet were optimized on the basis of p.m.
($Na^+ + K^+$)-activated ATPase recovery.

(c) The starting density of the Percoll and the composition of the
diluting medium were optimized with, as criteria, good p.m. recovery,
separation from other organelles and minimization of aggregation
of the membranes. Initially 0.25 M sucrose was used to dilute the
Percoll, and EDTA supplementation (to 0.1 mM) was found to reduce
membrane aggregation. Sorbitol gave further improvement. The scheme
subjected to closer study was as follows.-

Scheme B
1.   Centrifuge isolated acini for 3 min at 50 g.
2.   Re-suspend in 0.3 M sucrose, 2 mM $MgCl_2$, 1 mM benzamidine.
3.   Homogenize with 6 passes of a tight-fitting, all-glass Potter
homogenizer (protein concentration 1-3 mg/ml).
4.   Centrifuge for 10 min at 6000 g_{av}.
5.   Re-suspend pellet in 0.3 M sucrose, 0.1 mM EDTA (pH 7).
6.   Mix with Percoll in 6% (w/v) sorbitol (starting density
1.05 g/ml), 24 ml/ml of pellet, and underlay with 2.5 M sucrose.
7.   Centrifuge for 30 min at 60,000 g_{av} (then harvest the bands).

Table 2. Isolation of p.m. from a low-speed pellet: marker enzymes.
Isolated acini were homogenized and fractionated by Scheme **B**; LSP =
low-speed pellet obtained at step 5; PM = upper fraction from the
Percoll gradient. The % values and relative specific activity (rsa)
values refer to the homogenate activity; each is mean ±S.E.M. (with
no. of experiments).

	Na,K-ATPase	Cyt. oxidase	NADPH/Cyt. reductase	Amylase
LSP, %	28.0 ±3.0 (10)	73.3 ±3.5 (7)	N.D.	42.7 ±3.1 (10)
PM, %	22.7 ±5.3 (5)	17.0 ±6.4 (5)	8.2 ±2.0 (3)	1.1/0.4 (2)
LSP, rsa	0.72 ±0.08 (10)	2.10 ±0.22 (7)	N.D.	1.7 ±0.4 (3)
PM, rsa	5.6 ±0.8 (5)	3.4 ±1.0 (5)	1.7 ±0.4 (3)	0.2 /0.3 (2)

Three major bands were obtained from the gradient: **a)** on the
underlying cushion, **b)** from the upper quarter of the gradient, and
c) as for **b)** but above it. Band **a)** contained ~50% of the zymogen
granules from the 6,000 **g** pellet with no detectable mitochondria
or p.m. Band **b)** contained up to 70% of the mitochondria, 14-19%
of the p.m. and 13-17% of the zymogen granules.

For **c)**, which was somewhat aggregated, Table 2 gives detailed
marker-enzyme data. An average of 22.7% of the total p.m. in the
homogenate was recovered with a relative specific activity of $(Na^+ + K^+)$-
activated ATPase of 5.6. However, although zymogen granule contamination was ⊅1% ($vs.$ 6000 **g** pellet), contamination by e.r. (8%)   and mitochondria (17%)   was unacceptably high. The p.m. aggregation in
the Percoll gradient contributed to the disappointing purity, since
harvesting with a gradient-sampling device was difficult due to the
membranes sticking on the withdrawal tube's entrance or interior.
In addition, the harvested membranes were difficult to re-suspend
smoothly. In preliminary experiments to measure Ca^{2+} transport across
membranes vesiculated in the presence of the Ca^{2+}-activated photoprotein, aequorin, it was necessary to sonicate at lower power in
order to obtain vesicles with entrapped aequorin.

Entrapment was demonstrated (Fig. 1, panel A) by the low level
of luminescence of the aequorin in the presence of 1 mM Ca^{2+}. Unless
protected in an environment maintaining a lower Ca^{2+} concentration,
>99% of the aequorin would be utilized within 10 sec. Unreacted
aequorin was demonstrated after several minutes by solubilizing the
membranes with Triton X-100, The trace in panel B shows that the
bivalent cation ionophore A23187 caused an increased rate of aequorin
utilization inside the membrane vesicles. Thus, following injection
into the luminometer tube, the initial increase in luminescence and

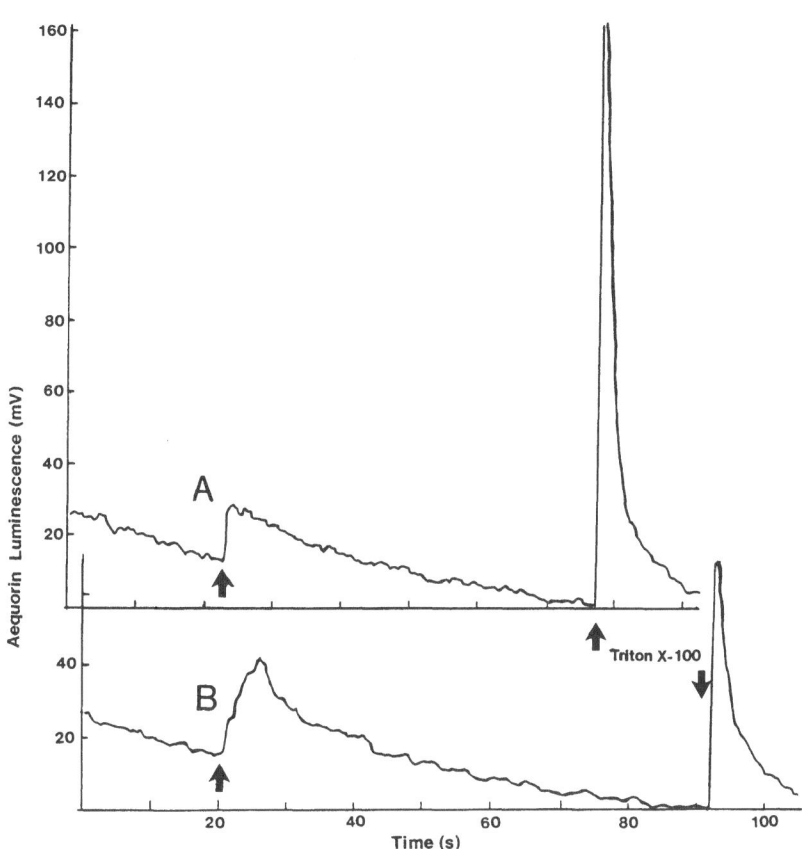

Fig. 1. Entrapment of aequorin in p.m. vesicles. The p.m. from Scheme **B** was diluted with 16 vol. 10 mM Tes, pH 7.4 (treated with Chelex-100 to reduce Ca^{2+} contamination [20]) containing 6% sorbitol and centrifuged (2 min, 1000 **g**). The pellet was washed twice in this medium and re-suspended in 10 mM Chelex-treated Tes, pH 7.4, containing 6 mM NaCl, 150 mM KCl, 2 mM $MgSO_4$, 10 mM PEP, 2 mM ATP, 20 u/ml pyruvate kinase and aequorin ($2-4 \times 10^8$ luminescent counts/ sec; assay as in [21]). After sonication (probe at 16 μm peak-to-peak) for 2×5 sec, membrane aliquots were taken and added to 0.5 ml 10 mM Chelex-treated Tes, pH 7.4, containing 140 mM NaCl, 5 mM KCl, 2 mM $MgCl_2$, 1 mM $CaCl_2$, and placed in the luminometer. The depicted traces began 3 min thereafter, to ensure that all non-entrapped aequorin had reacted. At 20 sec (*1st arrow*) an equal vol. of the same buffer + 0.1% (v/v) DMSO (**A**) or 8 μM A23187 (**B**) was injected. After ~1 min, Triton X-100 (final concentration 3% v/v) was injected (*2nd arrow*).

its subsequent decay was greater than in the control (panel A), indicating exposure of the photoprotein to a higher Ca^{2+} concentration. This is further demonstrated by the presence of less active aequorin following membrane solubilization with Triton X-100.

Two modifications of Scheme **B** reduced p.m. aggregation: (i) centri-fuging the homogenate at a lower speed so that the membranes were not pelleted, and applying the supernatant to the Percoll gradient; (ii) replacing sorbitol in the gradient by hypertonic sucrose (0.5 M) with 100 mM Tris-HCl, pH 7.0.

However, the resulting p.m. fraction still contained unacceptably high amounts of mitochondria (4.5% of the total) and e.r. (8.5%). Moreover, it is difficult to obtain accurate values for marker-enzyme specific activities since Percoll interferes in protein determin-ations. We encountered this in initial attempts to purify rough e.r. using Percoll gradients [13], and although the p.m. density position (1.05) in Percoll is lower than that of e.r. (1.14) as measured by density marker beads (Pharmacia), we considered it desirable to remove Percoll from the fractions if possible. Two methods were tested.-

(a) The p.m. fraction was underlaid with 40% (w/v) sucrose and centri-fuged for 90 min at 100,000 g_{av}. As previously described [13], 80% of the e.r. in the fraction passes through the sucrose but none of the p.m. In two experiments the values for the protein content of the fraction at the upper interface with the sucrose were 338 and 531 µg by the Coomassie Blue binding method [29] and 356 and 505 µg by the method of Kahn et al. [30] for protein determination in the presence of Percoll. This indicated that the procedure was effective in removing Percoll; in addition the contamination by mitochondria and e.r. was reduced by ~7-fold and 4-fold respectively.

(b) Belsham et al. [31] described a p.m. preparation from adipocytes in which Percoll was removed by washing in a NaCl-based solution. We found that diluting acinar-cell p.m. fractions with 10 vol. of 80 mM NaCl/0.15 mM KCl/0.1 mM ouabain in 31 mM imidazole-HCl buffer, pH 7.1, and centrifuging for 2 min at 10,000 **g** yielded up to 50% of the p.m. in the pellet but with much reduced contamination by mitochondria and e.r. Comparisons of protein content with the above two methods showed more variability in the effectiveness of Percoll removal by this procedure. However, the marked improvement in p.m. purity as shown in Table 3 made this the procedure of choice.

The contamination of this preparation by e.r. and mitochondria was <0.1% of the total in the homogenate. In terms of the relative amounts of membrane, with an average p.m. yield of 8.6% this represents only 5.6% of e.r. in the fraction. This is comparable to other preparations of p.m. from pancreas isolated either by centrifugation in a Ficoll-sucrose density gradient in a zonal rotor followed by differential and sucrose-gradient centrifugation [32], or by repeated

Table 3. Isolation of p.m. from a low-speed supernatant: marker enzymes. Pancreas chopped into fragments was homogenized by Polytron (4×5 sec) in a medium as in (**B**) (a) above + 50 mM Tris-HCl, pH 7.4. After centrifugation for 10 min at 300 g_{av}, 22.5 vol. of Percoll in 0.5 M sucrose, 100 mM Tris-HCl, pH 7.0 (starting density 1.11) was added to the supernatant. After centrifugation for 30 min at 30,000 **g**, the upper band in the gradient was harvested, diluted with 10 vol. 80 mM KCl + ouabain etc. and centrifuged - see (b) in text, opposite. LSS = low-speed supernatant; M = fraction harvested from Percoll; PM = final pellet. Values expressed as in Table 2.

		Na,K-ATPase	Cyt. oxidase	NADPH/Cyt. reductase
% homogenate activity	LSS:	48.4 ±6.8 (3)	80.6 ±2.9 (3)	83.1, 77.6 (2)
	M:	27.5 ±2.2 (4)	6.0 ±0.9 (4)	6.4 ±1.2 (4)
	PM:	8.6 ±3.0 (4)	0.09 ±0.02 (4)	0.04 ±0.01 (4)
Specific activity	LSS:	12.1 ±0.4 (3)	0.96 ±0.09 (3)	3.9, 3.9 (2)
	M:	72.4, 146.7 (2)	0.4, 1.6 (2)	3.2, 8.7 (2)
	PM:	316.0 ±25.7 (4)	0.64 ±0.28 (4)	0.57 ±0.24 (4)
Relative specific activity	LSS:	0.60 ±0,02 (3)	0.93 ± 0.09 (3)	1.0, 1.0 (2)
	M:	3.6, 7.3 (2)	0.4, 1.6 (2)	0.8, 2.2 (2)
	PM:	15.7 ±0.7 (4)	0.62 ±0.22 (4)	0.14 ±0.05 (4)

precipitation of the membranes by magnesium [6]. Each of the latter preparations had relative specific activities of up to 40 for $(Na^{+}+K^{+})$-activated ATPase; the value of 16 reported here is apparently lower, due to a higher contamination by mitochondria. Plasma membranes have also been isolated from rat pancreatic acinar cells using differential centrifugation and discontinuous sucrose-gradient centrifugation [33]. These preparations had a relative specific activity for $(Na^{+}+K^{+})$-ATPase of ~12-fold, although no data on contamination by e.r. were reported. The procedure reported here has the advantage of being rapid by comparison, taking only 50-60 min following homogenization.

The purity of this preparation has allowed us to demonstrate Ca^{2+}-activated Mg^{2+}-ATPase activity in the p.m. which is distinct from the rough-e.r. and mitochondrial ATPases. The p.m. enzyme is half-maximally active at 0.65 μM free Ca^{2+} and activated by calmodulin, whereas the rough-e.r. ATPase is half-maximally active at 0.16 μM [13, 14]. This suggests that the p.m. ATPase may be less important in maintaining the resting cytoplasmic free Ca^{2+} concentration, viz. 0.1-0.4 μM [1-3]. During stimulation of secretion, when calcium uptake by the cell is increased [34, 35], at the higher Ca^{2+} concentration the p.m. Ca^{2+}-Mg^{2+}ATPase may be fully activated by Ca-calmodulin and important in balancing calcium entry to prevent the free Ca^{2+} from rising to levels which would be injurious to the cell.

PATHOPHYSIOLOGICAL RELEVANCE OF THE APPROACHES

A knowledge of the Ca^{2+}-transporting properties of cellular membranes is important to our understanding of pathophysiological conditions which are thought to arise from a disturbance in intracellular Ca^{2+} [36-38]. The methods described here, for the rapid purification of p.m.'s from exocrine acinar cells, are of particular relevance to the study of derangements which occur in pancreatitis [39] and cystic fibrosis. Our evidence, from *in vitro* studies on human submandibular glands, has led us to propose that an abnormal intracellular regulator, linking Ca^{2+}- and cyclic ATP-activated pathways controlling exocrine function, underlies the basic defect in cystic fibrosis [40, 41]. Studies using readily accessible cells, e.g. blood cells and fibroblasts, have suggested an alteration in p.m. Ca^{2+} transport [37]. However, it is not clear whether changes observed in these cells are an expression of the primary, genetic defect or a secondary change due, for example, to alteration in the membrane lipid environment. Studies of the transport properties of membranes purified from the primarily affected exocrine cells will be required to resolve this question.

Acknowledgements

This work was supported by grants from the Medical Research Council, the Welsh Scheme for the Development of Health and Social Research, and King Faisal University, Saudi Arabia.

References

1. O'Doherty, J. & Stark, R.J. (1982) *Am. J. Physiol. 242*, G513-G521.
2. Ochs, D.L., Korenbrot, J.I. & Williams, J.A. (1983) *Biochem. Biophys. Res. Comm. 117*, 122-128.
3. Powers, R.E., Johnson, P.C., Houlihan, M.J., Saluja, A.L. & Steer, M.L. (1985) *Am. J. Physiol. 248*, C535-C541.
4. Clemente, F. & Meldolesi, J. (1975) *J. Cell Biol. 65*, 88-102.
5. Dormer, R.L. & Williams, J.A. (1981) *Am. J. Physiol. 240*, G130-G140.
6. Bayerdorffer, E., Eckhardt, L., Haase, W. & Schulz, I. (1985) *J. Membr. Biol. 84*, 45-60.
7. Ceccarelli, B., Clemente, F. & Meldolesi, J. (1975) *J. Physiol. 245*, 617-638.
8. Dormer, R.L. (1983) *Biosci. Rep. 3*, 233-240.
9. Dormer, R.L. (1984) *Biochem. Biophys. Res. Comm. 119*, 876-883.
10. Knight, D.E. & Koh, E. (1984) *Cell Calcium 5*, 401-418.
11. Williams, J.A. (1980) *Am. J. Physiol. 238*, G269-G279.
12. Case, R.M. (1984) *Cell Calcium 5*, 89-110.
13. Richardson, A.E. & Dormer, R.L. (1984) *Biochem. J. 219*, 679-685.
14. Richardson, A.E., Brown, G.R. & Dormer, R.L. (1984) *Biochem. Soc. Trans. 12*, 1066-1067.

15. Berridge, M.J. & Irvine, R.F. (1984) *Nature 312*, 315-321.

16. Doughney, C., Brown, G.R., McPherson, M.A. & Dormer, R.L. (1987) *Biochim. Biophys. Acta*, in press.

17. Streb, H., Bayerdorffer, E., Haase, W., Irvine, R.F. & Schulz, L. (1984) *J. Membr. Biol. 81*, 241-253.

18. Al-Mutairy, A.R. Dormer, R.L. (1985) *Biochem. Soc. Trans. 13*, 900-901.

19. Bolender, R.P. (1974) *J. Cell Biol. 61*, 269-287.

20. Campbell, A.K. & Dormer, R.L. (1975) *Biochem. J. 152*, 255-265.

21. Campbell, A.K. & Dormer, R.L. (1978) *Bioochem. J. 176*, 53-66.

22. Warren, L., Glick, M.C. & Nass, M.K. (1966) *J. Cell Physiol. 68*, 269-288.

23. McCollester, D.L. (1970) *Cancer Res. 30*, 2832-2840.

24. Bauer, H-C., Ferber, E., Golecki, J.R. & Brunner, G. (1979) *Hoppe-Seyler's Z. Physiol. Chem. 360*, 1343-1350.

25. Williams, J.A., Korc, M. & Dormer, R.L. (1978) *Am. J. Physiol. 235*, E517-E524.

26. Sottocasa, G.L., Kuylenstierna, B., Ernster, L. & Bergstrand, A. (1967) *J. Cell Biol. 32*, 415-438.

27. Cooperstein, S.J. & Lazarow, A. (1951) *J. Biol. Chem. 189*, 665-670.

28. Bernfeld, P. (1955) *Meths. Enzymol. 1*, 149-158.

29. Bradford, M. (1976) *Anal. Biochem. 72*, 248-254.

30. Kahn, M.N., Kahn, R.J. & Posner, B.I. (1981) *Anal. Biochem. 117*, 108-112.

31. Belsham, G.J., Denton, R.M. & Tanner, M.J.A. (1980) *Biochem. J. 192*, 457-467.

32. Milutinovic, S., Sachs, G., Haase, W. & Schulz, I. (1977) *J. Membr. Biol. 36*, 253-279.

33. Ansah, T-A., Molla, A. & Katz, S. (1984) *J. Biol. Chem. 259*, 13442-13450.

34. Kondo, S. & Schulz, I. (1976) *Biochim. Biophys. Acta 419*, 76-92.

35. Dormer, R.L., Poulsen, J.H., Licko, V. & Williams, J.A. (1981) *Am. J. Physiol. 240*, G38-G49.

36. Campbell, A.K. (1983) *Intracellular Calcium: its Universsal Role as a Regulator*, Wiley, Chichester.

37. Katz, S., Schoni, M.H. & Bridges, M.A. (1984) *Cell Calcium 5*, 421-440.

38. Postnov, Y.V. & Orlov, S.N. (1985) *Physiol. Rev. 65*, 904-945.

39. Powers, R.E., Saluja, A.K., Houlihan, M.J. & Steer, M.L. (1986) *J. Clin. Invest. 77*, 1668-1674.

40. McPherson, M.A., Dormer, R.L., Dodge, J.A. & Goodchild, M.C. (1985) *Clin. Chim. Acta 148*, 229-237.

41. McPherson, M.A., Dormer, R.L., Bradbury, N.A., Dodge, J.A. & Goodchild, M.C. (1986) *Lancet ii*, 1007-1008.

#NC(D)

COMMENTS related to

EPITHELIAL MEMBRANES IN DIGESTIVE TISSUES

Comments on #**D-1**: J.M. Wilkes et al. – GASTRODUODENAL APICAL MEMBRANE
　　　　　　#**D-2**: H. Sjöström & co-authors – INTESTINAL BRUSH BORDER

Comment (to Wilkes) by T.J. Peters.- If trypsin affected the Cl$^-$ channels in gastric microsomes, this could explain its dissociated effects, on H$^+$-ATPase and on H$^+$ secretion. **Question by T. Berg to Sjöström.**- Might the proteins found along with clathrin be proteins transported in the coated vesicle? **Reply.**- Possibly yes; they may also be receptors. **Questions by W.H. Evans.**- What is your view on the origin of the coated vesicles? Do they exist as independent entities, or do they originate by vesiculation of coated regions of membranes? **Sjöström's answer.**- We sometimes see coated vesicles well removed from membrane structures. We thus favour the view that they actually exist in the cell's cytoplasm.

G.J. Rucklidge (to Sjöström).- What yield of antigen do you obtain from the 5 ml column, and approximately what amount of Ab do you bind to the matrix? **Reply.**- We use CNBr-activated Sepharose with the IgG fraction coupled at 10 mg/ml. With 90% coupling efficiency and possibly 1-2% of the IgG fraction representing the required Ab, we have a yield of (conservatively) 2 mg of antigen finally purified from a 5 ml column, i.e. ~3 mg antigen/mg Ab. **Comments by A.W. Schram.**- To elute an immunoaffinity column in such a way that one ends up with an intact protein is very often a problem. We have developed a procedure for immunopurification of a membrane-associated protein using, as an eluent, ethylene glycol – which may be of general use for eluting hydrophobic proteins from such columns. When raising a panel of monoclonal Ab's for use in immunoaffinity chromatography it is advisable to include in the screening procedure the method favoured for eluting the affinity column; e.g. screen the panel in the absence and presence of ethylene glycol, and choose as affinity support those monoclonal Ab's that in the absence of ethylene glycol bind with the antigen but in its presence dissociate from it. **Reply (by Sjöström) to D.R. Headon.**- Ab's to gluten proteins (kM$_r$ 45 & 35) indeed occur in sera of all coeliac patients. We have no information concerning Ab reaction with proteolytic fragments of these proteins.

Comments on #**D-3**: R.J. Simpson & T.J. Peters – MUCOSAL IRON UPTAKE
          #**D-4**: R.L. Dormer & colleague – PANCREATIC ACINAR CELLS

Simpson, answering Sjöström.- (1) There are no grounds for believing that transferrin-mediated iron transport occurs in the enterocyte.   (2) The concentration of transferrin in the intestinal lumen is considerably lower than in serum. **H. Sjöström asked Dormer** whether the p.m. preparation contained both apical and basolateral membranes (**reply**: yes), and whether the two types might be separable; but (**Dormer's reply**) there are no described methods, and there would be problems with marker enzymes. **Question by S. Zucker.-** Why was a Percoll gradient used to isolate p.m.'s from pancreatic acinar cells? **Reply**: because of aggregation problems with these membranes it was important to isolate them rapidly with a Percoll gradient. **To Dormer: N. Crawford.-** Concerning the p.m. Ca^{2+} extrusion pump that acinar cells, like red cells, seem to possess, Na^+/Ca^{2+} exchange at the p.m. might participate in Ca^{2+} homeostasis. **Dormer's reply.-** Our data comparing Ca^{2+},Mg^{2+}-ATPase measurements in red-cell and acinar-cell p.m.'s suggest that the two enzymes are somewhat different; in particular, unlike the red cell, it is difficult to demonstrate activation of the acinar-cell p.m. ATPase by µM concentrations of Ca^{2+} at mM Mg^{2+} concentrations. With regard to Na^+/K^+ exchange I think that the evidence for its importance is not very convincing. The process has been demonstrated in purified p.m. vesicles from pancreatic acinar cells, by Schulz's group, but the activity is very low. In addition to whole cells, Schulz has shown that $^{45}Ca^{2+}$ efflux can be inhibited by 30-min pre-incubation with ouabain, whereas Williams showed that removal of extracellular Na^+ from isolated pancreatic acini had no effect on $^{45}Ca^{2+}$ efflux from resting or cholinergically stimulated cells.

Further point raised by Crawford.- There is the paradox that the p.m. Ca^{2+},Mg^{2+}-ATPase seems to be inhibited at high Ca^{2+} levels where one would reckon it had to work harder. **Dormer's reply.-** The p.m. Ca^{2+},Mg^{2+}-ATPase was not inhibited by up to 3 µM free Ca^{2+} which is about the maximum the cytoplasmic free Ca^{2+} is thought to reach in the stimulated cell. However, my point is that, in the resting cell, where the free Ca^{2+} is thought to be ~0.2 µM, the rough-e.r. Ca^{2+}-sequestering system is apparently much more active than the p.m. pump. Only when the free Ca^{2+} increases following stimulation of secretion, and the p.m. pump is activated by calmodulin, is it likely to take over from the e.r. in preventing the cell from being flooded with Ca^{2+} as a result of stimulated Ca^{2+} entry.

Supplementary refs. contributed by Senior Editor

Relevant to (but not citing) studies by Wilkes.- Nagaya, H., Satoh, H. & Maki, Y. (1987) *Biochem. Pharmacol. 36*, 513-519: "Actions of antisecretory agents on proton transport in hog gastric microsomes".
    Brush-border membranes from human small intestine (surgical specimens) have furnished subfractions, examined for hydrolases and polypeptides.- Turnbull, G. & Bailey, D. (1986) *Biochem. Soc. Trans. 14*, 783-784.

Section #E

STUDIES FOCUSED ON ENDOCYTOSIS, PROTEASES OR INVASIVENESS

#E-1

RECEPTOR-MEDIATED UPTAKE OF A TOXIN WHICH BINDS
TO PLASMA-MEMBRANE ION CHANNELS

[1]P.N. Strong and [2]W.H. Evans

[1]MRC Receptor Mechanisms Research   [2]National Institute for
Group, Department of Pharmacology   Medical Research
University College London   Mill Hill
Gower Street, London WC1E 6BT, U.K.   London NW7 1AA, U.K.

The toxin apamin, a specific marker for Ca^{2+}-activated K^+ channels, is internalized by a receptor-mediated uptake mechanism in guinea-pig liver cells. No such uptake is observed with the toxin on rat liver cells, where these channels are absent. Ca^{2+}-activated K^+ channels are present in purified endosomes, isolated from guinea-pig liver on sucrose density gradients. They are enriched in 'early' endosome subfractions. These results suggest that ion channels can be recycled and regulated in much the same way as has already been demonstrated for various receptor molecules. Endosomal K^+ channels activated by Ca^{2+} ions may play a role in controlling the ionic conditions necessary for sorting within the endocytic network.

Peptide hormones, serum proteins, viruses and bacterial toxins are examples of many different types of molecule that bind to specific receptors on cell surfaces and are subsequently internalized by receptor-mediated endocytosis [1]. For many ligands, this process involves a clustering of occupied receptors into clathrin-coated pits and transfer into the cell's endocytic compartment. This vesicular-tubule network is responsible for the 'sorting' of internalized ligands; some are returned to the surface of the cell whereas others are dispatched to the lysosome for degradation. Receptors, on the other hand, escape degradation and, after dissociation from ligands, are recycled to the cell surface. The low pH (pH ~5) in the endocytic network is thought to be responsible for ligand-receptor dissociation [2, 3] and is generated by an ATP-dependent proton pump [4, 5]. The membranes comprising the endocytic networks can be recovered, after subcellular fractionation, as low-density vesicles on sucrose

density gradients [6]. Endocytic vesicles have been shown to possess
a monensin-specific Mg-ATPase activity [5] which is most probably
responsible for the maintenance of the proton pump.

In this article we demonstrate that hepatic endosomes possess
a further mechanism for regulating transmembrane ion flow, namely
a Ca^{2+}-activated K^+ channel. This type of channel occurs widely
in both excitable and non-excitable cells and is thought to provide
a link between membrane permeability and cell metabolism. Recent
studies have shown that Ca^{2+}-activated K^+ channels can be modulated
by intracellular second messengers such as cyclic AMP and inositol-
1,4,5-triphosphate [7, 8].

Toxins which specifically block ion-channel function have become
established as essential tools for the identification and characterization of these transmembrane protein structures, and the bee venom
toxin, apamin, has been used as a marker of Ca^{2+}-activated K^+ channels.
Apamin has been shown to block these channels in a variety of cells,
including hepatocytes [9], and biologically active [^{125}I]monoiodo-
apamin has proved a useful ligand for the complementary pharmacological
characterization of these channels in binding studies on both isolated
and intact cells [10, 11]. The existence of apamin-specific Ca^{2+}-activ-
ated K^+ channels on liver cells and our ability to isolate endocytic
vesicles from the same tissue [12] enables us to study the internalization of these ion channels and their role in hepatic endosome
function.

APAMIN PURIFICATION AND IODINATION

Apamin was isolated from the crude venom of the European honey
bee *(Apis mellifera)* by chromatography on SP-Sephadex C-25. Final
traces of melittin (50% by wt. of the crude venom) were removed by
chromatography on a heparin-Sepharose Cl-6B affinity column [13].
The toxin was judged pure by HPLC analysis on a C-8 RP column, run
isocratically (2% propanol, 3 mM methanesulphonic acid, 10 mM tetra-
ethylammonium phosphate pH 3.0) [14].

Apamin was iodinated using the iodogen method, and ^{125}I$^-$, [^{125}I]-
monoiodo-, [^{125}I]diiodo- and unlabelled apamin were separated by ion-
exchange chromatography on SP-Sephadex C-25 [15]. [^{125}I]Monoiodo-
apamin was desalted on a disposable 1 ml C-18 RP column (Bond-Elut,
Analytichem], and subsequently stored at 4° in the presence of 1%
(w/v) serum albumin. Under these conditions, there was no deterioration
in biological activity for at least 8 weeks. Assuming 100% isotopic
abundance, the specific activity of [^{125}I]monoiodoapamin was
2000 Ci/mmol and its concentration could be directly determined in
a calibrated gamma counter.

IDENTIFICATION OF CALCIUM-ACTIVATED POTASSIUM CHANNELS
ON GUINEA-PIG LIVER PLASMA MEMBRANES (p.m.'s)

Saturable, high-affinity apamin binding sites (Ca^{2+}-activated K^+ channels) were characterized on the p.m.'s [16]. Livers were homogenized in ice-cold buffer (1 mM $NaHCO_3$, 0.5 mM $CaCl_2$); p.m.'s were preparred by the Neville method [17] and collected at the 43/48% (w/v) sucrose interface. Fractions were then resuspended in 1 mM EGTA/10 mM Tris buffer pH 7.4 containing 8% (w/v) sucrose and homogenized in a tight-fitting Dounce homogenizer. After centrifugation at 98,000 g_{av} for 30 min, the membranes were finally resuspended in 8% sucrose and stored in liquid N_2. Toxin binding was measured in low ionic strength medium using a filter assay [15]. Membranes (100 μg protein) were incubated with iodotoxin (0.9–50 pM) in a buffer containing 5.4 mM KCl, 0.1% bovine serum albumin and 10 mM Tris/Hepes pH 7.4. After incubation for 90 min at 0°, triplicate samples were rapidly filtered through cellulose acetate filters (0.45 μm; Sartorius type 11106) and washed with two 5 ml portions of ice-cold buffer. The binding of apamin to guinea-pig liver membranes was determined as a function of toxin concentration, in the presence and absence of a large excess (1 μM) of unlabelled toxin (Fig. 1).

Assuming that inhibitable binding could be described by a simple bimolecular model, values of apparent dissociation constants (K_D) and maximum binding capacities of the inhibitable binding components (B_{max}) were obtained (Table 1) by computer-aided curve-fitting of the data to a single component Langmuir isotherm, using an iterative least-squares method [18]. The linearity of a Scatchard plot (Fig. 2) derived from the data was in agreement with the chosen model of apamin binding to a single class of non-interacting binding sites.

Rat-liver p.m.'s had negligible levels of apamin binding sites (Table 1), in accord with both biochemical and physiological data indicating the absence of apamin-specific Ca^{2+}-activated K^+ channels on isolated rat hepatocytes [9, 19, 20].

INCORPORATION OF [^{125}I]MONOIODOAPAMIN INTO LIVER CELLS

Monoiodoapamin (1–20 pmol, in 0.5 ml 0.15 M NaCl, 0.001 M Na phosphate buffer, pH 7.4) was injected into the portal vein of anaesthetized Sprague-Dawley rats (pentabarbitone, 40 mg/kg) or small Hartley guinea pigs (60 mg/kg). At various time intervals, the liver was removed; after subcellular fractionation, radiolabel was sought (4 min after injection in the case of the Fig. 3 experiment) in both 'early' and 'late' endosomal vesicle subpopulations on sucrose density gradients. The guinea-pig liver was homogenized in ice-cold 0.25 M sucrose (loose- and then tight-fitting Dounce homogenizer, 10 and then 6 strokes respectively) and centrifuged at 1000 g_{av} for 10 min at 4°. After washing and re-centrifugation, the combined supernatants were centrifuged at 33,000 g_{av} for 8 min and the final supernatant

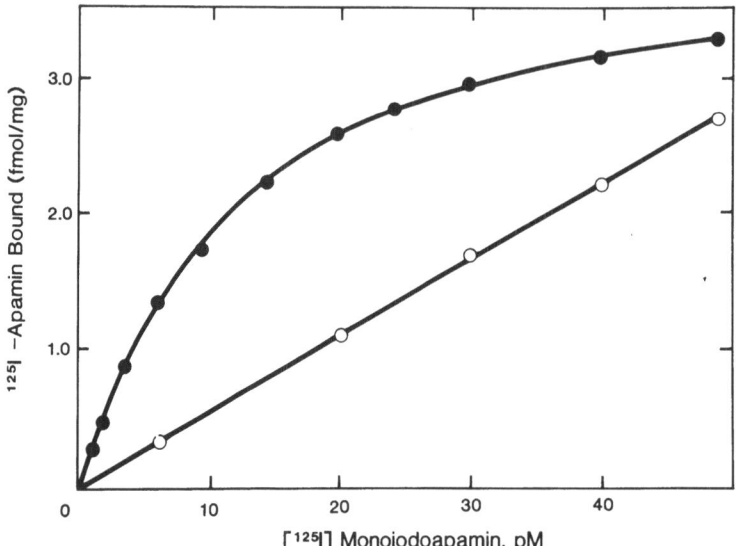

Fig. 1.   Binding of [125I]monoiodoapamin to guinea-pig liver p.m.'s. Inhibitable (●) and non-inhibitable (○) binding measured in the presence of 1 μM apamin.   Points represent the mean of 3 expts., each performed in triplicate (S.E.'s not shown).

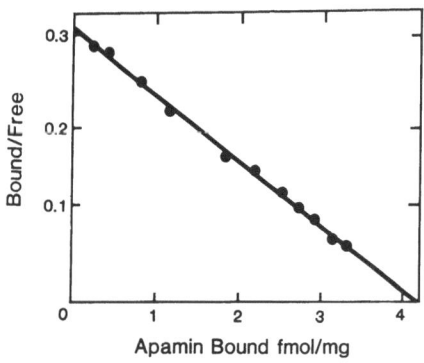

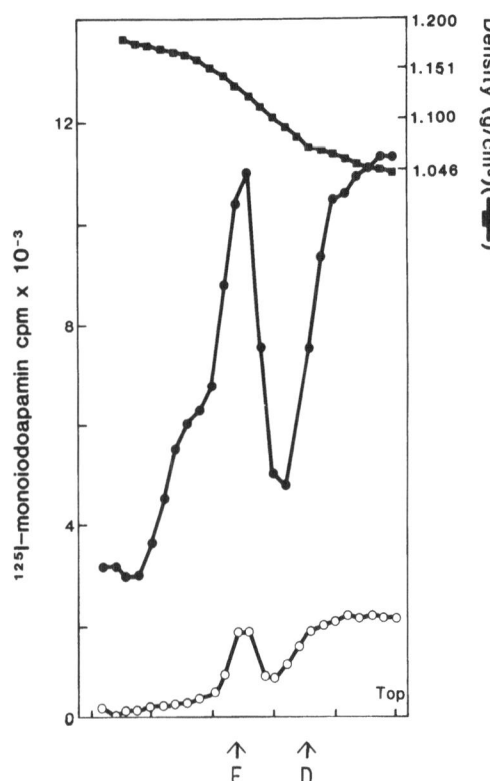

Fig. 2 *(above).*   Scatchard plot derived from the data shown in Fig. 1.

Fig. 3 *(right).*   Incorporation of [125I]monoiodoapamin, 4 min after injection, into endosome fractions isolated by gradient centrifugation (see text) from post-mitochondrial supernatants from liver: ●, guinea pig; ○, rat (means from 3 expts. varying by <10%). E, D: see text. *From [16], by permission.*

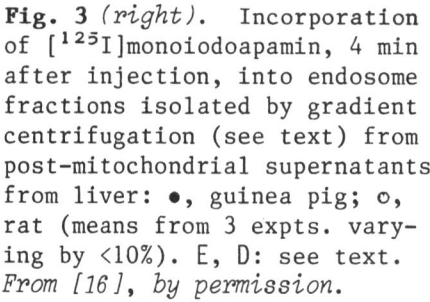

Table 1. [^{125}I]Monoiodoapamin binding parameters: K_D (pM) and B_{max} (fmol/mg protein). S.E.'s (n = 3) are based on the variance of residuals and the calculated normalizing elements for each parameter.

Liver preparation	Species	K_D	B_{max}
Plasma membrane	Guinea pig	12.6 ±0.8	4.2 ±0.2
'Early' endosome	Guinea pig	10.6 ±3.3	2.5 ±0.6
'Late' endosome	Guinea pig	10 ±4	0.7 ±0.3
Plasma membrane	Rat		<0.2

applied to a sucrose gradient (1 ml 70%, 5 ml 43%; 15 ml of a 15–40% continuous gradient) and centrifuged at 100,000 **g** at 4° in a Beckman SW 28 rotor for $3\frac{1}{2}$ h.

Almost all of the radiolabel appeared in the 'early' endosome fraction (fraction E, 1.12–1.14 d range), the radioactivity appearing at the top of the gradient representing free toxin that was released from lysed endocytic vesicles (Fig. 3). At later time intervals (data not shown) the amount of [^{125}I]monoiodoapamin found in 'early' endosomes decreased, and in contrast with many other internalized hormones and serum proteins [21] (cf. [6] in this series), the toxin could not be found in low-density, 'late' endosomal subfractions (fraction D, density range 1.095–1.117) on the sucrose gradients. However, the data with apamin resembled those obtained with the β-adrenergic antagonist iodocyanopindolol [6, 22] where the ligand was rapidly transferred to the lysosomes after appearing in 'early' endocytic vesicles and was difficult to interpret and recover in the endocytic network at physiological temperatures.

In comparison with the above data on toxin uptake by guinea-pig liver, only very small amounts of toxin were taken up by rat liver, and the corresponding radiolabel appearing in rat endosomal fractions was likewise extremely low (Fig. 3). The lack of uptake of [^{125}I]-monoiodoapamin into rat livers (which do not possess Ca^{2+}-activated K^+ channels) argued that internalization into guinea-pig livers was a specific process and did not proceed via a generalized fluid-phase endocytic mechanism.

The radiolabelled peak in both rat and guinea-pig fractions was identified as undegraded [^{125}I]monoiodoapamin by its elution characteristics in both SP-Sephadex C-25 ion-exchange chromatography [15] and TLC on silica gel in a butanol/pyridine/acetic acid/water (3:3:1:3 by vol.) solvent system [23]. The recovery of toxin in an undegraded form suggested that the 'early' endosome fraction was devoid of active peptidases, and provided further evidence for the subcellular distinction between endosomes and lysosomes.

Fig. 4. SDS-PAGE analysis of p.m.'s and endosomal membranes from guinea-pig liver, solubilized in 2% (w/v) SDS/1% β-mercapto-ethanol/50 mM Tris-HCl pH 7.4 and electrophoresed in 10% polyacrylamide gels. (a), p.m.'s. (b), 'early' endosomal membranes (fraction 'E'). (c), 'late' endosomal membranes (fraction 'D'). *From [16], by permission of European Journal of Biochemistry (as for Fig. 3).*

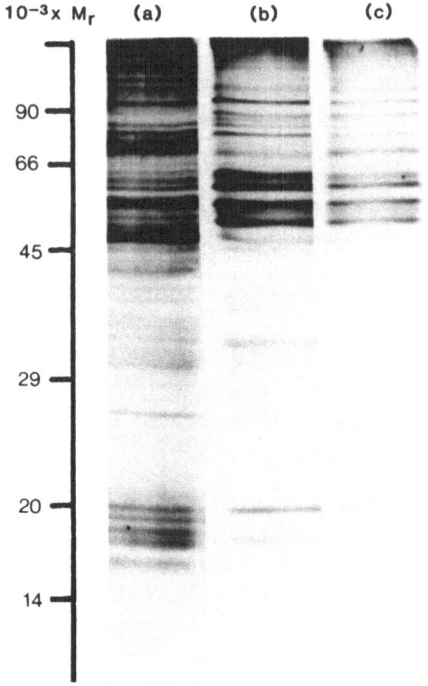

$10^{-3} \times M_r$ (a)   (b)   (c)

90 —
66 —
45 —
29 —
20 —
14 —

IDENTIFICATION OF CALCIUM-ACTIVATED POTASSIUM CHANNELS ON GUINEA-PIG LIVER PLASMA MEMBRANES

Since an enzymatic analysis and the polypeptide and phospholipid profiles of rat-liver endosomal membranes have been reported [21, 24] and guinea-pig endosomal membranes used in the present experiments were prepared by an identical method, a full characterization of guinea-pig endosomes was not carried out. Analyses of the protein composition of guinea-pig p.m. and endosome fractions (Fig. 4) showed them to correspond closely to the analysis reported for the corresponding rat fractions [24]. A limited enzymatic analysis ([16] and, for rat, [5]) showed that, like rat endosomes, both 'early' and 'late' guinea-pig endosomal fractions were enriched >30-fold in a monensin-activated Mg^{2+}-ATPase (expressed as μmol/h/mg protein):
- homogenate: guinea-pig, 0.2; rat, 0.77;
- 'early' endosome: 6.3   12.6;
- 'late' endosome: 6.5   15.1.
The guinea-pig endosomal fractions were very low in lysosomal marker enzymes and free of mitochondria (marker: succinate dehydrogenase [16].

[^{125}I]Monoiodoapamin bound to 'early' endosomal membranes from guinea-pig liver in a saturable manner (Fig. 5), very similar to that previously shown for liver p.m.'s [16]. A Scatchard plot (Fig. 6)

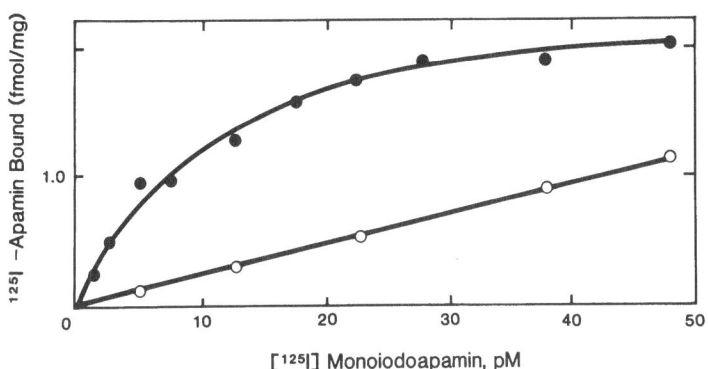

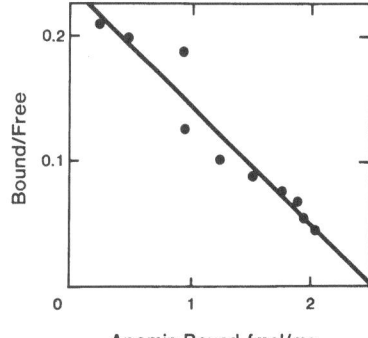

Fig. 5. Binding of [^{125}I]-monoiodo-
apamin to guinea-pig liver endosomal
membranes. Inhibitable (•) and non-
inhibitable (o) binding measured in
the presence of 1 μM apamin. See
Fig. 1 legend and Table 1 heading for
no. of observations and S.E. calculation.

Fig. 6 *(right).* Scatchard plot
derived from the data shown in
Fig. 5.

derived from the data supported the simple bimolecular model of toxin
binding to a single class of non-interacting binding sites. Although
the affinity of [^{125}I]monoiodoapamin for endosomal membranes was
very similar to that of the toxin for p.m.'s, the density of binding
sites was reduced (Table 1). The toxin also bound to the 'late'
endosome fraction in a saturable manner, displaceable by unlabelled
apamin, although the number of binding sites was even lower than
that seen with the 'early' endosome fraction. As with p.m.'s, no
binding sites could be detected on rat endosomal membrane fractions.

The existence of high-affinity intracellular binding sites for
apamin on endosomal membranes indicates that these sites, corresponding
to Ca^{2+}-activated K^+ channels, are present in the endocytic network,
irrespective of whether or not apamin is internalized. The results
suggest that these ion channels are present in 'late' as well as
'early' endosomes, which is in distinction to the radiolabelled toxin
which was observed only in the 'early' endosome fraction. Although
it is premature to identify specific pathways, the evidence suggests
that channel-toxin complexes are processed in the endocytic network,
affording recycling of the channel-forming protein and dispatch of
the toxin for degradation within the lysosome.

The results also indicate a wider generality for the trafficking of membrane proteins and suggest that channel-forming proteins, as well as receptor molecules, move between the cell surface and the endocytic compartment. The role of these K^+ channels (activated by Ca^{2+}) may relate to a mechanism controlling the ionic conditions necessary for the trans-membrane movement of ligands into the cytoplasm.

Acknowledgements

We thank Dr. M. Claret and Dr. J-P. Mauger for helpful advice, Mr. N. Flint and Miss J. Williamson for technical assistance and Miss V. Grant for preparing the manuscript. This work was supported in part by a Medical Research Council project grant to P.N.S.

References

1. Wileman, T., Harding, C. & Stahl, P. (1985) *Biochem. J. 232*, 1-14.
2. Tycko, B. & Maxfield, F.R. (1982) *Cell 28*, 643-651.
3. Geisow, M.J. & Evans, W.H. (1984) *Exp. Cell Res. 150*, 36-46.
4. Yamashiro, D.J., Fluss, S.R. & Maxfield, F.R. (1983) *J. Cell Biol. 97*, 929-934.
5. Saermark, T., Flint, N. & Evans, W.H. (1985) *Biochem.J. 225*, 51-58.
6. Evans, W.H., Flint, N. & Hadjiivanova, N. (1984) in *Investigation of Membrane-Located Receptors* (Reid, E., Cook, G.M.W.& Morré, D.J., eds.) [Vol. 13, this series], Plenum, New York, pp. 203-210.
7. Ewald, D.A., Williams, A. & Levitan, I.B (1985) *Nature 315*, 503-506.
8. Higashida, H. & Brown, D.A. (1986) *Nature 323*, 333-335.
9. Burgess, G.M., Claret, M. & Jenkinson, D.H. (1981) *J. Physiol. 317*, 67-90.
10. Lazdunski, M., Fosset, M., Hugues, M., Mourré, C., Romey, G. & Schmid-Antomarchi, H. (1985) *Biochem. Soc. Symp. 50*, 31-42.
11. Strong, P.N. & Castle, N.A. (1987) in *Metabolism and Development of the Nervous System* (Tucek, S., ed.), Wiley, Chichester, in press.
12. Evans, W.H. (1985) *Meths. Enzymol. 109*, 246-257.
13. Banks, B.E.C., Dempsey, C.E., Pearce, F.L., Vernon, C.A. & Wholley, T.E. (1981) *Anal. Biochem. 116*, 48-52.
14. Dotimas, E.M., Hider, R.C., Ragnarsson, U. & Tatham, A.S. (1984) *Proc. 18th Eur. Peptide Symp.*, 141-144.
15. Hugues, M., Duval, D., Kitabgi, P., Lazdunski, M. & Vincent, J.P. (1982) *J. Biol. Chem. 257*, 2762-2769.
16. Strong, P.N. & Evans, W.H. (1987) *Eur. J. Biochem. 163*, 267-273.
17. Neville, D.M. (1966) *Biochim. Biophys. Acta 154*, 540-552.
18. Marquards, D.W. (1963) *J. Soc. Indust. Appl. Math. 11*, 431-441.
19. Cook, N.S., Haylett, D.G. & Strong, P.N. (1983) *FEBS Lett. 152*, 265-269.
20. Cook, N.S. & Haylett, D.G. (1985) *J. Physiol. 358*, 373-394.

21. Evans, W.H. & Flint, N. (1985) *Biochem. J. 232*, 25-32.
22. Hadjiivanova, N., Flint, N., Evans, W.H., Dix, C. & Cooke, B. (1984) *Biochem. J. 222*, 749-754.
23. Habermann, E. & Fischer, K. (1979) *Eur. J. Biochem. 94*, 355-364.
24. Evans, W.H. & Hardison, W.G.M. (1985) *Biochem. J. 232*, 33-36.

#E-2

LYSOSOMAL PROTEOLYSIS IN CULTURED HEPATOCYTES

Peter Bohley, Gabriele Adam, Werner Hoch
and Jürgen Kopitz

Physiologisch-chemisches Institut der Universität Tübingen
Hoppe-Seyler-Str. 1, D-74 Tübingen, W. Germany (FRG)

We have investigated the degradation of labelled rat-liver cytosol proteins after their introduction ('micro-injection') into cultured hepatocytes by published procedures. These entailed incubation in a sucrose-PEG* medium, then incubations in a hypotonic medium, thereby facilitating selective lysis of the pinosomes without increasing the cytosolic activities of lysosomal enzymes. These treatments were without detriment to the survival rates of the hepatocytes in subsequent monolayer culture, during which the distribution and degradation of the introduced proteins was investigated by microscale cell fractionation and determination of protein-bound and free radioactivity in the subcellular fractions. Unexpectedly, lysosomes showed the highest selectivity for the uptake and degradation not only of long-lived but also of very short-lived cytosol proteins. Proteolysis was also investigated with isolated fractions.

Intracellular proteolysis is necessary for liver cells to adapt their enzyme content to environmental changes without a change in the total amount of protein. The half-lives of hepatocyte proteins differ from a few minutes to many weeks, though some cytosolic proteins are very short-lived, e.g. ornithine decarboxylase and tyrosine amino-transferase [review: 1]. As much as one-third of intracellular proteolysis is directed at nascent proteins in mammalian cells [2], but the cellular compartments as well as molecular mechanisms responsible for this rapid degradation are still unknown. One aim of our investigations is to find the main cellular compartments engaged in this process and to elucidate, for substrate proteins, the molecular properties which are important for their rapid degradation.

The use of cultured hepatocytes allows the isolation of highly labelled substrate proteins and their introduction into unlabelled cells so as to follow their cellular distribution and subsequent

---

*Abbreviations: PEG, polyethylene glycol; TCA, trichloroacetic acid; GDH, glutamate dehydrogenase; & see Fig. 2 legend.

degradation. We have preferred the hypertonic-hypotonic method of Okada & Rechsteiner [3] for introducing radiolabelled proteins into cultured cells, because this method avoids the use of additional membranes (as in, e.g., [4–7]) which may selectively bind the hydrophobic very-short-lived proteins.

The use of cultured hepatocytes also allows an investigation of the degradation of endogenous proteins in the different subcellular fractions. Here we compare the degradation of cellular proteins with the proteolysis of introduced cytosol proteins.

CULTURING AND LABELLING, AND RELATED PROCEDURES

Male Sprague-Dawley rats (body wt. 210–260 g) were kept for >6 days in a 12 h-dark/12 h-light cycle on the standardized diet of ALMA® and water *ad libitum*.

Liver parenchymal cells were isolated by a collagenase perfusion method [8]. The isolated cells were allowed to attach to collagen-coated Petri dishes in Williams E medium [9] with 10% (v/v) calf serum, 10^{-6} M insulin and 10^{-7} M dexamethasone for 2 h, and the medium was changed after a further 2 h to serum-free medium (and with radio-active amino acids if applicable; see below), and likewise every 24 h for long-term cultures unless otherwise indicated.

Double labelling of hepatocyte proteins in culture was effected with [^{14}C]leucine for 15 h followed by a 24 h chase and finally with [^{3}H]leucine for 30 min. Thereby ^{14}C-labelling was effected for long-lived proteins and ^{3}H-labelling mainly for shorter-lived proteins. The labelled cells were washed 3 times with Williams E medium and 3 times with 0.3 M sucrose, rapidly harvested from the monolayers with a rubber policeman, and homogenized in a Potter-type micro-homogenizer (80 μm clearance; 10 up-and-down strokes).

The cell-fractionation procedures were at 0–4° and essentially followed our earlier description [10] which gives details such as **g**-min values and purity checking by marker-assay (DNA, GDH, β-glycero-phosphatase, glucose-6-phosphatase) but not the following procedures for nuclei and lysosomes. Nuclei were purified by a re-spin in 2 M sucrose (100,000 **g**, 30 min). Lysosomes were selectively disrupted by homogenization of the crude mitochondrial-lysosomal fraction in water containing digitonin (0.3 mg/ml); since the supernatants contained <2% of the mitochondrial GDH, there was no need for no-digitonin blanks. All high-speed centrifugations were performed in an 'Airfuge' (Beckman Insts.); this made it possible to recover the cytosol fraction <2 h after the cell harvesting, which is very important for the yield of very-short-lived proteins.

To measure surviving and degraded labelled proteins, all isolated cell fractions (listed in Fig. 2 legend; also post-L supernatant,

and homogenates where applicable) were treated with 1 vol. of 10%
(w/v) TCA to precipitate the undegraded proteins. After centrifugation
(13,000 g, 10 min), aliquots of the supernatants, and of the undegraded
proteins solubilized by 1 M KOH (24 h, 37°), were mixed with Aqualuma-
plus (usually 20-200 µl sample to 4 ml scintillator) and counted
for at least 10 min in an LKB 1219 Rackbeta liquid scintillation
counter with automatic dpm calculation for ^3H and ^{14}C.

Before their introduction into unlabelled cells, the double-
labelled cytosol proteins were freed from labelled amino acids by
rapid passage through Sephadex G-25 gel. For some experiments (Fig.
2, below), in place of ^3H- and ^{14}C-labelling in culture, a purified
protein was ^3H-labelled by reductive methylation [11] to furnish
the substrate. All substrates had to be checked for possible autoproteolysis that could give artefactual results.

Checking for autoproteolysis.- The rate of autoproteolysis was
very low for culture-labelled cytosol proteins after the Sephadex
step: <0.4%/h for ^3H-labelled and <0.07%/h for ^{14}C-labelled proteins.
Therefore these substrate protein fractions suit well for investigating
intracellular proteolysis: endogenous proteinases that could interfere
are not co-introduced. Autodegradation was likewise minimal for
the reductively methylated ^3H-proteins.

The introduction of labelled proteins into cells

The hypertonic-hypotonic treatment [3] used for introducing the
cytosolic double-labelled or the ^3H-labelled proteins into unlabelled
cultured hepatocytes was as shown in Fig. 1. Monolayers (10^6 cells)
were incubated for 16 min at 37° with 3 ml of a solution of labelled
proteins in 0.5 M sucrose containing 10% (w/v) PEG-1000 in Williams
E medium, followed by a 3 min incubation in hypotonic (60%) medium
(Williams E medium/water, 6:4 by vol.). We prefer to repeat this
hypotonic step to facilitate the selective lysis of the newly formed
hypertonic pinosomes and thus to increase the uptake of labelled
proteins into the cytosol. This repetition did not increase the
cytosolic activities of the lysosomal enzyme β-glycerophosphatase
(determined as in [10]) or of proteinases (autoproteolysis in the
cytosol remained <0.5%/h even after two repetitions of the hypotonic
treatment). Moreover, the survival rates of the hepatocytes in monolayer were not changed by the hypertonic-hypotonic treatment or by
repetitions of the hypotonic step. In contrast, hepatocytes in suspension did not survive the hypertonic treatment (>85% dead cells
after 15 min).

After introduction of the labelled substrate proteins into the
cultured hepatocytes, they were harvested immediately or after further
incubation at 37° in unlabelled Williams E medium for 15, 30 or 45 min
or 24 or 48 h. Procedures described above were used for washing,
cell harvesting, homogenization, isolation of fractions and analysis

Incubation of cells in monolayer in a hypertonic medium:

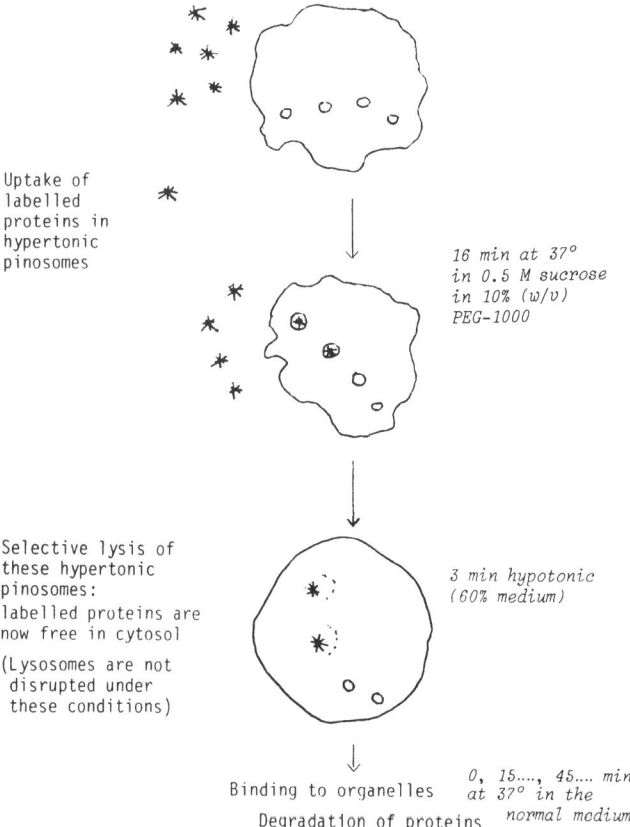

Uptake of
labelled
proteins in
hypertonic
pinosomes

*16 min at 37°
in 0.5 M sucrose
in 10% (w/v)
PEG-1000*

Selective lysis of
these hypertonic
pinosomes:
labelled proteins are
now free in cytosol

(Lysosomes are not
disrupted under
these conditions)

*3 min hypotonic
(60% medium)*

Binding to organelles
Degradation of proteins

*0, 15...., 45.... min
at 37° in the
normal medium*

Fig. 1. Introduction of labelled proteins into hepatocytes (method
of Okada & Rechsteiner [3]): selective lysis of hypertonic pinosomes.
See text for initial protein labelling in cultures and passage
through Sephadex, also for repetition of the hypotonic step and for
the final incubation medium (isotonic).

of these [10] for radiolabel (giving Q-values as in Fig. 2 legend)
and protein. Examination of subcellular fractions for proteolytic
capacity entailed a 2-h incubation, then the above TCA step.

PROTEOLYTIC CAPACITY OF CELL FRACTIONS

As shown in Fig. 2, the degradation of short-lived proteins
(ODC, TAT and ^{3}H-labelled cytosol proteins) and also of long-lived
proteins (LDH, ubiquitin and ^{14}C-labelled cytosol proteins) was by
far the greatest in the lysosomal fraction; the proteolytic capacity
of all other cell fractions was comparatively low. In particular
the cytosol is not able to degrade any of the substrate proteins

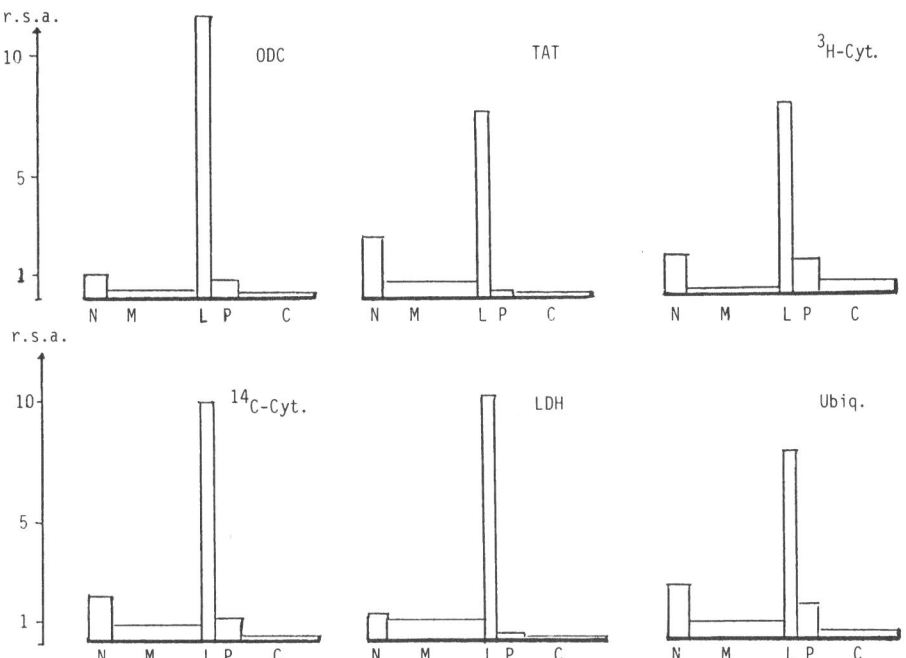

Fig. 2. Degradation of short-lived *(top)* and long-lived *(bottom)* proteins of rat-liver cytosol by isolated rat-liver cell fractions, as relative specific activity (r.s.a.; 'Q' value) for TCA-soluble material in each fraction compared with homogenate on a protein-N basis. N, nuclei (see text for 2 M sucrose re-spin); M, mitochondria; L, lysosomes; P, microsomes; C, cytosol (after rapid isolation in an 'Airfuge').
^{3}H-Cyt. = short- and ^{14}C-Cyt. = long-lived cytosol proteins.
ODC = ornithine decarboxylase, TAT = tyrosine aminotransferase; LDH = lactate dehydrogenase, Ubiq. = ubiquitin: all ^{3}H-labelled by reductive methylation, performed on purified proteins (as will be published for ODC and TAT).

(except perhaps short-lived cytosol proteins) at appreciable rates. This agrees well with the very low autoproteolysis of cytosol (see above) and also with many earlier investigations [1, 10]. This finding seems to warrant investigation of the extent to which the proteinase inhibitors in the cytosol might be responsible for this low proteolysis and whether cytosol proteins *in vivo* might be degraded in other organelles. Yet such experiments with isolated subcellular fractions cannot give full information on such processes in the living cells. For instance, the transport of the substrate proteins into the degradative organelles might be the rate-limiting step in intracellular proteolysis. Therefore we investigated this problem in more detail by introducing the double-labelled proteins from cultured hepatocytes into unlabelled hepatocytes in monolayer.

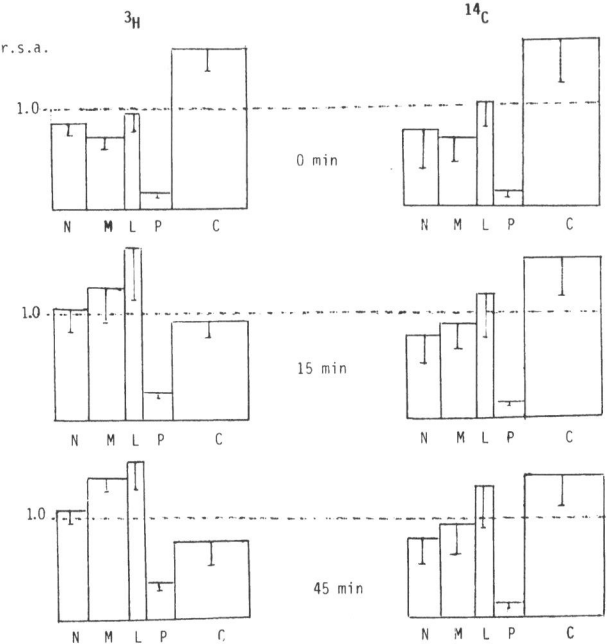

Fig. 3. Intracellular distributions of labelled cytosol proteins –
short-lived (^{3}H) and long-lived (^{14}C) – in hepatocyte monolayers
[2, 3 or 4 days of pre-exposure culture; no bearing on results]
following introduction of the proteins in hypertonic pinosomes and
then selective lysis of these pinosomes (as in [3]) and incubation
for the times indicated. See text for the preparation of the
labelled proteins including amino acid removal by Sephadex; Fig. 1
summarizes the Fig. 3 approach. TCA-insoluble radiolabel is represented as relative specific activity (r.s.a.; 'Q' values) compared
with homogenate on a protein-N basis; the vertical bars signify S.D.
(6–8 expts.). For degradation (TCA-soluble label) see text.
Subcellular fractions are designated as in Fig. 2.

INTRACELLULAR DISTRIBUTION AND DEGRADATION OF INTRODUCED CYTOSOL PROTEINS

As is shown in Fig. 3, immediately after the introduction of
the labelled cytosol proteins they were found mainly in the cytosol
fraction. Yet at this time substantial amounts were also found in
nuclei, mitochondria and especially lysosomes. This might be at
least partly due to fusion of the newly formed pinosomes (containing
labelled proteins) during the 16-min introduction period. However,
it is unclear why the introduced cytosol proteins also bind to nuclei
and mitochondria; possibly they have truly accumulated in the nuclei,
but it seems more likely that they have associated with structures
which co-sediment with nuclei, as has been suggested for a close
association of introduced proteins with the cytoskeleton before their
degradation by lysosomes [4].

Whereas the pattern of long-lived (^{14}C-labelled) introduced proteins did not change significantly during incubation of the recipient cultured hepatocytes for 15 or 45 min, a remarkable change in short-lived (^{3}H) proteins was observed after 15 min: their cytosolic specific activity decreased markedly , especially after 45 min, hand-in-hand with a rise in the lysosomal fraction in particular.

Concomitantly (not illustrated), the amount of TCA-soluble radioactivity increased in all subcellular fractions and in the extra-cellular incubation medium. From this increase, an overall degradation of 7% of the introduced ^{3}H-proteins after 15 min and of 19% after 45 min was calculated. The degradation of ^{14}C-proteins was <1% of all introduced proteins after 45 min. It was not possible to discriminate between translocation and degradation of these proteins in the cells by such methods.

The very small proteolytic activity in cytosol after its very rapid isolation in an 'Airfuge' (see Fig. 2) is not sufficient to explain the rapid loss of labelled proteins from this fraction. Thus we assume an intracellular cooperation in degradation of these proteins whereby lysosomes are responsible for their rapid uptake (see Fig. 3) and for their subsequent degradation. This assumption is supported by the very rapid degradation of endogenously labelled proteins in hepatocyte cultures: >30% of the very-short-lived proteins (^{3}H-labelling for 30 min) in the lysosomal fraction were degraded in the first hour after the labelling.

CONCLUDING REMARKS

Our results do not support the assumption that the selective degradation of introduced proteins occurs principally in the cytosol rather than in lysosomes (which has been based on experiments with [^{14}C]sucrose-labelled bovine serum albumin or pyruvate kinase [12]) nor on the assumption that lysosomes degrade mainly the long-lived proteins [13-15]. These assumptions were made before the discovery that a large proportion (up to 40%) of cellular proteins of mammalian cells are very short-lived. It is now important to identify the endogenous proteinases which are responsible for this rapid intra-cellular proteolysis. Experiments with cultured cells using specific inhibitors of lysosomal proteinases are in progress in our laboratory.

Proteins that turn over rapidly exhibit increased surface hydrophobicity [16-19] - a property that may be important for the control of proteolysis not only to permit protein entry into the degradative environment but because many cellular proteinases preferentially attack peptide bonds adjacent to hydrophobic residues [review: 1].

We plan to investigate whether changes in the surface hydrophobicity of substrate proteins may be important for the control of proteolysis of very-short-lived proteins in cultured hepatocytes.

Acknowledgements

This work was possible only with the help, which is gratefully acknowledged, of Prof. Dr. D. Mecke and Dr. R. Gebhardt and their group working with hepatocyte cultures, particularly W.A. Schultz, H. Burger, H. Heini, M. Landesvatter, M. Locher, K. Mayer and B. Ugele. Dr. A.J. Kenny (Leeds) kindly helped in preparing the manuscript. We are also grateful to the Bundesministerium für Forschung und Technologie for supporting this work by a research grant (BMFT 03 8655 1).

References

1. Bohley, P. (1987) in *New Comprehensive Biochemistry* (Neuberger, A. & Brocklehurst, K., eds.), Elsevier, Amsterdam, in press.
2. Wheatley, D.N. (1984) *J. Theoret. Biol. 107*, 127-149.
3. Okada, C.Y. & Rechsteiner, M. (1982) *Cell 29*, 33-41.
4. Doherty, F.J. & Mayer, R.J. (1985) *Biochem. J. 226*, 685-695.
5. Hendil, K.B. (1981) *Exp. Cell Res. 135*, 157-166.
6. Netland, P.A. & Dice, J.F. (1985) *Anal. Biochem. 150*, 214-220.
7. McElligot, M.A. Miao, P. & Dice, J.F. (1985) *J. Biol. Chem. 260*, 11986-11993.
8. Gebhardt, R. & Jung, W. (1982) *J. Cell Sci. 56*, 233-244.
9. Williams, G.M. & Gunn, J.M. (1974) *Exp. Cell Res. 89*, 139-142.
10. Bohley, P., Kirschke, H., Langner, J. & Ansorge, S. (1969) *FEBS Lett. 5*, 233-236.
11. Tack, B.F. & Wilder, R.L. (1981) *Meths. Enzymol. 73*, 138-147.
12. Bigelow, S., Hough, R. & Rechsteiner, M. (1981) *Cell 25*, 83-93.
13. Ballard, F.J. (1977) *Essays in Biochem. 13*, 1-26.
14. Khairallah, E.A., Bond, J.S. & Bird, J.W.C., eds. (1985) *Intracellular Protein Catabolism*, Liss, New York.
15. Pontremoli, S. & Melloni, E. (1986) *Ann. Rev. Biochem. 55*, 455-481.
16. Bohley, P. (1968) *Naturwiss. 55*, 211-219.
17. Bohley, P., Kirschke, H., Langner, J., Miehe, M., Riemann, S., Salama, Z., Schön, E., Wiederanders, B. & Ansorge, S. (1979) in *Biological Functions of Proteinases* (Holzer, H. & Tscheche, H., eds.), Springer-verlag, Berlin, pp. 16-34.
18. Mann, D.F., Shah, K., Stein, D. & Snead, G.A. (1984) *Biochim. Biophys. Acta 788*, 17-22.
19. Bohley, P., Hieke, C., Kirschke, H. & Schaper, S. (1985) *Prog. Clin. Biol. Res. 180*, 447-455.

#E-3

HEPATIC PROCESSING OF INSULIN:
CHARACTERIZATION AND MODULATION

G.D Smith and **J.R. Christensen**

Division of Clinical Cell Biology
MRC Clinical Research Centre
Watford Road, Harrow, Middlesex HA1 3UJ, U.K.

The liver is a major target for insulin, up to 50% of the circulating insulin being taken up after a single pass. Uptake is receptor-mediated. It has been postulated that the site of insulin degradation following internalization is lysosomal. We have studied the uptake and cell processing of radiolabelled insulin using a combination of ex vivo *liver perfusion and subcellular fractionation on sucrose density gradients. It appears that the processing of insulin is non-lysosomal. Various aspects of the processing pathway can be differentiated using weak bases: e.g. methylamine inhibits both uptake and processing whereas chloroquine affects only processing.*

Non-lysosomal processing was confirmed by the isolation of the vesicles concerned.* The processing of insulin continues within the isolated vesicles and is markedly inhibited by chloroquine (95% depression). The rate of processing in the vesicles can account for at least 25% of the observed hepatic insulin clearance. Inhibition of insulin processing is of clinical relevance since in a small clinical trial we showed significant improvement in glucose tolerance in Type II diabetics on short-term treatment with chloroquine.

The uptake and processing of ligands by various cell types is an area of cell biology that is currently under intense investigation [1]. Although cultured cell systems are often used to study this process, the liver offers an accessible means to study these phenomena in the intact organ. This article outlines some of the techniques employed in our laboratory to investigate this process, using as an example the uptake and processing of insulin by rat liver.

---

* In the ligand 'trafficking' context, W.H. Evans and co-workers have likewise studied vesicle isolation and features (# E-1 & Vol. 13).-*Ed.*

THE LIVER-PERFUSION APPROACH

One of the most common approaches is to use an iodinated ligand and to merely inject it into the animal's hepatic portal vein or jugular vein. This suffices if only qualitative data are required such as determining the cell type or subcellular locus to which the ligand is internalized. It does not allow control of the concentration or duration of exposure of the organ to the ligand. When this is required, perfusion of the liver is the preferred method. We have found that perfusion with oxygenated Krebs-bicarbonate Ringer containing 1 nM insulin and 4.4×10^5 Bq/ml of A14-iodo-insulin for 2 min followed by perfusion with Ringer alone for between 2.5 and 20 min represents suitable conditions to follow the uptake and sub-cellular processing of insulin.

In the first instance, examination of the perfusate itself may give valuable information on the uptake and processing of the ligand under investigation. Thus the effluent perfusate can be monitored to detect the appearance of radioactive label during the processing phase. In the case of insulin, both total radioactivity and that soluble in trichloroacetic acid (TCA) were measured to give an estimate of ligand degradation. These experiments showed that 50% of the insulin in the ingoing perfusate was taken up by the liver. Of the material taken up, very little emerged as intact insulin; a peak of degraded material comprising ~65% of the internalized material was detected at 13 min. The perfusion system also allows examination of the effect of various agents on uptake and processing by including them in the perfusion medium. In this instance addition of chloroquine (to 1 mM) to the perfusion medium not only prevented degradation, but also caused the ligand to be retained in the liver (Fig. 1).

Detailed kinetic analysis of binding and uptake parameters is feasible if the volume of the bolus of radioactive material is minimal (only 0.5 ml) and, moreover, if radioactive markers for blood and extracellular spaces are used and if perfusate effluent is collected at 1 sec intervals. We have used this technique in conjunction with a mathematical modelling package (MLAB) to construct a model of the perfused liver. The model consists of a double catenary of compartments and has shown that in the perfused rat liver endocytosis of insulin is very rapid, having a half-life of 19 sec [2].

SUBCELLULAR STUDIES WITH PERFUSED LIVER

These perfusion experiments can not only provide detailed infor-mation on the amount of ligand taken up by the liver but also can illuminate the subcellular processing pathway of the ligand within the hepatocytes. In our investigations of insulin processing, we have combined *in vivo* perfusion with analytical subcellular fractiona-tion of whole-liver homogenates on sucrose density gradients.

CONTROL   CHLOROQUINE

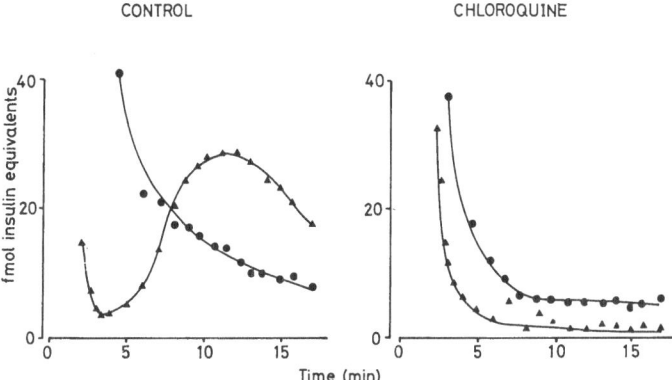

Fig. 1. Time course of the appearance of insulin degradation products in rat liver perfusate. Livers were perfused for 2.5 min with Krebs-Ringer containing 1 nM insulin and [A-14]-iodo-tyrosyl-insulin as a tracer, in the presence and absence of 1 mM chloroquine. The liver was then perfused with Ringer, with and without chloro-quine but in the absence of insulin. Perfusate samples were assayed for TCA-soluble (▲) and TCA-precipitable (●) radioactivity.

Fractionation following perfusion at 4° such that binding can occur but not internalization showed that the insulin was associated with the plasma membrane [3]. After perfusion at 37°, fractionation showed that the label was associated with low-density membranes (1.12 g/ml). These membranes have a similar equilibrium density to the Golgi as marked by galactosyl transferase, but may be dis-tinguished by the action of digitonin [4]. We have termed these membranes *ligandosomes*, and they probably represent a class of endo-cytic vesicles.

Another factor that can be varied in a perfusion model is the time allowed for the liver to process the ligand. This is done by altering the time allowed for the wash period with Ringer alone. In our experiments this interval was increased to 15 min. Under these conditions, fractionation showed that although the amount of label associated with the liver decreased by >60% over this time, most of the label was still associated with ligandosomes.

The integrity of the ligand associated with the gradient fractions can also be estimated from the TCA-solubility of the radiolabel. For insulin, although no evidence was found at any time point for association of insulin with the lysosomes, examination of its integ-rity, as described above, showed that even at early time points at least 30% of the label in the ligandosome fraction was in a degraded form. Inclusion of chloroquine in the perfusion medium as before, followed by subfractionation, showed that not only did chloroquine

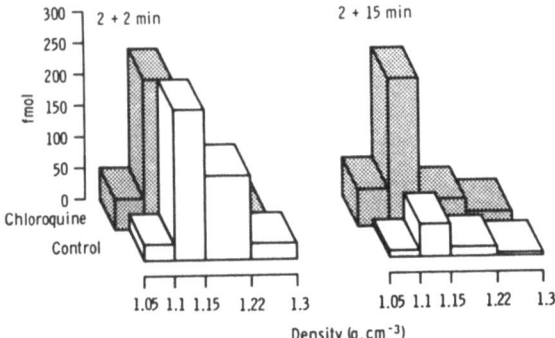

Fig. 2. Variation with time in the subcellular distribution of
TCA-precipitable insulin: effect of chloroquine. Perfusion was
as in Fig. 1 with or without chloroquine, for 2 min with the wash
period indicated. The liver was removed, a 2 g sample homogenized,
and then subjected to analytical subcellular fractionation on a
sucrose density gradient. The gradient fractions were assayed for
TCA-precipitable radioactivity as a marker for intact insulin. The
data were reprocessed mathematically to show the total amount of
intact insulin over 4 discrete density intervals.

cause the ligand to be retained in ligandosomes, but in addition the
amount of degraded material associated with these membranes was
drastically reduced (Fig. 2).

These results suggested that insulin was not only being internalized in endocytic vesicles, but was also undergoing at least partial
degradation therein. However, the analytical fractionation experiments cannot rule out a small but rapid flux through the lysosomes.
One way of testing the hypothesis that insulin can undergo degradation
in endocytic vesicles is to attempt to prepare vesicles that contain
the internalized ligand but are relatively uncontaminated with lysosomes.

The equilibrium density of the ligandosome fraction obtained
in the analytical fractionation experiments can be used as a starting
point to determine optimum conditions for preparation of the membrane..
Thus a membrane preparation, enriched in ligandosomes, was prepared
from liver perfused with radiolabelled insulin. The vesicles are
obtained by layering the liver homogenate onto a two-step discontinuous
sucrose gradient (d 1.09-1.14) and centrifuging at 100,000 **g** for
60 min. The ligandosome-enriched membranes are obtained at the interface and contain 60% of the internalized insulin but <4% of the lysosomal marker enzymes [5], meeting the aim of a high insulin recovery
with minimal lysosomes rather than a purified endosomal preparation.
(The vesicles contained 76% of a putative endosomal marker, latent
NADH pyrophosphatase, and 72% of the Golgi marker galactosyl transferase; there was <10% p.m. and e.r., and <1% mitochondria.)

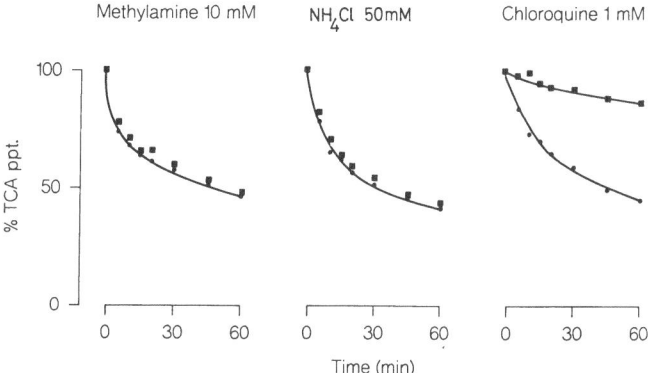

Fig. 3. Intravesicular degradation of insulin: effect of weak bases. Low-density vesicles enriched in endocytosed insulin were prepared as described in the text. The vesicles were incubated in isotonic sucrose containing 10 mM imidazole buffer, pH 7.2, in the presence (-■-■-) or absence (-●-●-) of the indicated concentrations of weak bases. Samples were removed at various time intervals and assayed for TCA-precipitable radioactivity.

INSULIN PROCESSING IN ISOLATED VESICLES

Incubation of the ligandosome-enriched vesicles in isotonic medium and assay of TCA-solubility of the radiolabel at various time intervals confirmed that degradation of the insulin continued *in vitro* [6]. This isolated vesicle preparation provides a very useful means of investigating the properties of the insulin-processing system. Thus, use of detergents (Brij) and inhibitors of the proteolysis (Ag^+, bacitracin) confirmed that degradation was indeed intravesicular [6]. Incubation in the presence of weak bases showed a differential effect in that methylamine and ammonium chloride had no effect on the rate of degradation, whereas chloroquine produced a marked inhibition (Fig. 3). The inhibition was unaffected by the presence of detergents, indicating that this was not an effect produced by the 'acidotrophic' mechanism normally associated with chloroquine.

The rate at which insulin is degraded in the vesicles also provides valuable information. Kinetic analysis of the degradation curves showed that the process can be described as two concurrent first-order reactions, one of which is ~20 times faster than the other. The faster process has a rate constant of 0.08 min^{-1} [6]. This is of the same order as that described for insulin degradation in intact hepatocytes [7] and for insulin clearance by the liver [8], suggesting that the process being observed is capable of supporting physiological rates of insulin clearance. Furthermore, the effect of chloroquine seemingly is to reduce the degradation process to a single first-order reaction, equal to the slower process observed in the absence of an inhibitor. The mechanism of this inhibition is not at present clear.

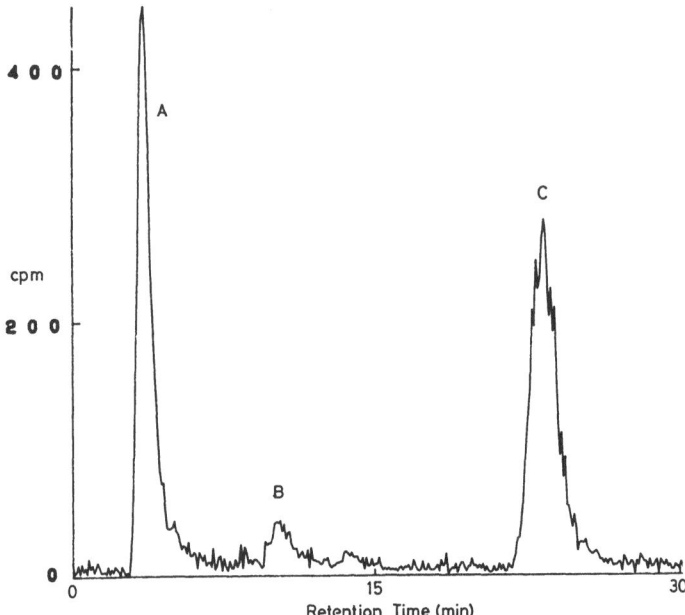

Fig. 4. HPLC of intravesicular degradation products of endocytosed insulin. Low-density vesicles wre prepared from the liver of a rat 2.5 min after portal vein injection of radiolabelled insulin. The vesicles were extracted with an equal volume of 40 mM HCl containing 0.1% Triton X-100 and then centrifuged at 100,000 **g** for 30 min. A 200 μl sample of the freeze-dried supernatant dissolved in elution buffer was chromatographed [9]. Peak A is TCA-soluble, and elutes with the solvent front; B is TCA-precipitable, and C corresponds in elution position to intact [A14]-iodo-tyrosyl-insulin.

The vesicle preparation may also be used to study the products of ligand processing. In this instance we can use HPLC of a detergent extract of the vesicles to follow the appearance of iodinated degradation products with continuing incubation. HPLC with the conditions previously described [9] showed the conversion of intact ^{125}I-insulin to TCA-soluble products with the appearance of a small amount of an intermediate product that was still TCA-precipitable, indicating only minimal degradation (Fig. 4). In the presence of chloroquine, all the radiolabel was in the form of intact native insulin, suggesting that the step inhibited by chloroquine is early in the processing pathway.

APPLICABILITY OF THE CHLOROQUINE EFFECT TO TREATING DIABETICS

The potent inhibition of insulin processing shown by chloroquine may prove to be of value in the treatment of diabetes. A recent

small-scale clinical trial [10] showed that the short-term administration of chloroquine to normal subjects had little effect on glucose
and insulin homeostasis on glucose challenge.   However, in Type-II
diabetics there was a marked improvement in the glucose tolerance
that was parallelled by an increase in plasma insulin concentration.
These changes were not accompanied by an alteration in the plasma
C-peptide, suggesting that the effect of chloroquine was not mediated
by an increase of pancreatic output but was a result of increased
bioavailability of insulin, perhaps by decreased hepatic breakdown.
Although chloroquine is not the drug of choice in view of the possible
side-effects during long-term administration, the *in-vitro* isolated
vesicle test system offers a relatively easy means of screening other
compounds for enhanced inhibitory activity.

References

1. Willingham, M.C. & Pastan, I. (1984) *Rec. Prog. Hormone Res. 40,*
    569-587.
2. Smith, G.D.. Hammond, B.J., Christensen, J.R. & Peters, T.J.
    (1986) *Biochem. Soc. Trans. 14,* 324.
3. Christensen, J.R., Smith, G.D. & Peters, T.J. (1985) *Cell
    Biochem. Function 3,* 13-19.
4. Smith, G.D. & Peters, T.J. (1982) *Biochim. Biophys. Acta 716,* 24-30.
5. Pease, R.J., Sharp, G., Smith, G.D. & Peters, T.J. (1984)
    Biochim. Biophys. Acta 774, 56-66.
6. Pease, R.J., Smith, G.D. & Peters, T.J. (1985) *Biochem. J. 228,*
    137-146.
7. Juul, S.M. & Jones, R.H. (1982) *Biochem. J. 206,* 295-299.
8. Jones, R.H., Sonksen, P.H., Boroujerdi, M.A. & Carson, E.R.
    (1984) *Diabetologia 27,* 207-211.
9. Rideout, J.M., Smith, G.D., Lim, C.K. & Peters, T.J. (1985)
    Biochem. Soc. Trans. 13, 1225-1226.
10. Jones, R.H., Amos, T.A.S., Mahler, R. & Peters, T.J. (1987)
    Br. Med. J., in press.

#E-4

RECEPTOR-MEDIATED ENDOCYTOSIS IN RAT LIVER PARENCHYMAL AND NON-PARENCHYMAL CELLS, STUDIED BY MEANS OF SUBCELLULAR FRACTIONATION IN DENSITY GRADIENTS

G.M. Kindberg, W. Eskild, K-J. Andersen[⊗],
K.R. Norum and T. Berg

Institute for Nutrition Research [⊗]Section for Clinical
School of Medicine   Research and Molecular
University of Oslo   Medicine, University of
P.O. Box 1046   Bergen, Haukeland Sykehus
0316 Oslo 3, Norway   5016 Bergen, Norway

In order to evaluate the influence of cellular derangements on hepatic endocytosis, its pathways have to be elucidated. Ways are now described to separate different cell types concerned, and to follow the endocytosis of different labelled ligands by subcellular fractionation techniques employing sucrose, Nycodenz or Percoll density gradients. The fractions were analyzed for degraded (acid-soluble) and undegraded ligands and for relevant marker enzymes (more meaningful because of the use of purified cell types). Different cell types differed in the intracellular processing of the ligands. Endocytosis has been studied in cultures also. The steps seem to be: → small, then larger, endosomes → (degradation) lysosome progression.

The liver removes by endocytosis a variety of macromolecules from blood. Many of them are proteins that function as transporters of other molecules: transferrin [1, 2], lipoproteins [3], ceruloplasmin [4], α_1-macroglobulin [5], RBP[*] [6] and haptoglobin [7]. In addition, the liver removes from the blood peptide hormones [8], intracellular enzymes that have leaked out of cells [9, 10], denatured proteins [11], glycosaminoglycans [12] and various glycoproteins [13]. A derangement of any of these uptake processes may have profound homeo-static consequences.

---

[*] Abbreviations: RBP = retinol-binding protein, ASOM = asialooroso-mucoid, CMR = chylomicron remnants; HDL, LDL: normal lipoprotein connotation. Liver cells: PC = parenchymal, NPC = non-parenchymal - EC = endothelial, KC = Kupffer, SC = stellate. See below (LIGANDS) for FSA, TC. Specific activity denoted s.a.; p.m., plasma membrane.

At least 4 different cell types, viz. PC (hepatocytes), EC, KC and SC, are involved in endocytosis [14]. In each cell type the process depends on several subcellular structures: p.m., various endosome types, lysosomes, microtubules and microfilaments [15]. The endocytic process has a thermodynamic requirement for input of external energy [16] and may therefore also depend on proper mitochondrial function.

Hepatic endocytosis consists of several steps, each of which may be a target for detrimental influences from drugs, carcinogens, reduced temperature, nutritional step-down conditions, etc. In order to determine the step(s) at which a given factor acts it is necessary to have methods available to follow the sequence of events in the endocytic process. We have followed the hepatic endocytosis of various ligands by means of subcellular fractionation techniques, with density gradients of sucrose, Nycodenz (Nyegaard) or Percoll (Pharmacia).

Subcellular fractionation involves the use of marker enzymes, and this presents a problem when working with whole liver homogenates: the distribution of a marker enzyme may not be representative of the cell in which a specific ligand is endocytosed. We have tried to solve this problem by purifying the various liver cell types prior to subcellular fractionation. We have also studied the endocytic processes in isolated cells *in vitro*, though the processes may of course deviate from those taking place *in vivo*. However, the use of isolated cells makes it possible to measure the effects of detrimental influences under precisely controlled conditions, and in homogeneous cell cultures.

The purpose of this article is to describe (a) methods for preparing purified, isolated rat-liver cells, and (b) cell fractionation methods for studying the intracellular transport and degradation of labelled ligands in rat-liver PC and NPC *in vivo* and *in vitro*. [Arts. #E-1, -2 & -3 in this vol. are also pertinent to (b), and Vol. 8 to (a).- *Ed.*].

PREPARATION OF PURIFIED RAT-LIVER CELLS

The starting material for separating liver cells is a cell suspension prepared by collagenase perfusion of the liver [17-19]. PC can be separated from NPC by differential centrifugation of the total cell suspension (50 **g**, 30 sec) [11]. If highly purified PC are needed, this can be achieved by centrifugal elutriation of the PC obtained by differential centrifugation ([20], & Vol. 8, this series, e.g. #B-1.) Of the three methods that have been used to obtain a preparation of NPC (EC, KC and SC), the simplest is to start with the supernatant after sedimenting the PC (see above) [11]: it is centrifuged for 3.5 min at 400 **g**. The pellet is resuspended and similarly recentrifuged 3 times, then resuspended and centrifuged for 20 sec at 50 **g** to remove remaining PC. This procedure furnishes ~25% of the hepatic NPC.

Two methods for obtaining highly purified NPC in high yield involve selective destruction of the PC, either with pronase [19] or with enterotoxin from *Clostridium perfringens* [20]. The initial suspension (~10^6 PC/ml) is incubated at 37° with 0.25% pronase - which disintegrates all PC after 45 min - or 10 µg/ml of the entero-toxin, which renders the PC leaky (as judged by trypan blue uptake) after ~15 min. After pronase the NPC can be separated from the damaged PC by repeated centrifugations (5 × 3.5 min) and washings. After enterotoxin the NPC can be separated from the leaky PC by flotation in a solution containing 18% (w/v) Nycodenz, in which the leaky cells sediment while the viable NPC move to the top of the solution. This method may be applied in general for the efficient separation of viable from permeabilized cells. Separation of EC and KC can be achieved by centrifugal elutriation [20] or by the selective adherence method described by Munthe-Kaas et al. [21] and Smedsrød et al. [22].

HOMOGENIZATION OF LIVER AND ISOLATED CELLS

A Dounce homogenizer was used for whole liver and PC, with 0.25 M sucrose medium. For liver this contained 1 mM Hepes/1 mM EDTA (pH 7.2): after 5 strokes with a loose-fitting and 10 with a tight-fitting pestle, the homogenate was centrifuged for 2 min at 2000 **g**. The nuclear fraction, resuspended in sucrose solution, was rehomogenized (5 strokes, tight pestle) and similarly centrifuged, and the super-natants combined for gradient separation.

Homogenates of suspended PC (tight pestle, 25 strokes) were centrifuged; the pellet containing unbroken cells and nuclei was rehomogenized (15 strokes), and the supernatants combined.

For NPC homogenization several methods have been tested, including Dounce, Potter-Elvehjem and sonication; we favoured the LoX-Press, used essentially according to Tagesson et al. [23]. NPC suspensions in sucrose/Hepes/EDTA (as above) were exposed to an increasing pressure up to 10-15 Kp/cm^2 in the cylindrical chamber, with monitoring in the presence of trypan blue by counting apparently intact cells in a Bürker chamber before and after extrusion; 80-90% of the cells were disintegrated at 15 Kp/cm^2 operating pressure. Then, after 5 strokes (Dounce, tight pestle), the homogenate was centrifuged (2 min, 2000 **g**) and the crude nuclear fraction resuspended and rehomo-genized; the supernatants were pooled.

SUBCELLULAR FRACTIONATION

In preparing the linear density gradients, tubes fitting the Sorvall TST 41-14 (10 ml) or Beckman SW 27 (34 ml) rotor were used for sucrose or Nycodenz. Tubes (14 ml) fitting the Sorvall SS-34 rotor were used for the Percoll gradients, self-generated from 10 ml of 30% (v/v) Percoll in 0.25 M sucrose. The sucrose gradients were prepared by mixing 20% and 58% (w/w) solutions supplemented with

Hepes/EDTA as above. The Nycodenz gradients were prepared by mixing
0.25 M sucrose (containing Hepes/EDTA) and 40% (w/v) Nycodenz in
1.2 mM KCl/0.1 mM Ca,Na-EDTA/0.1 mM NaCl/0.2 mM Tris-HCl, pH 7.5.
Onto the gradients in the 34 and 10 ml tubes were placed 4 and 2 ml
of post-nuclear fraction respectively; to the 14 ml tubes containing
Percoll, 2 ml was added. The sucrose and Nycodenz gradients were
centrifuged at 85,000 **g** for 4 h and 45 min respectively, and the
Percoll tubes at 40,000 **g** for 1 h.

LIGANDS [See foot of title p. for some abbreviations]

We have studied the intracellular transport of the following
ligands: asialofetuin [18, 24], ASOM [25-27], CMR containing labelled
retinyl ester [28-30], HDL [31], LDL [32], invertase [33] and formal-
dehyde-treated serum albumin (FSA) [34, 35]. Polyvinylpyrrolidone
has been used as a marker for fluid-phase endocytosis [36].

Here we consider the applicability of subcellular fractionation
to ASOM and FSA, which are taken up exclusively in PC and EC respectively. They can be labelled directly with [125]I. However, there
is the advantage of trapping labelled degradation products at the
site of formation if the ligand is labelled with either ^{14}C-sucrose
or ^{125}I-tyramine-cellobiose (^{125}I-TC), and we have applied both
labelling procedures [37] to ASOM and FSA. Only labelling with
^{125}I-TC is considered here, since it gives higher s.a. and eliminates
quenching problems.

LIVER FRACTIONATION AFTER ASOM AND FSA INJECTION

^{125}I-TC-ASOM or ^{125}I-TC-FSA was injected i.v. into rats. At
varying times the liver was perfused with ice-cold 0.25 M sucrose.
Tissue samples were homogenized and fractionated by gradient centrifugation: Nycodenz gradients were preferable for the study of intracellular ASOM transport, while sucrose gradients seemed more efficient
in separating the steps in FSA endocytosis.

Fig. 1 shows the distribution of degraded and undegraded ligands
in density gradients at different times. Evidently with Nycodenz
gradients ^{125}I-TC-ASOM was in a relatively slow-sedimenting vesicle
during the first minute and subsequently in denser endosomes (d =
1.09). The degradation products were detected in the gradient after
~15 min and accumulated later in the same gradient region as the
lysosomal enzyme β-acetylglucosaminidase. Undegraded ASOM remained
relatively long in the denser endosomes, and the rate-limiting step
in transport is probably from the endosomes to the lysosomes.

The kinetics of the endocytosis of ^{125}I-TC-FSA was distinctly
different from that of ^{125}I-TC-ASOM (Fig. 1). Fractionation in sucrose
gradients revealed three steps in the intracellular transport. The
ligand was first at d = 1.14 and then rapidly transferred to an endosome

Fig. 1. Density-gradient fractionations of post-nuclear fractions from rats given a labelled ligand i.v. For conditions, see text. Radioactivity of fractions, in relation to density: •, acid-precipitable; o, acid-soluble.

Upper panels: ^{125}I-TC-ASOM, at times ranging from 45 sec (A) to 60 min (E). Linear Nycodenz gradient.
Panel F: distribution of β-acetylglucosaminidase.

Lower panels: ^{125}I-TC-FSA, at times ranging from 1 min (A) to 24 min (F). Linear sucrose gradient.

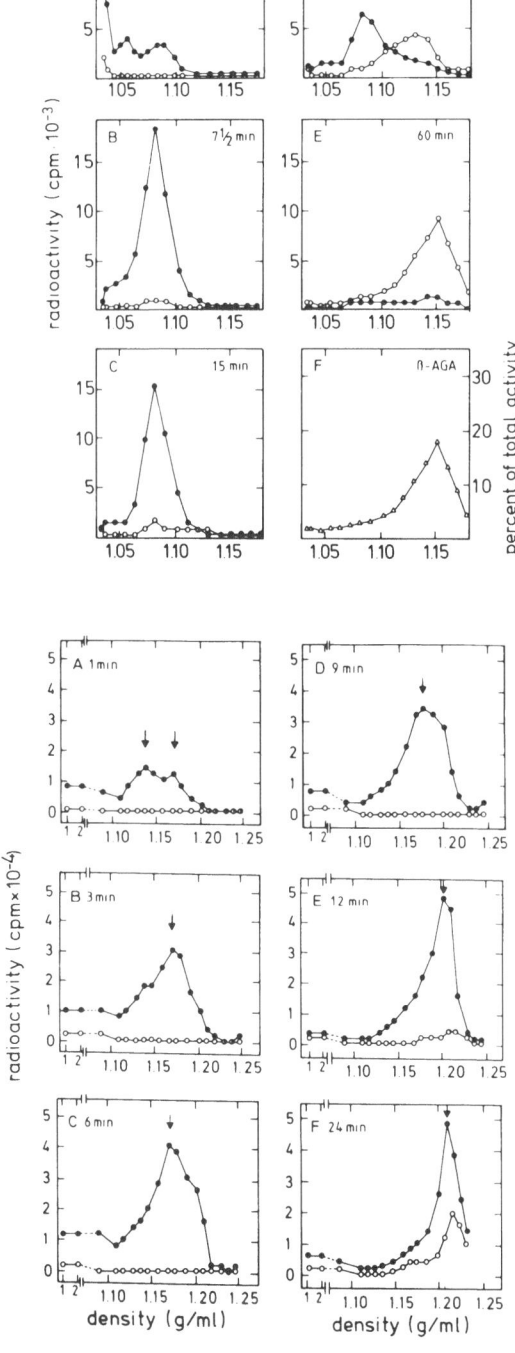

of d = 1.17. After 6 min it started to be transferred to a lysosomal
compartment (d = 1.21) and already after 12 min this transfer was
almost quantitative. Owing to this rapid transfer, undegraded
ligand accumulated in the lysosomes, which therefore became rate-
limiting in the intracellular degradation of this ligand.

DISTRIBUTION OF A LYSOSOMAL ENZYME IN FRACTIONS FROM PC AND NPC

If hepatic endocytosis is studied by means of subcellular fractionation it is crucial to know if the lysosomal marker enzymes are
representative for the lysosomes of the cells in which the endocytosis
under study takes place. We have fractionated PC and NPC in sucrose
and Nycodenz gradients and measured a lysosomal enzyme. Moreover,
we have labelled the lysosomes with two ligands which are selectively
taken up in PC and EC and whose degradation products, as Fig. 1 showed
for whole liver, eventually accumulate in the lysosomes.

The results for β-acetylglucosaminidase distribution (Fig. 2)
indicate that in sucrose gradients the EC contain slightly denser
lysosomes than the PC, the peaks being at d = 1.19 and 1.21 respectively. Conversely, the peak activities are at 1.11 and 1.14 respectively in Nycodenz gradients. Taking account of Fig. 1, the subcellular
distribution of enzymes in the isolated cells probably reflects the
situation for the two cell types *in situ*.

INTRACELLULAR TRANSPORT OF ^{125}I-TC-ASOM IN ISOLATED PC

In order to synchronize the intracellular transport of this
ligand, isolated PC may first be incubated at 4° in the presence
of the ligand. The cells are then centrifugally washed 3 times and
re-incubated in fresh medium at 37°. At various time points, cell
aliquots are homogenized and fractionated in the sucrose, Nycodenz
and Percoll gradients, Fig. 3 shows the distributions of degraded
and undegraded ^{125}I-TC-ASOM in the gradients. After 30 min of incubation the labelled degradation products are mostly in a low-density

Fig. 2. Distribution of the
lysosomal enzyme β-acetyl-
glucosaminidase following
fractionation of PC and NPC
in sucrose and Nycodenz
gradients: % of total
gradient activity *vs.* density
for each fraction.

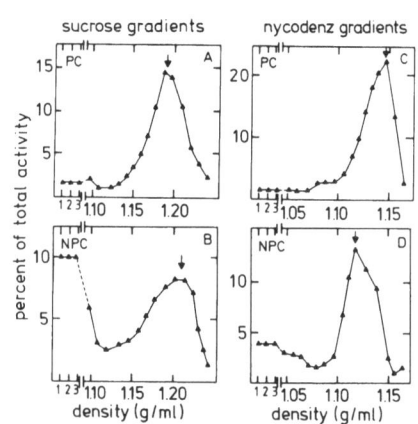

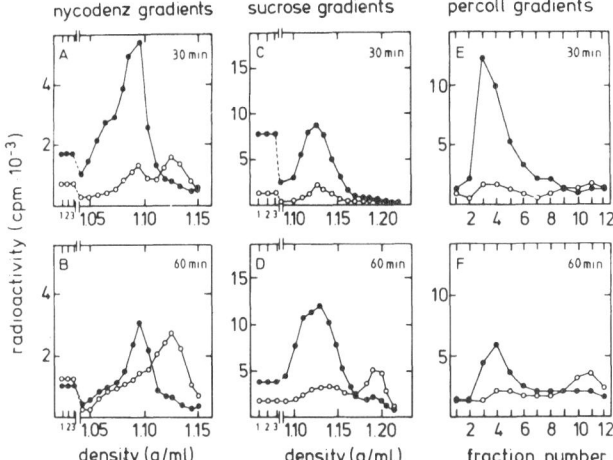

Fig. 3. Fractionation of isolated hepatocytes in Nycodenz, sucrose and Percoll gradients. The cells had been incubated for 1 h at 4° in the presence of 50 nM ^{125}I-TC-ASOM and then, after removing extracellular ligand, at 37°: cell aliquots were removed at 30 and 60 min, homogenized, and fractionated. Radioactivity: •, acid-precipitable; o, acid-soluble.

region in all gradient systems, but after 1 h degradation products accumulate in a compartment of higher density.

The data are, then, compatible with the notion that degradation is initiated in a pre-lysosomal/endosomal vesicle [27]. Very little undegraded ^{125}I-TC-ASOM is found in the lysosomal region of the gradient, suggesting that degradation of the glycoprotein in the secondary lysosomes is very efficient. The rate-limiting step in the intracellular transport in the isolated cells - as in the intact liver - is therefore the transfer from the endosomes to the lysosomes.

We believe that Nycodenz gradients are particularly well suited for separating the steps in the endocytic process. Fig. 4 shows the distribution of degraded and undegraded ^{125}I-TC-ASOM for cells incubated for 1-90 min after starting the 37° incubation. Evidently the ligand is first (<1 min) in a slowly sedimenting vesicle and subsequently in vesicles banding at a higher density (1.08). Degradation products are first found at d = 1.09 but subsequently in denser structures coinciding with the lysosomal enzyme β-acetylglucosamini-dase. The 'early' vesicle containing ligand <1 min after the start of the uptake has an equilibrium density of ~1.10. However, because of its small size it does not reach density equilibrium during 45 min of centrifugation [26].

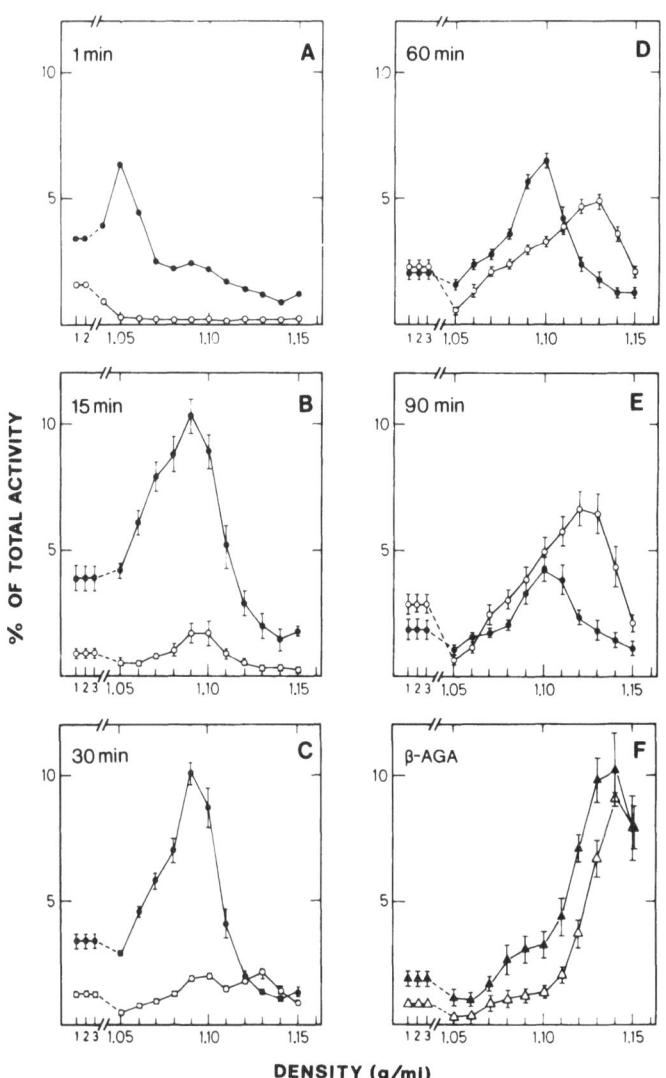

Fig. 4. Subcellular fractionation of hepatocytes containing acid-soluble (o; degraded) and acid-precipitable (●; undegraded) ^{125}I-TC-ASOM, after incubation in its presence (50 nM) for 1 h at 4°, then at 37° after removing extracellular ligand: at the indicated times cell aliquots were removed, homogenized, and fractionated in linear Nycodenz gradients. **A-E:** radioactivity distributions, as % of total cell-associated radioactivity at start of 37° incubation. **F:** distri-of β-acetylglucosaminidase after incubation for 0 (Δ) or 90 (▲) min, as % of toal activity in the gradient. **A:** results from one typical experiment; **B-F:** results are means ±S.E. from 6 experiments.

Fig. 5. Subcellular fractionation of endothelial cells (EC) containing degraded (o) and undegraded (•) ^{125}I-TC-FSA, fractionated in a sucrose gradient after incubation for 60 min at 37° in the continued presence of the ligand (cf. Fig. 4). Also shown (▲): distribution of β-acetyl-glucosaminidase.

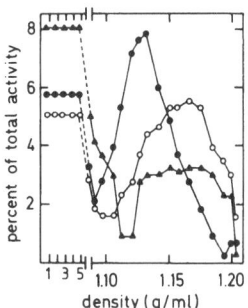

Altogether these data indicate that 4 steps in the intracellular transport can be discerned: (a) uptake in small endosomes, (b) transfer to large endosomes, (c) initiation of degradation in a prelysosomal vesicle, and (d) final degradation in a secondary lysosome. In addition the binding and internalization of ligand can easily be studied in the isolated cells [25]. [For 'lysosomal shift', see also #NC(E)-3.]

In earlier work [38] the effect of monensin on the subcellular distribution of ^{125}I-TC-ASOM indicated that the ligand may be internalized into two separate compartments, 'acid' and 'neutral'. When monensin was added to cells that had been incubated at 37°, e.g. for 10 min, it led to a dual distribution of undegraded ligand in the Nycodenz gradients: part remained at the original density (1.09) but the rest was diverted into a lighter vesicle (d = 1.06) [38]. In other words, there exist monensin-sensitive and monensin-insensitive compartments; this may relate to the finding by Weigel [39] that the ligand is internalized into two separate compartments in the cells.

INTRACELLULAR TRANSPORT OF ^{125}I-TC-FSA IN EC

We have previously shown that isolated EC in suspension effectively take up and degrade ^{125}I-FSA [35]. Fig. 5 illustrates preliminary experiments in which EC incubated with ligand have been fractionated in a sucrose gradient. The EC bind very little ligand at 0°, which precluded pre-binding of the ligand at 0° and following its intracellular transport at 37°. In Nycodenz gradients the distributions of degraded and undegraded ^{125}I-TC-FSA coincided, suggesting that the endosomes and lysosomes do not separate in this medium. In sucrose gradients acid-soluble radioactivity accumulated in the density range 1.15-1.20, as did β-acetylglucosaminidase. The lysosomes became progressively lighter with time of incubation. This may be due to autophagy which in PC is accompanied by a reduction in the buoyant density of the lysosomes [40].

These preliminary data show that receptor-mediated endocytosis of ^{125}I-TC-labelled ligands may be followed in isolated rat-liver NPC by means of subcellular fractionation. The method may be improved by testing the cell culture conditions, by employing monolayers rather than cell suspensions, and by using other cell culture media.

Acknowledgements

This research was financially supported by the Norwegian Cancer Society, Anders Jahre Foundation and the Nansen Foundation.

References

1. Morgan, E.H., Smith, G.D. & Peters, T.J. (1986) *Biochem. J. 237*, 163-173.
2. Kishimoto, T. & Tavassoli, M. (1985) *Biochim. Biophys. Acta 846*, 14-20.
3. Pittman, R.D. & Steinberg, D. (1984) *J. Lipid Res. 25*, 1577-1585.
4. Tavassoli, M., Kishimoto, T. & Kataota, M. (1986) *J. Cell Biol. 102*, 1298-1303.
5. Bergsma, J., Boelen, M.K., Duursma, A.M., Schutter, W.G., Bouma, J.M.W. & Gruber, M. (1985) *Biochem. J. 226*, 75-84.
6. Blomhoff, R., Norum, K.R. & Berg, T. (1985) *J. Biol. Chem. 260*, 13571-13575.
7. Higa, Y., Oshiro, S., Kino, K., Tsunoo, H. & Nakajima, H. (1981) *J. Biol. Chem. 256*, 12322-12328.
8. Posner, B.I., Patel, B.A., Khan, M.N. & Bergeron, J.J.M. (1982) *J. Biol. Chem. 257*, 5789-5799.
9. De Jong, A.S.H., Duursma, A.M., Bouma, J.M.W., Gruber, M., Brouwer, A. & Knook, D.L. (1982) *Biochem. J. 202*, 655-660.
10. Kamimoto, Y., Horiuchi, S., Tanase, S. & Morino, Y. (1985) *Hepatology 5*, 367-375.
11. Nilsson, M. & Berg, T. (1977) *Biochim. Biophys. Acta 497*, 171-182.
12. Smedsrød, B., Pertoft, H., Ericsson, S., Fraser, J.R.E. & Laurent, T.C. (1984) *Biochem J. 223*, 617-626.
13. Ashwell, G. & Harford, J. (1982) *Ann. Rev. Biochem. 51*, 531-554.
14. Berg, T., Blomhoff, R., Eskild, W. & Norum, K.R. (1986) in *Cells of the Hepatic Sinusoids*, Vol. 1 (Kirn, A., Knook, D.L.& Wisse, E., eds.), Kupffer Cell Foundation, Rijswijk, pp. 137-138.
15. Kolset, S.O., Tolleshaug, H. & Berg, T. (1979) *Exp. Cell Res. 122*, 159-167.
16. Clarke, B.L. & Weigel, P.H. (1985) *J. Biol. Chem. 260*, 128-133.
17. Seglen, P.O. (1976) in *Methods in Cell Biology*, Vol. 13 (Prescott, D.M., ed.), Academic Press, New York, pp. 29-59.
18. Tolleshaug, H., Berg, T., Nilsson, M. & Norum, K.R. (1977) *Biochim. Biophys. Acta 499*, 73-84.
19. Berg, T. & Boman, D. (1973) *Biochim. Biophys. Acta 321*, 585-596.
20. Blomhoff, R., Smedsrød, B., Eskild, W., Granum, P.E. & Berg, T. (1984) *Exp. Cell Res. 150*, 194-204.
21. Munthe-Kaas, A.C., Berg, T., Seglen, P.O. & Seljelid, R. (1975) *J. Exp. Med. 141*, 1-10.
22. Smedsrød, B., Johansson, S. & Pertoft, H. (1986) *Biochem. J. 228*, 415-424.

23. Tagesson, C., Stendahl, O., Magnusson, K.E. & Edebo, L. (1973) *Acta Path. Microbiol. Scand. Sect. B 81*, 464-472.

24. Tolleshaug, H., Berg, T., Frølich, W. & Norum, K.R. (1979) *Biochim. Biophys. Acta 585*, 71-84.

25. Berg, T., Blomhoff, R., Naess, L., Tolleshaug, H. & Drevon, C.A. (1983) *Exp. Cell Res. 148*, 319-330.

26. Kindberg, G.M., Ford, T., Blomhoff, R., Rickwood, D. & Berg, T. (1984) *Anal. Biochem. 142*, 455-462.

27. Berg, T., Kindberg, G.M., Ford, T. & Blomhoff, R. (1985) *Exp. Cell Res. 161*, 285-296.

28. Blomhoff, R., Helgerud, P., Rasmussen, M., Berg, T. & Norum, K.R. (1982) *Proc. Nat. Acad. Sci. 79*, 7326-7330.

29. Blomhoff, R., Holte, K., Naess, L. & Berg, T. (1984) *Exp. Cell Res. 150*, 186-193.

30. Berg, T., Blomhoff, R. & Norum, K.R. (1982) in *Sinusoidal Liver Cells* (Knook, D.L. & Wisse, E., eds.), Elsevier, Amsterdam, pp. 37-44.

31. Ose, L., Ose, T., Norum, K.R. & Berg, T. (1980) *Biochim. Biophys. Acta 620*, 120-132.

32. Ose, T., Berg, T., Norum, K.R. & Ose, L. (1980) *Biochem. Biophys. Res. Comm. 97*, 192-199.

33. Tolleshaug, H., Berg, T. & Blomhoff, R. (1984) *Biochem. J. 223*, 151-160.

34. Blomhoff, R., Eskild, W. & Berg, T. (1984) *Biochem. J. 218*, 81-86.

35. Eskild, W., Smedsrød, B. & Berg, T. (1986) *Internat. J. Biochem. 7*, 647-651.

36. England, I.G., Naess, L., Blomhoff, R. & Berg, T. (1986) *Biochem. Pharmacol. 35*, 201-208.

37. Pittman, R.C., Carew, T.E., Glass, C.K., Green, S.R., Taylor, C.A. & Attie, A.D. (1983) *Biochem. J. 212*, 791-800.

38. Berg, T., Kindberg, G.M., Ford, T. & Blomhoff, R. (1986) in *Receptor-mediated Uptake in the Liver* (Greten, H., Windler, E. & Beisiegel, V., eds.), Springer-Verlag, Berlin, pp. 174-182.

39. Weigel, P. (1987) in *Vertebrate Lectins* (Olden, K. & Paxent, J.B., eds.), Van Nostrand/Reinhold, New York, in press.

40. Seglen, P.O. & Solheim, A.E. (1985) *Exp. Cell Res. 157*, 550-555.

#E-5

USE OF CHROMOGENIC PEPTIDE SUBSTRATES IN THE STUDY OF PROTEOLYTIC ENZYMES AND MECHANISMS REGULATING THEM

P. Friberger

KabiVitrum Haematology AB, Diagnostica
S-431 33 Mölndal, Sweden

The idea of mimicking the structure of the protein metabolite cleaved by enzymes of the humoral defence systems has been very fruitful. Synthetic peptide substrates with a chromophore at the cleaved bond often prove to be more selective than the natural substrate. Chromogenic peptide substrates exist not only for most coagulation, fibrinolysis, kallikrein and complement enzymes but also for chymotrypsin-, elastase- and papain-type enzymes. Various venoms and cell enzymes also show high sensitivity towards certain peptide substrates. Using well-defined substrates and other reagents, it has often been easy to opimize assay conditions. In a few cases inhibitors have been used to improve specificity.

Assay simplicity has also enabled factors regulating such enzymes to be determined. By adding a purified enzyme to a biological sample, the amount or activity of inhibitors therein can be determined. Proenzymes can be quantified after total activation. Activators and various moderators can be determined by having all factors except the analyte in excess. Such methods can be made very specific and can notably aid proteolytic enzyme research. The rapid results from manual as well as automated procedures make it possible to follow various processes in biological fluids. Both research and clinical practice may benefit, because proteolysis regulates many metabolic and defence reactions.

CHROMOGENIC PEPTIDE SUBSTRATES

The first synthetic peptide substrates, designed to mimic the natural protein substrates (Fig. 1), were found to be more sensitive and selective than the previously used amino acid derivatives. Since then there has been a rapid development in this area. Synthetic peptide substrates can be varied in three ways (Table 1): in respect

Fig. 1. The synthetic peptide substrate is made to mimic the natural protein substrate. In this example, pNA denotes *p*-nitroaniline.

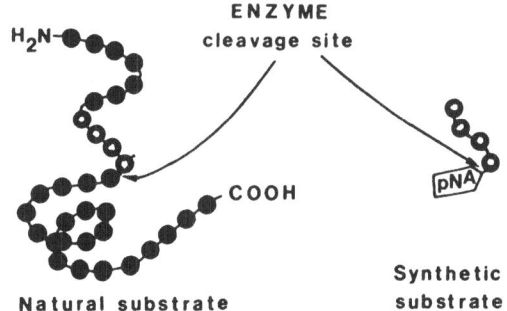

of the marker group, of the peptide sequence (important for selectivity), and of the N-terminal protecting group. Changes in any of these three parts of the peptide substrate can significantly influence several properties important for their usefulness.

Chromogenic and fluorogenic substrates can be used in direct kinetic assays. This holds also for certain measurements based on electrochemical properties. Luminogenic*, radiometric and certain indirect chromogenic procedures need an extra stage for the measurement, after the marker ('reporter') group has been split off. The fluorogenic and luminogenic assay techniques can be made more sensitive than the chromogenic. Then it is necessary, however, to consider technical difficulties that can appear when handling very low concentrations of enzymes, especially in purified systems [1].

Regarding the peptide sequence, the trypsin-like enzymes split on the C-terminal side of Arg or Lys, chymotrypsin-like enzymes split C-terminal to Phe, Tyr or Trp, while elastase-like enzymes prefer Ala or Val. There is virtually no cross-reactivity between these types of serine proteases. Collagenase-like enzymes do not seem to split the present type of peptide substrates. However, certain bacterial enzymes have been found to split C-terminal to Pro while others split C-terminal to Arg, Lys, Phe, Val, Met or Cit [2-4]; further delineation of specificities has yet to be made.

Among the trypsin-like proteases, some including plasmin and plasma kallikrein prefer the Phe-Arg sequence, while trypsin, factor X_a, acrosin and urokinase prefer Gly-Arg, and thrombin, factor XI_a and activated Protein C prefer the Pro-Arg sequence. Further selectivity information has appeared elsewhere [1, 5]. From such data it can be concluded that selectivity is usually of the order of 10- to 100-fold on a molar basis. The reaction conditions are of course selected to favour selectivity. Besides, a number of more or less selective inhibitors can be used to quench interfering activity.

---

* 'Luminogenic' here connotes phosphorescence.

Table 1. Different types of synthetic peptide substrates consisting of three moieties as tabulated (also, see text). Cbo = carbobenzoxy.

N-terminal group	Peptide sequence	Marker-group feature
Hydrophobic: Bz, Cbo	X-X-Arg,Lys	Chromogenic, ester or amide
Aliphatic: Ac	X-X-Phe,Tyr	Fluorigenic
Hydrophilic: Suc	X-X-Ala,Val	Luminogenic
D-amino acid	X-Phe-Arg	Radioactive
Protected D-amino acid	X-Phe-Arg	Electric property
Unprotected	X-Pro-Arg	Indirect chromogenic

The N-terminal group of the peptide substrate can be varied to fit the demands of the enzyme as well as to increase substrate solubility. This group can also protect the substrate from degradation, e.g. by aminopeptidase which is present in many biological samples.

Besides the obviously necessary requirements such as being well defined, pure [1] and stable, the kinetic data and the solubility are the most important parameters for synthetic peptide substrates. The kinetic constants determine the selectivity of the substrates, but from the practical viewpoint the concentrations of the various proteases in the reaction mixture as well as their proteolytic efficiency are equally important. All these factors should be considered each time a substrate for a certain enzyme is chosen as a reagent in a particular assay. Furthermore, it could be argued that a substrate should contain amino acids on both sides of the bond to be split, making it more like a protein substrate. The amino acids next to pNA (cf. Fig 1) would afterwards be split off by another protease, e.g. aminopeptidase, present in excess. This protease, however, should not split the intact substrate to any significant extent. Such a system has to be used for enzymes which really need certain amino acids on both sides of the susceptible bond, e.g. renin and pepsin.

Spatial conformation and charge distribution within the substrate are important for the enzyme to recognize its substrate. Some affinity sites of the enzymes may be situated at a distance from the active site and thus cannot influence the reaction with substrates of low mol. wt. There are a couple of cases reported where defects in the enzyme are undetectable with small peptide substrates but detectable with the natural protein substrates [1]. Yet more information about the enzyme is obtainable by using the two types of substrate in parallel. The possibility of being thus misled must be kept in mind.

The synthesis of >500 peptide substrates and their testing on available enzyme preparations has been very informative about these

Fig. 2. Comparison of u.v. absorbance curves for intact (pNA-containing) substrate (S) and pNA as split off from it (both 50 μM; pH 5-10). Liberated pNA as the index of enzyme activity is measured at 405 nm, where the substrate contribution is <1%.

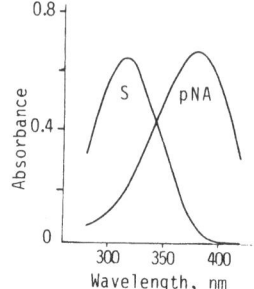

reactions. We are, however, only at the beginning of a new era of investigation of limited proteolysis in metabolic and defence mechanisms.

METHODOLOGY

The basis for use of a peptide-pNA substrate is the difference in absorbance between the pNA split off from the peptide and the intact substrate.   At 405 nm, the wavelength most commonly used, substrate absorbance is <1% of that of an equimolar amount of pNA (Fig. 2).   As photometers can be of various types and are widely available, such assays are facile.

Chromogenic substrates are now available for numerous proteases:- thrombin, trypsin and chymotrypsin; $C\bar{1}s$, $C\bar{1}r$; factors X_a, XI_a and XII_a; plasma and tissue kallikreins; plasmin, urokinase and tissue plasminogen activators; protein C_a; cathepsin G; leucocyte and pancreatic elastase; acrosin; snake venoms; Limulus and crayfish lysates; papain-type enzymes; bacterial and fungal proteases.   The wide availability of these substrates has spurred the ever-increasing study of limited proteolysis.

Protease determination being simple, the approach can readily be extended to enzyme-inhibitor assay merely by adding a known amount of the purified enzyme to the sample, incubating it for a suitable time, and then measuring the residual activity.   The specificity of inhibitory assays depends on the choice of enzyme as well as the use of supplements such as heparin which potentiates antithrombin, or methylamine which ensures that α_2-macroglobulin-enzyme complexes are not formed.   Proenzymes can be determined after activation. Usually it has been feasible to use selective activators, and different strategies have been used to avoid inhibition.   Dilution of the test sample has been effective in several cases.

Activators have been determined as such by using an excess of the proenzyme which they activate.   This type of assay can be both specific and highly sensitive.   If a single-stage procedure is used the method is also very simple to perform.   Other types of moderators of protease systems can be determined using all reagents except the

analyte in excess. Numerous articles [e.g. 1, 2, 6] give methodological
details.   For automated operation, which reduces error and assay
cost, special precautions may be needed [1, 6].

     Available assays are exemplified by the following.-
Enzyme standardization: plasmin, urokinase.
Enzyme activity in biological samples: urinary kallikrein; plasma
plasmin-like activity.
For the following, kits are available.-
Enzyme inhibitors: antithrombin, antiplasmin; Cl-esterase inhibitor;
α_2-macroglobulin; α_1-antitrypsin; t-PA inhibitor.
Proenzymes: prothrombin, factor X, plasminogen, prekallikrein,
protein C.
Enzyme activators: t-PA, endotoxin.
Moderators: factor VIII, heparin.

ASSAY APPLICATIONS

     Numerous physiological and pathological systems are regulated
by limited proteolysis, involving proteolytic enzymes, inhibitors
and various types of cofactor. Examples include: regulation of digestive enzymes; coagulation and fibrinolysis; blood pressure regulation;
cell lysis; chemotaxis; inflammation; toxin release; hormonal regulation; fertilization; tumour development; aggregation of blood cells;
activation of various protein-degrading enzymes; pathological protein
degradation; bacterial defence systems (Ig-ase); food metabolism
and energy conversion.

     Proteases in the saliva, stomach and intestines are important
for the absorption of various food elements. Trypsin and chymotrypsin
are secreted as proenzymes and, once activated, are also regulated
by intestinal inhibitors. Blood coagulation and fibrinolysis are
two cascade oxidants consisting of several proenzymes, inhibitors,
and cofactors for both activation and inhibition. Recently a negative
feedback process regulated by a proteolytic enzyme (protein C) which
destroys two key factors has been discovered within coagulation.
Such regulation has long been known in the complement system.

     There exists a haemostatic balance consisting of four forces,
viz. coagulating enzymes, fibrinolytic enzymes and inhibitors of
the two oxidants. If there is imbalance a trigger in any direction
will lead to thrombo-embolic or bleeding complications.

     Blood pressure is regulated by proteolytic enzymes and connected
factors, viz. kallikreins, renin and angiotensin-converting enzyme
as well as kininogens and angiotensinogen.

     Protection of the host when attacked by foreign cells involves,
besides antibodies, complement factors and, for bacterial invasion,
leucocyte and macrophage factors.

Regulation similar to that *in vivo* is prevalent in isolated
cells too. Proteolytic enzymes may be discharged from the cell and
found either on the cell surface or in the extracellular space, or
else the release may be significant only when the plasma membrane
is broken. These three cases call for different techniques of measurement. Discharged enzymes can be measured in the medium after centrifugation or, if cell-bound, in a cell suspension with a suitable
medium; intracellular enzymes can be measured after dispersing and
centrifuging down the plasma membrane.

Bacterial protease assays may be used for bacterial typing as
well as for investigating virulent properties of the species.
Proteolytic enzymes can be studied in various cultures or various
cells including microorganisms such as yeasts and bacteria. Moreover,
genetic manipulations can be used to study the consequences of defective
proteases, inhibitors and other regulators, e.g. in the areas of
fertilization/reproduction, metabolism, and the defence system under
different conditions. Synthetic peptide substrates obviously aid
such studies, and existing assay procedures can furnish adequate
sensitivity.

With enzyme systems from Limulus or crayfish it is possible
to determine very small amounts of endotoxins and β-1,3-glycans
respectively. These substances from cell membranes of gram-negative
bacteria and fungi respectively are potent triggers of mammalian
protease systems. For study of histological sections and detection
of separated proteases in gels, both chromogenic and fluorogenic
substrates have been used [7, 8].

Concluding comments.- Chromogenic peptide substrates will be
increasingly exploited in research work. Many are routinely used
in the pharmaceutical industry where specific procedures are
appreciated. There is increasing routine use, especially in diagnosis
and therapeutic monitoring, in medical areas including the following.-
Internal medicine: ischaemic heart disease, allergic diseases, acute
leukaemia, diabetes, hypertension, liver disease, RES dysfunction,
haemodialysis, cancer.
Haematology: coagulation profile, liver cirrhosis, DVT[@], vascular
disorders, angina pectoris, thromboembolism, cancer.
Obstetrics/Gynaecology: pre-eclampsia, DIC, DVT.
Nephrology: nephrotic syndrome, transplant rejection.
Surgery: major surgery, CPB, abdominal surgery, tumour surgery, major
trauma, sepsis, vascular surgery, embolism.
Orthopaedics: DVT, embolism.
Anaesthesiology/Intensive care: severe burns, pancreatitis, peri-
tonitis, respiratory distress, DIC, shock.
Infectious diseases: sepsis, bacteraemia, viraemia.
Geriatrics: circulatory diseases, rheumatoid arthritis.

[@] DVT, deep venous thrombosis; DIC, disseminated intravascular
coagulation; CPB, cardiopulmonary bypass

References

1. Friberger, P. (1982) *Scand. J. Clin. Lab. Invest. 42, suppl. 162.*
2. Berdal, B.P., Bøvre, K., Olsvik, Ø. & Omland, T. (1983) *J. Clin. Microbiol. 17*, 970–974.
3. Øgaard, A.R., Berdal, B.P. & Bøvre, K. (1984) *Acta Path. Microbiol. Immunol. Scand., sect. B, 92*, 31–37.
4. Gray, C.J., Barker, S.A., Pharimal, M.S. & Sullivan, J.M. (1985) *Biotech. Bioengin. 27*, 1717–1720.
5. Claeson, G. & Aurell, L. (1981) *Ann. N.Y. Acad. Sci. 370,* 798–811.
6. Friberger, P. (1983) *Seminars in Thrombosis & Hemostasis 9,* 281–300.
7. Blasini, R., Stemberger, A., Wriedt-Lübbe, I. & Blümel, G. (1978) *Thrombosis Res. 13*, 585–590.
8. Wagner, O.F., Bergmann, I. & Binder, B.R. (1985) *Anal. Biochem. 151*, 7–12.

#E-6

PEPTIDASES IN CONNECTIVE TISSUE DEGRADATION AND REMODELLING IN REPRODUCTIVE AND INVASIVE TISSUES

[1]J. Ken McDonald, [1]Christian Schwabe and [2]Noel O. Owers

[1]Department of Biochemistry   [2]Department of Anatomy
Medical University of   Medical College of Virginia
  South Carolina   Virginia Commonwealth University
Charleston, SC 29425, U.S.A.   Richmond, VA 23219, U.S.A.

An outline is given of the properties and substrate specificities of three lysosomal peptidases, and of efforts to elucidate their potential role in the intracellular degradation of collagen. Use of specific fluorogenic substrates such as Gly-Phe-NNap*, Lys-Ala-NNap and Gly-Pro-Met-NMec allows direct assay of DPP I, DPP II and TPP I respectively. They may hydrolyze macromolecules as exemplified by insulin A and B chains and by poly(Gly-Pro-Ala) as a model collagen α-chain. These are extensively hydrolyzed at acidic pH by the coupled action of the three exopeptidases, which could explain the degradation of proline-rich, collagen-derived polypeptides within lysosomes lacking any known endopeptidases active at prolyl bonds.

With suitable derivatives of specific peptide sequences, the above exopeptidases can be localized by light- and electron-microscopy in cryostat sections of a range of connective tissues and invasive reproductive and metastatic cells. A method for detecting and localizing endopeptidases of general or undefined specificity, in relation to specific cells and tissue structures, is also illustrated. It involves incubating cells or cryostat tissue sections on a glutaraldehyde-fixed, India ink-containing gelatin membrane that is sensitive to a wide range of proteinase activities including that of mammalian collagenase. Examples include the remodelling (relaxin-induced) pubic symphysis, spermatozoa, and the implanting blastocyst.

Collagen is the major extracellular protein of connective tissue and the most abundant vertebrate protein. Its primary function is to provide a stabilizing architecture for cell support, attachment

---

* *Abbreviations.*- In amide context: NNap = 2-naphthylamine, NMec = 7-(4-methyl)coumarylamine, NNapOMe = 2-(4-methoxy)naphthylamine. DPP, dipeptidyl peptidase; TPP, tripeptidyl peptidase.

and migration. Whereas the interstitial collagens, types I, II and III, form fibrils that provide the major source of mechanical strength for the extracellular matrix [1], type IV collagen forms the mat-like network of basement membranes [2] that functions as a filtration and structural barrier between epithelial cells and underlying connective tissue. Despite their chemical and functional differences, all of the known collagen types display repeating Gly-X-Y triplet sequences (where X is frequently proline) as well as non-triplet regions that are presumably globular in nature [3].

The catabolism of insoluble (cross-linked) interstitial collagen fibrils appears to follow two possible pathways: an extracellular route that occurs primarily at neutral pH and involves either metalloendopeptidases (physiological) or serine endopeptidases (pathological), and an intracellular route that occurs primarily at acidic pH and involves lysosomal cysteine and aspartic endopeptidases [4]. Apparently the accelerated rate of connective tissue breakdown seen during rapid physiological remodelling (simultaneous degradation and synthesis), as well as that associated with pathological damage, is a local exaggeration or derangement of a normal process. Diseases such as arthritis and periodontitis have long been associated with excessive (uncontrolled) rates of interstitial collagen breakdown, and the metastatic potential of tumour cells correlates well with their ability to destroy basement membrane (type IV) collagen [5, 6].

Fibroblasts involved in the rapid remodelling of connective tissues, as in wound repair [7-9], as well as in fibroblastic and osteoblastic cells present in invasive human sarcomas [10, 11], contain native (banded) collagen fibrils within secondary lysosomes (phagolysosomes). Although lysosomal endopeptidases such as cathepsins B, H, L and N are commonly cited as being responsible for the degradation of endocytosed collagen fibrils [12-16], none of these enzymes is capable of catalyzing the hydrolysis of the prolyl and hydroxyprolyl linkages [17] that constitute about one-quarter of the polypeptide (α-chain) residues [18, 19]. On the other hand, studies summarized here suggest that certain lysosomal exopeptidases could make an important contribution to the intracellular degradation of collagen-derived, proline-rich α-chains. The properties and substrate specificities of these exopeptidases are summarized herein. Specific fluorimetric and histochemical methods are illustrated for their assay and localization, and a general detection method is described for localizing endopeptidase (gelatinase) activity in cells and cryostat sections of fresh tissues.

CLASSIFICATION OF THE PEPTIDASES ('PROTEASES')

Whereas the **endopeptidases** (proteinases) are classified according to their catalytic mechanism as 'serine', 'cysteine', 'aspartic' or 'metallo' [17, 20], the **exopeptidases** are classified according to their specificity [20, 21], as illustrated in Fig. 1. Although

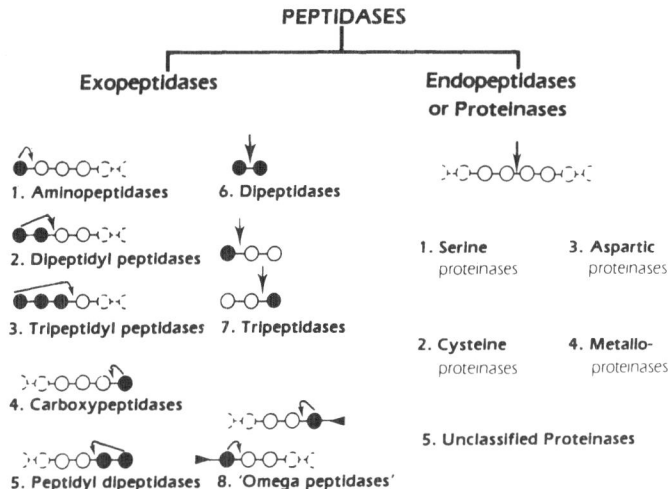

Fig. 1. Trivial terms currently recommended by the I.U.B.'s Nomen-clature Committee for **peptidase** classes – based on specificity determinants for **exopeptidases** but on catalytic mechanism for the **endopeptidases.**

the I.U.B. favours replacing the term 'protease' by 'peptidase' [22], the term 'proteinase' has been retained, and is considered to be synonymous with 'endopeptidase'. A rationale for these nomenclature changes has been offered [23].

Insofar as the terms 'endopeptidase' and 'exopeptidase' were introduced ~50 years ago [24], it is rather surprising that they have not enjoyed the same widespread usage as have the analogous terms applied to the glycosidases, viz. endoglycosidase and exoglycos-idase.

Herein, emphasis is given to the lysosomal members of classes #2 and #3 on left of Fig. 1. They include DPP I, DPP II and TPP I, generally described as N-terminal exopeptidases because their specifi-cities are characteristically for the unsubstituted N-termini of polypeptide substrates (as surveyed: [21, 25]). The DPP's ('dipeptidyl peptide hydrolases'), originally termed 'dipeptidyl aminopeptidases' [26], remove two amino acids at a time, while the TPP's remove three at a time, and were named by analogy with the DPP's [27].

DIPEPTIDYL PEPTIDASE I (DPP I)

DPP I (EC 3.4.14.1) was initially recognized as an SH-activated pituitary-gland enzyme having an absolute halide requirement [28], a lysosomal localization [29, 30] and an exceptionally broad specifi-city [26]. It can degrade large polypeptides, exemplified by insulin

A CHAIN OF REDUCED (CARBOXYMETHYLATED) BOVINE INSULIN

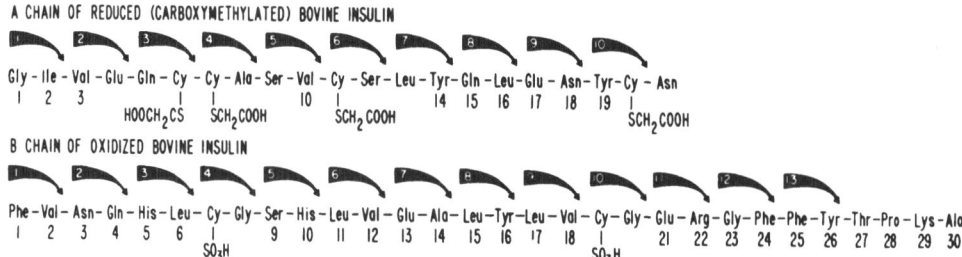

Gly - Ile - Val - Glu - Gln - Cy - Cy - Ala - Ser - Val - Cy - Ser - Leu - Tyr - Gln - Leu - Glu - Asn - Tyr - Cy - Asn
 1 2 3 | | 10 | 14 15 16 17 18 19 |
               HOOCH$_2$CS SCH$_2$COOH   SCH$_2$COOH   SCH$_2$COOH

B CHAIN OF OXIDIZED BOVINE INSULIN

Phe - Val - Asn - Gln - His - Leu - Cy - Gly - Ser - His - Leu - Val - Glu - Ala - Leu - Tyr - Leu - Val - Cy - Gly - Glu - Arg - Gly - Phe - Phe - Tyr - Thr - Pro - Lys - Ala
 1 2 3 4 5 6 | 9 10 11 12 13 14 15 16 17 18 | 21 22 23 24 25 26 27 28 29 30
               SO$_3$H   SO$_3$H

Fig. 2. Degradation of the A and B chains of bovine insulin by DPP I at pH 5.5 in the presence of -SH and Cl⁻ as activators.

chains (Fig. 2), through rapid and successive release of a wide range of dipeptides [30]; although attack is confined to the unprotected N termini, in fact, as Barrett [31] pointed out, the 13 bonds cleaved in the insulin B chain (Fig. 2) exceed the number cleaved by the combined actions of trypsin, chymotrypsin and pepsin. Polypeptide fragmentation is, however, halted by either a prolyl residue or by the emergence of an N-terminal Arg or Lys. Hence when DPP I is used as a sequencing reagent, it has been recommended [32] that tryptic fragments be employed as a means of placing all arginyl and lysyl residues in C-terminal positions. Penultimate basic residues, on the other hand, are among the most rapidly hydrolyzed linkages. Some DPP I properties (above; general surveys in [21, 33]) were unprecedented when first reported; but reinvestigation of cathepsin C [26] (detected in 1948 in porcine kidney by H.R. Gutmann & J.S. Fruton) disclosed a common identity.

Fluorimetric assay of DPP I in tissue extracts can be performed specifically (pH 6.0, 37°) with Gly-Phe-NNap or, with 10- to 20-fold sensitivity gain, Gly-Arg-NNap, plus Cl⁻ and -SH activators [26]. A chart record of 410 nm emission (336 nm excitation) gives directly the rate of NNap release. Colorimetry of NNap is also feasible [34].

Localization by fluorescence enzyme histochemistry

Pro-Arg-NNapOMe, a substrate originally developed for DPP I azo dye histochemistry [30], offers maximal sensitivity and selectivity because of its penultimate arginyl residue and terminal prolyl residue. The latter endows the substrate with resistance to breakdown by various aminopeptidases. The 4-methoxy substituent on the naphthalene ring reduces the aqueous solubility of the NNapOMe leaving group and, in the detection step, enhances its rate of coupling to 5-nitrosalicyl-aldehyde and increases the substantivity of the resulting Schiff-base – which fluoresces at 595 nm (365 nm excitation) and can be readily visualized by fluorescence microscopy [35]. Because the coupling agent is unaffected by the presence of thiols included in incubations with cysteine peptidases such as DPP I, the procedure is an excellent

Table 1. Substrate specificities and properties of bovine DPP I and DPP II. *Data collated from refs. [26, 29, 36, 37].*

Feature	DPP I	DPP II
Assay substrates	Gly-Phe-NNap Pro-Arg-NNap	Lys-Ala-NNap Lys-Pro-NNap
pH optimum	6.0	5.5
Activators	-SH, Cl⁻	None
Inhibitors	SH reagents Gly-Phe-CHN₂	Dip-F Lys-Ala-CH₂Cl
Catalytic class	Cysteine	Serine
M_r	200,000	130,000
Subcellular localization	Lysosomal	Lysosomal

alternative to azo-dye techniques that involve the use of reactive diazonium salts in simultaneous coupling procedures (as described below for DPP II histochemistry) which thiols disrupt.

Illustrative studies, e.g. on remodelling.- We have used this approach to follow and localize DPP I in fresh-frozen sections of cartilagenous connective tissues that were hormonally induced to undergo remodelling. As depicted elsewhere [38], large increases in activity were attributable to regions of the mouse pubic symphysis that were rapidly remodelling in response to relaxin administered to oestrogen-primed animals. Parallel, quantitative determination of DPP I activity were conducted by excising and extracting symphyseal tissue for assay by direct fluorimetry as above. The histochemistry approach has been used by others to demonstrate DPP I in skeletal, cardiac and vascular smooth muscles [39].

The visualization, by immunofluorescent staining, of collagen types I and II in the remodelling pubic symphysis revealed that type II was degraded following relaxin administration. These unpublished studies employed a method [40] that utilizes antibodies raised against specific collagen types.

DIPEPTIDYL PEPTIDASE II (DPP II)

Properties and specificity (summarized in Table 1).- DPP II (EC 3.4.14.2) is the other known lysosomal member of the exopeptidase class being considered. It was first distinguished from DPP I on the basis of its indifference to -SH and Cl⁻ activators and its ability, which DPP I lacks, to release Lys-Ala from Lys-Ala-NNap [29]. It was subsequently shown to be a cation-sensitive, lysosomal peptidase [29, 30] whose range of action on NNap derivatives, although very narrow [41], included the unique ability to hydrolyze prolyl arylamide bonds as readily as alanyl arylamide bonds, as in Lys-Ala-NNap and Lys-Pro-NNap [36].   This unusual substrate specificity was first encountered during studies employing preparations of DPP II derived from a bovine connective tissue source [36].   Later [42] a novel (porcine) form was uncovered that displayed a remarkable (7-fold greater) preference for the prolyl linkage in Phe-Pro-NNap as compared to that in Lys-Pro-NNap.

Whereas DPP I could act on polypeptides of virtually any length, the bovine DPP II was manifestly restricted to tripeptides, especially those having Ala or Pro in the central position, as shown by the following relative rates [36, 41] (10 mM substrate; pH 5.0, 37°).-
  Ala-Ala-Ala *100%*,   Lys-Ala-Pro or Lys-Ala-Ala 125%;
  Lys-Pro-Ala 44%,   Gly-Pro-Ala 36%,   Ala-Pro-Ala 31%;
  Nil for N-Acetyl-Ala-Ala-Ala, Gly-Pro-Ala-Gly, Ala$_4$, Ala$_5$, Ala$_6$ or
    Poly(Ala).
DPP II purified from rat kidney [43] and porcine ovary [42] displayed the same overriding preference for tripeptides.   A fuller summary, including distributions, has been published recently [21].

Fluorimetric assay.- DPP II in tissue extracts can be assayed specifically with Lys-Ala-NNap or Lys-Pro-NNap at pH 5 (37°) similarly to DPP I (above; details in [29, 36]).

Localization by azo-dye histochemistry

Since NNap, the leaving group in commonly used substrates for fluorimetry and colorimetry, displays appreciable aqueous solubility and a relatively slow coupling rate to diazonium salts, NNapOMe was adopted, in accordance with Nachlas et al. [44], as an improved leaving group for localizing peptidase activity in tissue sections.   Reaction-product diffusion was greatly reduced owing to the lower aqueous solubility and the faster (40-fold) rate of coupling (at pH 6.5) to the diazonium salt, fast blue B (tetrazotized diorthoanisidine) which gives a bright-red insoluble azo dye readily seen by light microscopy.   Even the colour intensity of the resulting azo dye is ~3-fold greater [45].

To exploit these findings, Lys-Ala-NNapOMe was prepared and tested as a possible histochemical substrate for DPP II localization. The success of this approach was manifested by the appearance of an intense (bright-red) granular staining in the follicular (secretory) epithelium of the rat thyroid [30, 46], a tissue known to be a rich source of DPP II from a survey employing the foregoing fluorimetric assay [29].   More recently, the guinea pig epididymis was found to be a rich source of DPP II.   The acrosomal cap on the guinea pig spermatozoon, a structure believed to mediate ovum penetration, was clearly delineated (Fig. 3) when stained for DPP II by a published procedure [46].   A lysosomal localization has since been shown for

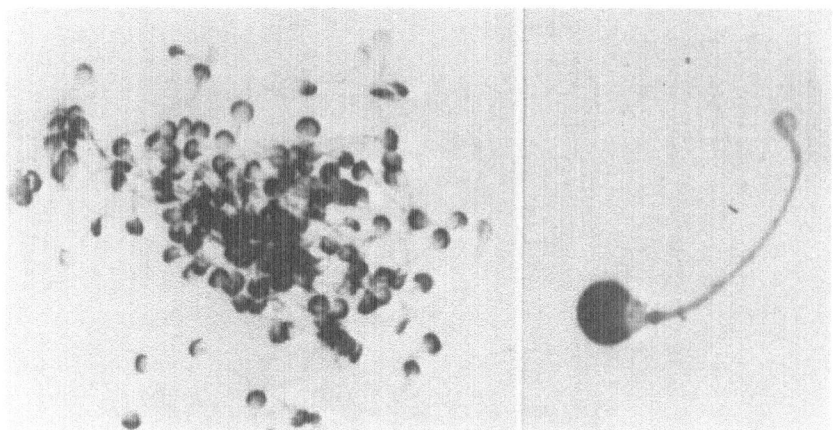

Fig. 3. Formalin-fixed spermatozoa from guinea pig epididymis, stained for DPP II by simultaneous incubation with Lys-Ala-NNapOMe and fast blue B at pH 5.5 Note azo-dye reaction product, viz. dark crescents [bright red], in the acrosomes of individual sperm. ×155 and (inset) ×390.

DPP II in diverse cells and tissues and a subcellular lysosomal distribution has consistently been shown. The prominence of DPP II in the guinea pig acrosome therefore appears to support the contention that this invasive organelle is a lysosomal derivative [47].

Other biological material.- DPP II is a prominent lysosomal enzyme in most cell types of bone and associated connective tissues, including osteocytes, osteoblasts, chondrocytes, chondroblasts, fibroblasts and macrophages. Furthermore, since DPP II is stable to the harsh conditions of fixation and decalcification, which are pre-requisite to the sectioning and staining of bone, the discrete, bright-red reaction product generated by this exopeptidase has made it possible to visualize and identify cells present in these connective tissues [48]. Cells located in skeletal tissues, as well as in the fibroblasts of ligaments, were highly reactive for DPP II, especially in the transition regions between ligament and fibrocartilage - a zone where matrix turnover is believed to be particularly high. Chondrocytes of articular cartilage, and the cells of synovium and periosteum, were also very reactive, as were macrophages in all sites [48].

P.L. Sannes and B.H. Schofield at the Johns Hopkins Medical School (personal communication) have utilized DPP II histochemistry to identify metastatic myeloma cells in bone biopsies, and to visualize invasion routes associated with matrix erosion. Similar observations were made for bone invaded by metastatic cells arising from large-cell tumours of the lung.

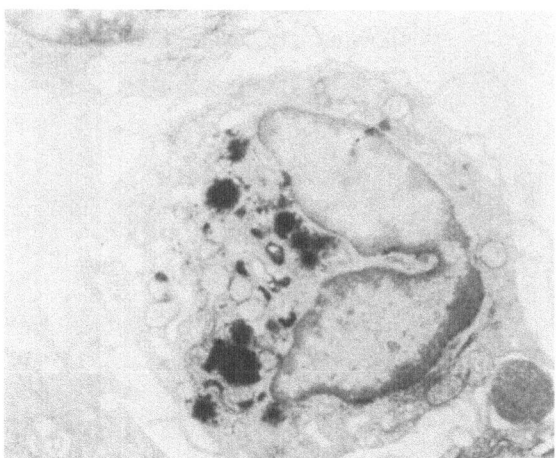

Fig. 4. Electron micrograph of a rat peritoneal macrophage fixed with glutaraldehyde and incubated simultaneously with Lys–Ala–NNapOMe and hexazotized pararosaniline at pH 5.5, showing DPP II reaction product (as the azo dye shown on left) principally confined to lysosomal structures. Procedures as in [46]. ×5,500.

Localization by electron microscopy

Attempts to develop an azo–dye metal chelate useful for the ultrastructural localization of peptidases led to the adoption of hexazotized pararosaniline, a diazonium salt that yields a large azo–dye complex containing three NNapOMe leaving groups (Fig. 4, left). This dye was found in Dr. Robert Smith's laboratory to show a substantivity and osmiophilia (3 atoms of Os/mol of azo dye) which significantly exceeded that obtained with fast blue B and several other diazonium salts. Furthermore, there were no extra-lysosomal artifacts, as found with fast blue B due to displacement of reaction product [46].

What is believed to be the first example of a discrete ultra-structural demonstration of a lysosomal peptidase was accomplished when rat-thyroid follicular cells were found to be the locus of DPP-II. Glutaraldehyde-fixed sections were incubated at pH 5.5 with Lys-Ala-NNapOMe and freshly prepared hexazotized pararosaniline, as amplified and illustrated elsewhere [30, 46]. Biochemical and cyto-chemical procedures were later used jointly to demonstrate that DPP II was also a prominent peptidase in macrophages (Fig. 4, right). Interestingly, DPP II was seen to be present in many but not all of the cytoplasmic dense bodies recognized as secondary lysosomes or phagolysosomes. Such evidence of lysosomal heterogeneity was seen in peritoneal and bone-marrow macrophages [49] and in pulmonary alveolar macrophages [50] as well.

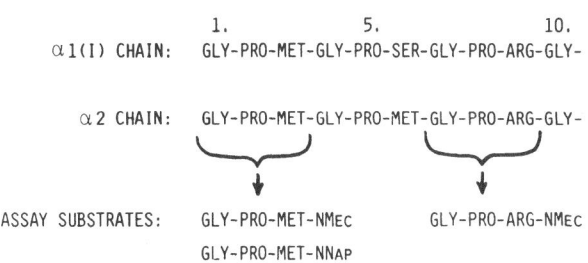

ASSAY SUBSTRATES: GLY-PRO-MET-NMEC   GLY-PRO-ARG-NMEC

                     GLY-PRO-MET-NNAP

Fig. 5. Model fluorigenic substrates based on repeating triplet sequences present in the α-chains at the N terminus of the triple helical regions of type I collagen.

TRIPEPTIDYL PEPTIDASE I (TPP I)

A description [51] of a bovine pituitary exopeptidase that can release tripeptides sequentially from the N terminus of growth hormone prompted us to look in the ovary, where collagenolytic activity is hormonally regulated [52], for a similar peptidase able to depolymerize the repeating triplet sequence (Gly-X-Y) found in the collagen α-chains. For this purpose, specially designed fluorigenic substrates were used as probes. They included two tripeptide NMec's (Fig. 5) corresponding to the first and third tripeptides found (in both α1 and α2 chains) at the start of the triple helical regions of type I collagen [19]. Thereby an exopeptidase, now termed TPP I [21], was found in hog ovary [53]. Assays performed at pH 5.0 (37°; 0.15 M substrate) [53] gave the following values for enzyme specific activity as mU/mg (and *relative activity*) with different substrates.-
    Gly-Pro-MetNMec 1660 *(100%)*, Gly-Pro-Arg-NMec 160 *(10%)*.
    Nil activity: Pro-Met-NMec, Met-Nec, Suc(MeO)-Gly-Pro-Met-NMec;
      Z-Arg-Arg-NMec, Arg-NMec. ['Suc(MeO)' = succinic mono-methyl ester.]
Evidently TPP I requires a free N-terminal α-amino group and a tripeptidyl moiety attached to the NMec leaving group. Its ability to hydrolyze the arginyl bond in GlyProArg-NMec was indicative of a relatively broad specificity. Like DPP II, TPP I exhibited a serine catalytic mechanism [53]. As summarized elsewhere [21], TPP I was found to be a lysosomal peptidase (M_r ~55,000) whose pH optimum was 4.0-4.5, with a wide tissue distribution.

Action on poly(Gly-Pro-Ala).- The ability of purified TPP-I to depolymerize this model collagen α-chain to its constituent Gly-Pro-Ala tripeptides [53] was of special interest, in relation to novel mechanisms of collagen-chain degradation. When the reaction was coupled to DPP II, the fragments were further reduced to Gly-Pro and free Ala [53], products that would be expected to diffuse to the cytosol [54] where proline dipeptidase ('prolidase') could complete the breakdown of collagen to its constituent amino acids. Thus,

Fig. 6. Proposed
contribution of a
coupled exopeptidase
mechanism (TPP I/DPP
II/'prolidase') to
intracellular degra-
dation of α-chain
sequences (amplified
in text).

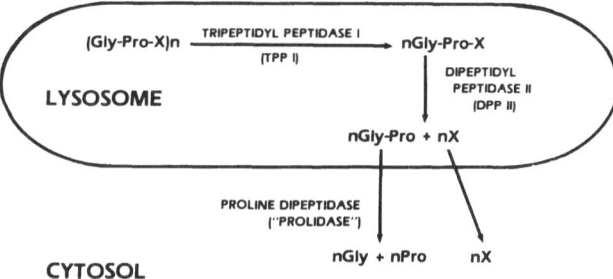

a coupled TPP I/DPP I mechanism, as illustrated in Fig. 6 for
$(Gly\text{-}Pro\text{-}Ala)_n$ breakdown, could substantially contribute to the task
of reducing proline-rich, collagen-derived oligopeptides to free
amino acids. The terminal action of DPP II, like the action of DPP I
on tripeptides lacking proline serves to reduce tripeptides (which
are believed to be incapable of permeating the lysosomal membrane)
to dipeptides and free amino acids, products which are able to permeate
it [54].

Fluorimetric assay.- A version [53] of a method outlined above
allows TPP I in extracts to be assayed specifically with
Gly-Pro-Met-NMec or Gly-Pro-Met-NNap, in pH 5.5 Na acetate buffer
at 37°.

Azo-dye histochemistry.- The above readily hydrolyzable fluorigenic substrates were resynthesized as NNapOMe derivatives as potential
histochemical substrates. Disappointingly, the hydrolysis rates
were only ~2% of those for the corresponding NNap derivatives. This
unprecedented lack of activity on the slightly bulkier NNapOMe derivatives was recently circumvented by substituting l-naphthol (ONap)
as the leaving group, giving ~10-fold faster cleavage than for the
NNap derivatives. Preliminary trial of these substrates in TPP I
histochemical reactions coupled to fast blue B has yielded promising
results with fresh-frozen rat ovary sections.

LOCALIZATION OF TISSUE PROTEINASES USING FIXED GELATIN MEMBRANES

We utilize a glutaraldehyde-denatured India ink-impregnated gelatin membrane fixed to a microscope slide as a general substrate for
detecting the activity of a range of tissue proteinases and simultaneously assigning them to particular cells and structures. The membrane
(which resists boiling water and acids) is exquisitely sensitive
to proteinases possessing a range of specificities, as illustrated
in Fig. 7 for trypsin and Fig. 8 for two mammalian collagenases.
Areas of digestion of the gelatin are seen as clear zones (with the
aid of back lighting), attributable to release of the India ink (carbon)
particles. Digestion (area of lysis) of the membrane by trypsin
was time- and concentration-dependent over the trypsin range 0.3-300 ng
(Fig. 7).

Figs. 7 & 8. Sensitivity of
gelatin–India ink membranes
to a range of enzyme concentrations per 3 µl drop.
 7.- Trypsin: (from top to
 bottom) 300, 30, 3 & 0.3 ng.
 Buffer: 1 mM CaCl₂/150 mM
 glycerol/0.001% Triton X-100
 /30 mM Tris–HCl pH 8.0.
 Sensitivity is increased
 >1000-fold if incubation
 (37°) 18 h rather than 1 h.

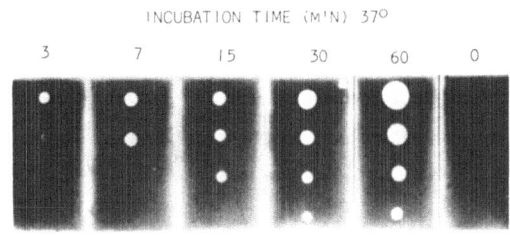

INCUBATION TIME (M'N) 37°

3   7   15   30   60   0

8.- Collagenases. Trypsin as
used for procollagenase acti-
vation was inactivated (with
soybean trypsin inhibitor,
SBTI) in controls.
Buffer: 0.1 M NaCl/10 mM
CaCl₂/0.05% Brij 35/50 mM
Tris–HCl pH 7.5.

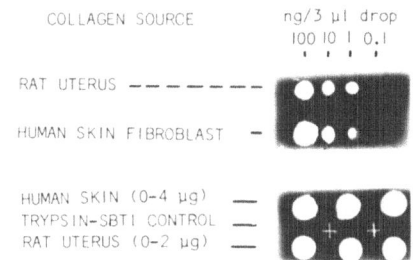

COLLAGEN SOURCE   ng/3 µl drop
                       100 10 1 0.1

RAT UTERUS — — — — — —

HUMAN SKIN FIBROBLAST —

HUMAN SKIN (0-4 µg)   —
TRYPSIN–SBTI CONTROL   —
RAT UTERUS (0-2 µg)   —

     Evidently (Fig. 8) two trypsin-activated collagenases, one
purified from rat uterus and the other from cultured human skin fibroblasts, were readily detected at the 1-µg level on the gelatin membrane.
Lysis of the membrane was attributed to the intrinsic gelatinolytic
activity displayed by mammalian collagenase [55]. Fig. 8 shows no
action on the membrane in controls where the trypsin used for procollagenase activation was pre-treated with SBTI as normally used to
remove residual trypsin. The metallopeptidase nature of the two
collagenases was evidenced by Ca²⁺ activation and EDTA inhibition.

     Preparing the membranes, with denatured gelatin (area covered:
~25 × 45 mm) on microscope slides (25 × 75 mm).- Working solution of
gelatin: mix 10 ml 6.8% (w/v) aqueous stock solution (37°) with 10 ml
water (37°) containing 0.55 ml Higgins India ink (#4465, waterproof)
pre-shaken vigorously by hand for 1 min and left for 15 min before
drawing from the supernatant. Gently mix to suspend the ink in the
gelatin solution, leave 10-30 min at 37°, then allow to come to 29°
(or the lowest temp. at which still a sol) just prior to adding 0.1
ml to each slide. Spread drop over the set area (using the edge
of another slide or a glass rod), place slide on an absolutely level
surface and allow film to air-dry at room temp. (not >23°). Ideally
there should be very little retraction of the gelatin film along
the edges of the slide. **Fixation.-** Submerge the slides (on edge in
a 20-place dipping rack) in freshly prepared 0.05% glutaraldehyde

Fig. 9. Pubic symphysis cryostat section, incubated at pH 5.5 for 18 h at 37° on a gelatin/India ink membrane. Clear (white) areas indicate proteolytic action; note regions of acid protease activity associated with the cartilage caps and bone undergoing remodelling.

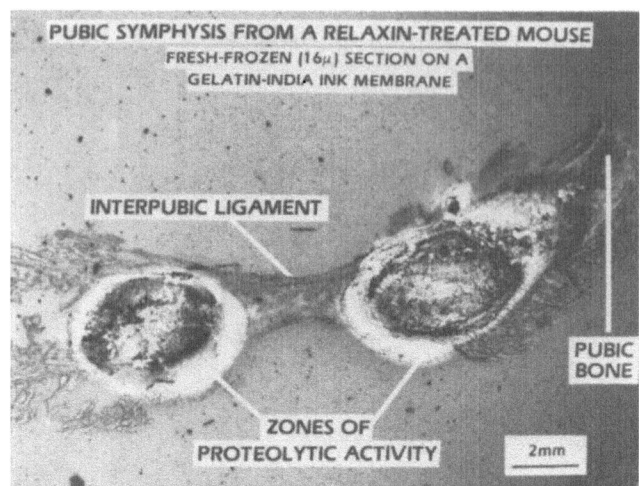

in 25 mM Na phosphate buffer pH 7.0, at ambient temp. After 2 min, rinse (2 × 10 sec) in phosphate buffer and again, when dry, in water (3 × 2 min). Shake off excess water, allow to dry, and store.

Slide testing.- To check thermal stability, submerge representative slides in 95° water for 10 min and 37° water for 24 h; the membranes should not dissolve if gelatin fixation is complete. To check sensitivity, apply 3-μl drops of serial (1:10) dilutions of trypsin in buffer (Fig. 7 legend). Sensitivity (clear zones of lysis) should be 0.3–300 ng trypsin with a 1-h incubation (Fig. 7), 0.03–30 pg with 18 h; use a moist chamber (to prevent drops from drying) at 37°. Slide storage: at room temp. on edge in a covered slide box; sensitivity to proteases survives >1 year.

Applicability to remodelling and invasive tissues.- Fluorescence histochemistry for DPP I as outlined earlier, with mouse pubic symphysis undergoing relaxin-induced remodelling, was for localization of a specific peptidase. In contrast, detection of general proteinase activity was achieved with the gelatin membrane in a similar (adjacent) tissue section (Fig. 9). The 10 to 20-fold widening of the symphysis to ~3 mm and the replacement of the firm cartilagenous attachments by an interpubic ligament manifest extensive connective remodelling. The clear zones (Fig. 9 legend) displayed little or no acid (lysosomal) proteinase activity prior to inducing remodelling.

In contrast with the azo-dye method that showed specific DPP II location in spermatozoan acrosomes (Fig. 3), where DPP I could not be detected, with the gelatin membrane sperm acrosomes were shown (Fig. 10A) to be loci for general (neutral) proteinase activity. The lysis zones appear as 'halos' around the acrosomal caps.

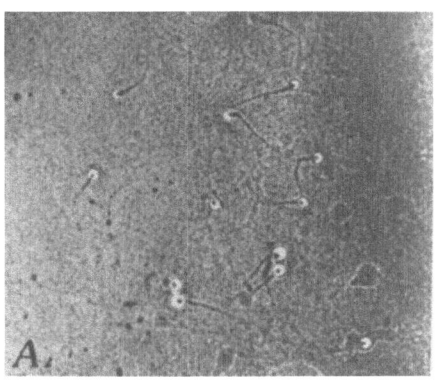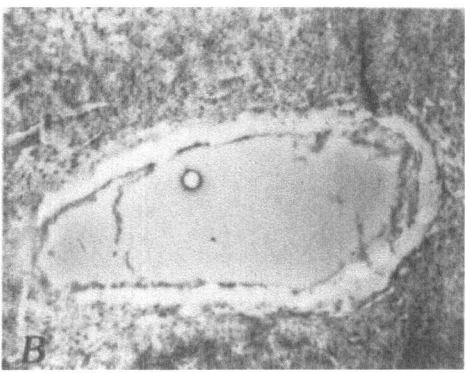

Fig. 10. Localization of proteolytic activity in cells and tissue sections following incubation for 18 h at 37° on the gelatin membranes. **A:** a smear of fresh epididymal sperm (in pH 7 veronal acetate buffer) from a *Cebus* monkey, showing zones of neutral proteinase activity (clear zones) associated with heads, but not with the midpiece or tail segments. ×170. **B:** a cryostat section of fresh tissue conceptus taken from a rat following implantation of the invading blastocyst in the uterine wall, showing a clear (white) zone of proteolytic activity (at pH 5.0) surrounding the oval-shaped (late-stage) blastocyst. Maternal (background) tissue shows little or no proteolytic activity. ×60.

The gelatin membrane may also be used to visualize 'invasion fronts' by exploiting the presence of associated proteinase activity. The clear zone of lysis seen in Fig. 10B surrounds a rat blastocyst (in cross-section) that is in the process of implanting in the uterine wall.

Acknowledgements

The authors thank Drs. Sannes and Schofield (Baltimore) for permitting the mention of their unpublished peptidase studies on multiple myeloma samples, and Dr. John J. Jeffrey (of Washington University Medical School, St. Louis) for generously gifting human and rat collagenase. Our thanks also go to Mrs. Anne R. Hoisington for valuable technical assistance and to Mrs. Burnett Bryant for typing the manuscript. These studies were supported, in part, by USPHS grant HD-16106 from NICHD.

References

1. Miller, E.J. (1985) *Ann. N.Y. Acad. Sci. 460*, 1-13.
2. Timpl, R., Wiedemann, H., van Delden, V., Furthmayr, H. & Kühn, K. (1981) *Eur. J. Biochem. 120*, 203-211.

3. Miller, E.J. (1984) in *Extracellular Matrix Biochemistry* (Piez, K.A. & Reddi, A.H., eds.), Elsevier, New York, pp. 41-81.

4. Sellers, A. & Murphy, G. (1981) *Internat. Rev. Connect. Tiss. Res. 9*, 151-190.

5. Liotta, L.A., Thorgeirsson, U.P. & Garbisa, S. (1982) *Cancer Metastasis Rev. 1*, 277-288.

6. Woolley, D.E., Tetlow, L.C., Mooney, C.J. & Evanson, J.M. (1980) in *Proteinases and Tumor Invasion* (Staüli, P., Barrett, A.J. & Baici, A., eds.), Raven Press, New York, pp. 97-115.

7. Ten Cate, A.R. (1972) *J. Anat. 112*, 401-414.

8. Deporter, D.A. & Ten Cate, A.R. (1973) *J. Anat. 114*, 457-461.

9. Ten Cate, A.R. & Freeman, E. (1974) *Anat. Rec. 179*, 543-546.

10. Welsh, R.A. & Meyer, A.T. (1967) *Arch. Path. 84*, 354-362.

11. Levine, A.M., Reddick, R. & Triche, T. (1978) *Lab. Invest. 39*, 531-540.

12. Dingle, J.T. (1976) in *Proteolysis and Physiological Regulation* (Ribbons, D.W. & Brew, K., eds.), Academic Press, New York, pp. 339-355.

13. Woessner, J.F., Jr. (1976) in *as for* 12., pp. 357-369.

14. Singh, H., Kuo, T. & Kalnitsky, G. (1978) in *Protein Turnover and Lysosome Function* (Segal, H.L. & Doyle, D.J., eds.), Academic Press, New York, pp. 315-331.

15. Etherington, D.J. (1980) in *Protein Degradation in Health and Disease, Ciba Found. Symp. 75*, 87-103.

16. Murphy, G. & Reynolds, J.J. (1985) *BioEssays 2*, 55-60.

17. Barrett, A.J. & McDonald, J.K. (1980) *Mammalian Proteases: A Glossary and Bibliography*, Vol. 1: *The Endopeptidases*, Academic Press, New York, 357 pp.

18. Piez, K.A., Eigner, E.A. & Lewis, M.S. (1963) *Biochemistry 2*, 58-66.

19. Fietzek, P.P. & Kühn, K. (1976) *Internat. Rev. Connect. Tiss. Res. 7*, 1-60.

20. Nomenclature Committee, Internat. Union of Biochem. (1984) in *Enzyme Nomenclature 1984*, Academic Press, New York, pp. 330-366.

21. McDonald, J.K. & Barrett, A.J. (1986) *Mammalian Proteases: A Glossary and Bibliography*, Vol. 2: *The Exopeptidases*, Academic Press, New York, 416 pp.

22. Nomenclature Committee, Internat. Union of Biochem. (1985) *Biochem. J. 231*, 808 (cf. 807).

23. Barrett, A.J. & McDonald, J.K. (1986) *Biochem. J. 237*, 935.

24. Bergmann, M. & Ross, W.F. (1936) *J. Biol. Chem. 114*, 717-726.

25. McDonald, J.K. (1985) *Histochem. J. 17*, 773-785.

26. McDonald, J.K., Zeitman, B.B., Reilly, T.J. & Ellis, S. (1969) *J. Biol. Chem. 244*, 2693-2709.

27. McDonald, J.K., Hoisington, A.R. & Eisenhauer, D.A. (1985) *Biochem. Biophys. Res. Comm. 126*, 63-71.

28. McDonald, J.K., Ellis, S. & Reilly, T.J. (1966) *J. Biol. Chem. 241*, 1494-1501.

29. McDonald, J.K., Reilly, T.J., Zeitman, B.B. & Ellis, S. (1968) *J. Biol. Chem. 243*, 2028-2037.

30. McDonald, J.K., Callahan, P.X., Ellis, S., & Smith, R.E. (1971) in
    Tissue Proteinases (Barrett, A.J. & Dingle, J.T., eds.), North-
    Holland, Amsterdam, pp. 69-107.
31. Barrett, A.J. (1975) in *Proteases and Biological Control* (Reich, E.,
    Rifkin, D.B. & Shaw, E., eds.), Cold Spring Harbor Laboratory, New
    York, pp. 476-477.
32. McDonald, J.K., Callahan, P.X. & Ellis, S. (1972) *Meths. Enzymol.*
    25, 272-281.
33. McDonald, J.K. & Schwabe, C. (1977) in *Proteinases in Mammalian
    Cells and Tissues* (Barrett, A.J., ed.), North-Holland, Amsterdam,
    pp. 311-391.
34. Barrett, A.J. & Heath, M.F. (1977) in *Lysosomes: A Laboratory
    Handbook*, 2nd edn. (Dingle, J.T., ed.), Elsevier/North-Holland,
    Amsterdam, pp. 19-145.
35. Dolbeare, F.A. & Smith, R.E. (1977) *Clin. Chem. 23*, 1485-1491.
36. McDonald, J.K. & Schwabe, C. (1980) *Biochim. Biophys. Acta 616*,
    68-81.
37. Green, G.D. & Shaw, E. (1981) *J. Biol. Chem. 256*, 1923-1928.
38. McDonald, J.K. & Schwabe, C. (1982) *Ann. N.Y. Acad. Sci. 380*, 178-
    186.
39. Stauber, W.T. & Ong, S-H. (1982) *J. Histochem. Cytochem. 30*, 162-164.
40. Timpl, R. (1982) *Meths. Enzymol. 82*, 472-498.
41. McDonald, J.K., Leibach, F.H., Grindeland, R.E. & Ellis, S. (1968)
    J. Biol. Chem. 243, 4143-4150.
42. Eisenhauer, D.A. & McDonald, J.K. (1986) *J. Biol. Chem. 262*, 8859-
    8865.
43. Fukasawa, K., Fukasawa, K.M., Hiraoka, B.Y. & Harada, M. (1983)
    Biochim. Biophys. Acta 745, 6-11.
44. Nachlas, M.M., Monis, B., Rosenblatt, D. & Seligman, A.M. (1960)
    J. Biophys. Biochem. Cytol. 7, 261-264.
45. Barrett, A.J. (1973) *Biochem. J. 131*, 809-822.
46. Smith, R.E. & Van Frank, R.M. (1975) in *Lysosomes in Biology
    and Pathology* (Dingle, J.T. & Dean, R.T., eds.), North-Holland,
    Amsterdam, pp. 193-249.
47. Hartree, E.F. (1975) *J. Reprod. Fert. 44*, 125-126.
48. Sannes, P.L., Schofield, B.H. & McDonald, J.K. (1986) *J. Histochem.
    Cytochem. 34*, 983-988.
49. Sannes, P.L., McDonald, J.K. & Spicer, S.S. (1977) *Lab. Invest.
    37*, 243-253.
50. Sannes, P.L. (1983) *J. Histochem. Cytochem. 31*, 684-690.
51. Doebber, T.W., Divor, A.R. & Ellis, S. (1978) *Endocrinology 103*,
    1794-1804.
52. Curry, T.E., Jr., Dean, D.D., Woessner, J.F. Jr. & LeMaire, W.J.
    (1985) *Biol. Reprod. 33*, 981-991.
53. McDonald, J.K., Hoisington, A.R. & Eisenhauer, D.A. (1985) *Biochem.
    Biophys. Res. Comm. 126*, 63-71.
54. Ehrenreich, B.A. & Cohn, Z.A. (1969) *J. Exp. Med. 129*, 227-243.
55. Welgus, H.G., Grant, G.A., Sacchettini, J.C., Roswit, W.T. &
    Jeffrey, J.J. (1985) *J. Biol. Chem. 260*, 13601-13606.

#E-7

BRAIN TUMOUR CELL INVASION: POSSIBLE ROLE OF LYSOSOMAL ENZYMES

[1]Knut-Jan Andersen, [2]Rolf Bjerkvig and [2]Ole Didrik Laerum

[1]Section for Clinical Research   [2]The Gade Institute
and Molecular Medicine   Department of
Medical Department A   Pathology
                    University of Bergen
                    N-5016 Haukeland Sykehus
                    Bergen, Norway

A syngeneic invasion system where cell aggregates from foetal rat brain are used as a target for glioma cell invasion is described. With this model system we have been able to demonstrate different invasion patterns for different continuous glioma cell lines. When grown in a serum-free chemically defined medium the glioma cell lines studied secreted glycosidases and also collagenolytic peptidases, for which increased activity was recorded in the medium from the cell line causing the more severe tissue destruction.

Tumour cell invasion into adjacent normal tissue, and metastatic spread to other sites, are important features of malignant behaviour. These properties may depend on the intrinsic abilities of tumour cells to penetrate into the normal tissue, and on the composition of the surrounding normal tissue. Primary brain tumours are regarded as highly invasive with little or no ability to metastasize. The clinical experience is that these tumours cannot be radically excised due to the fact that malignant cells are already beyond the stage of surgical respectability.

At present the relationship between invasiveness and other biological properties of brain tumour cells is unclear. The specific mechanisms involved in the invasive behaviour of brain tumours are still largely unknown. Since primary brain tumours consist of heterogeneous tumour cell populations, it is likely that several different mechanisms are involved in tumour invasion. As will be shown, the development and modification of cell culture techniques for normal as well as malignant cells may give valuable information on the mechanisms involved during invasive growth.

EXPERIMENTAL SYSTEMS FOR STUDYING BRAIN TUMOUR INVASION

In vivo **experiments.-** Selective induction of malignant tumours of the nervous system by resorptive carcinogens has given valuable information regarding the pathogenesis and morphology of gliomas in rodents [1-3]. Frequently the induced tumours have been cultured *in vitro* and transplanted intracranially for invasion studies [3]. However, there are several limitations of such transplantation studies. Firstly, it is difficult to obtain a dynamic picture of the invasive process since observations are done on static images of histopathological preparations. Secondly, the implantation procedure may induce wound healing as well as inflammatory and other reactions, which in turn may influence the invasive process. Thirdly, inter-experimental variations frequently occur even though inbred animals are used.

In vitro experiments.- Several *in vitro* systems have been described for tumour cell invasion studies [4, 5]. Although data obtained from *in vitro* studies should be interpreted with some reservation, *in vitro* models of invasion may give a dynamic picture of the processes taking place and also allow direct observations of interactions between the tumour cells and the normal tissue. By the use of 3-dimensional confrontation culture systems it has been demonstrated that both rat and human neurogenic cell lines can invade and destroy embryonic chick heart fragments *in vitro* [6, 7]. In addition, recent studies have shown that invasive behaviour *in vitro* correlates closely with *in vivo* observations [2] (cf. M. Mareel et al., #E-8 in this vol.).

Recently we have described a syngeneic invasion system (Fig. 1) where both foetal rat brain fragments and cell aggregates from foetal rat brain have been used as target in glioma cell invasion experiments [8-12]. With this model system we have been able to demonstrate individual differences in the invasion patterns for two different continuous glioma cell lines [11]. One line (BT_5C) invaded the brain tissue by massive replacement of the normal tissue, causing severe tissue destruction of normal structures (Figs. 2 & 3). The other cell line (BT_4Cn) showed single-cell invasion with little initial tissue destruction of normal tissue (Fig. 4). Moreover, when normal brain aggregates were incubated in BT_5C cell-conditioned media, the normal cells were observed to degenerate [9]. This was not the case for the BT_4Cn cell line. These observations indicated that the BT_5C cells secrete soluble factors into the medium that directly or indirectly cause destruction of the normal tissue. These *in vitro* results were verified by intracranial implantation of the same cell lines; two different invasion patterns were manifest [11].

EXTRACELLULAR ACCUMULATION OF 'LYSOSOMAL' ENZYMES

It has been proposed by several groups that malignant cells secrete matrix-degrading enzymes, usually found within cellular

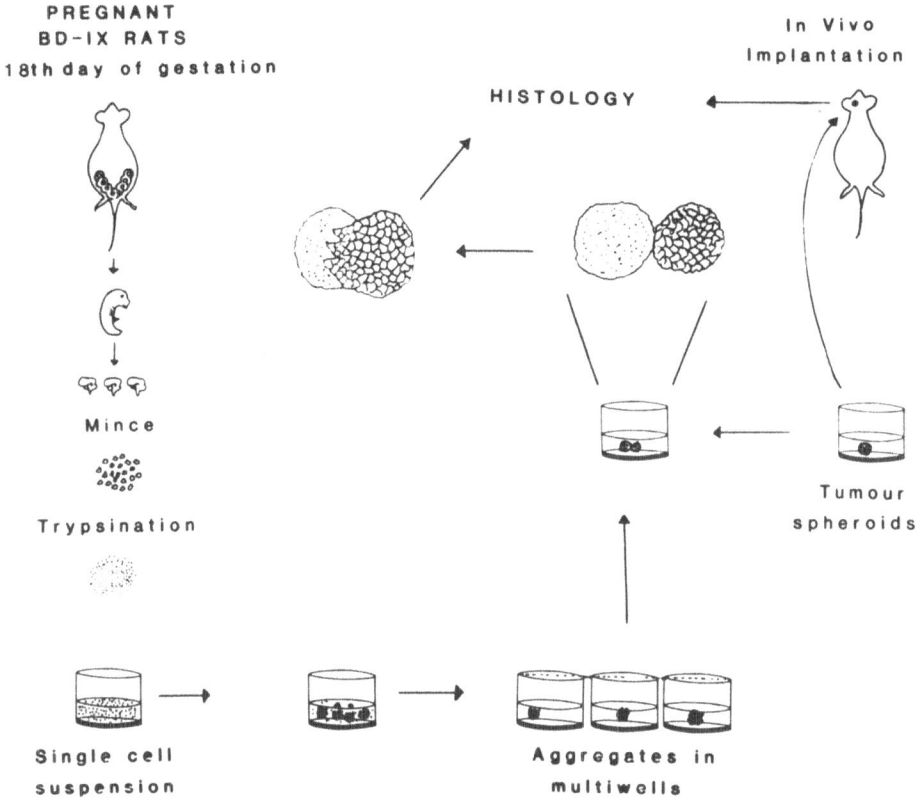

Fig. 1. Brains from foetal rats are minced and trypsinized to obtain a single-cell suspension, which is seeded into medium-agar coated multi-wells. During 48 h the cells will form multicellular aggregates, which are then cultured individually for 20 days. At this time the cellular differentiation is complete, and the aggregates are confronted with tumour cells, which are also implanted intracranially into rats. Both *in vivo* and *in vitro* invasion patterns are evaluated by histology.

lysosomes, to facilitate their invasion into surrounding tissues [12-15]. From our morphological observations (Figs. 2-4) it is evident that different mechanisms may be involved in the invasive process. We have therefore been screening culture media obtained from monolayer cultures of several human and rat glioma cell lines, and have found that these cells secrete variable amounts of glycosidases as well as endo- and exo-peptidases involved in extracellular matrix degradation in addition to cell surface modulation [14]. Due to the presence of hydrolytic activity and also natural peptidase inhibitors in serum, cells in the late exponential phase were transferred to a serum-free

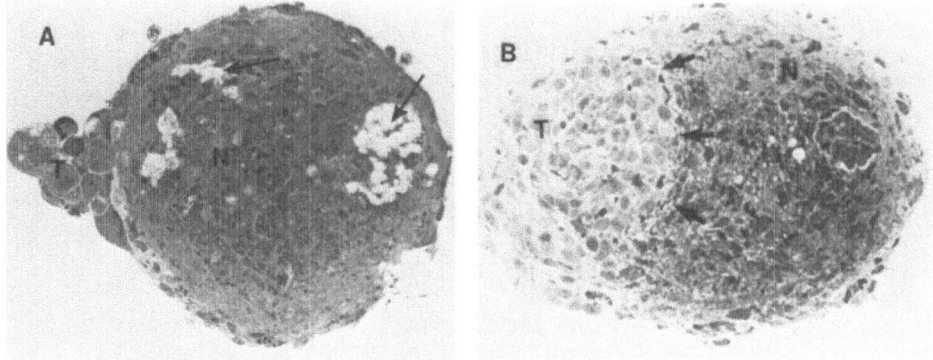

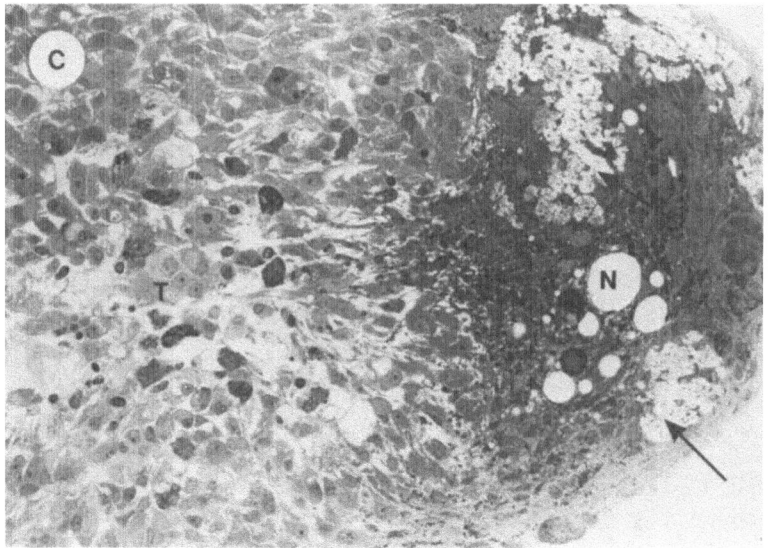

Fig. 2. A. Semi-thin toluidine-stained section of a foetal rat brain aggregate (N) 72 h after confrontation with BT$_5$C cells (T). Tissue destruction is observed in the normal tissue distant from the .tumour cells (*arrows*). ×400.
B. After 144 h a distinct border (*arrows*) can be seen between the tumour cells (T) and normal tissue (N). Much normal tissue has been replaced by tumour cells, in **C** likewise. ×400.
C. After 292 h, normal-tissue destruction is severe (*arrows*). ×700.

chemically defined medium found to support growth of a broad variety of glioma cell lines. After 48 h the medium was assayed for glycosidases [16] and collagenolytic peptidases [17]. With both cell lines (Table 1) an extracellular accumulation of acid hydrolases was observed, although activity levels differed. We have reason to believe that these enzymic activities reflect secreted or released

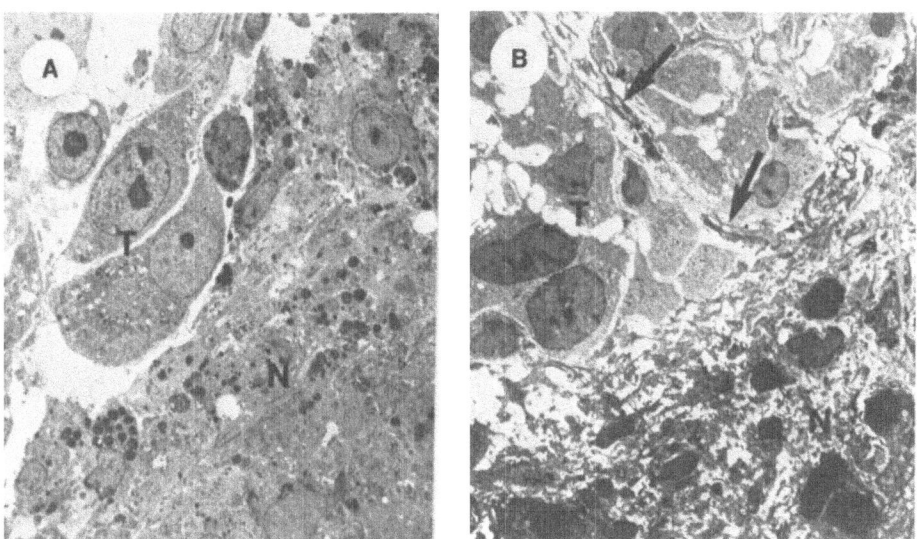

Fig. 3. Transmission electron micrographs (× 3200) showing the contact zone between BT$_5$C cells and normal tissue, **(A)** 24 h, and **(B)** 96 h after confrontation. No tissue destruction is observed after 24 h, while after 96 h the neutrophil has fallen completely apart. *Arrowed:* normal tissue within the tumour cell population.

enzymes and not cell death, as the DNA content [16] of exposed media was found to be <1 µg/ml.

Of particular interest to us was the presence of both tripeptidyl peptidase I and dipeptidyl peptidase II in the medium, as Gly-Pro-X tripeptides released from collagen-like structures by tripeptidyl peptidase I would be favoured substrates for attack by dipeptidyl peptidase II [17]. Also noteworthy is the observed increased activity of both enzymes in medium from the BT$_5$C cells found to cause severe tissue destruction of normal structures (Figs. 2 & 3).

However, it would be premature to correlate the proteolytic activity of some tumour cells with invasive behaviour, pending more studies including several cell lines.

Acknowledgement

Thanks for support are expressed to the Norwegian Society for Fighting Cancer.

Table 1. Extracellular accumulation (mU/ml) of 'lysosomal' enzymes.

Enzyme	BT$_5$C cell line	BT$_4$Cn cell line
Acid phosphatase	0.18	0.49
Acid β-galactosidase	0.03	0.03
Acid β-glucuronidase	0.10	0.10
N-Acetyl-β-glucosaminidase	1.01	0.36
Cathepsin B	–	0.06
Tripeptidyl peptidase I	0.18	0.04
Dipeptidyl peptidase II	1.0	0.19

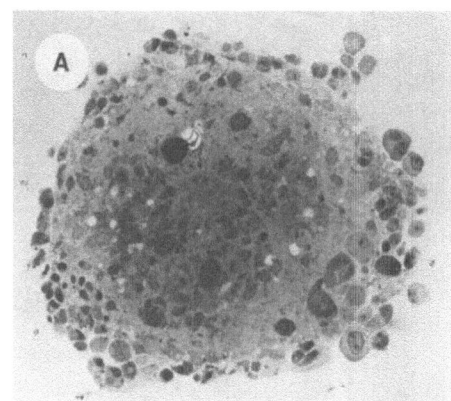
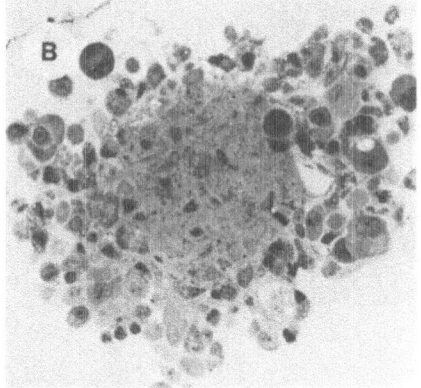
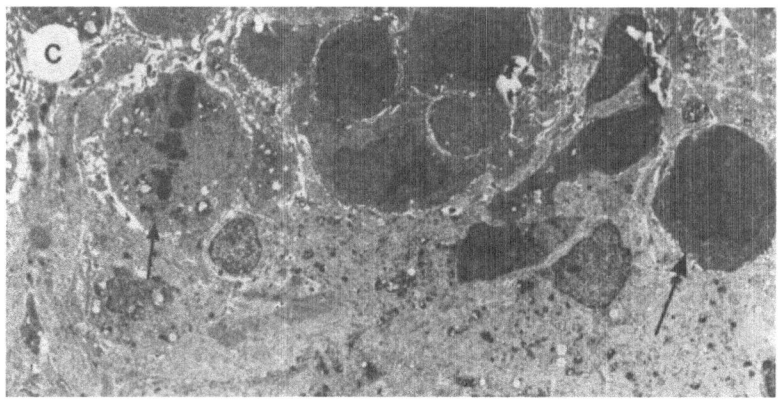

Fig. 4. Brain cell aggregates confronted with BT$_4$Cn cells. (A), (B): semi-thin sections after 96 and 144 h respectively. Single-cell invasion is observed with little initial destruction of normal tissue. × 460.   (C): a transmission electron micrograph of the contact zone between tumour cells and normal tissue. Single-cell invasion is observed *(arrows)*; no massive tissue destruction. (D): After 292 h of co-culture only remnants of normal tissue are evident.   × 700.

References

1. Druckrey, H., Ivankovic, S., Preussmann, R., Zulch, K.J. & Mennel, H.D. (1972) in *The Experimental Biology of Brain Tumors* (Kirsch, W.M., Paoletti, E.G. & Paoletti, P., eds.), Thomas, Springfield, IL, pp. 85-147.
2. Mörk, S.J., de Ridder, L. & Laerum, A.D. (1982) *Anticancer Res.* 2, 1-10.
3. Pilkington, G.J. & Lantos, P.L. (1979) *Acta Neuropath.* 45, 177-185.
4. Easty, G.C. & Easty, D.M. (1984) in *Invasion: Experimental and Clinical Implications* (Mareel, M.M. & Calman, K.C., eds.), Oxford University Press, Oxford, pp. 24-62.
5. Laerum, A.D., Steinsvag, S.K. & Bjerkvig, R. (1985) *Acta Neurol. Scand.* 72, 529-549.
6. de Ridder, L. & Laerum, A.D. (1981) *J. Nat. Cancer Inst.* 66, 723-728.
7. de Ridder, L., Laerum, A.D., Mörk, S.J. & Bigner, D.D. (1987) *Acta Neuropath.*, in press.
8. Steinsvag, S.K. & Laerum, A.D. (1985) *Experientia 41*, 1517-1524.
9. Bjerkvig, R., Steinsvag, S.K. & Laerum, A.D. (1986) *In Vitro 22*, 180-192.
10. Steinsvag, S.K., Laerum, A.D. & Bjerkvig, R. (1985) *J. Nat. Cancer Inst.* 74, 1095-1104.
11. Bjerkvig, R., Laerum, A.D. & Mella, O. (1986) *Cancer Res. 46*, 4071-4079.
12. Woolley, D.E. (1984) in *Invasion: Experimental and Clinical Implications* (Mareel, M.M. & Calman, K.C., eds.), Oxford University Press, Oxford, pp. 228-251.
13. Liotta, L.A., Thorgeirsson, U.P. & Garbisa, S. (1982) *Cancer Metastasis Rev. 1*, 277-288.
14. Jones, P.A. & DeClerk, Y.A. (1982) *Cancer Metastasis Rev. 1*, 289-317.
15. Bernacki, R.J., Niedbala, M.J. & Korytnyk, W. (1985) *Cancer Metastasis Rev. 4*, 81-102.
16. Andersen, K-J., Schjönsby, H. & Skagen, D.W. (1983) *Scand. J. Gastroenterol. 18*, 241-249.
17. Andersen, K-J. & McDonald, J.K. (1987) *Am. J. Physiol.*, in press.

#E-8

INVESTIGATION OF TUMOUR-INVASION MECHANISMS

M. Mareel, [*]F. Van Roy, L. Messiaen,
M. Bracke, E. Boghaert and P. Coopman

Laboratory of Experimental	[*]Laboratory of Molecular
Cancerology, Department	Biology
of Radiotherapy and	State University Ghent
Nuclear Medicine	K.L. Ledeganckstraat 35
University Hospital	(F.V.R.)
De Pintelaan, B-9000 Ghent	B-9000 Ghent, Belgium

Methods currently used in our laboratories to analyze the invasive and metastatic capabilities of cell populations in vitro *and* in vivo *are as follows: confrontation of cell clusters or cellular aggregates with fragments of embryonic chick heart in organ culture; implantation of cells aggregated to collagen sponges under the renal capsule of syngeneic mice and rats; s.c., i.p. or i.v. injection of cell suspensions and s.c. implantation of cellular aggregates in syngeneic and immunodeficient animals. Qualitative and quantitative techniques, mainly morphological, are used to score for invasion and metastasis. These methods cover various aspects of the multi-step process of invasion and metastasis from 'natural' tumours. We discuss here those aspects of our methods that influence the experimental results and determine the relevance of these results to the 'natural' situation.*

Invasion and metastasis are hallmarks of malignant tumours. Methods of analyzing these activities are rather crude, because little is known about the molecular mechanisms underlying them. Invasion is defined as the loss of tissue barriers with break-through of cells into surrounding tissue domains. Metastasis comprises invasion of cells into the vasculature (intravasation), transport and survival in the vasculature, adhesion to the capillary wall, egression from the capillary lumen (extravasation), and growth at a distant site. Most methods entail confronting a 'tumour' with a 'host' either *in vivo* or *in vitro*. Such confrontations are different from the natural situation where a tumour develops, possibly from a single cell, inside and in continuous interaction with its host. Experimental procedures unquestionably simplify this interaction, with the risk of producing results irrelevant to the behaviour of natural tumours.

In respect of the 'tumour', major concerns are sampling and
selection. The weight of these factors increases from taking a biopsy
from a heterogeneous tumour, through preparing a cell suspension
and establishing a cell line, to isolating a clone. In respect of
the 'host', major concerns are lack of neurohumoral and immunological
influences and loss of histiotypical structure. The weight of these
factors increases from use of immunosuppressed animals *in vivo*, to
isolation and maintenance *in vitro* of organs, tissues, cells and
extracellular matrices. For the two partners in conjunction, the
main concern is the unwitting production of false images of invasion
and metastasis.

We discuss here some aspects of the methods currently used in
our laboratories. An extensive review of methods for studying invasion
and metastasis was published by Easty & Easty [1].

SEPARATE ASSAYS FOR INVASION AND METASTASIS

There are arguments in favour of the acquisition of invasiveness
and of metastatic capability being discrete steps in tumour progression
towards increasing malignancy (discussed in [2]; cf. S. Zucker, p. 374).

So far, we have been able to predict in most cases the invasiveness
in vivo of cells on the basis of an assay *in vitro*. However, we
found no correlation between metastatic capability (*in vivo*) and
invasiveness *in vitro*. One low-metastatic and two high-metastatic
B16 mouse melanoma sublines were equally invasive *in vitro* [3].
Similar results were obtained with 2 non-metastatic, 3 high-metastatic
and 3 low-metastatic Lewis lung carcinoma sublines (unpublished work,
with G. Vaes & A. Van Lamsweerde), with 2 non-metastatic and 3 metastatic sublines from a Balb-3T3/A31 cell family (unpublished work,
with A. Raz), and with the non-metastatic DX3 *vs.* the metastatic
DX3-ara C melanoma cell lines (unpublished work, with I. Hart &
J. Ormerod). These findings accord with those of Waller et al. [4]
who described an Eb mouse lymphoma subline that was low-metastatic
but highly invasive *in vitro*. The interpretation of these results
is that invasion is needed for metastasis but not necessarily accompanied by it, or sufficient to cause it.

To characterize cell populations in our current experiments,
we use a series of methods that score for either invasion or metastasis
or for both. One advantage is that correlations between the results
from separate assays strengthen the relevance of each assay.

CURRENT TECHNIQUES

To characterize matched cell populations [e.g. 5], to evaluate
potential anti-invasive drugs [e.g. 6-8] and to study the role of
certain molecules in invasion [9], we use confrontations of tumour
cells with fragments of embryonic chick heart in organ culture [10].

Scheme 1.
The organ-culture
assay for invasion.

(Source of Gyrotory
shaker: New
Brunswick
Scientific Co.,
New Brunswick, NJ.)

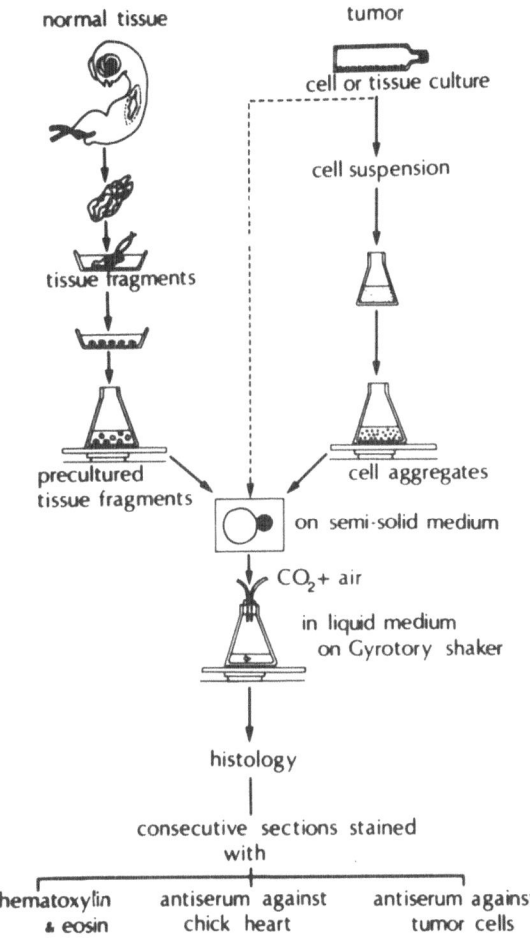

The relevance of this assay *in vitro* to invasiveness *in vivo* has been discussed [11]. The 'tumour' usually consists of a cell aggregate (Scheme 1). When cells do not produce aggregates, we use a drop of a dense cell suspension or a cluster of cells mechanically removed from the tissue culture substrate. The latter is also used when we suspect from morphotypic heterogeneity that pre-cultured aggregates contain a subpopulation of cells that might not be representative of the whole population. The volume of the 'tumour' can be standardized in the case of aggregates but not of suspensions or cell clusters.

As 'host' we routinely use 9-day old embryonic chick heart because the viability of heart fragments (0.4 mm diam.) can be easily judged from their beating function; but most other organs serve as well [12]. In order to conserve the organotypic structure of the heart, it is either kept on top of a semi-solid agar medium or suspended by Gyrotory shaking in liquid medium. Adhesion of the fragment to any solid tissue culture substrate leads to outgrowth of cells and loss of organotypic structure.

The invasive and metastatic capabilities of cell populations are also examined *in vivo* through transplantation into syngeneic or immunodeficient animals in four ways distinguished by the injection sites, and by the organization and volume of the inoculum (Scheme 2).

Injection of tumour cells i.v. leads to formation of lung colonies, also termed artificial metastases. This method bypasses some of the steps of spontaneous metastasis, namely release of cells from the primary tumour and invasion into the vessels (intravasation). It is, therefore, not surprising that the ability of cells to form artificial metastases does not always correlate with their ability to form spontaneous metastases.

Injection s.c. into the flank permits large inocula, but with possible troubles: irregular distribution of cells in natural cleavage planes, producing false images of invasion at least at the onset; accidental i.v. injection, leading to artificial metastases in the lungs; accidental penetration of the injection needle into the abdominal cavity, producing false images of invasion through the abdominal wall.

Immediately after i.p. injection, cells are in a non-invasive state with respect to the tissues to be invaded, except for those cells that are trapped in the abdominal wall upon withdrawal of the injection needle. With this method, nodules inside the thoracic cavity are not unquestionably metastases, since direct invasion through the diaphragm is difficult to exclude.

Implantation of an aggregate (0.3 mm diam.) s.c. into the tail is our method of choice for studying the metastatic capability of a cell line. The spontaneous character of metastases from tail tumours is verified through amputation of the tail from 1 day to 3 weeks after implantation. For the tumour cells examined so far, metastases are found mainly in the gluteal, ileac and para-aortic lymph nodes and in the lungs.

In experiments *in vivo*, evaluation is as follows. All animals are examined twice per week and calliper measurements of the smaller (a) and larger (b) diameters of flank tumours and tail tumours are made. The choice of the time of killing is subjective, since it depends on the observer's judgement of the general state of health of the tumour-bearing animal. We do not consider spontaneous death as a usable end-point because it hampers autopsy and histological examination. For flank and tail tumours the following values are considered: (i) volumes of the primary tumour calculated from the

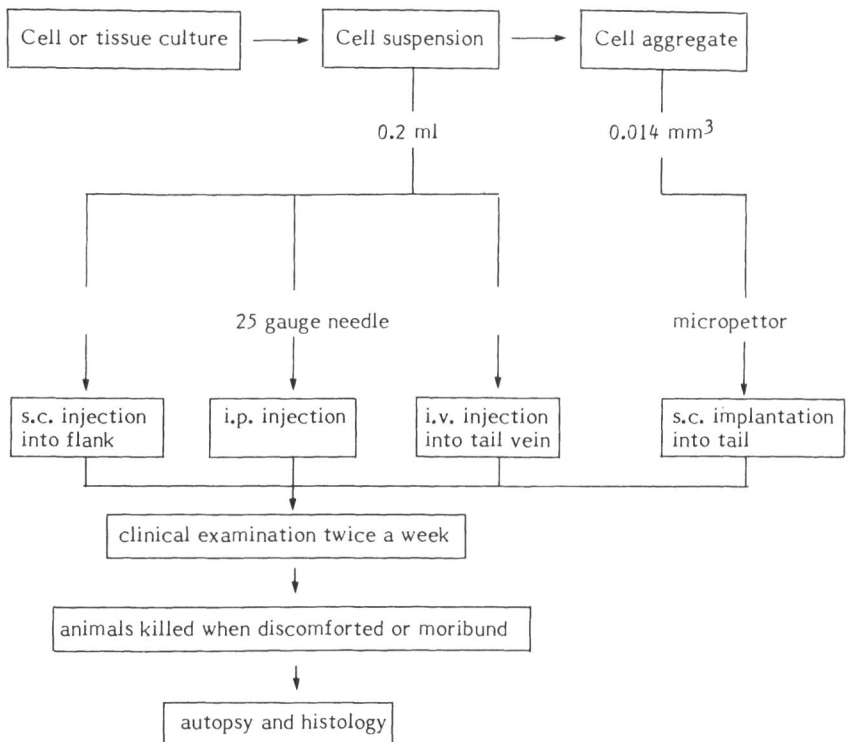

Scheme 2. Transplantation of cell populations to investigate metastatic and invasive capabilities, in four ways differing in respect of the character and volume of the inoculum and of the site (and mode) of injection.

diameters: $V = 0.4 \times a^2 \times b$ (a<b) according to Attia & Weiss [13]; (ii) time of observation, i.e. the time between injection and sacrifice (also for i.p. tumours); (iii) latency period, i.e. time between injection and appearance of palpable tumours; (iv) minimum tumour-bearing period, i.e. the time of observation minus the latency period. Obviously, for individual tumours these values are related to the occurrence of metastases. However, variations of these values could not entirely account for differences in metastasis between different types of tumour [2].

Careful macroscopic observation is most important when scoring for metastasis, since systematic histological examination of even a few organs is too time-consuming. So far [2, 5] we have found metastatic and non-metastatic variants within single cell families. All primary tumours of mesenchymal origin examined by us were invasive, and non-invasive variants were not observed *in vivo*. Possible explanations for this finding are that invasiveness is acquired by these cell lines before or simultaneously with tumorogenicity or that invasion is needed for tumour formation by these cells.

STANDARDIZING THE INOCULUM

Amongst the *in vivo* methods, s.c. implantation of an aggregate in the tail gave the most reproducible results, presumably because an aggregate is a standardized well-delineated implant from which invasion and growth starts in a reproducible way [14]. Aggregates have two main limitations: many cell types do not produce viable aggregates of a useful size; the number of cells in an aggregate is limited to $\sim 1 \times 10^4$ because of central necrosis, so that cells with a higher TD50 (no. of cells required to produce a tumour in 50% of the animals) do not form tumours from a single implanted aggregate. We therefore sought an alternative technique that should be more widely applicable but retain the advantages of aggregates. From a limited number of experiments including cell types that form aggregates only poorly, it appears that cells aggregated to a collagen sponge as suggested by Kraemer et al. [15] meet the need. The steps are as follows.-

(i) Cut cubes of ~ 1 mm^3 from dry (~ 8 mm^3 when wet) collagen sponges (Spongostan standard, Ferrosan, Denmark).
(ii) Put into 50 ml Erlenmeyer flasks 6 ml of culture medium (suitable for cell type used), 10 or 20 sponge cubes, and 0.5 or 1×10^6 cells, and incubate on a Gyrotory shaker (70 rpm) at 37° in 5% CO_2/95% air for 2-3 days.
(iii) Pour flask contents into a Petri dish and transfer individual sponges with 1 ml fresh culture medium into the wells of a 24 multiwell dish (Nunc, Roskilde, Denmark). Incubate in 5% CO_2/95% air at 100% humidity.
(iv) Sample 3 sponges, measure their sides under a Macroscope (Wild,

Heerbrugg, Switzerland; ×25), calculate their outer surface, and incubate them for ~2 h with 25 ml Puck's saline containing 50 mg bacterial collagenase and 100 mg BSA [16].
(v) Count the cells in a haemocytometer, and determine cell viability.
(vi) Infer no. of cells on other sponges from no. on the 3 samples with correction for sponge surface.
(vii) Implant sponges in 2 μl serum-free medium with a micropipettor as described for aggregates [14].

All 16 cell types tested so far aggregated to sponges, the no. of cells per sponge varying from 1×10^3 to 5×10^5. The presence of the sponge did not affect tumorigenicity, invasiveness or metastatic capability of the cell types tested. Sponges cannot be used in the organ-culture assay since cells from the normal tissue colonize the sponge with loss of histiotypic structure.

METASTASIS AND THE SITE OF INOCULATION

Few authors [review:. 17] provide us with the reason why they chose a given cell site for the inoculation of cells in tumorigenicity tests. The following recent reports demonstrate the necessity of taking the inoculation site into account in the analysis of invasion and metastasis. ECA109 human oesophageal carcinoma cells produced non-invasive tumours after s.c. injection into nu/nu mice, whereas tumours invading most abdominal organs were found after i.p. injection [18]. MO_4 cell aggregates implanted s.c. either in the pinna or in the tail of syngeneic C_3H mice produced invasive tumours of comparable size at both sites but only tail tumours were metastatic [14]. Site-dependence of metastasis has also been observed with human colorectal carcinoma cells in nude mice [19] and with mammary carcinoma cells in syngeneic mice [20].

Table 1 shows data about tail and flank tumours from our recent experiments on the role of oncogenes in invasion and metastasis [2, 5 & unpublished results]. The overall frequency of metastasis from tail tumours is ~2-fold higher than that from flank tumours. Cell types that produce metastatic flank tumours invariably give rise to metastases from tail tumours. In 5 cell types there was a clear-cut difference between flank and tail tumours. Flank tumours were not metastatic, tail tumours were always metastatic.

In the experiments shown in Table 1, the inocula giving rise to tail rather than flank tumours are relatively fewer, leading to longer latency periods and longer periods of observation for tail tumours. Nevertheless, these data confirmed that s.c. implantation in the tail is a specific test for metastasis. We do not know whether the tail environment favours the proliferation of a small pre-existing metastatic subpopulation (selection) or non-metastatic cells become metastatic under the influence of tail factors (adaptation).

Table 1. Frequency of metastasis (as no. of tumour-bearing rats with metastasis/no. of rats examined) from flank tumours as compared with tail tumours in F.344 Fischer rats. The cited refs. describe the origin of the cell lines.

Cell type: code [& ref.]	from flank tumours	from tail tumours
PyT21 [21]	0/3	0/2
Rat1W [22]	3/4	6/7
Rat1pEJ6.6 [23]	0/2	2/2
Rat2pT24B4 [5]	0/6	4/6
REFpEJmcyN7 [23]	10/11	3/3
Rat2TD2A [5]	0/4	4/4[+]
BPV3 [24]	2/4	3/3
BPV6 [24]	1/4	3/3
BPV1TD3 [24]	5/5	6/6
BPV3TD1 [24]	3/3	3/3
RVLC12 [25]	0/4	2/2
RVLC53 [25]	4/4	3/3
RV691-TD1 [25]	0/4	2/2
Total	28/58 (48%)	41/46 (89%)

[+] Cells aggregated onto a collagen sponge were used for implantation instead of a cellular aggregate.

SCORING INVASION

With 3-dimensional confrontations in organ culture, as with most assays *in vivo*, invasion has to be reconstructed from histological sections of tissues fixed after various time intervals. Since the critical activity of invasive cells is unknown, we have to score invasion on the basis of the results of this activity: occupation of the normal tissue, which is usually accompanied by degeneration of the normal tissue. In the chick-heart organ-culture assay (Scheme 1) the analysis of occupation and degeneration is facilitated by immuno-staining of the sections with an antiserum that recognizes the heart tissue but not the confronting cells [3].

To compare the invasiveness *in vitro* of various cell types and to examine the anti-invasive activity of drugs, a scale (grading from 0 to IV) was established [26] as follows.-

Grade 0.- Serial sections of the entire culture showed no confronting cells, as when these attached poorly to the pre-cultured heart fragments so that they were released into the culture medium, or they died under the culture conditions of the assay. With one cell-line family, derived from a mouse mammary carcinoma, grade 0 was scored in cultures fixed after 14 days. Surprisingly, a number of cultures fixed after 28 days showed invasion of the heart tissue by mouse cells. Immuno-

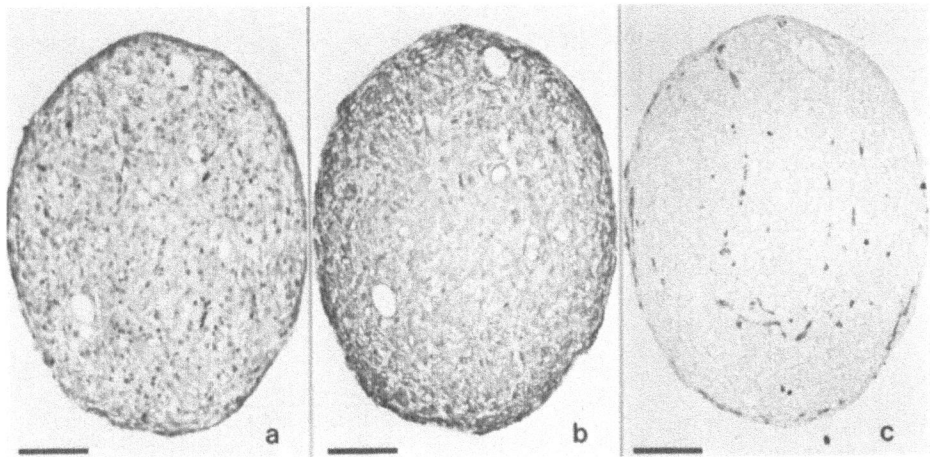

Fig. 1. Consecutive sections from a confrontation of mouse mammary tumour cells with chick-heart fragments in organ culture. Fixation after 14 days; staining with **(a)** haematoxylin-eosin, **(b)** an anti-serum against chick heart, and **(c)** an antiserum against mouse cells. Scale bar = 0.1 mm. (Unpublished work with A. Sonnenberg and J. Hilgers.)

staining of sections from the grade 0 cultures with an antiserum against mouse cells revealed within the heart a few mouse cells that were not manifest in sections stained with haematoxylin-eosin or with an anti-chick heart antiserum (Fig. 1). This observation demonstrated the need, at least in grade 0 cultures, for a double immuno-staining either on the same or on consecutive sections. Poor attachment of confronting cells to the heart tissue can be overcome by prolonged incubation on semi-solid medium before transfer of the confronting pair to fluid medium. For only a few cell types tried in this assay has non-survival hampered testing.

Grade I.- Confronting cells surrounded the heart fragment, and were separated from the core of muscle cells by a connective-tissue layer which, pre-confrontation but after pre-culturing, was manifest at the fragment's periphery and probably represented some kind of wound healing. Grade I cultures are rare. So far, they have been seen with cells from the canine kidney-cell line MDCK [27] and from the human mammary carcinoma cell line MCF-7 (unpublished results), and after treatment with the anti-invasive flavonoid (+)-catechin [26]. We interpret grade I cultures as modelling the natural situation of epithelia in the case of these cells, and for (+)-catechin as a drug-mediated stabilization of the fragment's peripheral layers.

Grade II.- As reviewed [12], when two normal tissues are confronted with each other in organ culture either they fuse at the site of attachment, or one of the confronting partners surrounds the other. Since in our assay for invasion one of the partners is always the

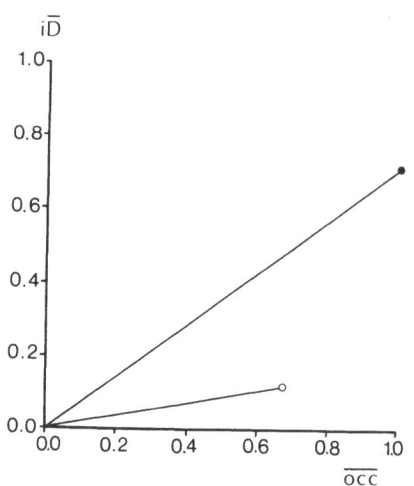

Fig. 2. Cell occupation (OCC)
and degeneration (iD) of chick
heart in organ culture:
●, MO$_4$ mouse cells; o, *ras*-
transfected Fischer rat cells.
Analytical details in ref. [28].

same (chick heart), we have distinguished between grades IIa and
IIb. In neither is the outer layer of chick connective tissue still
visible. Grade IIa indicates that the confronting cells have formed
a cap at the pole of attachment or have surrounded the cardiac muscle.
Grade IIb indicates that the cardiac muscle has surrounded the nodule
of confronting cells with a minimal area of contact (circular on
sections) between both partners. Grade IIa and IIb cultures were
observed when non-invasive cells were confronted with heart fragments.
With both grades the disposition of the partners is acquired during
the first 2 days and remains stable during further culture. Treatment
with most anti-invasive agents resulted in Grade IIa cultures.

Grades III and IV.- Here the confronting cells have occupied $<\frac{1}{2}$
(grade III) or $>\frac{1}{2}$ (grade IV) of the cardiac muscle; with most such
cell types there was accompanying degeneration and disappearance
of the heart tissue. Most of the invasive cells produced grade III
after 4-7 days, progressing towards grade IV during further incubation, and thus fulfilling all criteria of invasiveness. With some
cell types progression from grade III to IV was not demonstrated;
these were anchorage-dependent for their growth and produced invasive
tumours in syngeneic animals with long latency periods. Improved
organ-culture conditions are needed to model more closely the behaviour
in vivo of the latter cell types.

Analysis of invasion through this grading, although subjective,
gave reproducible results and showed a good correlation between scores
obtained by independent observers on blindly coded preparations.

As Fig. 2 illustrates, using a computer-assisted image-analysis
program, De Neve et al. [28] have quantitated separately two aspects
of invasion in the organ culture assay. Occupation of heart tissue
by MO$_4$ cells was scored on the basis of distances between MO$_4$ areas

Fig. 3. Invasion rate *vs.*
time after implanting mouse
cells under the renal
capsule of syngeneic mice:
●, MO_4 mouse cells;
□, Rac5E mammary cells;
△, B16B16 melanoma
isolates. Best fit curves;
each point represents one
tumour.

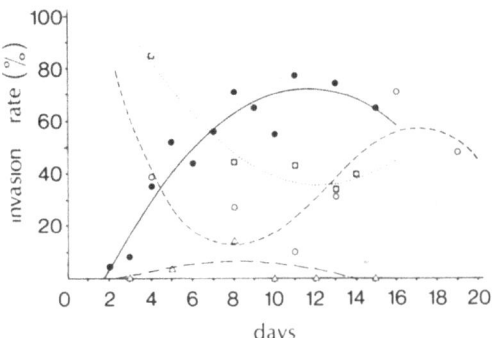

separated by heart tissue.   Degeneration was scored as a decrease
in the amount of immunoreactive heart material. Complex combinations
of occupation and degeneration showed a new qualitative aspect of
invasion that was different in various invasive cell types (exemplified
in Fig. 2).

Numerical assessment of invasion *in vivo* can be done with the
subrenal capsule assay (E. Boghaert, W. Distelmans, R. Van Ginckel
& M. Mareel, in preparation).   In this assay, cells aggregated to
sponges as described above are transplanted under the renal capsule
of male rats or mice, and then the animals are killed at regular
intervals.   The kidneys are hemisectioned, and the total thickness
of the tumour (T) as well as its invasive part (I = distance between
the original site of contact and the deepest point of invasion) are
measured.   Cell types are characterized by an invasion rate (I_r = T/I)
plotted against time after implantation (exemplified in Fig. 3).

Acknowledgements

This work was supported by grants from the Kankerfonds van de
A.S.L.K. and from the F.G.W.O. (20.093, 39.000.983 & 3.000.584),
Belgium.   P.C. and L.M. are beneficiaries of a Ph.D. grant from the
I.W.O.N.L., Belgium.

References

1. Easty, G.C. & Easty, D.M. (1984) in *Invasion* (Mareel, M.M. &
    Calman, K.C., eds.), Oxford University Press, Oxford, pp. 24–62.
2. Mareel, M.M. & Van Roy, F. (1986) *Anticancer Res. 6*, 419–436.
3. Mareel, M.M., De Bruyne, G.K., Vandesande, F. & Dragonetti, C.
    (1981) *Invasion Metastasis 1*, 195–204.
4. Waller, C.A., Braun, M. & Schirrmacher, V. (1986) *Clin. Exp.
    Metastasis 4*, 73–89.
5. Van Roy, F.M., Messiaen, L., Liebaut, G., Gao, J.,
    Dragonetti, C.H., Fiers, W.C. & Mareel, M.M. (1986) *Cancer
    Res. 46*, 4787–4795.

6. De Mets, M. & Mareel, M.M. (1984) *Int. Rev. Cytol. 90*, 125-167.

7. Geuens, G.M.A., Nuydens, R.N., Willebrords, R.E., Van de Veire, R.M.L., Gossens, F., Dragonetti, C.H., Mareel, M.M. & De Brabander, M.J. (1985) *Cancer Res. 45*, 733-742.

8. Bracke, M.E., Van Cauwenberge, R.M-L., Mareel, M.M., Castronovo,V. & Foidart, J-M. (1986) in *Plant Flavonoids in Biology and Medicine: Biochemical, Pharmacological and Structure-Activity Relationships* (Middleton, E., Jr., ed.), Liss, New York, pp.441-444.

9. Mareel, M.M., Dragonetti, C.H., Hooghe, R.J. & Bruyneel, E.A. (1985) *Clin. Exp. Metastasis 3*, 197-207.

10. Mareel, M.M., Kint, J. & Meyvisch, C. (1979) *Virchows Arch. B Cell Path. 30*, 95-111.

11. Mareel, M.M. (1982) in *Tumor Invasion and Metastasis* (Liotta, L.A. & Hart, I.R., eds.), Martinus Nijhoff, The Hague, pp. 207-230.

12. Mareel, M.M. (1983) *Cancer Metastasis Rev. 2*, 201-218.

13. Attia, M.A.M. & Weiss, D.W. (1966) *Cancer Res. 26*, 1787-1800.

14. Meyvisch, C. & Mareel, M.M. (1985) *Invasion Metastasis 5*, 185-192.

15. Kraemer, P.M., Travis, G.L., Saunders, G.C., Ray, F.A., Stevenson, A.P., Bame, K. & Cramm, L.S. (1984) in *Immune-Deficient Animals* (Sordat, B., ed.), Karger, Basel, pp. 214-219.

16. Kerkof, P.R. (1982) *J. Tissue Cult. Meth. 7*, 23-26.

17. Meyvisch, C. (1983) *Cancer Metastasis Rev. 2*, 295-306.

18. Gao, J., Xue, K., Li, B., Dong, H., Wei, L., Zhang, Z., Tian, S. & Xu, Z. (1984) *Clin. Exp. Metastasis 2*, 205-212.

19. Giavazzi, R., Campbell, D.E., Jessup, J.M., Cleary, K. & Fidler, I.J. (1986) *Cancer Res. 46*, 1928-1933.

20. Unemori, E.N., Ways, N. & Pitelka, D.R. (1984) *Br. J. Cancer 49*, 603-614.

21. Seif, R. & Cuzin, F. (1977) *J. Virol. 24*, 721-728.

22. Freeman, A.E., Gilden, R.V., Vernon, M.L., Wolford, R.G., Hugunin, P.E. & Huebner, R.J. (1973) *Proc. Nat. Acad. Sci. 70*, 2415-2419.

23. Land, H., Parada, L.F. & Weinberg, R.A. (1983) *Nature 304*, 596-602.

24. Grisoni, M., Meneguzzi, G., de Lapeyrière, O., Binétruy, B., Rassoulzadegan, M. & Cuzin, F. (1984) *Virology 135*, 406-416.

25. Cuzin, F., Meneguzzi, G., Binétruy, B., Cerni, C., Connan, G., *Molecular and Clinical Aspects* (Howley, P.M. & Broker, T.R., eds.), Liss, New York, pp. 473-486.

26. Bracke, M.E. & Van Cauwenberghe, R.M-L. (1984) *Clin. Exp. Metastasis 2*, 161-170.

27. Schroyens, W., Bruyneel, R., Tchao, R., Leighton, J., Dragonetti, C. & Mareel, M.M. (1984) *Invasion Metastasis 4*, 160-170.

28. De Neve, W.J., Storme, G.A., De Bruyne, G.K. & Mareel, M.M. (1985) *Clin. Exp. Metastasis 3*, 87-101.

#NC(E)

NOTES and COMMENTS related to

STUDIES FOCUSED ON ENDOCYTOSIS, PROTEASES OR INVASIVENESS

Comments on #**E-1**: P.N. Strong & W.H. Evans – UPTAKE OF A TOXIN
#**E-2**: P. Bohley et al. – LYSOSOMAL PROTEOLYSIS [*]
and #**E-4**: T. Berg & co-authors – ENDOCYTOSIS BY LIVER CELLS

H. Sjöström/W.H. Evans.– Some workers have claimed 2-way transport between Golgi and endosomes. Pulse-chase experiments with an endosome preparation could help in examining whether newly synthesized material passes the endosome compartment. **T. Berg asked Bohley** whether, in view of his biochemical evidence that exogenous proteins are seques-tered mainly in lysosomes, the content of such proteins in lysosomal fractions is increased by inhibitors of lysosomal proteases. **Reply.–** This indeed happens, as shown by adding leupeptin.

J.K. McDonald (to Bohley).– In view of your surprising finding of more cathepsin L in hepatocytes than in Kupffer cells, the cell-type distribution of other cysteine proteases, e.g. cathepsins B and H, would be of interest. **Reply.–** Kupffer-cell enrichment is known for B (Katunuma & co-workers) and D (s.a. ~30-fold higher than for paren-chymal cells); hence D seems to be the proteinase mainly responsible for heterophagy, and L for autophagy.

Remark by J.K. McDonald to T. Berg.– Your unexplained observation of a Ca^{2+} requirement for endocytosis could relate to a conceivable role of a p.m. transglutaminase, which has been reported to be involved in the binding and uptake of α_2M-protease complexes by (I believe) pulmonary macrophages and to be Ca^{2+}-dependent.

Comment on #**E-5**: P. Friberger – CHROMOGENIC PEPTIDE SUBSTRATES

J.K. McDonald asked Friberger whether methods or special sub-strates have been developed to assay plasma proteases such as plasmin or plasminogen activator in the presence of the plasma protease inhibi-tors. **Reply.–** Indeed one has to consider the effect of inhibitors in the assay of enzymes and proenzymes. Success with plasminogen has come from use of a technique to activate it with streptokinase

[*] Relevant to #E-2: *a citation on p. 393*

and obtain an enzyme which is not inhibited by plasmin or other inhibitors in plasma.   For tissue plasminogen activator, the plasma or
blood sample has to be treated with 1 M acetic acid, pH 3.9, to destroy
the inhibitors.

Comment on #**E-7**: K-J. Andersen et al. - BRAIN TUMOUR-CELL INVASION
        and #**E-8** M. Mareel et al. - TUMOUR-INVASION MECHANISMS

   Remarks by S. Zucker.- It might be inferred from the assay systems
in the two presentations that similar proteolytic enzymes mediate
neoplastic destruction of host connective tissue and normal host
cells. Alternatively, through a mechanism independent of proteolytic
activity, cancer cells might, as suggested by our investigations,
produce a cytolytic factor which although needing calcium is not
a protease, but is similar to complement in activity as has been
described for lymphocyte 'cytolysin-perforin'.

Supplementary refs. contributed by Senior Editor

Various observations on proteases or connective tissue degradation

   Tissue and organelle localization of the enzymes featured in
a thorough review of control of proteolysis.- Holzer, H. & Heinrich, P.C.
(1980) *Annu. Rev. Biochem. 49*, 63-91; few derangements are mentioned.
To help searches for neuropeptide-specific peptidases, a useful survey
has been published on the well characterized microvillar peptidases
found in intestine and especially in kidney.- Turner, A.J., Matsas, R.
& Kenny., A.J. (1985) *Biochem. Pharmacol. 34*, 1347-1356.

   Endogenous protein degradation in resting hepatic cells - largely
lysosomal - has been studied by lysosomal morphometry during liver
perfusion with the proteolytic inhibitor leupeptin present, and by
study of subcellular fractions including lysosomes plus functional
inhibitors.- Glaumann, H., Ahlberg, J., Berkenstam, M., Falk, M. &
Henell, F. (1985) *Biochem. Soc. Trans. 13*, 1010-1012.
   The effect of swainsonine on glycoprotein degradation by lysosomes
has been examined.- Segal, H.L. & Winkler, J.R. (1985) *Prog. Clin.
Biol. Res. 180*, 491-494.

   Assessment of collagen breakdown, especially in arthritic
patients, is the focus of studies involving the Strangeways Laboratory
in relation to ubiquitous molecules: interleukin-1 [J.T. Dingle et
al.] and TIMP, viz. tissue inhibitor of metalloproteinases [G. Murphy,
J.J. Reynolds; see Docherty, A.J.P. et al. (1985) *Nature 318*, 66-69.]
   In a review (from the Kennedy Inst. of Rheumatology; with a
sketch of lysosome biochemistry and storage diseases), pre-1980 work
is considered to furnish circumstantial evidence for a major role
of extracellular lysosomal enzymes in connective tissue diseases.-
Bitensky, L. (1978) *J. Clin. Path. 31, Suppl. 12*, 105-116.

[CONTINUATION: p. 393

#NC(E)-1

A Note on

PLASMA MEMBRANE DISTRIBUTION OF METALLOPROTEINASES AND CYSTEINE PROTEINASES IN METASTATIC HUMAN CANCER CELLS

Stanley Zucker, Rita M. Lysik, Janine Wieman and Bernard Lane

Departments of Medicine and Pathology
Veterans Administration Medical Center
Northport, NY 11768 and State University of New York
at Stony Brook, Stony Brook, NY 11794, U.S.A.

There has been considerable interest in the mechanism by which cancer cells invade and destroy normal host tissues and cells during the process of cancer invasion [1, 2]. Initial studies with tumours were based on the observation that cancer cells release plasminogen activator, a serine proteinase that is responsible for the initiation of fibrin digestion [3]. More recently, it has been demonstrated that cancer cells release other types of proteinase, besides hydrolases other than proteinases, that appear to be responsible for the digestion of connective tissue structural proteins, e.g. interstitial and basement-membrane collagens, fibronectin and laminin, and proteo-glycan-like ground substances [1, 2] (cf. J.K. McDonald et al., #E-6, this vol.-*Ed.*). Fig. 1 provides an overview of the steps postulated to be requisite for cancer cells to invade and destroy normal tissue. They must be able not only to digest host tissue but also to migrate from the primary tumour mass to penetrate capillaries and venules, thereby gaining access to the bloodstream.

Our studies concern whether the crucial proteinases required for cancer invasion are bound to the surface of cancer cells or are secreted into the surrounding environment [4-7]. We have employed subcellular fractionation techniques to validate the localization of proteinases within the p.m.[*] Recently we have focused our attention on highly invasive and metastatic human pancreatic ductal and small-cell lung cancers. Here we describe experiments showing that the p.m.'s of these cancer cells are highly enriched in proteolytic enzymes that we propose are responsible for the capacity of cancer cells to invade and destroy normal tissues.

---

[*]Abbreviations: p.m., plasma membrane; e.m., electron microscopy

CANCER CELL ACTIVITY   HOST INJURY AND RESPONSES

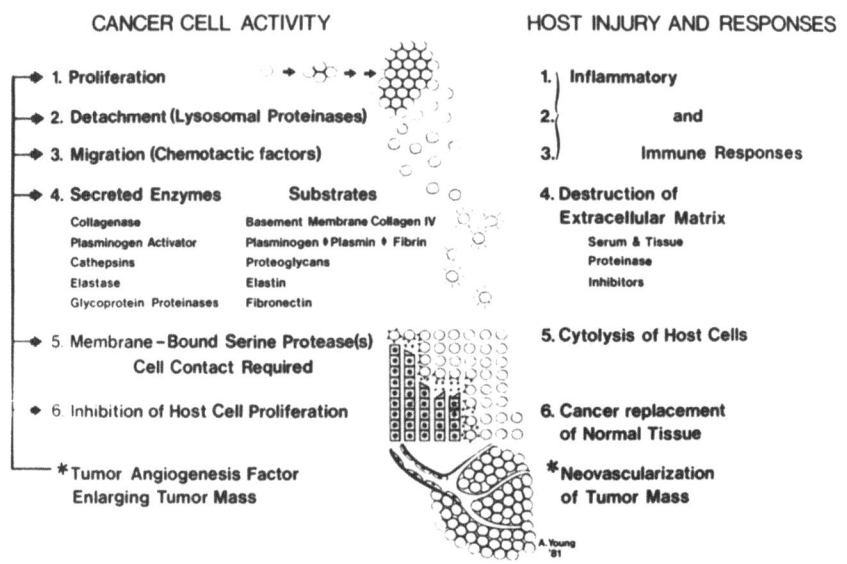

1. Proliferation   1. ⎫ Inflammatory

2. Detachment (Lysosomal Proteinases)   2. ⎬   and

3. Migration (Chemotactic factors)   3. ⎭   Immune Responses

4. Secreted Enzymes   Substrates   4. Destruction of
                                                        Extracellular Matrix
   Collagenase   Basement Membrane Collagen IV
   Plasminogen Activator Plasminogen ⬦ Plasmin ⬦ Fibrin   Serum & Tissue
   Cathepsins   Proteoglycans   Proteinase
   Elastase   Elastin   Inhibitors
   Glycoprotein Proteinases Fibronectin

5. Membrane–Bound Serine Protease(s)   5. Cytolysis of Host Cells
   Cell Contact Required

6. Inhibition of Host Cell Proliferation   6. Cancer replacement
                                                        of Normal Tissue

*Tumor Angiogenesis Factor   *Neovascularization
 Enlarging Tumor Mass   of Tumor Mass

                                                    A. Young
                                                    '81

Fig. 1. Postulated 6 sequential steps in cancer-cell invasion and destruction of normal tissues and in the host responses. The exact order of the stages is uncertain, and in fact may vary according to the cancer type and the organ invaded. The requisite Tumour Angio-genesis Factor participates at several stages *(arrowed)*. Although the secreted enzymes are listed separately from membrane-bound proteinases, there is evidence to suggest that secreted plasminogen activator, collagenase, and cathepsin B have their origins in the p.m. For further details, see refs. [1, 2, 5 & 6].

PREPARATION AND SUBCELLULAR FRACTIONATION OF CANCER CELL LINES

The human NCIH69 and NCIH82 small-cell lung cancer lines were generously provided by Dr. J.D. Minna and propagated *in vitro* as floating aggregates of cells [8]. The human RWP-1 pancreatic cancer cell line was kindly provided by Dr. D.L. Dexter and propagated as an adherent cell line [5]. Cancer cells (10^7/flask) propagating *in vitro* in flasks or *in vivo* in tumour masses in nude mice were collected non-enzymatically, re-suspended in cold, hypotonic lysis buffer containing 25 mM sucrose, 5 mM $CaCl_2$ and 5 mM Tris-HCl, pH 7.4. For nitrogen cavitation they were placed in a Parr bomb which was exposed to 500 psi (lung) or 200 psi (pancreatic) of N_2 at 4° for 30 min. The homogenate was centrifuged at 770 **g** for 15 min at 4°. The nuclei-enriched pellet was discarded and the supernatant was centrifuged at 50,000 **g** for 20 min, resulting in a pellet enriched in cell organelles and a supernatant containing soluble cytosol proteins. The re-suspended pellet was layered onto a 20-50% (w/v) discontinuous sucrose density gradient and centrifuged at 100,000 **g** for 3 h; visible membrane bands were removed, washed by dilution and centrifugation, and examined by marker-enzyme assay and by e.m. [4, 5]. Band 1, overlying the 25% sucrose layer, showed the greatest enrichment in 5'-nucleotidase (16- to 18-fold) and by e.m. showed abundant round profiles consistent with p.m. [5].

METHODOLOGY FOR METALLOPROTEINASES AND MEMBRANE PROTEINS

From the proteins in the tumour cell cytosol, metalloproteinases were isolated by sequential steps: 25-60% (w/v) ammonium sulphate precipitation; anion-exchange chromatography on a Mono Q H5/5 column (attached to a Pharmacia FPLC system); Superose-12 gel filtration chromatography (FPLC - Pharmacia); and zinc-chelated Sepharose column chromatography as we recently described [9]. After each chromatography step the fractions rich in collagenase-gelatinase activity were combined, dialyzed against sodium cacodylate buffer, pH 7.2, and concentrated by ultrafiltration. Apparent mol. wts. were estimated using known mol. wt. standards. A discontinuous system for non-reduced SDS-PAGE was employed [10].

Degradation of native collagen and heat-treated collagen (gelatin) was measured using acid-soluble lathyritic rat-skin type I [^{3}H-*methyl*]- collagen labelled *in vitro* with [^{3}H]formaldehyde [5]. Fluorograms showing the digestion of this substrate by tumour subcellular fractions were prepared by the method of Sodek et al. [11]. Cysteine proteinase activity was determined using benzoyloxcarboxyl-phenylalanyl-L-arginyl-7(4-methyl)-coumarylamide, which is degraded by cathepsins B and L as previously described [4]. Enzyme inhibitor assays [4] entailed a 2-h pre-incubation with drugs.

Membrane proteins.- To remove non-integral membrane proteins, crude tumour-cell membranes derived by pelleting of the post-nuclear supernatant at 100,000 **g** for 2 h were repeatedly washed with 10 mM Hepes-buffered NaCl (150 mM), pH 7.4, and then kept in 2 M KCl or buffer (as control) for 30 min at 4°, finally with washing steps (buffer or KCl) entailing centrifugation at 50,000 **g** for 20 min. Integral membrane proteins were solubilized by treatment with 1% n-octylglucoside (non-ionic detergent) for 1 h. The resulting pellets and washes (after dialysis against Hepes buffer to remove KCl and detergent) were then tested for proteinase activity. Metalloproteinase was associated with solubilized integral membrane proteins.

Partial purification of collagenase-gelatinase.- Starting with lung-cancer cytosol, 900-fold enrichment was achieved (Fig. 3) by ammonium sulphate precipitation (to 60%) and chromatography (zinc-chelated column, anion-exchange, then gel filtration). The partially purified enzyme, of kM_r 59 by SDS-PAGE, demonstrated both collagenolytic and gelatinolytic activity and was inhibited by metal chelators and diluted (× 10) human serum.

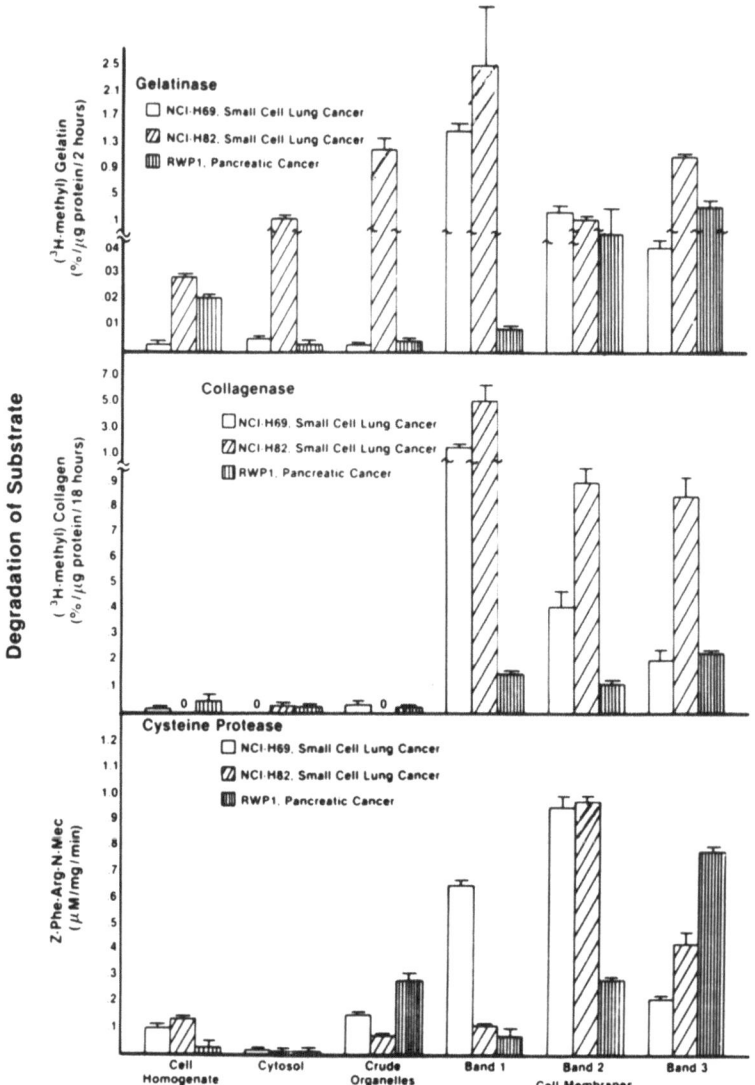

Fig. 2. Comparison of gelatinolytic, collagenolytic (trypsin-activated) and Cys proteinase activities in small-cell lung cancers and pancreatic cancer: means ±S.D. for subcellular fractions. Values are specific activities, expressed as % degradation of 1 µg of substrate or, for Cys protease, µM degraded.

See text for pre-activation (DTT and EDTA) and substrate in the Cys proteinase assay. and for the organelle bands. There was significant enrichment ($p < 0.05$) of Cys proteinase activity (compared with homogenate) in the crude organelles[†] and, along with collagenolytic and gelatinolytic activities, in membrane bands 1-3; the latter activities were most enriched in p.m. bands 1 & 2 with the lung cell lines and in the microvilli-containing band 3 with the pancreatic cells.

[†50,000 **g** pellet (see text)

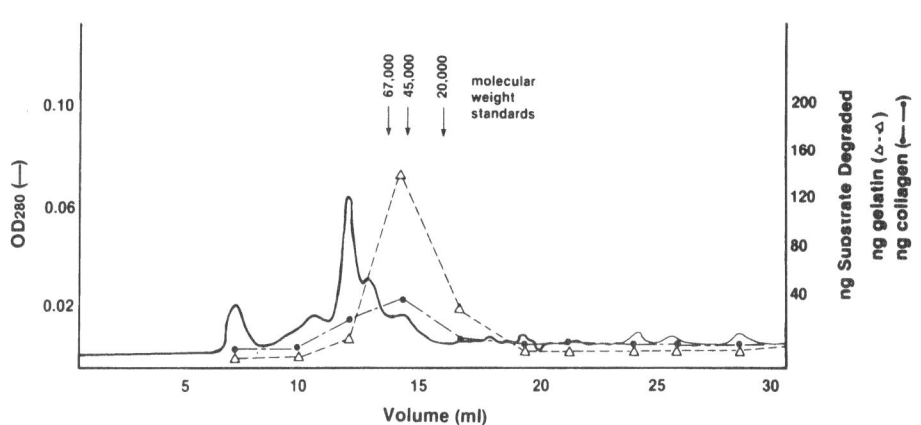

Fig. 3. Purification of pancreatic cancer (RWP-1) metalloprotein-ases: pattern from a gel filtration column (Pharmacia Superose–12HR) loaded with pooled cytosol proteins from anion–exchange (Mono Q) separation.   Fractions (1.5 ml) were pooled for assay.   Mol. wt. calibration was with BSA, ovalbumin and soybean trypsin inhibitor.

DISCUSSION AND CONCLUSIONS

We have shown that: (1) human lung and pancreatic cancer-cell p.m.'s contain proteinases with collagenolytic, gelatinolytic and cathepsin B-like activity; (2) metalloproteinases are integral p.m. proteins; (3) different types of human cancer cells differ in the localization of proteinases in the p.m. domains; and (4) collagen-ase-gelatinase purification from cancer cytosol gave a protein of M_r 59,000.   Concerning (2), earlier studies [12, 13] had not sought to determine whether the metalloproteinases isolated from tumour cell extracts resided in a subcellular organelle or the cytosol; our p.m. attribution is supported by morphological and biochemical evidence of connective tissue degradation occurring primarily at the tumour-cell surface, especially at focal sites of pseudopodial projections [14-16].

Our findings (2) and (4) tally with our recent localization and/or purification of similar metalloproteinases in mouse malignant melanoma cells and rat breast-cancer cells [5, 17, 18].   The data from different tumour-cell lines and species suggest that these proteinases are a common component of invasive cancer cells, and are not limited to specific tumour types.   Of pertinence to cathepsin B being an extrinsic membrane protein in human cancer cells is its presence in mouse-melanoma p.m. too [17, 18].   Our observation that the partially purified human-lung metalloproteinase has both collageno-lytic and gelatinolytic activity is contrary to most studies of collagen-ases [2] but not to those on a mouse melanoma metalloproteinase [4]. In summary, then, with membrane extraction procedures we have been able to characterize a metalloproteinase and a cysteine proteinase and demonstrate localization on the surface of invasive cancer cells.

Acknowledgements

We thank Drs. Ramamurthy and Golub for help in the fluorographic studies, Dean Wilkie, Clara Krause and Hsi-Ming Lee for technical assistance, and Drs. Minna and Dexter for providing the human cancer cell lines. The research was supported by Merit Review funds from the Veterans Administration.

References

1. Liotta, L.A., Rao, C.N. & Barsky, S.H. (1983) *Lab. Invest. 49*, 636-649.
2. Nicolson, G.L. (1982) *Biochim. Biophys. Acta 695*, 113-176.
3. Unkeless, J.C., Tobia, A., Ossowski, L., Quigley, J.R., Rifkin, D.R. & Reich, E. (1973) *J. Exp. Med. 37*, 85-111.
4. Zucker, S., Lysik, R.M., Ramamurthy, N.S., Golub, L.M., Wieman, J.M. & Wilkie, J.M. (1985) *J. Nat. Cancer Inst. 75*, 517-525.
5. Zucker, S., Lysik, R.M., Wieman, J.M., Wilkie, J.M. & Lane, B. (1985) *Cancer Res. 45*, 6168-6178.
6. DiStefano, J.F., Beck, G., Lane, B. & Zucker, S. (1982) *Cancer Res. 42*, 207-218.
7. Zucker, S., Beck, G., DiStefano, J.F. & Lysik, R.M. (1985) *Br. J. Cancer 52*, 223-232.
8. Carney, D.N., Gazdar, A.F., Bepler, G., Guccion, J.G., Marangos, P.J., Moody, P.J., Zweig, M.H. & Minna, J.D. (1985) *Cancer Res. 45*, 2913-2923.
9. Zucker, S., Wieman, J.M., Lysik, R.M., Wilkie, J.M., Ramamurthy, N.S. & Lane, B. (1987) *Biochim. Biophys. Acta*, in press.
10. Laemmli, U.K. (1970) *Nature 227*, 680-695.
11. Sodek, J., Hurum, S. & Feng, J. (1981) *J. Peridontal Res. 1*, 425-433.
12. Tane, N., Hashimoto, K., Kanzaki, T. & Ohyama, H. (1978) *J. Biochem. (Tokyo) 84*, 1171-1176.
13. Wirl, G. & Frick, J. (1979) *Urological Res. 7*, 103-108.
14. Kramer, R.H., Bensch, K.G. & Wong, J. (1986) *Cancer Res. 46*, 1980-1989.
15. Yee, C. & Shiu, R.P. (1986) *Cancer Res. 46*, 1835-1839.
16. Jones, P.A. & DeClerk, Y.A. (1980) *Cancer Res. 40*, 3222-3227.
17. Zucker, S., Wieman, J.M., Ramamurthy, N.S., Lysik, R.M., Wilkie, J.M., Liotta, L.A. & Golub, L. (1985) *Clin. Res. 33*, 461.
18. Sloane, B.F., Rozhin, J., Johnson, K., Taylor, H., Crissman, J.D. & Honn, K.V. (1986) *Proc. Nat. Acad. Sci. 83*, 2483-2487.

#NC(E)-2

A Note on

PROTEASES AND OXIDANTS IN INJURY TO ALVEOLAR EPITHELIAL CELLS CAUSED BY LUNG-DERIVED NEUTROPHILS *IN VITRO*

Kenneth Donaldson, Joan Slight and Robert E. Bolton

Institute of Occupational Medicine
Roxburgh Place, Edinburgh EH8 9SU, U.K.

This contribution concerns leucocyte-mediated injury in the alveolar region of the lung. Inflammatory leucocytes accumulate in this region in a range of chronic and acute conditions [1]. Whilst undoubtedly performing a defensive function in normal inflammatory responses within the lung, persistent or extensive accumulation of activated leucocytes within the fragile structure of the lung parenchyma is associated with damage to the alveolar septa which may be mediated by leucocyte products [2]. The outcome of this damage may be either fibroplasia within the septal interstitium or destruction of septa [1].

We set out to develop a model for detecting injurious effects of inflammatory leucocytes on one element of the alveolar septa, the alveolar epithelial cells. Leucocytes which accumulate in the alveolar spaces during alveolitis [1] are in close contact with these cells, and we sought to reproduce this situation in the assay. Here we describe preliminary results from experiments aimed at elucidating the mechanism of inflammatory cell-mediated injury to epithelial cells *in vitro*. The benefits of the assay system are also discussed.

THE ASSAY SYSTEM

The target cells were a Type II alveolar epithelial cell line A549 [3] which we ascertained by e.m. to be virus- and mycoplasma-free. The cells were cultured under standard conditions in Eagle's Minimum Essential Medium + 10% Foetal calf serum. For assay, 5×10^4 cells were seeded into microtitre plate wells with 100 µl of complete medium, and incubated overnight in the presence of 74 KBq of ^{51}Cr. The cells were washed to remove free ^{51}Cr, and effector cells were added, viz. two different populations of cells obtained by bronchoalveolar lavage of inbred, SPF, PVG rats: (i) control cells (98% alveolar macrophages);

Fig. 1. Dose-dependent detachment
of A549 alveolar epithelial cells
caused by co-culture with control
and inflammatory cells *(see text)*.
The latter caused detachment (P<0.001)
at ratios other than the lowest (*vs.*
no effector); none occurred with
control cells at any ratio.
Neither cell population caused
significant lysis.
Vertical bars denote S.E.M.'s.
Data from triplicate wells in three
separate experiments.

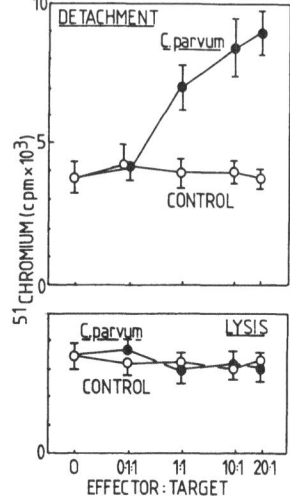

(ii) 16 h after instillation of 1.4 mg of heat-killed *Corynebacterium parvum* into the lung (~90% neutrophils). Effectors and targets were co-cultured for 4 h at effector/target ratios of 0.1 to 20. Both lytic injury and non-lethal detachment injury were assessed [4].

OBSERVATIONS ON EPITHELIAL CELL DETACHMENT

Depending on the effector/target ratio, inflammatory leucocytes differed significantly from control bronchoalveolar cells in producing detachment of epithelial cells *in vitro* (Fig. 1). Neither of these populations caused lytic damage to the epithelial cells.

The most obvious candidates for causing epithelial detachment amongst the leucocyte products are oxidants and proteases, both of which have already been implicated in tissue damage [2]. We therefore used exogenous proteases and oxidants for attempted mimicking of the detachment injury caused by inflammatory leucocytes.

Exogenous proteases.- The four proteases used all caused statistically significant dose-dependent detachment (Fig. 2), much more so with trypsin or elastase than with pronase or microbial collagenase. None of the proteases caused lytic damage.

Hydrogen peroxide or superoxide anion.- Dose-dependent detachment of significant extent resulted from H_2O_2 addition (Fig. 2). The decrease in detachment at the highest concentration (500 µM) was due to extensive lysis which decreased the available detachable cells. Increased detachment (215% of control; P<0.001) likewise occurred with acetaldehyde/xanthine oxidase, a system which generated superoxide anion (~30 µmol). However, we later discovered that the system also generated 20 µmol of H_2O_2 which itself caused some detachment (Fig. 2).

Fig. 2. Dose-dependent detachment
of epithelial cells caused by
exogenous proteases or H_2O_2. All
data normalized to control (100%);
error bars omitted for clarity, but
S.E.'s were all <10% of the mean.
All treatments caused significant
detachment (P <0.05 to P <0.001).
No lysis was caused by any treatment except for 500 µM H_2O_2.

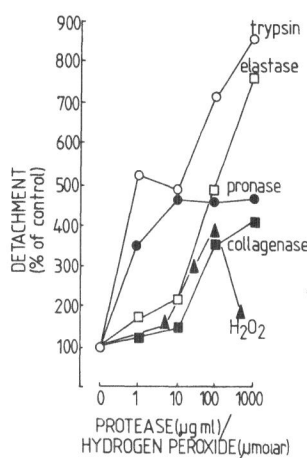

Use of anti-oxidants or anti-proteases.- The use of specific
inhibitors gave the following results for % inhibition of the detachment
caused by inflammatory bronchoalveolar cells: αl-protease inhibitor,
100% (P <0.001); catalase,   17.4%; superoxide   dismutase, 67.8%
(P <0.001).

Conclusion.- Evidently the *in vitro* detachment injury caused
to alveolar epithelial cells by leucocytes from the bronchoalveolar
space of acutely inflamed lung is mediated both by leucocytic proteases
and by superoxide anion.   These agents could also damage the other
cellular or matrix elements of the alveolar septum, and subsequent
re-modelling may lead to alveolar fibrosis or destruction of septa.

METHODOLOGICAL CONSIDERATIONS

The co-culture model used has several points of advantage for
studies of leucocyte-mediated cellular injury.-
1. **Sub-lethal injury *in vitro*** - The loss of integrity of cellular
barriers such as epithelium and endothelium are important factors
in many pathobiological processes, but cell killing does not have
to occur for the integrity of such a barrier to be threatened.   Sub-
lethal injury, detectable as loss of attachment to substratum, is
detectable in the assay described here, while lytic injury can also
be measured if it occurs.
2. **Role of cell attachment factors.-** All cells, and   particularly
mesenchymal   and   epithelial   cells,   have important associations
with their surrounding extracellular matrix, and these may be of
particular importance during cell injury.   Coating of wells with
elements of the extracellular matrix before culturing cells on them
enables their role in any subsequent injury to be assessed.
3. **Triggering of leucocytes.-** Leucocytes exist in a range of
secretory states which can be altered by both exogenous agents and
endogenous signals generated during inflammation and immunity. These

triggers can be included in the assay to assess their significance for leucocyte-mediated injury during inflammation or immune responses.

4. **Mechanism of the injury.**- The system easily allows the inclusion of agents which block particular effector pathways (e.g. anti-oxidants or anti-proteases) so that the mechanism of leucocyte-mediated injury can be dissected.

Acknowledgement

The work was carried out with funding from the Colt Foundation.

References

1.   Gee, J.B.L. & Fick, R.B. (1980) *Thorax 35*, 18.
2.   Fantone, J.C. & Ward, P.A. (1984) *Am. Rev. Resp. Dis. 130*, 484-491.
3.   Lieber, M., Smith, B., Szakal, A., Nelson-Rees, W. & Todaro, G. (1976) *Int. J. Cancer 17*, 62-70.
4.   Ayars, G., Altman, L.C., Rosen, H. & Doyle, T. (1984) *Am. Rev. Resp. Dis. 130*, 964-973.

#NC(E)-3

A Note on

AUTOPHAGY AND PROTEIN DEGRADATION IN RAT HEPATOCYTES, STUDIED BY MEANS OF SUBCELLULAR FRACTIONATION IN NYCODENZ GRADIENTS

T. Berg, K.R. Norum and G.M. Kindberg

Institute for Nutrition Research
School of Medicine, University of Oslo
P.O. Box 1046, 0316 Oslo 3,   Norway

Cellular autophagy is inducible by reduction in nutrient supply or by the accumulation of macromolecules which are rendered ineffective or potentially toxic [1].   The process may be a response to cell derangements and may serve to degrade cytoplasmic constituents during remodelling of tissues.

Overt autophagy in rat hepatocytes is induced by 'nutritional step-down' conditions [2]. Autophagosomes formed initially (possibly from the e.r.) fuse with lysosomes which thereby are rendered more buoyant in gradients of Nycodenz (Nyegaard) [3].   We have attempted to analyze autophagic activity in cultured rat hepatocytes by sub-cellular fractionation in Nycodenz gradients.   Before the cells were prepared and seeded, the rats were injected i.v. with ^{125}I-tyramine cellobiose-asialoorosomucoid (TC-ASOM): labelled degradation products from the ^{125}I-TC serve as a marker for lysosomes, being trapped therein ([4], & #E-4, this vol.).

Liver parenchymal cells were prepared from male Wistar rats (200 g) by collagenase perfusion [5]. The cells were seeded (6-8 × 10^4 /cm^2) in Costar plastic flasks.   The complete culture medium used (0.15 ml/cm^2) was bicarbonate-buffered Dulbecco's modified Eagle medium (DMEM) supplemented with horse serum (2.5%), penicillin (100 U/ml), streptomycin (0.1 mg/ml), dexamethasone (0.25 µM) and insulin (400 nM). The flasks were gassed with 95% air/5% CO_2,   and kept overnight in culture prior to each experiment.

To induce autophagy the cells were incubated in a minimal salt medium (MSM; gassed as above): 7.0 g NaCl, 0.4 g KCl, 60 mg Na_2HPO_4. $2H_2O$, 47 mg KH_2PO_4, 0.2 g $MgSO_4.7H_2O$, 2.5 g $NaHCO_3$, 0.29 g $CaCl_2.2H_2O$

and water to 1000 ml; pH adjusted to 7.5 by NaOH addition. After repeated washes (0.9% NaCl, × 3; 0.25 M sucrose/1 mM Hepes/1 mM EDTA pH 7.3, × 2), the cells were removed using a rubber policeman, and ~5 × 10⁶ cells in 3 ml of the sucrose solution were homogenized (tightly fitting Dounce, 6 strokes).

The homogenate was centrifuged for 115 sec at 2000 g, and the pellet was re-suspended, re-homogenized (6 strokes) and re-centrifuged. From the combined supernatants, 2 ml was layered onto 10 ml (in a 14 ml tube) of a linear Nycodenz gradient, prepared by mixing 0.25 M sucrose solution with 35% (w/v) Nycodenz (dissolved in 3 mM KCl/0.3 mM (Ca,Na)-EDTA and 5 mM Tris-HCl, pH 7.5). After centrifuging in a Sorvall TST 41-14 rotor (85,000 g, 45 min), 16 fractions were collected, using Maxidenz for upwards displacement; their densities were obtained from the refractive index at 20° (n) by the formula [6] $d = (n \cdot 3.41) - 3.55$. The acid-soluble radioactivity (unprecipitated by TCA to 10% w/v; Kontron gamma counter) was the measure of degraded ¹²⁵I-TC-ASOM. Its distribution closely coincided with that of lysosomal markers, notably acid phosphatase which served to identify lysosomes in the gradient (Fig. 1, B).

CONVERSION OF DENSE INTO LIGHT LYSOSOMES DURING AUTOPHAGY

During the 3-h post-injection period the ¹²⁵I-TC-ASOM was completely degraded. The cells then isolated were pre-incubated overnight in DMEM; their whole lysosomal population was therefore assumed to have a homogeneous distribution of the ¹²⁵I-TC marker. After 3 washes with 0.9% NaCl, the cells were incubated in MSM. After 10-240 min (Fig. 1, A) the cells were removed, homogenized and fractionated in Nycodenz gradients. For control cells (0 and 240 min) the medium was DMEM. During incubation in MSM the cells gradually became more buoyant, and after 4 h the radioactivity and enzyme-activity peaks had moved from d = 1.14 to d = 1.08. Seemingly there is both a dense lysosomal population and a light one which is formed from it during the period of increased autophagy.

To test the possibility that the formation of light lysosomes was a sign of irreversible cell damage, we incubated cultured hepatocytes for 2 h in the MSM and then replaced this 'poor' medium by DMEM, with before-and-after determination of ¹²⁵I-TC distribution in Nycodenz gradients (Fig. 1, C). There was complete reversal of the change in density distribution of the lysosomal marker, which after the 2-h period resembled that for cells incubated continuously in DMEM.

INSULIN PREVENTS THE LYSOSOMAL DENSITY SHIFT

To determine the medium composition required to prevent the density shift, parallel incubations of cells were performed (Fig. 2) in five media: MSM, culture medium (DMEM), MSM with amino acids,

Fig. 1. Distributions (% of total in the gradient) of intralysosomal ^{125}I-TC (formed by ligand removal *in vivo*; see text) and lysosomal marker enzyme in hepatocytes incubated for up to 4 h, usually in a minimum salt medium (MSM), and fractionated in Nycodenz gradients.

A: acid-soluble radioactivity with different incubation times.
B: complete medium (DMEM; Δ,▲) for the 4-h incubation, compared with MSM (o,●), showing acid phosphatase and, giving very similar distributions, acid-soluble radioactivity (the MSM curve, o, being the same as in **A**).
C: acid-soluble radioactivity distributions which show that the lysosomal density changes induced by incubation in deficient medium are reversible:- after 2 h in MSM ('KRB'; distribution denoted □) either this incubation was continued to 4 h (o) or the cells were washed and DMEM added for the 2 h continuation (■); cells were also incubated with DMEM throughout the 4 h (●).

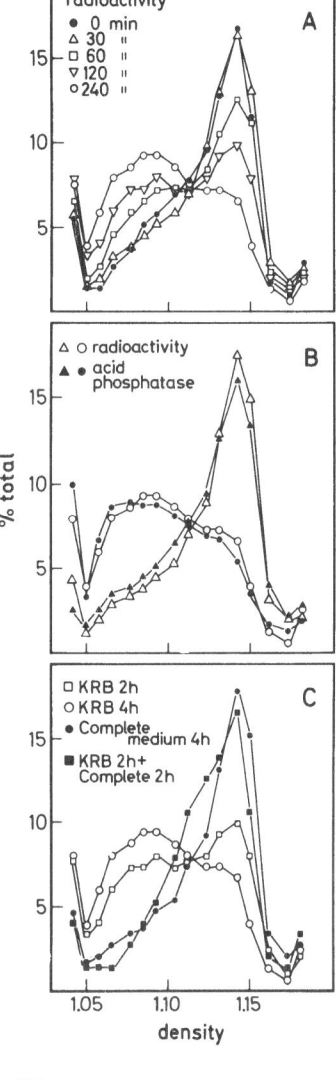

Fig. 2. Insulin effect on lysosomal distribution, in Nycodenz gradients, in hepatocytes initially containing ^{125}I-TC (as in Fig. 1) and incubated for 4 h. Acid phosphatase (not shown) and acid-soluble radioactivity curves were identical. The curve ● is for complete medium (DMEM). Otherwise the incubation was in MSM, alone (▲) or with supplements: Δ, insulin; □, amino acids; o, both insulin and amino acids.

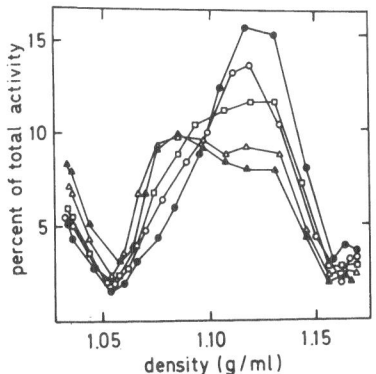

MSM with insulin and MSM with both amino acids and insulin. Since lysosomal marker-enzyme results were identical with those for acid-soluble radioactivity, only the latter is shown (Fig. 2) in the 4-h distributions. Evidently the addition of insulin to MSM supplemented with amino acids is sufficient to prevent the density shift completely. This insulin effect is compatible with the finding by Pfeifer [7] that insulin inhibits cellular autophagy.

Acknowledgements

This research was financially supported by the Norwegian Cancer Society, Anders Jahres Foundation and the Nansen Foundation.

References

1. Glaumann, H., Ericsson, J.L.E. & Marzella, L. (1981) *Int. Rev. Cytol.* *73*, 149–182.
2. Mortimore, G.E. (1982) *Nutrit. Rev. 40*, 1–91.
3. Seglen, P.O. & Solheim, A.E. (1985) *Exp. Cell Res. 157*, 550–555.
4. Berg, T., Kindberg, G.M., Ford, T. & Blomhoff, R. (1985) *Exp. Cell Res. 161*, 285–296.
5. Seglen, P.O. (1976) in *Methods in Cell Biology* (Prescott, D.M., ed.), Vol. 13, Academic Press, New York, pp. 29–83.
6. Rickwood, D. (1983) in *Iodinated Density Gradient Media* (Rickwood, D., ed.), IRL Press, Oxford, pp. 173–174.
7. Pfeifer, U. (1978) *J. Cell Biol. 78*, 152–167.

#NC(E)-4

A Note on

IMMUNOCHEMICAL TECHNIQUES IN THE STUDY OF
PROTEIN DEGRADATION WITH PARTICULAR REFERENCE TO COLLAGEN

Garry J. Rucklidge and Valerie Dean

Rowett Research Institute
Bucksburn, Aberdeen AB2 9SB, U.K.

Type I collagen is a highly organized rod-like molecule consisting of three α-chains arranged in a triple helix in the native form. Proteolytic degradation *in vivo* is initiated by collagenase cleaving the three α-chains at a precisely defined location which results in the helix unwinding. A combination of intra- and extra-cellular cross-links prevents the immediate removal of the disrupted helices. The possibility that antigenic determinants unique to denatured collagen may be exposed during proteolysis makes collagen a particularly suitable candidate for immunolocalization in tissue in a degraded form.

The routine production of polyclonal antibodies (Ab's) is performed by the intradermal injection of an emulsion of adjuvant and antigen of choice [1, 2]. The isolation of Ab's directed against the injected antigen is then achieved by chromatography with an immunoabsorbent [3]. The Ab's thus produced can be used for a variety of immunochemical procedures including immunocytochemical localization of antigens within tissues [4], estimation of antigen in body tissues or hydrolysates of tissue by ELISA or RIA [5, 6], immunodetection of Western-blotted proteins [7] and isolation of antigens from complex mixtures using the Ab covalently bound to a rigid support [8].

In this article we demonstrate how collagen antigens in a native form injected into sheep can elicit a response to both native and denatured states of the antigen, and describe the separation of these polyclonal Ab's and their application to ELISA and immunolocalization procedures, exemplified by kidney.

MATERIALS

Isolation of Type I collagen from skin.- The 0.5 M acetic acid soluble fraction from adult rat skin was subjected to differential salt precipitation at 4° [9]. A fraction which precipitated between 2.2 M and 2.5 M NaCl in 50 mM Tris–HCl (pH 7.4) was found to contain highly purified type I collagen as judged by SDS–PAGE. This material was dialyzed against water and freeze–dried.

   Injection schedule.- The type I collagen (4 mg) was suspended in phosphate–buffered saline (PBS; 1 ml) and emulsified with 1 vol. of Freund's complete adjuvant. Sheep were immunized intradermally at 2–week intervals and blood was collected 10 days after the third injection.

Preparation of antibodies by column chromatography

   Column preparation (2 forms).- (**i**) Rat skin type I collagen was dissolved (1 mg/ml) in 0.1 M acetic acid at 4° to preserve native structure. This material was dialyzed against 0.1 M $NaHCO_3$-0.2 M NaCl and then linked to Sepharose 4B–CNBr (Sigma Chemical Co.) for 16 h at 4° at a concentration of 10 mg collagen/ml settled gel. Before use the column was washed with 0.2 M glycine/HCl-0.5 M NaCl (pH 2.8) and re-equilibrated with PBS. (**ii**) Denaturation of rat skin type I collagen was performed by heating a solution in coupling buffer (10 mg/ml) at 60° for 30 min. After coupling to the Sepharose at 37° for 4 h, washing and re-equilibration were performed as in (i), and the column stored at 37° in PBS containing 0.2% NaN_3 to retain denatured structure.

   Antibody isolation.- To a 5 ml column, (i) or (ii), 3–5 ml of serum was applied at 5 ml/h. The column was washed with PBS until the absorbance (280 nm, 1 cm) was <0.01, then eluted with pH 2.8 buffer [see (i)]. The eluate was immediately pH–adjusted with 1.0 M Tris–HCl (pH 7.6) and then dialyzed against PBS. The Ab's thus obtained were tested by ELISA to verify specificity.

ELISA PROCEDURES

   Type I collagen was dissolved at 4° in 0.1 M acetic acid at 1 mg/ml and diluted in cold PBS to 5 µg/ml. A similar solution was heated at 60° for 30 min to denature the collagen. These solutions were used to coat 96–well microtitre plates (Gibco–BRL, Paisley, Scotland) at 200 µl/well either at 4° for 16 h (native) or at 37° for 4 h (denatured), after which time the plates were washed 3 times with PBS containing 0.05% Tween–20 before assay using the appropriate Ab. After Ab isolation and ELISA screening it was found that cross–reacting Ab's were present. To ensure monospecificity, material initially absorbed to the native type I collagen–Sepharose column was passed through the denatured type I collagen column, and conversely. Unbound material in each case was tested by ELISA. The preparations were then specific for the particular antigen form tested (Fig. 1).

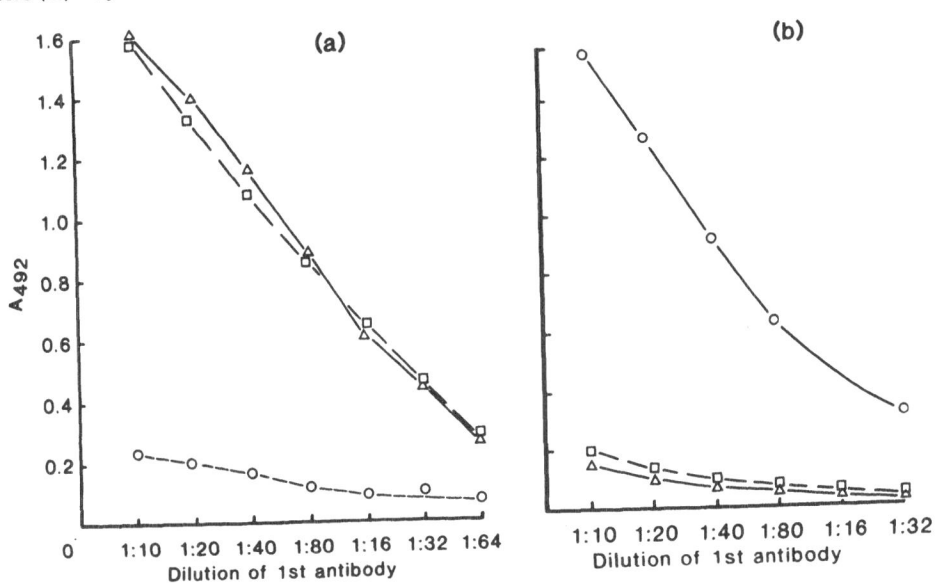

Fig. 1. ELISA test of anti-collagen type I specificity. Ab directed against (a) denatured type I collagen or (b) native type I collagen reacted with: Δ, denatured type I collagen □, type I collagen CNBr peptides; or o, native type I collagen (all 0.1 μg/200 μl well).

From these preliminary results it is apparent that some degradation of the collagen molecule must occur after intradermal injection of the antigen in its native form. It is probable that endogenous collagenase can cleave the α-chains of the native molecule, causing it to unwind and subsequently produce an immunogenic response to the denatured protein. Our unpublished follow-up of this finding suggests that in the sheep the response to the denatured type I collagen is more pronounced than that towards the native molecule. Furthermore, the positive ELISA response towards cyanogen bromide peptides of type I collagen [10] by the Ab against denatured collagen (but not by Ab to native collagen) indicated that the Ab's can be distinguished on the basis of sequence- (denatured) or conformation-dependent (native) antigenic determinants.

TISSUE STUDIES

The kidneys from 10 week-old rats were fixed with 2.5% paraformaldehyde in 30 mM Pipes buffer (pH 7.2) for 24 h at room temperature and processed to paraffin blocks. Sections (6 μm) were cut and de-waxed in preparation for immunostaining.

Immunostaining

Immunolocalization of different collagen antigens was performed using the indirect immunoperoxidase method as already described [11].

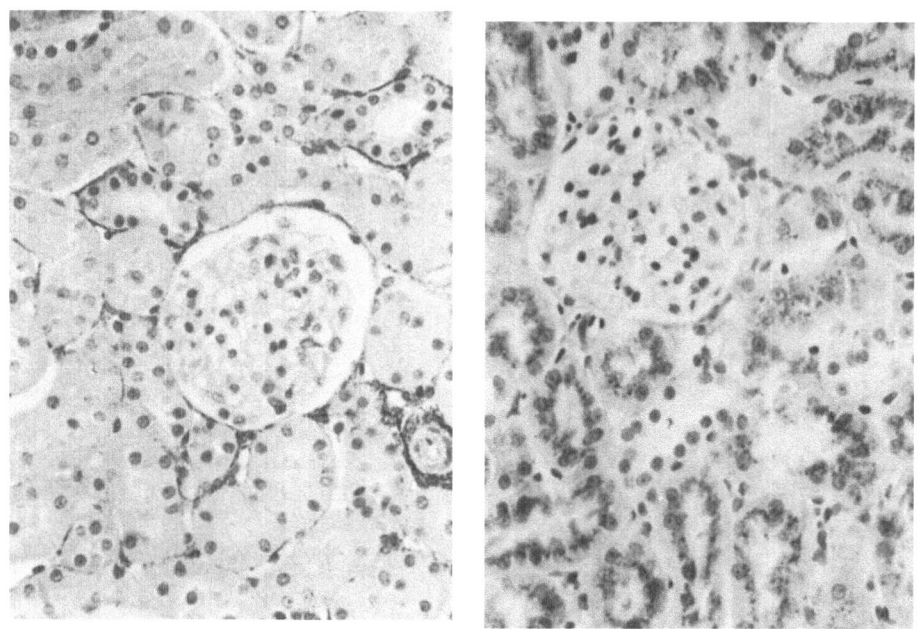

Fig. 2. Immunolocalization of type I collagen in normal rat kidney: *left*, native type I collagen; *right*, denatured type I collagen. Mag. × 70.

Ab's directed against native type I collagen in normal rat kidney showed staining around the glomeruli, proximal and distal tubules and blood vessels. The staining was localized in the extracellular interstitium, but was not associated with the basement membrane in the regions specified above or within the glomerular tuft (Fig. 2a). The capsule surrounding the kidney also stained strongly for this Ab.

The Ab's to denatured type I collagen exhibit a different staining pattern compared with the native state (Fig. 2b). The staining occurred around renal blood vessels and in the capsule surrounding the kidney. However, the interstitial connective tissue surrounding the tubules was unstained, but intracellular staining was seen in the proximal tubule region and appears to be localized in the brush border region of these epithelial cells. Since collagen is found in the extracellular space in the native form (and in the denatured form transiently during normal degradative processes) it is probable that the immunopositive material present in the proximal tubule cells represents fragments of degraded collagen molecules arising from the large body pools of type I collagen – skin, bone and tendon. These fragments, normally degraded to component amino acids by concerted proteolytic action at the site of initial cleavage, may evade these catalytic procedures

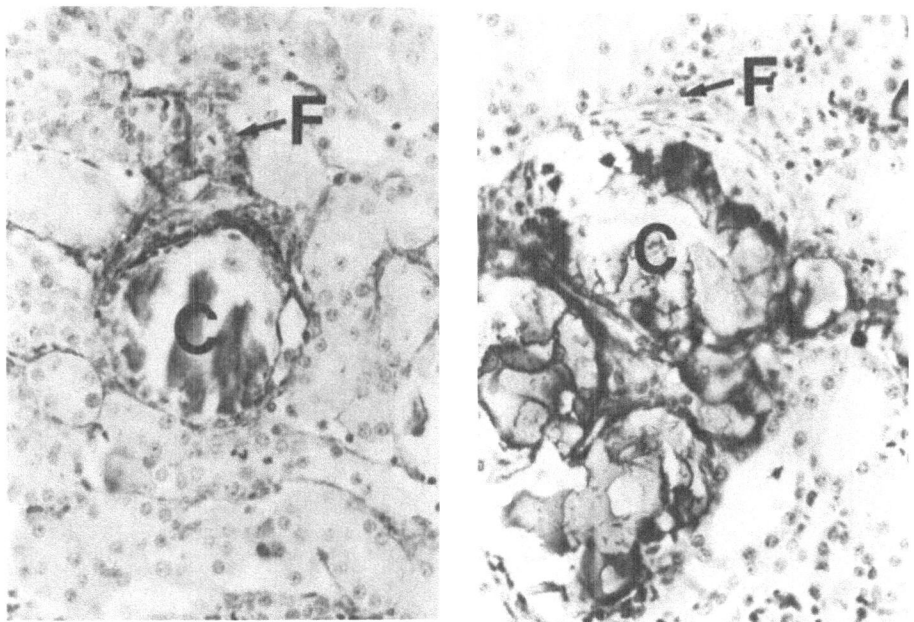

Fig. 3. Immunolocalization of type I collagen in rat kidney containing calcareous deposits: *left*, native type I collagen; *right*, denatured type I collagen. F, fibrosis; C, calcareous deposit. × 70.

and are resorbed by the proximal tubule cells after glomerular filtration. The plethora of lysosomal proteases present in kidney proximal tubule epithelia completes the degradative process, as considered elsewhere in this volume (K-J. Andersen & M. Dobrota, #F-4; J.K. McDonald et al., #E-6).

The staining pattern found in an example of a kidney containing calcareous deposits was also of interest. The very extensive fibrosis surrounding these deposits contained both native and denatured type I collagen (Fig. 3), indicating that degradation and/or remodelling of the fibrotic region was in progress. The absence of staining with the Ab against denatured type I collagen in interstitial areas of the normal kidney may indicate much lower rates of degradation in these regions.

The above findings have prompted an extension to our studies, and currently we are attempting to characterize the tumour invasion process in rat brain by glioma cell lines using these Ab's. The brain neuropil has a very low collagen content which appears to be confined to the vascular tissue [12]; but the response of the extracellular matrix to tumour invasion is not known. The degradation

of collagen as detected by the immunochemical techniques described
here may provide an insight into the invasion process by these highly
destructive tumours.

References

1. Herbert, W.J. (1978) in *Handbook of Experimental Immunology*,
    3rd edn., Vol. 3 (Weir, D.M., ed.), Blackwell, Oxford, pp. A3.1-A3.15.
2. Dresser, D.W. (1986) *Immunochemistry*, 4th edn., Vol. 1 (Weir, D.M.,
    ed.), Blackwell, Oxford, pp. 8.1-8.21.
3. Fuchs, S. & Sela, M. (1986) in *Immunochemistry*, 4th edn., Vol. 1
    (Weir, D.M., ed.), Blackwell, Oxford, pp. 16.1-16.5.
4. Bullock, G.R. & Perusz, P. (1982) *Techniques in Immunocytochemis-
    try* (Bullock, G.R. & Perusz, P., eds.), Acad. Press, London , 306 pp.
5. Nakamura, R.M., Voller, A. & Bidwell, D.E. (1986) in *Immunocyto-
    chemistry*, 4th edn. Vol. 1 (Weir, D.M., ed.), Blackwell, Oxford,
    pp. 27.1-27.20.
6. Bolton, A.E. & Hunter, W.M. (1986) in *Immunochemistry*, 4th edn.,
    Vol. 1 (Weir, D.M., ed.), Blackwell, Oxford, pp. 26.1-26.56.
7. Gershoni, J.M. & Palade, G.E. (1983) *Anal. Biochem. 131*, 1-15.
8. Chase, H.A. (1984) *Chem. Eng. Sci. 39*, 1099-1125.
9. Chung, E. & Miller, E.J. (1974) *Science 183*, 1200-1201.
10. Reiser, K.M. & Last, J.A. (1983) *Conn. Tiss. Res. 12*, 1-16.
11. Rucklidge, G.J., Riddoch, G.I. & Robins, S.P. (1986) *Coll. Rel.
    Res. 6*, 41-49.
12. Maxwell, W.L., Duance, V.C., Lehto, M., Ashurst, D.E. & Berry, M.
    (1984) *Histochem. J. 16*, 1219-1229.

Comments on #**NC(E)-2**: K. Donaldson et al. - ALVEOLAR CELL INJURY
         and #**NC(E)-4**: G.J. Rucklidge & V. Dean - COLLAGEN DEGRADATION

Iona Pratt asked Donaldson whether the effect of activated macrophages on his cell detachment system had been looked at, in the context
of elastase (microbial) being a strong promoter of detachment, as
also caused by a mixed inflammatory cell population but not by normal
alveolar macrophages; both the neutrophil and the activated macrophage
are known to produce elastase. **Reply.-** We have not yet studied a
pure population of activated macrophages; but time studies on the
inflammatory response to *Corynebacterium* in the lung showed a mixed
macrophage/neutrophil response; seemingly macrophage elastase also
contributes to the observed detachment phenomena.

Rucklidge was asked whether possibly the tumour cells produce
collagen, possibly explaining the presence of native collagen in
the brain in invaded tissue. **His reply:** no, the collagen could not
come from the tumour cells; but **Mareel remarked** that the tumour cells
do produce collagen. **R.G. Price remarked** on the interest of the
observation concerning the distribution in the proximal tubule of
peptidases with collagenolytic activity, particulary membrane-bound
tripeptidyl peptidase. One wonders about the role of these enzymes,
taking into account that the availability of the substrate in the
tubule will depend on molecular size and its ability to pass the
glomerular filter (cut-off 68,000 KDa). **Reply.-** Enzymes play a
scavenging role but may be involved in fibrosis in interstitial tissue.

Supplementary refs. contributed by Senior Editor

Proteases/Connective tissue degradation, *continued from p. 372*

Genetic deficiencies involving β-galactosidase may be attributable to lack of certain proteins that normally protect against intra-
lysosomal proteolysis of the enzyme; fibroblast studies have shown
that in galactosialidosis the loss of the enzyme (and of sialidase)
is due to degradation of the monomeric form which normally becomes
aggregated, with participation of the protective proteins, to a high
mol. wt. form with enzyme activity.- Hoogeveen, A., Verheijen, F.W.
& Galjaard, H. (1983) *J. Biol. Chem.* 258, 12143-12146.

─────────

Citation relevant to #E-2 (P. Bohley et al.): González-Cadavid, N.F.,
Bravo, M. & Campbell, P.N. (1968) *Biochem. J. 107*, 523-529.- In homogenates, mitochondrial cytochrome **c** partly leaks into the 'cytosol' and
and then becomes adsorbed to microsomes (fragments of e.r., the site
where cytochrome **c** biosynthesis occurs).

Section #F

RENAL INVESTIGATIONS

#F-1

A NUCLEAR MAGNETIC RESONANCE APPROACH TO INVESTIGATE
THE BIOCHEMICAL AND MOLECULAR EFFECTS OF NEPHROTOXINS

J.K. Nicholson & K.P.R. Gartland

Department of Chemistry
Birkbeck College, University of London
Malet Street, London WCiE 7HX, U.K.

^{1}H-NMR* can facilitate study of the metabolism and biochemical effects of foreign compounds, with no need for tailoring analytical conditions to particular metabolite classes and with minimal pretreatment of the sample. Here, NMR and conventional biochemical analyses have been used to investigate the physiological effects of ACZ (a renal carbonic anhydrase inhibitor) and the nephrotoxic effects of Hg^{2+} and pAP in the rat. ACZ caused a rapid reduction in excretion of citrate cycle intermediates, reflecting increased mitochondrial utilization. Hg^{2+} affected their excretions differently, as did pAP which also caused glycosuria, amino aciduria and lactic aciduria. In comparison with biochemical assessment including GGT and NAG determinations, NMR gave, with little effort, a much more complete picture of the nephrotoxic lesions, characteristic of the nephrotoxin type.

Knowledge of the detailed composition of the urine can illuminate the functional state of the kidney and the nature, severity and location of structural or biochemical lesions. Many urinalysis procedures are, however, rather insensitive and may detect only gross functional changes associated with significant structural damage. Despite recent major developments in biochemical toxicology and analytical biochemistry, there have been few advances in methods for detecting nephrotoxic lesions or studying the molecular actions of nephrotoxins. Here we describe a novel non-invasive approach to renal function assessment, using ^{1}H-NMR to measure urinary metabolite excretion profiles.

---

*^{1}H-NMR signifies proton NMR spectroscopy; Hg^{2+} = mercury II (chloride). *Abbreviations (some by Editor):* GGT, γ-glutamyltransferase; NAG, *N*-acetyl-β-D-glucosaminidase; αKG, α-ketoglutarate; ACZ, acetazolamide; pAP, p-aminophenol; FID, TSP - *see text.*

Many low mol. wt. compounds present in plasma, urine, cells and cell extracts can be measured simultaneously by ^{1}H-NMR [1-6] (& J.K. Nicholson et al. in Vol. 16, this series). Advantages over HPLC and GC are that little or no sample pre-treatment is necessary and a wide screen of many ^{1}H-containing metabolites in the mM range can be run within a few minutes. Only 0.3 ml of sample is required, and the technique is non-destructive and does not perturb equilibrium mixtures of compounds. These features are invaluable in metabolism studies on small animals, where only limited volumes of sample are available that must serve for a number of biochemical assays, or in experiments that involve the production of unstable metabolites that would decompose under conventional sample 'clean-up' or isolation procedures.

Specific conventional assays are often more sensitive for the detection of metabolites than NMR; but in toxicology the question is often "Which metabolite or biochemical change should be studied?". Many metabolites are suffiently abundant in biological materials to be measured by ^{1}H-NMR directly. Thereby information on many bio-chemical pathways can be obtained, without pre-selection of detection conditions for 'expected changes' in metabolic profiles due to the effects of a given xenobiotic, and toxicological mechanisms can be elucidated. Our quantitative analytical studies suggest that ^{1}H-NMR methods offer a sensitive multi-parametric indicator of renal damage, and other metabolic information that may relate to basic mechanisms of toxicity or renal disease [4, 5]. Furthermore, in the early stages of drug development, information on potential toxic side-effects is sparse, and could be forthcoming from suitable NMR techniques in toxicological screening.

MATERIALS AND METHODS

Animals and urine samples.- Animals (3 per group) were housed individually in metabolic cages (with water *ad lib.*) to readily get timed urine samples including a pre-treatment 24-h sample, collected over ice. In **ACZ** and **Hg^{2+}** studies, each with 5 treatment groups, Sprague-Dawley rats (~250 g) were used, with post-dose feeding *ad lib.* (ACZ) or withdrawn (Hg^{2+}). The ACZ injection (15 mg/kg in 0.9% NaCl) was given i.m., and urine collected during 0-4 h and 4-24 h, noting pH and volume for each collection. HgCl$_2$ was given i.p. as a single dose:- 0 (control), 0.5, 1, 1.5 or 2 mg/kg. Urine was collected for 24 h (and assayed for GGT [7]), then all animals were killed and their kidneys prepared for histology by standard techniques.

In **pAP** studies, with male Fischer 344 rats (~200 g) allocated at random to 4 groups, a single i.p. injection of a 100 mg/ml solution in 0.9% NaCl was given: 0, 25, 50 or 100 mg/kg. The urine collected 8, 24, 32 and 48 h after dosing was centrifuged and then frozen prior to NMR assay; assays for NAG [8] were performed. Post-dose feeding was *ad lib.*

Histopathology after Hg^{2+} or pAP dosing.- Small cubes of kidney were immersed in 2% Karnovsky's fixative, dehydrated, resin-embedded, sectioned (1 μm) and stained with toluidine blue. For electron micros-copy (e.m.) ultrathin sections mounted on formvar-coated grids were stained with uranyl nitrate and lead citrate.

^{1}H-NMR.- Spectra were recorded on a Bruker (WH400) spectrometer operating at 400 MHz with quadrature detection. To 0.4 ml urine was added 0.1 **ml** ^2H_2O (to provide a field frequency signal) containing Na 3-(trimethylsilyl)-[2,2,3,3-^{2}H]-1-propionate (TSP) which provided a concentration standard (2 mM) and the internal chemical shift reference (δ = 0 ppm). Typically, for each sample, 64 free induction decays (FID's) were collected into 16384 computer points (acquisition time 1.7 sec) using a 50° pulse angle, and a total pulse recycle time of 5.6 sec to allow complete T_1 relaxation of solute protons. A continuous secondary irradiation field was applied at the resonance frequency of water in order to suppress the signal and allow optimal digitization of the solute proton signals [2].

RESULTS AND DISCUSSION

To illustrate the potential of NMR urinalysis in the study of perturbed renal function, we describe three NMR case studies of agents which significantly modify the renal physiology and biochemistry of experimental rats after a single injection.

Biochemical effects of acetazolamide (ACZ)

ACZ is a commonly used drug which produces a marked osmotic diuresis in man and other mammals. Its molecular action is inhibition of renal tubular carbonic anhydrase [10, 11], preventing reabsorption of bicarbonate (Fig. 1) whose tubular concentrations therefore rise and exert an osmotic 'drag' on tubular water, producing a dilute alkaline urine. The reduced bicarbonate reabsorption results in

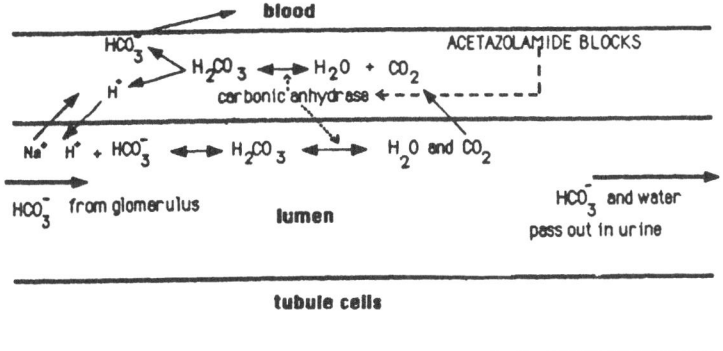

Fig. 1. Processes connected with bicarbonate reabsorption and excretion in the nephron.

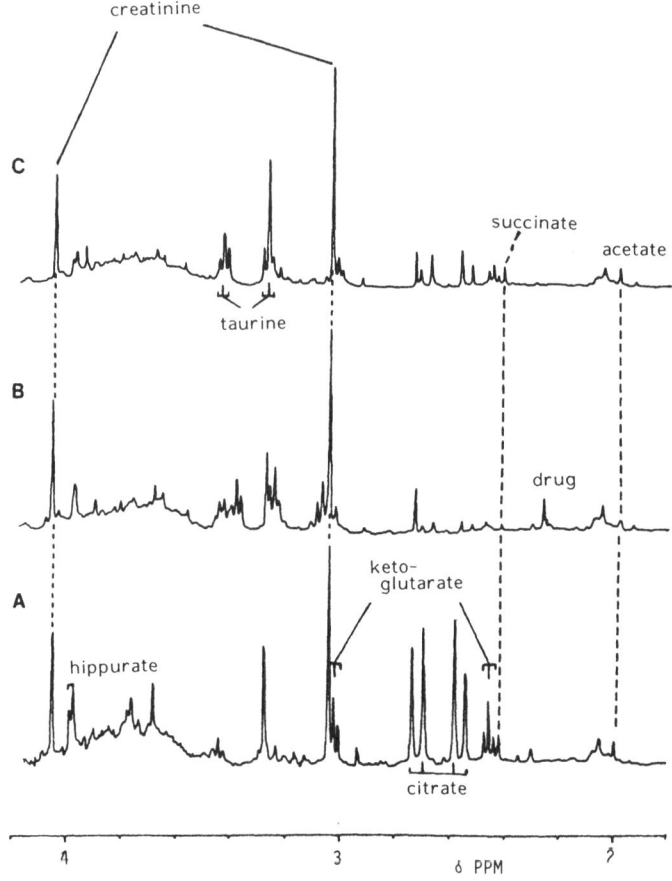

Fig. 2. Aliphatic regions of 400 mHz ^{1}H-NMR spectra of rat urine: control (A), and during 0-4 h (B) or 4-24 h (C) after 15 mg/kg ACZ.

a fall in intracellular pH and classical renal tubular acidosis occurs, causing significant changes in renal substrate metabolism. We have used NMR urinalysis to investigate the metabolic effects of ACZ, as nephrotoxic states may often be accompanied by changes in renal acid-base balance; if NMR is to be useful in elucidating toxic mechanisms, biochemical changes due to altered acid-base balance must be distinguishable from changes due to the direct toxic effects of the xenobiotic.

In aqueous solution ACZ has only one NMR-observable proton signal, that of the *N*-acetyl methyl group ($\delta = 2.01$ ppm). This resonance can be detected in ^{1}H-NMR spectra of urine from animals and subjects treated with the drug, and rats also show major changes in the excretion patterns of NMR-detectable metabolites (Fig. 2). These changes in excretion profiles of low mol. wt. compounds reflect drug-induced subtle alterations in renal physiology and biochemistry, most notably

Table 1. Effect of ACZ on excretion of citrate cycle intermediates and creatinine, as µmol/kg body wt. per h (mean ±S.E.M.; n = 3). For spin-spin coupling pattern (proton signal): s, singlet; t, triplet; AB, 2nd-order spin system; approx. chemical shifts at pH 7 (TSP δ = 0 ppm) are given. Control = pre-dose 24-h collection.

Metabolite	NMR signal		δ	Control rate	0-4 h	4-24 h
Succinate	CH_2	[s]	2.42	1.86 ±0.27	1.80 ±0.11	0.94 ±0.11
αKG	$(CH_2)_2$	[t]	3.02	2.15 ±0.21	1.21 ±0.32	not detected
Citrate	$(CH_2)_2$	[AB]	2.64	3.40 ±0.24	2.47 ±0.29	0.36 ±0.08
Creatinine	CH_2	[s]	4.06	19.3 ±1.25	21.9 ± 1.30	21.3 ±0.52

a sudden rapid decline in the excretion rate of citrate and two other cycle intermediates, succinate and αKG. Their concentrations can be measured directly from the proton NMR spectra of urine, quantitation being achieved by integration of the area of the metabolite resonances relative to that of the TSP standard. Excretion rates can then be easily calculated, taking into account the collection volume and the molar equivalent number of protons contributing to the measured resonance. Table 1 shows metabolite excretion data for rats given 15 mg/kg of ACZ. Creatinine excretion being unchanged, we conclude that the glomerular filtration rate is hardly affected. Citrate excretion data obtained by conventional biochemical assays [9] agreed well with the NMR values.

The biochemical basis of these observations is as follows. Intracellular pH and bicarbonate concentration are important factors in the mitochondrial utilization of substrates and in particular citrate cycle intermediates such as citrate [10]. Recent evidence suggests that the effects of acid-base changes in the kidney are mediated by alteration in the pH gradient across the inner mitochondrial membrane. Metabolic acidosis causes intracellular bicarbonate to decrease, with an associated increase in the mitochondrial pH gradient. This stimulates the tricarboxylic carrier and increases the flux of citrate into the mitochondrial matrix compartment [10]. The cytoplasmic citrate level is therefore reduced and so tubular and peritubular uptake is increased with a consequent reduction in the urinary citrate clearance [10]. Our data also show that αKG and succinate reabsorption from the urine follows the same pattern as citrate, when the animal is treated with ACZ (Fig. 2 & Table 1). Here we have shown that NMR quantitative data on citrate cycle intermediates can provide direct information on renal carbonic anhydrase activity. Understanding the changing excretion profiles produced in situations of perturbed acid-base balance is important if alterations in excretion profiles caused by nephrotoxins are to be understood and distinguished from indirect physiological and biochemical effects (see below).

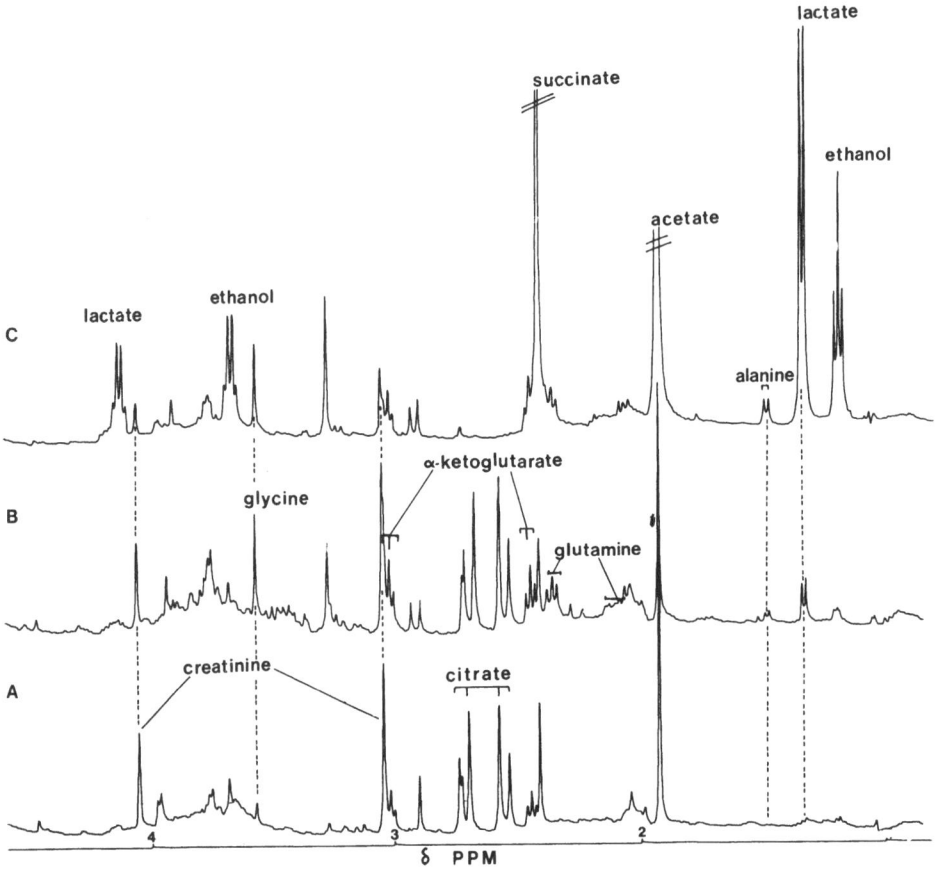

Fig. 3. Aliphatic regions of 400 MHz ^{1}H-NMR spectra of 24 h rat urines: control (A), and after treatment with 1 mg/kg (B) or 2 mg/kg (C) HgCl$_2$.

Biochemical effects of exposure to mercury II chloride

HgCl$_2$ is a potent nephrotoxin reproducibly causing severe necrosis of the pars recta of the proximal tubule after an i.p. injection of >1.5 mg/kg [12, 13]. As for other authors ([14] & this vol.) who have used Hg^{2+} to model acute renal failure, we have chosen this agent for the present critical assessment of the merits of NMR urinalysis for detecting nephrotoxic lesions. The damage to the proximal tubule results in blocked reabsorption from the tubular lumen and hence produces an acquired Fanconi syndrome with amino aciduria, glycosuria, calciuria and low mol. wt. proteinuria [15]. Uncoupling of oxidative phorphorylation and decreased respiratory control in proximal tubule mitochondria has been thought to account for the renal damage [16]. Following proximal tubular dysfunction, secondary anoxic damage to the tubular epithelium follows because of the reduction

in renal blood flow, due to massive disruption of the renin/
angiotensin-mediated glomerulotubular feedback mechanism [14]; the
main degenerative changes are cloudy swelling and hydropic degeneration
followed by necrosis of the pars recta.   The renal histopathology
of Hg^{2+} poisoning has been fully described [13, 17, 18].   At the
e.m. level the most striking degenerative change prior to tubular
necrosis and autolysis is severe mitochondrial swelling [13].

We previously reported profound changes in the excretion profiles
of 14 out of 24 NMR-detectable endogenous urinary metabolites in
Hg^{2+}-treated rats [4].   Fig. 3 illustrates the wealth of biochemical
data that can be derived from ^1H-NMR urinalysis in toxic episodes.
The NMR fingerprints in Hg^{2+}-NMR poisoning are grossly abnormal, demonstrating that NMR is potentially a good detector of acute renal damage.
Since, moreover, the metabolite excretion patterns relate to functional
changes in the renal tubules after exposure to the metal, interpretation
of these patterns can also illuminate the basic mechanism of the
Hg^{2+}-induced renal damage.

Alanine and glycine levels were greatly increased after Hg^{2+}
treatment (Fig. 3).   These appear to be good NMR markers for proximal
tubular toxicity as they are abundant in plasma and so readily overflow
into the urine after tubular damage; they also give well resolved,
simple NMR signals and therefore are easy to detect.   Their presence
in the urine at high concentration is an immediate indication of
proximal tubular damage.

Other striking changes in the NMR spectra included an increase
in the intensity of the succinate and αKG signals and the complete
absence of the citrate signal at high $HgCl_2$ doses   (>1.5 mg/kg).
Table 2 shows the effect of various Hg^{2+} doses on NMR-derived excretion
rates of these metabolites and creatinine. The data seemingly conflict
with those from the ACZ experiments, where metabolite excretion
changes were always parallel and in the same direction (Table 1).
We interpreted this in terms of the combined effects of carbonic
anhydrase inhibition (resulting in enhanced citrate catabolism and

Table 2.   Effect of Hg^{2+} on excretion of citrate cycle intermediates
and lactate, as μmol/kg body wt. per h (mean ±S.E.M.; n = 3), at
different $HgCl_2$ dose levels expressed as mg/kg.

Metabolite	HgCl$_2$: 0	0.5	1.0	1.5	2.0
Succinate	2.75 ±0.58	1.12 ±0.27	4.12 ±1.4	4.04 ±1.3	10.6 ±1.79
αKG	0.85 ±0.08	0.27 ±0.06	1.83 ±0.37	2.29 ±0.50	2.58 ±0.54
Citrate	2.02 ±1.18	0.58 ±0.16	0.83 ±0.45	none detected	
Lactate	0.38 ±0.18	0.09 ±0.01	1.12 ±0.08	1.50 ±0.58	8.42 ±2.67

reduced renal clearance) and the inhibition by Hg^{2+} of mitochondrial malate and succinate dehydrogenases, leading to accumulation of αKG and succinate in the cell and consequently to increased renal clearance of these compounds. The reduced activity of the proximal tubular citrate cycle pathway is also reflected in the massive increase in lactate excretion (Fig. 3 & Table 2) due to increased utilization of anaerobic pathways [4].

Interestingly, high doses of Hg^{2+} also cause ethanol excretion due to inhibition of liver alcohol dehydrogenase, this in turn leading to accumulation of ethanol derived from fermentation reactions in the gut [4]. This unexpected finding indicates the power of NMR in detecting novel markers of toxicity and enzyme inhibition. The significance of raised urinary acetate levels (Fig. 3) is not known, although preliminary data suggest a relationship to altered renal fatty acid metabolism (unpublished).

Biochemical effects of p-aminophenol (pAP)

pAP, which causes severe proximal tubular necrosis after a single injection, is a metabolite of phenacetin and paracetamol and has been used to study experimental analgesic nephropathy [19, 20]. It causes a dose-dependent depletion of renal (but not hepatic) reduced GSH, and it has been suggested that pAP-induced nephrotoxicity results from the covalent binding to protein of a reactive intermediate, possibly the benzoquinone imine [21]. Like Hg^{2+}, pAP causes degenerative changes in proximal tubular mitochondria prior to cellular necrosis [21]. However, the molecular mechanisms of toxicity are likely to be very different from those of Hg^{2+}. Hence the patterns of toxin-induced metabolite excretion should be significantly different, despite the similarity of the histopathological manifestations of toxicity and the topographical localization of the pAP- and Hg^{2+}-induced renal lesions.

The 400 MHz ^{1}H-NMR spectra (aliphatic region only) of rat urine collected 0–8 h after different pAP doses are shown in Fig. 4. Examination of the changes in pattern of excreted metabolites (shown by NMR spectra) indicates that the threshold toxic dose is between 25 and 50 mg/kg in the Fischer 344 rat. This is consistent with the findings of previous workers [19, 20]. Profound dose-related glycosuria, lactic aciduria and amino acidurias were observed with >25 mg/kg (Fig. 4).

With only 25 mg/kg (Fig. 4, B) no significant changes were seen in the profiles for low mol. wt. metabolites. The presence of the N-acetyl singlet of paracetamol sulphate (δ = 2.18 ppm) confirms that metabolism (i.e. N-acetylation) of the pAP has occurred. With 50 mg/kg (Fig. 4, C) significant elevations were seen in urinary levels of glucose, lactate, alanine, glutamine and valine, during the first 8 h after dosing, indicating a functional proximal tubular

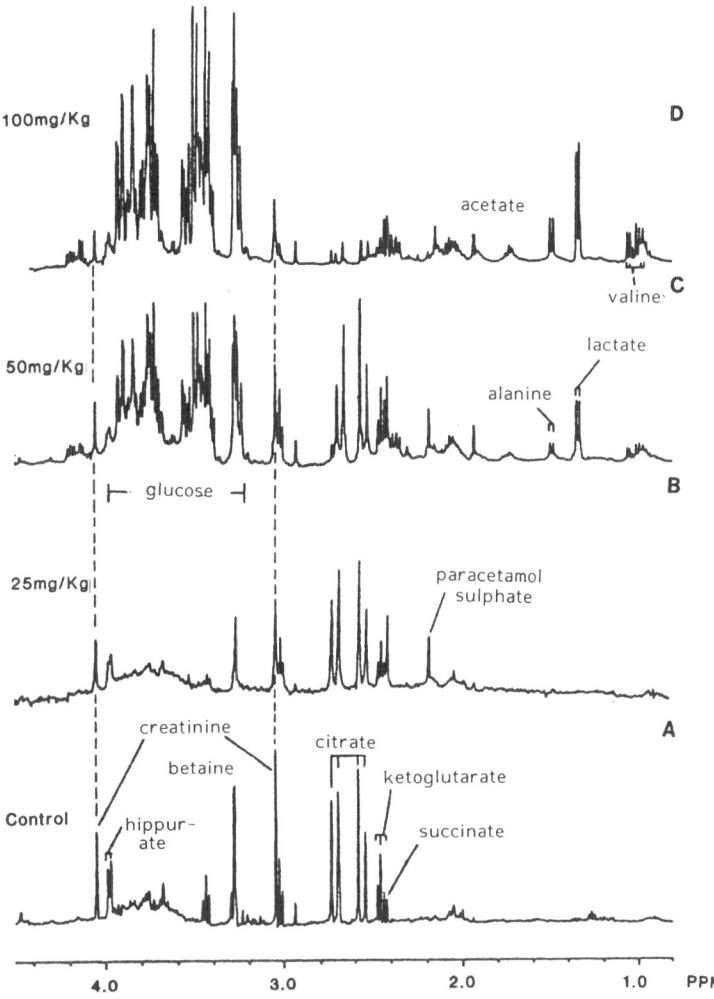

Fig. 4. Aliphatic regions of 400 MHz ^{1}H-NMR spectra of rat urine: control (A) or after injection of pAP: 25 (B), 50 (C) or 100 (D) mg/kg. Post-dose period 8 h.

defect and hence reduced solute reabsorption efficiency. At this dose level the glycosuria, lactic aciduria and alaninuria are most elevated over the period 8-24 h after dosing; but after 24 h the levels fell, approaching those of controls at 48 h (Fig. 5).

The highest dose of pAP (100 mg/kg, Fig. 4, D) resulted in gross and sustained elevations in urinary glucose, lactic acid, alanine and other amino acids, the increases being greatest during 8-24 h and persisting until at least 48 h (data not shown), suggesting that more severe or irreversible structural damage had occurred. Conventional assays for glucose and lactate confirmed the NMR values.

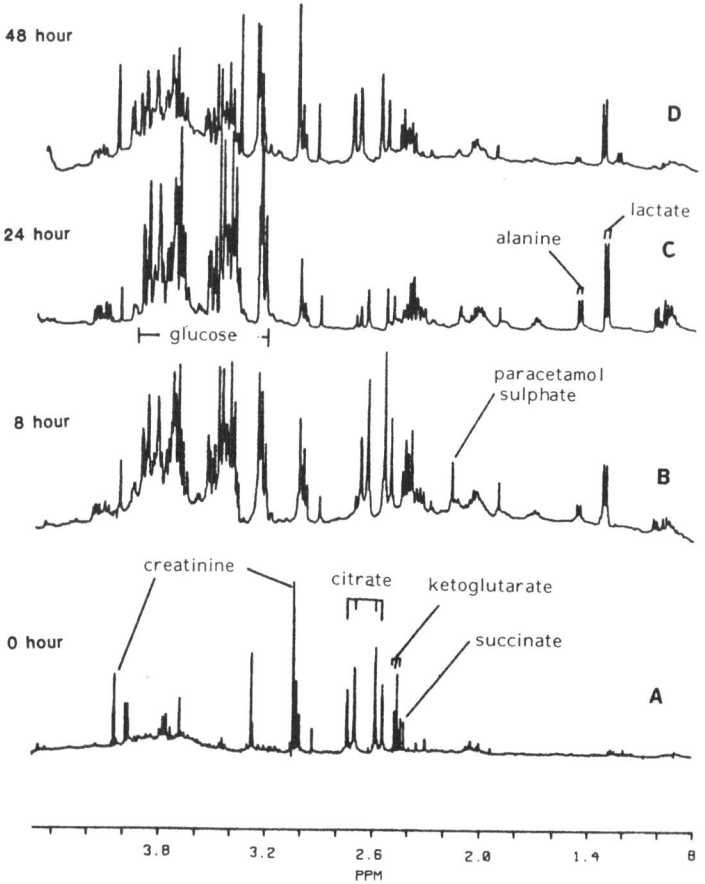

Fig. 5. Aliphatic regions of 400 MHz ^{1}H-NMR spectra of rats given 50 mg/kg pAP: 24 h pre-dose (A), and 0-8 (B), 8-24 (C) and 24-48 h (D) post-dose.

The urinary excretion of damage marker enzymes (NAG, GGT) was significantly elevated, maximally at ~8 h, with 50 or 100 mg/kg, but not with 25 mg/kg - supporting the indications from NMR that there was no toxicity.

Comparison of pAP and Hg^{2+}, both producing acute proximal tubular lesions, in respect of urinary metabolites shows significant differences although there were features in common, e.g. the alaninuria reflecting a generalized reduction in solute reabsorption effiency in the proximal tubule. Toxins such as pAP and Hg^{2+} cause lactic aciduria, especially severe with Hg^{2+}. Besides having reduced reabsorptive capacity (for ultrafiltered lactate), the degenerating proximal cells also over-produce lactate through more use of anaerobic pathways. With pAP the excretion of citrate cycle intermediates was less severely disrupted. But glycosuria was much more apparent (as confirmed by conventional assays). This suggested that besides a gross pathological lesion in the proximal tubule's pars recta, there was also a functional defect in the pars convoluta, the main site of urinary glucose reabsorption. Sodium chromate, a toxin that specifically damages the pars convoluta, produces an even more severe glycosuria (unpublished).

CONCLUSIONS

^{1}H-NMR urinalysis was found to be at least as sensitive as conventional methods of analytical biochemistry, histopathology and urinary enzyme assays for the detection of renal lesions. In the case of Hg^{2+}, NMR measurements on urine provided novel multi-parametric indices of renal function, and also enabled information on the basic mechanisms of nephrotoxic action to be derived by interpreting changes in metabolite excretion patterns. Different classes of nephrotoxin produce characteristic changes in the patterns, closely relating to specific biochemical lesions of mechanistic importance.

Acknowledgements

The work was supported by the S.E.R.C. and the National Kidney Research Fund.

References

1. Nicholson, J.K., Buckingham, M.J. & Sadler, P.J. (1983) *Biochem. J. 211*, 605-615.
2. Bales, J.R., Higham, D.P., Howe, I., Nicholson, J.K. & Sadler, P.J. (1984) *Clin. Chem. 30*, 426-432.
3. Nicholson, J.K., O'Flynn, M., Sadler, P.J., Juul, S., MacLeod, S. & Sonksen, P. (1984) *Biochem. J. 217*, 365-375.
4. Nicholson, J.K. & Timbrell, J.A. (1985) *Mol. Pharmacol. 27*, 644-651.
5. Gartland, K.P.R., Nicholson, J.K., Timbrell, J.A. & Bonner, F. (1986) *Human Toxicol. 5*, 122-123.
6. Nicholson, J.K., Timbrell, J.A., Bales, J.R. & Sadler, P.J. (1985) *Mol. Pharm. 27*, 634-643.
7. Naftalin, L., Sexton, M., Whitaker, J.F. & Tracy, D. (1969) *Clin. Chim. Acta 26*, 293.
8. Maruhn, D. (1976) *Clin. Chim. Acta 73*, 453-461.
9. Warty, V.S., Busch, R.P. & Virji, M.A. (1984) *Clin. Chem. 30*, 1231-1233.
10. Grollman, A.P., Harrison, H.C. & Harrison, H.E. (1961) *J. Clin. Invest. 40*, 1290-1296.
11. Hager, S. & Simpson, D. (1982) *Mol. Physiol. 2*, 203-211.
12. Haagsman, B.H. & Pound, A.W. (1979) *Br. J. Exp. Path. 60*, 341-352.
13. Nicholson, J.K., Kendall, M.D. & Osborn, D. (1983) *Nature 304*, 633-635.

14. Barnes, J.L., McDowell, E.M., Nagle, R.B., McNeill, J.S. &
    Flamenbaum, V. (1980) *Virchow's Arch. B Cell Path. 32*, 201-232.
15. Foulkes, E.L. (1979) *Toxicol. Appl. Pharmacol. 71*, 445-450.
16. Southard, J.H. & Nitisewojo, P. (1973) *Biochem. Biophys. Res.
    Comm. 32*, 201-232.
17. Cuppage, F.E. & Tate, A. (1967) *Am. J. Path. 51*, 617-625.
18. Gritza, T.A. & Trump, B.F. (1968) *Am. J. Path. 52*, 1255-1258.
19. Carpenter, H.M. & Mudge, G.H. (1981) *J. Pharmacol. Exp. Ther.
    38*, 161-167.
20. Newton, J.F., Kuo, C.H., Gemborys, M.R., Mudge, G.H. & Hook, J.B.
    (1982) *Toxicol. Appl. Pharmacol. 65*, 336-344.
21. Crowe, C.A., Young, A.C., Calder, I.C., Ham, K.N. & Tange, J.D.
    (1979) *Chem. Biol. Interact. 27*, 235-243.

#F-2

ENZYME RELEASE IN NEPHROTOXICITY AND RENAL DISEASE

P.N. Boyd, J. Halman, S.A. Taylor and R.G. Price

Department of Biochemistry
King's College, Kensington Site
Campden Hill, London W8 7AH, U.K.

A combination of non-invasive and histopathological techniques is particularly valuable in assessing the nephrotoxicity of drugs and in the diagnosis of renal disease. Morphological evaluation and microdissection techniques have been used to establish the site of the twin tubular lesions in streptozotocin-diabetic rats. A feature of enzymuria following the initiation of diabetes was the excretion of brush-border and lysosomal enzymes, suggesting that these organelles are perturbed early in renal damage. However, no biochemical change was apparent in the brush border, suggesting that in this case the lesion may be at the functional level.

Changes in renal brush border resulting from the injection of mercuric chloride were accompanied by an immediate elevation of NAG* and LAP in the urine; the levels then fell to normal, whereas urinary NAG activity in diabetic rats remained elevated for >6 months. A complete understanding of the biochemical mechanisms which underlie renal disease will require use of a wide range of subcellular and morphological techniques. However, the excretion of renal enzymes provides a convenient way of monitoring the development of renal disease and evaluating the nephrotoxicity of drugs.

The kidney is particularly vulnerable to toxins because of its high blood supply and metabolic activity, while renal involvement is often an important feature of major disorders, e.g. diabetes. A characteristic of renal damage is that it often occurs initially in a specific region of the nephron. The release of enzymes into the urine is a sensitive indicator of renal disease and the pattern of enzyme release reflects the site and severity of damage [1].

Abbreviations: NAG, *N*-acetyl-β-D-glucosaminidase; LAP, leucine aminopeptidase; PAGE, polyacrylamide gel electrophoresis; e.m., electron microscopy.

A wide range of procedures is used to characterize renal damage but the final diagnosis has usually depended on morphological evidence. However, since histopathological procedures are carried out on biopsy material these investigations are justifiable only when the disease is well established. In animal studies the histopathological investigations are usually carried out on renal material obtained by autopsy. A further problem is that as the kidney has a large functional reserve, up to 65% of the capacity of the nephrons can be lost before changes are observed with the parameters used to assess renal function, e.g. glomerular filtration rate and tubular reabsorptive capacity.

Because of the insensitivity of physiological techniques and the difficulties encountered in obtaining material for pathology, considerable interest has developed in non-invasive tests which allow continuous monitoring of renal damage. Initially these procedures attracted interest as simple screening methods, but the ultimate aim will be to use them in differential diagnosis. The kidney has only a limited number of ways of responding to damage, and as many types of renal disease are classified on the basis of morphology it may not be possible to achieve a specific diagnosis on the basis of urine data alone. However, the combination of non-invasive tests with histopathology and simple tests of renal function has great potential both for monitoring the development of disease and in differential diagnosis. Careful selection of the tests to be used can allow the determination of the initial site of damage, allowing glomerular and tubular damage to be readily distinguished; the urinary pattern will also change if damage affects a second region of the nephron.

Understanding the basis of renal damage and disease calls for studies at the cellular and molecular level. Since the kidney contains numerous cell types [2], besides the nephron region the cells initially affected should be determined by histochemical and/or e.m. techniques.

Once the region and target cells have been identified, the next step is to determine the organelle or biochemical system that is perturbed. A wide range of organelles and subcellular systems are affected by drugs, and the mechanisms involved are little understood, although molecular probes can now be used to determine whether changes have occurred in protein synthesis. Drugs are known to affect the cytochrome P-450 system [3] and lysosomes [4]; the luminal membrane is a target for heavy metals [5], while other agents affect the mitochondria [6]. The glomerulus is adversely affected in many types of renal disease (as discussed in #F-3 later in this vol.). To illustrate how the urinary excretion of enzymes can be related to renal disease and nephrotoxicity, the effect of mercuric chloride and streptozotocin diabetes on the cells of the proximal tubule of rat kidney will be discussed.

THE NON-INVASIVE APPROACH

Urinary enzymes are sensitive indicators of renal damage or disease, and are elevated in urine prior to changes in classical renal function tests [7] and usually before gross proteinuria. The enzymes assayed are usually chosen on the basis of their distribution in the cell:
- NAG, widely used as a lysosomal enzyme;
- LAP or alanine aminopeptidase, commonly used as a brush-border enzyme;
- lactate dehydrogenase (LDH) or β-glucosidase (cytosolic), indicative of loss of material into the tubular lumen, and often measured.

Stability in urine is a crucial factor and has limited the number of enzymes which can be conveniently used [1]; brush-border, lysosomal or cytosolic enzymes are preferred to mitochondrial enzymes which are inherently unstable in urine. The metabolism and functional specialization of the cells along the nephron is associated with a high degree of specialization at the biochemical level, and each segment has a characteristic complement of enzymes [8]. The ratio of different enzymic activities in urine gives an indication of the initial site of damage [9].

Data obtained on urinary enzyme excretion is often enhanced if other simple tests of renal function are also performed. A wide range of different tests can be used, the following being the most useful.-
Renal function:
- volume, specific gravity, osmolarity;
- creatinine (index of concentration);
- protein, glucose (indices of reabsorption);
- pH; electrolytes - Na^+, K^+, Mg^{2+}, Ca^{2+}.
Celluria:
- bacteria, blood cells.
Proteins:
- *specific* (RIA or ELISA available): albumin [10] (diabetes); retinol binding protein [11] (proximal tubular damage); $β_2$-microglobulin [12], $α_1$-microglobulin [13]; Tamm-Horsfall glycoprotein [14] (distal tubule);
- *general:* electrophoresis (mol. size separation).
Isoenzymes: NAG, LDH.

Routinely a small number of tests are usually performed, the choice depending on the circumstances - normally including one or two urinary enzymes and protein. Further tests can be carried out to confirm or extend diagnostic information: e.g. urinary volume and glucose and protein excretion, reflecting tubular reabsorptive capacity, and changes in pH and electrolytes - likewise easy to measure and reflecting the functional efficiency of the tubular cells. The qualitative nature of proteinuria can be characterized by SDS-PAGE [15] or by 2-D electrophoresis [16]; the molecular size profile may indicate glomerular, tubular or mixed proteinuria [17]. Immunological procedures can be used to determine the amount of an individual protein present in urine, as indicated above.

THE INVASIVE APPROACH

Previous studies [5] established that necrosis of the proximal tubule cells was a characteristic feature of acute proximal-tubule damage following $HgCl_2$ administration. Early loss of microvilli was seen by e.m., and the necrotic process was parallelled by a fall in the activity of brush-border marker enzymes in preparations of brush-border membranes. These changes occurred concomitantly with an increase in urinary enzyme excretion, exemplified in Fig. 1 by LAP (brush border), and also by NAG which increased as a result of damage to the lysosomes as well as to the luminal membrane. Both activities returned to normal as the tubular cells recovered from the nephrotoxin.

Morphological studies using light and electron microscopy demonstrated two tubular lesions in streptozotocin-diabetic rat kidney (P.N. Boyd, unpublished data). A marked accumulation of glycogen-like, PAS-positive material was observed in the distal tubule, and its location between the start of the cortical straight tubule and the macula densa was confirmed by microdissection studies. Although the morphology of the proximal tubule cells appeared normal there was a faster rate of thickening of the tubular basement membrane over the 6 months. Both lesions were fully developed after 2 months of diabetes. Although the proximal tubule cells were normal, both NAG and LAP activities showed a sustained elevation in urine (Fig. 2). Indeed, increased excretion of urinary NAG is a feature of diabetes [18], markedly contrasting with its return to normal following damage by $HgCl_2$. The molecular size of both enzymes is too great to allow them to pass through the filtration barrier of the glomerulus. Hence they must originate from cells of the nephron; but as no morphological damage was discernible they must be excreted at an accelerated rate owing to a functional abnormality which may be related to the thickening seen in the tubular basement membrane.

Figs. 1 and 2 exemplify two possible modes of expressing enzyme activity data. One mode is units/mmol creatinine: it allows for urine flow variation. This mode is preferable when random urine samples are assayed. When accurately timed samples can be obtained it is convenient to express enzyme activity as a daily excretion rate (u/24 h).

THE ORGANELLE LEVEL

Plasma membranes were isolated from the renal cortices of normal and $HgCl_2$-injected rats by rate-zonal centrifugation in a B-XIV rotor, and compared at different stages of the necrotic process [5]. The

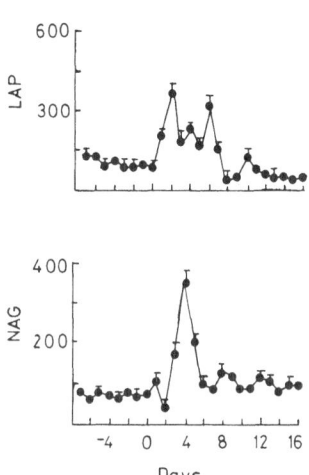

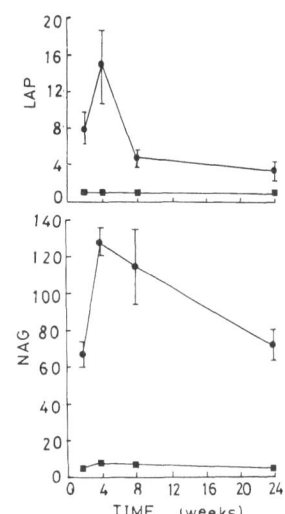

Fig. 1. Comparison of NAG and LAP activities in the urine of rats given $HgCl_2$ (3 mg/kg, i.p.) on Day 0 (normal excretion levels before then). Values are means ±S.D. (6 rats), as μmol MNP released/h/mmol creatinine [MNP = 2-methoxy-4-(2'-nitrovinyl)phenol.]

Fig. 2. Urinary NAG and LAP over a 6-month period in rats injected i.v. with a placebo (■) or streptozotocin, 65 mg/kg; ●). Values are means ±S.E.M. (10 rats), as units/24 h.

recovered activities of brush-border marker enzymes (alkaline phosphatase, LAP, maltase and 5'-nucleotidase) were all decreased, while that of a basolateral membrane marker enzyme, (Na-K)-ATPase, was unchanged.

Brush-border membranes were prepared from diabetic and control rat renal cortices using a modification of the Mg^{2+} precipitation method of Booth & Kenny [19]. The following modifications improved the yield of membranes and increased the relative specific activities of the marker enzymes. (1) The homogenate was 5% rather than 10%. (2) Homogenization was carried out in an Atomix blender at full speed for 8 × 5 sec with appropriate cooling. (3) The initial low-speed centrifugation step [19] was omitted. (4) No separation of the P2 pellet into layers was made. (5) The final centrifugation step was increased from 180,000 to 1,500,000 **g**-min to recover microvilli remaining in the P4 fraction. Renal brush-border preparations thus isolated showed enrichments of 12-fold for maltase and 17-fold for LAP, with ~30% recoveries.

No differences in LAP activity were observed when membrane isolated from rats 2 weeks after diabetes onset was compared with control preparations (Table 1). Indeed, no significant differences in the protein, lipid or enzymic activity of isolated microvillar

Table 1. LAP and NAG in 'P4' preparations of kidney brush border from normal and 2 week-diabetic rats: means ±S.D. (n = 10). Enzyme activities expressed as μmol/h, protein as mg.

	Total act./amount	Relative specific act.	% Recovery
LAP: control	1277 ±399	16.72 ±3.34	33.4 ±5.8
diabetic	*777 ±317*	*16.25 ±4.02*	*27.3 ±8.1*
NAG: control	2.0 ±1.2	0.45 ±0.26	1.9 ±0.7
diabetic	*1.9 ±0.7*	*0.26 ±0.13*	*0.6 ±0.1*
Protein: control	10.4 ±2.4		2.1 ±0.5
diabetic	*13.0 ±6.5*		*2.0 ±0.4*

preparations was found over the 6-month period of the study (unpublished data). However, differences in the (Na-K)-ATPase activity of baso-lateral membrane preparations have been reported [20], and (P.N. Boyd, G. Mann and R.G. Price, unpublished) we were able to demonstrate in perfusion experiments with isolated kidney that streptozotocin diabetes caused a decrease in glutamine and serine uptake at both the brush border and basolateral surfaces.

These experiments confirm that functional abnormalities do occur in the proximal tubular cells in diabetes. The elevated excretion of NAG may result from an increased rate of exocytosis which may also involve the loss of enzymes from the brush border. Why such a loss is not reflected in the composition of the membrane fractions is not clear at present. Further understanding of this problem will require more detailed biochemical studies at the organelle and functional level.

CONCLUSIONS

The elevation of enzyme activities in urine is a sensitive indicator of renal cellular damage: activities rise rapidly to a peak in acute damage and return to normal levels with recovery. In long-standing conditions such as diabetes, enzyme levels may remain abnormally high over an extended period, indicating that functional abnormalities have occurred in the kidney. When the rises in urinary enzyme activity occur in conjunction with other parameters such as microalbuminuria, they provide an excellent means of monitoring the progress of the disease.

Maximum information is obtained regarding the sequence of events in renal damage when urinary enzyme excretion is correlated with subcellular studies and sequential histopathology. Isolation of the target organelle from normal and pathological tissue should provide information on the basic pathological mechanism involved. Urinary enzymes are most effective when assayed in conjunction with other indices of renal function [21]. The enzymes of choice are those which are highly active in the kidney, exist as isoenzymes, are stable in urine, and have a discrete localization in a defined region of the nephron. Urinary enzymes have now found a place in human and animal chemical pathology, in occupational medicine, and in the screening for early signs of renal disease. The next phase should involve the assessment of the molecular perturbations occurring in the kidney that give rise to well defined excretion patterns.

Acknowledgements

These studies were supported by the British Diabetic Association, S.E.R.C., and a Queen Elizabeth College Scholarship (P.N.B.).

References

1. Price, R.G. (1982) *Toxicology 23*, 99–134.
2. Price, R.G. (1979) in *Cell Populations* [Vol. 8, this series] (Reid, E., ed.), Horwood, Chichester, pp. 105–110.
3. Rush, G.F., Newton, J.F. & Hook, J.B. (1985) in *Renal Cell Heterogeneity and Target Cell Toxicity* (Bach, P.H. & Lock, E.A., eds.), Wiley, Chichester, pp. 107–118.
4. Wellwood, J.M., Simpson, P.M., Tighe, J.R. & Thompson, A.E. (1975) *Br. Med. J. iii*, 278–291.
5. Kempson, S.A., Ellis, B.G. & Price, R.G. (1977) *Chem.-Biol. Interact. 18*, 217–234.
6. Trump, B.F., Berezesky, I.K., Lipsky, M.M. & Jones, T.W. (1985) *as for* 3., pp. 31–42.
7. Ellis, B.G., Price, R.G. & Topham, J.C. (1973) *Chem.-Biol. Interact. 7*, 101–113.
8. Ross, D.A. & Guder, W.G. (1982) in *Metabolic Compartmentation* (Sies, H., ed.), Academic Press, New York, pp. 363–409.
9. Ellis, B.G., Price, R.G. & Topham, J.C. (1973) *Chem.-Biol. Interact. 7*, 131–142.
10. Fielding, B.A., Price, D.A. & Houlton, C.A. (1983) *Clin. Chem. 29*, 355–357.
11. Beetham, R., Dawnay, A., Landon, J. & Cattell, W.R. (1986) *Clin. Chem. 31*, 1364–1367.
12. Carlier, Y., Colle, A., Tachon, P., Bout, D. & Capron, A. (1981) *J. Immunol. Meths. 40*, 231–238.
13. Kusano, E., Suzuki, M., Asano, Y., Itoh, Y., Takagi, K. & Kawai, T. (1985) *Nephron 41*, 320–324.
14. Dawnay, A.B., McLean, C. & Cattell, W.R. (1980) *Biochem. J. 185*, 679–687.
15. Balant, L. & Fabre, J. (1979) in *Diagnostic Significance of Enzymes and Proteins in Urine* (Dubach, U.C. & Schmidt, U., eds.), Hans Huber, Bern, pp. 216–234.
16. Marshall, T., Williams, K.M. & Vesterberg, O. (1985) *Electrophoresis 6*, 47–52.

17. Fowler, J.S.L. (1987) in *Nephrotoxicity, Assessment and Pathogenesis* (Bach, P.H., Grasso, P., Molland, E.A. & Bridges, J.W., eds.), Wiley, Chichester, pp. 66-77.
18. Whiting, P.H., Ross, I.S. & Borthwick, L. (1979) *Clin. Chim. Acta 92*, 459-463.
19. Booth, A.G. & Kenny, A.J. (1974) *Biochem. J. 158*, 581-591.
20. Ku, D.D. & Meezan, E. (1984) *J. Pharmacol. Exp. Ther. 229*, 664-670.
21. Yuen, C-T., Kind, P.R.N., Price, R.G., Praill, P.F.C. & Richardson, A.C. (1984) *Ann. Clin. Biochem. 21*, 295-300.

#F-3

THE INVESTIGATION OF SUBCELLULAR DERANGEMENTS IN NEPHROTOXICITY

Peter H. Bach

The Nephrotoxicity Research Group, Robens Institute
of Industrial and Environmental Health and Safety
University of Surrey, Guildford GU2 5XH, U.K.

Particular chemicals in vivo *specifically damage the glomeruli, the medullary cells or the proximal tubular cells. Elucidation of the mechanism(s) of such unique interactions between the target cell and the toxin has proved most difficult using conventional biochemical methods. Histochemical methods may be helpful, e.g. to define toxin distributions or the location of enzymes that effect requisite activation. Isolated renal cells may be used to study the molecular basis of target cell toxicity, provided that - as we have shown for several such chemicals - the same cell type is damaged* in vitro.

Freshly isolated proximal tubular fragments and cultured proximal tubular cells need an intact renal anion-transport system for the toxicity of the haloalkene conjugates. Cultured renal medullary interstitial cells* (and various cell lines) have been used to show that the toxicity of 2-bromoethanamine† is closely related to the presence of peroxidative enzyme activity and intracellular lipid droplets, suggesting a role for lipid peroxidation. The selective toxic effects of adriamycin towards the epithelial cells of freshly isolated and cultured glomeruli appears to be related to the selective uptake into epithelial cells, where peroxidative activation or redox cycling may explain the toxicity.

Many medicines and both industrial and agricultural chemicals cause renal lesions [1-5] that are similar to naturally occurring diseases. These chemicals often target selectively for a group of cells or a single morphologically distinguishable cell type [4].

---

* termed RMIC's below; † BEA. *Other abbreviations include:* HCBD, hexachloro-1:3-butadiene; PAN, puramycin aminonucleoside; PG, prostaglandin; SOD, superoxide dismutase.

There is a very important need to identify the mechanism(s) of injury, but this depends on the ability to differentiate between the primary pathophysiological events that led from insult to cell death, as opposed to those processes that are an indirect consequence.   The complexity of the kidney has limited the identification of most molecular mechanisms. *In vivo* the number of affected cells is very small compared to the total cell population, which makes conventional biochemical methods of subcellular investigation inappropriate. Thus special subcellular techniques have to be used to study the biochemical changes.

Histochemistry has been applied to define the biochemical properties of individual cells in terms of their micromolecular and macromolecular components and how endogenous and exogenous molecules are distributed within the cell. Isolated cells from the affected region offer a simple biological system in which the interaction between a cell and a chemical can be studied, and identify the primary molecular mechanism(s) and the cascade of degenerative changes that follow. Provided that isolated cells are sensitive to the same range of chemicals that cause lesions *in vivo*,   then it is likely that whatever is learned about the mcehanism(s) will be directly applicable to the intact organ.

Any attempt to define the mechanism of selective target–selective toxicity *in vitro* depends on ability to show *a priori* that the injury mechanism continues to operate in isolated cells, and furthermore that the cells are not sensitive to toxic substances that do not damage them in the living animal.

KIDNEY TOXICITY

There is a marked biochemical and morphological heterogeneity both along and between nephrons [4]. It is not certain what chemicals reach the kidney because of extra–renal and renal metabolism. Further-more, compartmentalization [6] within the kidney is very complex due to a series of transcellular pH gradients, solute and/or solvent permeability,   intracellular binding of molecules and the counter-current concentration mechanism. Each of these processes can trap or exclude metabolites selectively in (or from) one of the different cell types. Some filtered chemicals are actively reabsorbed from the nephron lumen, whereas others are secreted into it. The combination of these physiological factors contributes to the distribution of chemicals within the kidney and may also explain target selectivity. The accumulation or exclusion of substances from renal cells also makes it difficult to choose test concentrations of chemicals for *in vitro* studies that are realistic to the *in vivo* situation, and not excessive and therefore non–specifically cytotoxic.

RENAL METABOLISM AND METABOLIC ACTIVATION OF CHEMICALS

The molecular basis of renal cell injury [1-5, 7] may hinge on direct interaction between the chemical and the target cell. These interactions may entail an accumulation of the xenobiotic or endogenous molecules, interference with normal intermediary metabolism, and especially detoxification or metabolism to toxic products such as electrophilic intermediates, unstable reactive intermediates or free radicals [7]. Cellular constituents – free-radical scavengers (e.g. carotenoids and retinoids), anti-oxidants (e.g. α-tocopherol and ascorbic acid) and the nucleophile, reduced GSH – combine with the reactive species to protect cells. Reactive oxygen species such as the superoxide anion ($O_2^{\cdot-}$), hydroxy radicals ($OH^{\cdot}$) or hydrogen peroxide (H_2O_2) may also be generated, directly or as a secondary consequence of the presence of other unstable intermediates [8].

The inactivation of these reactive oxygen (and other) species depends on the compartment in which they are generated, and on the presence of the protective anti-oxidants, nucleophiles, etc., and/or those enzyme systems that can change each intermediate. Neither the superoxide anion nor H_2O_2 *per se* represents the ultimate toxic oxygen species, provided that the enzymes are present to deactivate them by conversion, respectively, to H_2O_2 + water (SOD) and to water + oxygen (catalase) [9]. Hydroxy radicals do not have an enzyme system that converts them to less active species, and once formed their toxic potential will be reduced only by anti-oxidants, free-radical scavengers and nucleophiles.

It is not certain how cell death is caused by the reactive intermediate(s); but if these protective mechanisms are absent or depleted the reactive intermediates may bind to essential small molecules and/or macromolecules, e.g. lipids, proteins and nucleic acids, or may cause metabolic injury, etc. [7]. In the absence of anti-oxidants, reactive species can also interact with polyunsaturated fatty acids to form peroxylipid free radicals, which give rise to a chain reaction. Most of the lipid peroxidation processes linked to cell degeneration are envisaged to be confined to cellular membranes, but free polyunsaturated phospholipid droplets would also support lipid peroxidation [10].

PROXIMAL TUBULAR INJURY

A number of chemicals damage the proximal tubules, and many of these injuries are thought to be mediated by reactive intermediates [11-16]. The biotransformation capacity of the proximal tubules is qualitatively similar to that of the liver (but quantitatively less), as shown by the presence of the cytochrome P-450 mixed-function oxidase system [11]. Antibodies to cytochrome P-450 show localization of this enzyme in the proximal tubule [17]. Paracetamol (acetaminophen) damages the proximal tubule following P-450-mediated metabolic activation to a quinolimine, and causes GSH depletion and covalent binding of material derived from the parent compound [12].   There is now significant evidence to suggest that the deacetylated metabolite of paracetamol, *p*-aminophenol, is the proximate toxin [13] which is also activated to a quinolimine.

High doses of paracetamol decrease GSH and increase covalent binding [18].   The formation of the GSH conjugate has been shown by the presence of both the paracetamol-mercapturic acid and the cysteine conjugate in urine and by its formation in isolated renal cells [19].   The essential protective role of GSH was also shown by the marked exacerbatory effect of pre-dosing with any compound that depletes GSH [20].   Relatively little is known about the renal distribution of protective molecules such as   free-radical scavengers and anti-oxidants,   GSH [21], SOD [22] and catalase [23] are all located in the proximal tubule.

There are other oxidative systems in the kidney that may form reactive intermediates from chemicals.   For example, peroxisomes are confined to the P3 portion of the proximal tubule, being absent from the glomeruli and the distal nephron [24].   These organelles contain urate oxidase, catalase, and D-amino acid oxidase [25]. The renal roles of the peroxisomes remain ill-defined. It is generally assumed that urate oxidase is responsible for the conversion of uric acid to urea [26], but the physiological function of the enzyme is far from certain.   Reddy [14] has speculated on the role of D-amino acid oxidase in the genesis of the highly localized necrosis confined to the P3 region of the nephron following the administration of D-serine, because of the release of H_2O_2 when the D-amino acids are oxidized to their α-pyruvate derivatives.   Normally deactivation by catalase would be expected, and a mechanistic explanation for this specifically localized and rapidly developing lesion is still needed. (*Editor's note*.- Ref. [26] deals with chick embryo.)

There are also non-oxidative mechanisms that may mediate proximal tubular injury.   A single injection of HCBD selectively necroses the pars recta in the rat [15].   HCBD is metabolized in the liver to its GSH conjugate; this is excreted via the kidney, where it is metabolized by γ-glutamyl transferase and cysteinyl glycinase, giving the cysteine conjugate.   The transferase is sharply confined to the brush border of the proximal tubule. Before this conjugate is metabolized to the mercapturate (by acetylation) in the pars recta, β-lyase cleaves the C-S bond (generating pyruvate and ammonia), and thiochlorobutadiene -   a potent alkylating sulphur mustard - is formed [16].   At present there is no known physiological role for renal β-lyase.

Recently   the   organic   anionic   transport   blocking   agent, probenecid, has been shown to inhibit the uptake of the cysteine conjugate of HCBD and protect the kidney both *in vivo* and in renal

slices [27].   The proximal tubule is selectively damaged *in vitro*
by the proximate metabolite HCBD-**N**-acetylcysteine (HCBD-NaC) which
effectively inhibits the incorporation of amino acids into proximal
tubular macromolecules [28, 29]. Probenecid protects animals exposed
to HCBD or HCBD-NaC *in vivo* [27, 30], and tubular fragments pre-exposed
to HCBD-NaC almost completely regain their protein-synthetic capacity
when exposed to 400 μM probenecid [29].   HCBD-NaC also selectively
targets for cultured proximal tubular cells   (C.P. Ketley, unpublished),
but has no adverse effect on fresh or cultured glomeruli (I. Ahmed,
unpublished).

MEDULLARY INJURY

Renal papillary necrosis is a consequence of long-term mixed
analgesic abuse [31], and can be induced in experimental animals
using analgesics and non-steroidal anti-inflammatory drugs over
6-12 months or acutely by a single dose of BEA hydrobromide. A variety
of mechanisms have been proposed [31], and chronically induced renal
papillary necrosis has been explained in terms of metabolic activation
[31]. The fact that the medulla is very low in mixed-function oxidase
enzymes [32] excludes P-450-mediated reactive intermediates, because
the short half-life of the 'ultimate' toxic intermediates means that
they can only exert their toxic effects at or very close to the site
of formation [7].

PG synthase, an enzyme with two discrete and biochemically different activities (cyclo-oxygenase and PG endoperoxidase), also activates
both paracetamol [33-36] and *p*-phenetidine [35, 36] to produce biologically reactive quinolimine species which bind protein [35, 37],
GSH [36] and/or nucleic acids [38].   This is due to arachidonic acid-
dependent PG endoperoxidase-mediated activation.   Cyclo-oxygenase
activity occurs in the medulla [39], where its loci are the interstitial
cells, collecting ducts, arteriolar endothelial cells and glomerular
endothelial cells as shown by use of an antibody raised to this enzyme
[40]. Individual PG's show a discrete distribution [41]. Arachidonic
acid-dependent diaminobenzidine oxidation shows intense staining of
the medullary collecting ducts and interstitial cells [42] which
may represent PG hydroperoxidase or a lipoxygenase.

Role of renal medullary interstitial cells (RMIC's)

The high levels of peroxidative enzymes in the medulla may,
then, be pertinent to the generation of reactive intermediates; but
the widespread localization of PG synthesis or peroxidation in the
medulla (or elsewhere in the kidney) makes it difficult to explain
why the papillotoxic compounds target selectively for the medullary
interstitial cells, unless other predisposing factors are also
involved.

The RMIC's have very low GSH levels [43], and both catalase
[23, 25] and SOD [22] activities are absent.   This suggests that

reactive intermediates formed in the medulla would not be quenched. More importantly, the RMIC's contain very numerous lipid droplets, the distribution of which has been well characterized light-microscopically [44] and ultrastructurally [45]. Ultracentrifugation [46] and chemical analysis [47] of these lipid droplets has shown that they are particularly rich in polyunsaturated fatty acids, which would provide an ideal fuel to sustain lipid peroxidation, if free radicals were generated by the peroxidation of papillotoxic chemicals. Thus, the RMIC's have an extensive quantity of fuel for oxidation and would have a propensity to form free radicals and peroxidative chain reactions, once such an event was initiated [31].

A single dose of BEA hydrobromide damages the renal medulla within a few days. The RMIC's undergo the earliest degenerative changes [48], which parallel the target selectivity of other papillo-toxins including analgesics and non-steroidal anti-inflammatories. At present there is no full explanation for this highly selective targeting for interstitial cells, but BEA is amenable to oxidation to a reactive intermediate via its halide group, or it forms an alkyla-ting agent which could generate reactive oxygen species and/or initiate lipid peroxidation [31].

We have cultured rodent RMIC's and shown that they are sensitive to established papillotoxins *in vitro* [49]. Fluorescent probes have shown high levels of peroxidase activity and/or peroxides, and Nile Red confirms the presence of lipid droplets [50]. The mechanism of medullary necrosis is thought to be linked to the peroxidative activ-ation of analgesics to form reactive intermediates [51]. In an attempt to confirm this association, we have also studied the sensitivity of cell lines to papillotoxins. Both RMIC's and 3T3 fibroblasts are sensitive to BEA [50], but there were no cytotoxic changes in MDCK or HaK cells exposed to a 10-fold increased concentration of BEA for 4 times as long. RMIC's are thought to be fibroblastic in origin, hence the comparison with 3T3 cells, whereas HaK and MDCK represent renal epithelial cells [50]. The absence of lipid droplets and peroxidase activity from HaK and MDCK cells may explain the lack of BEA cytotoxicity.

GLOMERULAR INJURY

Puramycin aminonucleoside (PAN) selectively damages glomerular epithelial cells *in vivo* and causes loss of the basement membrane-associated polyanion, retraction of the podocytes and a proteinuria [52]. PAN also damages the epithelial (but not mesangial) cells in cultured glomeruli [53]. We are ignorant about the molecular changes that develop within the epithelia. The glomeruli contain both PG synthase and lipoxygenase activity [54], but little is known about the presence of protective substances, although both SOD [22] and catalase [23] are absent from this region of the kidney.

PAN is broken down to hypoxanthine, a substrate for xanthine oxidase (or perhaps other oxidases) which forms uric acid and the superoxide anion. The level of xanthine oxidase is lower in kidney than in other organs (liver, spleen), but its distribution is unknown. A role for the superoxide anion seems likely, because the co-administration of SOD, but not catalase or dimethylsulphoxide, protects *in vivo*. Allopurinol, a selective inhibitor of xanthine oxidase, also inhibited the PAN-induced proteinuria [55].

Adriamycin also damages the glomerular epithelial cells [56], causing fusion of podocytes, loss of basement membrane polyanions, and a proteinuria. Its toxicity has also been related to the NAD(P)H-dependent reduction to a free radical that is easily oxidized back to the parent compound [57], due to the comparatively high O_2 tension. This redox cycling produces the superoxide anion radical ($O_2^{\cdot-}$) making the parent compound available for reduction to the radicals. The superoxide anion may be the reactive oxygen species that causes the cell lesion, because of the absence of catalase and SOD [22, 23], which would facilitate the generation of other reactive oxygen species - especially the hydroxyl radical ($OH^{\cdot}$), which in the absence of anti-oxidants, free-radical scavengers, etc., has the potential to cause cell death.

Whereas adriamycin at levels of < 0.05 mM inhibits amino acid incorporation into freshly isolated rat and pig glomerular protein and basement membrane, similar effects are observed in proximal tubular fragments when concentrations are >0.5 mM (L. Dela Cruz, C.P. Ketley & I. Ahmed, unpublished data). If glomeruli are cultured to allow the individual outgrowth of epithelial and mesangial cells, it is only the former that are affected. The use of fluorescent microscopy has suggested that adriamycin is taken up selectively into epithelial but not mesangial cells [50].

CONCLUSION

Exploiting the subcellular approach for elucidating the mechanisms of target cell injury can be effected using histochemistry and isolated cells from different regions of the kidney, and offers the potential to understand intracellular events that lead to cell death or injury. This approach also could lead to development of rational *in vitro* tests for evaluating chemical safety and facilitate drug design at an early stage.

Acknowledgements

The research was supported by the Wellcome Trust, the Humane Research Trust, the Johns Hopkins Center for Alternatives to Animals in Testing, the Dr. Hadwen Trust for Humane Research, and the Cancer Research Campaign. I am grateful to M.E. van Ek for preparing the manuscript.

References

1. Hook, J.B., ed. (1979) *Toxicology of the Kidney*, Raven Press, New York, 276 pp.
2. Porter, G.A., ed. (1982) *Nephrotoxic Mechanisms of Drugs and Environmental Toxins*, Plenum, New York, 466 pp.
3. Bach, P.H., Bonner, F.W., Bridges, J.W. & Lock, E.A., eds. (1982) *Nephrotoxicity: Assessment and Pathogenesis*, Wiley, Chichester, 528 pp.
4. Bach, P.H. & Lock, E.A., eds. (1985) *Renal Heterogeneity and Target Cell Toxicity*, Wiley, Chichester, 571 pp.
5. Bach, P.H. & Lock, E.A., eds. (1987) *Nephrotoxicity in the Experimental and the Clinical Situation*, MTP Press, Lancaster,
6. Mudge, G.H. (1985) *as for* 4., pp. 1-12.
7. Aldrich, N.W. (1981) *Trends Pharmacol.Sci. 2*, 228-231.
8. Seis, H., ed. (1985) *Oxidative Stress*, Academic Press, London, 507 pp.
9. Flohe, L. (1982) in *Free Radicals in Biology* (Pryor, W.A., ed.), Academic Press, New York, pp. 223-254.
10. Bus, J.S. & Gibson, J.E. (1979) *Rev. Biochem. Toxicol. 1*, 125-149.
11. Anders, M.W. (1980) *Kidney Internat. 18*, 636-647.
12. McMurtry, R.J., Snodgrass, W.R. & Mitchell, J.R. (1978) *Toxicol. Appl. Pharmacol. 46*, 87-100.
13. Carpenter, H.M. & Mudge, G.H. (1981) *J. Pharmacol. Exp. Ther. 218*, 161-167.
14. Reddy, J.K., Rao, M.S., Moody, D.E. & Qureshi, S.A. (1976) *J. Histochem. Cytochem. 24*, 1239-1248.
15. Ismael, J., Pratt, I. & Mudge, G.H. (1984) *J. Path. 142*, 195-203.
16. Lock, E.A. (1982) *as for* 3., pp. 396-408.
17. Dees, J., Parkhill, L.K., Okita, R.T., Yasukochi, Y. & Masters, B.S. (1982) *as for* 3., 246-249.
18. Newton, J.F., Kuo, C-H., Gemborys, M.W., Mudge, G.H. & Hook, J.B. (1982) *Toxicol. Appl. Pharmacol. 65*, 336-344.
19. Moldeus, P., Jones, D.P., Ormstad, K. & Orrenius, S. (1978) *Biochem. Biophys. Res. Comm. 83*, 195-200.
20. Newton, J.F., Yoshimoto, M., Bernstein, J., Rush, G.F. & Hook, J.B. (1983) *Toxicol. Appl. Pharmacol. 69*, 291-306.
21. Chieco, P. & Boor, P.J. (1983) *J. Histochem. Cytochem. 31*, 975-976.
22. Thaete, L.G., Crouch, R.K. & Spicer, S.S. (1985) *J. Histochem. Cytochem. 33*, 803-808.
23. Gilloteaux, J. & Steggles, A.W. (1983) *Cell Biol. Internat. Rept. 7*, 31-33.
24. Novikoff, A.B. & Goldfischer, S. (1969) *J. Histochem. Cytochem. 17*, 675-680.
25. Roels, F., Wisse, E., de Prest, B. & van der Meulen, J. (1974) *Histochem. J. 6*, 91-92.
26. Essner, E. (1970) *J. Histochem. Cytochem. 18*, 80-91.
27. Lock, E.A. (1985) *as for* 4., pp. 149-152.
28. Ketley, C.P. & Bach, P.H. (1987) *Toxicol. Appl. Pharmacol.* (submitted).

29. Ketley, C.P. & Bach, P.H. (1987) *Toxicol. Lett.* (submitted).
30. Lock, E.A., Odum, J. & Ormond, P. (1987) *Arch. Toxicol. 59*, 12–15.
31. Bach, P.H. & Bridges, J.W. (1985) *CRC Crit. Rev. Toxicol. 15*, 217–439.
32. Zenser, T.V., Mattammal, M.B., Herman, C.A., Joshi, S. & Davis, B.B. (1978) *Biochim. Biophys. Acta 542*, 486–495.
33. Zenser, T.V., Mattammal, M.B., Brown, W.W. & Davis, B.B. (1979) *Kidney Internat. 16*, 688–694.
34. Mohandas, J., Duggin, G.G., Horvath, J.C. & Tiller, D.J. (1981) *Toxicol. Appl. Pharmacol. 61*, 252–259.
35. Boyd, J.A. & Eling, T.E. (1981) *J. Pharmacol. Exp. Ther. 219*, 659–664.
36. Moldeus, P. & Rahimtula, A. (1980) *Biochem. Biophys. Res. Comm. 96*, 469–475.
37. Anderssen, B., Larsson, R., Rahimtula, A. & Moldeus, P. (1983) *Biochem. Pharmacol. 32*, 1045–1050.
38. Kadlubar, F.F., Fredrick, C.B., Weiss, C.C. & Zenser, T.V. (1982) *Biochem. Biophys. Res. Comm. 108*, 253–258.
39. Davis, B.B., Mattammal, M.B. & Zenser, T.V. (1981) *Nephron 27*, 187–196.
40. Smith, W.L. & Bell, T.G. (1978) *Am.J. Physiol. 235*, F451–F457.
41. Perex, G. & McGuckin, J. (1972) *Prostaglandins 2*, 393–398.
42. Litwin, J.A. (1977) *Histochemistry 53*, 301–315.
43. Mudge, G.H., Gemborys, M.W. & Duggin, G.G. (1978) *J. Pharmacol. Exp. Ther. 206*, 218–226.
44. Burry, A.F. (1968) *Nephron 5*, 185–201.
45. Bohman, S-O. (1980) in *The Renal Papilla and Hypertension* (Mandal, A.K. & Boham, S-O., eds.), Plenum, New York, pp. 7–34.
46. Bojesen, I. (1974) *Lipids 9*, 835–843.
47. Bojesen, I.N. (1980) in *The Renal Papilla and Hypertension* (Mandal, A.K. & Boham, S-O., eds.), Plenum, New York, pp. 121–147.
48. Gregg, N., Courtauld, E.A. & Bach, P.H. (1987) *Br. J. Exp. Path.* (submitted).
49. Benns, S.E., Dixit, M., Ahmed, I., Ketley, C.P. & Bach, P.H. (1985) in *Alternative Methods in Toxicology*, Vol. 3 (Goldberg, A., ed.), M.A. Liebert, Baltimore, pp. 435–447.
50. Bach, P.H., Ketley, C.P., Dixit, M. & Ahmed, I. (1986) *Food Chem. Toxicol. 24*, 775–779.
51. Bach, P.H. & Bridges, J.W. (1984) *Prostagland. Leuk. Med. 15*, 251–274.
52. Fajaro, L.F., Eltringham, H.R., Stewart, J.R. & Klauger, M.R. (1980) *Lab. Invest. 43*, 242–253.
53. Kreisberg, J.I. & Karnovsky, M.J. (1983) *Kidney Internat. 23*, 439–447.
54. Sraer, J., Rigaud, M., Bens, M., Rabinovitch, H. & Ardaillou, R. (1983) *J. Biol. Chem. 258*, 4325–4330.
55. Diamond, J.R., Bonventre, J.V. & Karnovsky, M.J. (1985) *Kidney Internat. 29*, 478–483.

56. Bertani, T., Poggi, A., Pozzoni, R., Delaini, F., Sacchi, G.,
    Thoua, Y., Mecca, G., Remuzzi, G. & Donati, M.B. (1982) *Lab.
    Invest. 46*, 16-23.
57. Gianna, L., Corden, B.J. & Meyers, C.E. (1983) *Rev. Biochem.
    Toxicol. 5*, 1-87.

#F-4

HETEROGENEITY AND FUNCTION OF KIDNEY CORTEX LYSOSOMES

Knut-Jan Andersen and Miloslav Dobrota

Section for Clinical Research and Molecular Medicine
Medical Department A, University of Bergen
N-5016 Haukeland Sykehus, Bergen, Norway   and
Robens Institute of Industrial and Environmental Health
and Safety, University of Surrey, Guildford GU2 5XH, U.K.

Reliable distribution profiles have been obtained for rat cortical lysosomal populations, fractionated by rate and isopycnic sedimentation and characterized comprehensively by employing marker enzyzes and labelled proteins as population probes. The known distributions of lysosomal enzymes manifest along the intact nephron and the morphologically observed variety of lysosomes in the different cortical cell types suggest considerable lysosomal heterogeneity. With cortical homogenates we have shown heterogeneous lysosomal populations, reflecting the variety of lysosomes in the major cell type of the cortex and cell-type heterogeneity, viz. (1) large lysosomes, which band at d = 1.235 in sucrose gradients and probably represent protein droplets originating from the proximal tubule cells; (2) small dense lysosomes, also banding at d = 1.235 which are also probably from proximal tubule cells; (3) small light lysosomes, banding at d = 1.20, which are probably a mixture of lysosomes from distal and proximal regions of the nephron.

Experiments with ^{125}I-lysozyme and ^{109}Cd-thionein administered i.v. suggest that all the lysosomal populations identified are at some stage involved with uptake and catabolism of these two proteins. Their preferential accumulation in the small lysosomes may reflect their rapid digestion in the large protein droplets, whereas they remain for longer periods in the small lysosomes whose content of proteases is relatively lower.

From the earliest work on the subfractionation of kidney organelles [1, 2] it was clear that cortical lysosomes were heterogeneous in terms of their enzyme content and morphology. This heterogeneity may partly be due to the complex cellular heterogeneity of the kidney

[3] where cellular hydrolases are present, in varying proportions, in all the segments of the nephron [4] suggesting considerable heterogeneity of lysosomal populations attributable to different physiological functions of the various cell types of the nephron.   In addition, lysosomal heterogeneity such as differences in lysosomal size and density most certainly reflect dynamic changes within the cell which take place as a result of membrane fusion following endocytosis, membrane recycling, and the fusion of membranes involved in the formation of autophagic vacuoles.   For several years we have been studying heterogeneity in relation to kidney cortex lysosome function. This article outlines the experimental approach used [5-7] to delineate the most likely cellular origin of the different lysosomal populations found in the kidney cortex homogenate, and also to study the lysosomal involvement in nephrotoxicity and in renal diseases.

EXPERIMENTAL APPROACH

Through the experimental approach outlined in Scheme 1 we have been able to separate three distinct populations of lysosomes from the rat kidney cortex [5, 7], namely:
(1) large lysosomes, at d 1.235, probably protein droplets from proximal tubule cells;
(2) small dense lysosomes, at d 1.235, probably from proximal tubule cells likewise;
(3) small 'light' lysosomes, at d 1.20, probably from both proximal and distal regions.
In addition our microsomal fraction at d 1.18-1.20 contained (4) some very small particles containing acid hydrolases, in a heterogeneous mixture (cf. Fig. 1 legend).

The properties of lysosomal populations isolated from rat kidney cortex are summarized in Fig. 1 as a classical S-ρ diagram, showing the relationship of approximate sedimentation rate or size to banding density in sucrose [8].

LYSOSOMAL FUNCTION IN RELATION TO HETEROGENEITY

Renal accumulation of low mol. wt. proteins occurs mainly in the convoluted proximal tubule and also, to a lesser extent, in the straight proximal tubule. Following i.v. injection of ^{125}I-lysozyme and [109]Cd-thionein which are both avidly taken up by the proximal tubule cells, we have shown that these two proteins are indeed rapidly incorporated into all the lysosomal populations identified [7, 9, 10], demonstrating that they are involved with uptake and catabolism of proteins.   However, recent observations have demonstrated that cathepsin B and cathepsin D, which have similar distributions, are located mainly in the dense lysosomes [11].   This may explain the significant accumulation of ^{125}I-lysozyme and [109]Cd-thionein in the small lysosomes [7, 9] as these are relatively poorer in proteases.

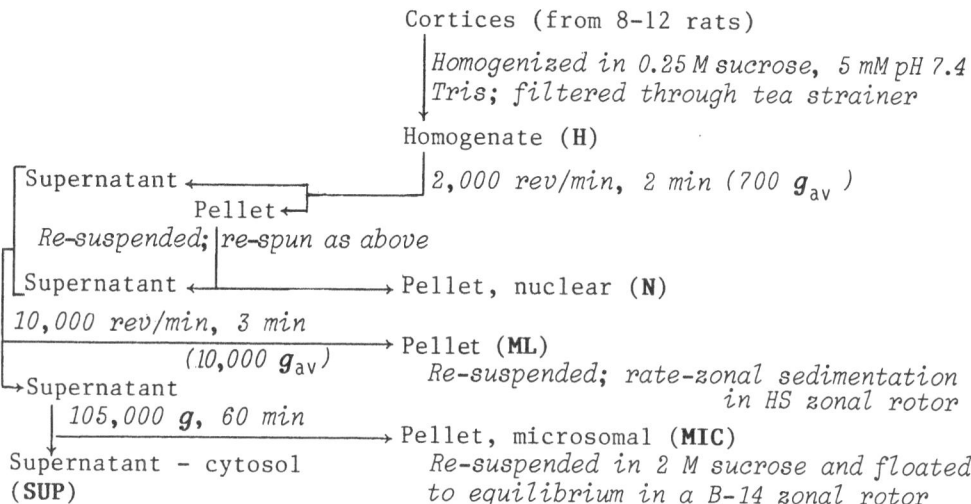

Cortices (from 8-12 rats)

Homogenized in 0.25 M sucrose, 5 mM pH 7.4
Tris; filtered through tea strainer

Homogenate (**H**)

2,000 rev/min, 2 min (700 g_{av})

Supernatant ←
    Pellet ←
Re-suspended; re-spun as above

Supernatant ← ————————→ Pellet, nuclear (**N**)

10,000 rev/min, 3 min
        (10,000 g_{av}) ————————→ Pellet (**ML**)
→Supernatant   *Re-suspended; rate-zonal sedimentation*
                                in HS zonal rotor

105,000 g, 60 min
              ————————→ Pellet, microsomal (**MIC**)
Supernatant – cytosol   *Re-suspended in 2 M sucrose and floated*
(**SUP**)   *to equilibrium in a B-14 zonal rotor*

Scheme 1. Preparation of subcellular fractions, providing reliable distribution profiles for cortical 'lysosomal' populations which have been further fractionated by rate and isopycnic sedimentation and characterized by employing a comprehensive range of marker enzymes, by using labelled proteins as probes for various lysosomal populations, and by morphological examination [5-7].

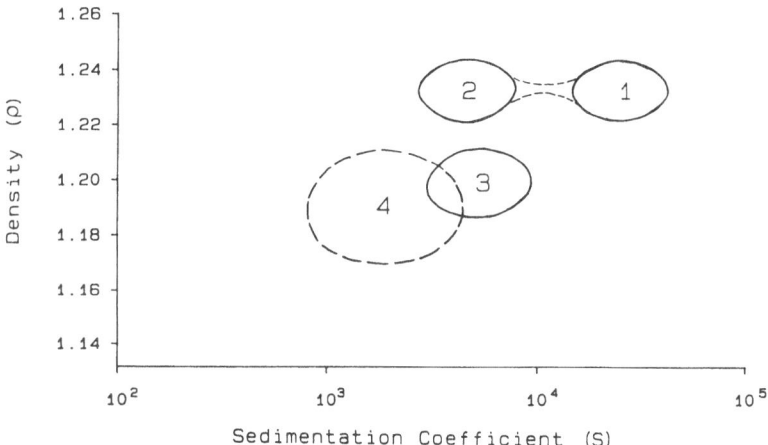

Fig. 1. S-ρ diagram, showing approx. S values (based on values given by Hinton & Dobrota [8]) and isopycnic (equilibrium) banding of cortical lysosomal populations fractionated in sucrose gradients. See text for nephron attributions of (1)-(3). Both (1), 'protein droplets', and (2), possibly a subpopulation of large lysosomes, contain all assayed lysosomal enzymes; (3) contain acid glycosidases but not cathepsin B or D, nor acid RNase. (4) = very small (microsomal) 'light' lysosomes, probably including lysosomal fragments and endocytic vesicles from various cell types of the nephron.

Table 1. Time course of uptake and processing of ³H-labelled
Cd-thionein: % distribution of ³H label in cortex subcellular
fractions following i.v. injection into rats. The labelled thionein
was injected at the stated times before sacrifice and sub-fractiona-
tion of the cortical homogenate (H; Scheme 1). ³H % values for small
and large lysosomes relate to total activity in **ML** as subjected to
rate-sedimentation to furnish these two regions; the peaks were
distinguished by scrutiny of acid hydrolase distributions.

Time	% of **H**: **ML**	**MIC**	% of **ML**: Small lysosomes	Large lysosomes
10 min	25.7	26.4	53.8	3.5
30 min	31.1	16.0	43.3	9.1
1.5 h	79.8	<0.1	40.9	26.4

The loss, with time of ³H-labelled thionein from 'microsomes'
and the corresponding increase of ³H in fraction 'ML' of Scheme 1,
as also observed for the large lysosomes (Table 1), strongly suggest
transfer of some undegraded Cd-thionein from endocytic vesicles to
small lysosomes and then to large lysosomes (further discussed in
#F-5, which follows).

From the time-course studies with ³H-labelled Cd-thionein [7]
it is apparent that this protein is broken down exremely rapidly.
This breakdown probably starts in the endosome compartment, which
in macrophages has been reported [12] to possess proteolytic activity.
Application of the procedure of W.H. Evans [13] for preparing the
liver 'endosome' fraction did not reveal any endo- or exo-peptidases
in the endosome subfractions (Fig. 2). However, the recovery of a
significant amount of latent N-acetyl-β-glucosaminidase (NAG) activity
in the 'late' endosomal fraction, banding at d = 1.10 in sucrose medium,
may explain the early increase in the urinary excretion of this
particular enzyme under pathological conditions [15].

Acknowledgement

This work was supported by research grants from the Norwegian
Research Council for Science and the Humanities.

References

1. Straus, W. (1956) *J. Biophys. Biochem. Cytol.* 2, 513-521.
2. Straus, W. (1964) *J. Cell Biol.* 20, 497-507.
3. Tisher, C.C. (1981) in *The Kidney*, 2nd edn., Vol. 1
    (Brenner, B.M. & Rector, F.C., eds.), Saunders,
    Philadelphia, pp. 3-75.
4. LeHir, M., Dubach, U.C. & Schmidt, U. (1979) *Histochemistry 63*,
    245-251.

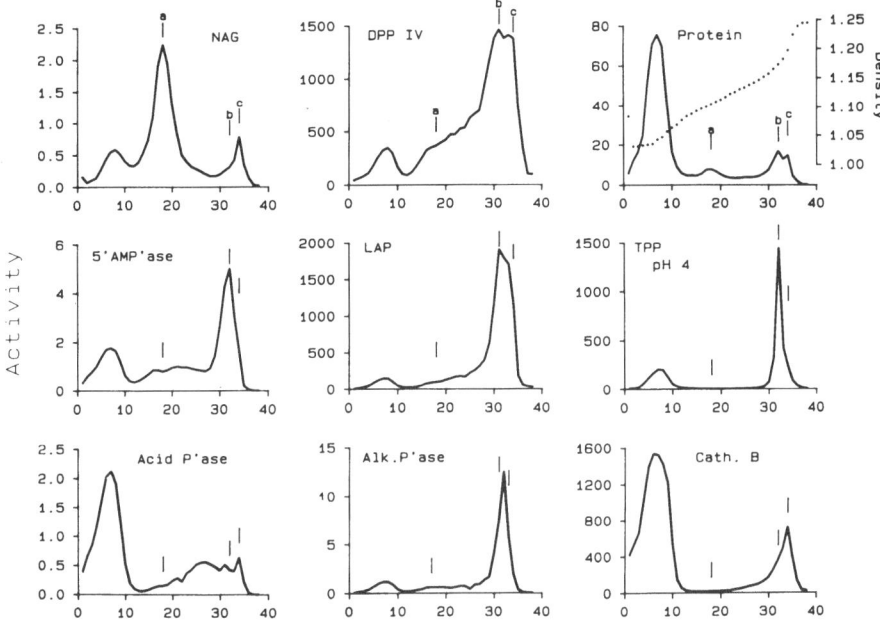

Fig. 2. Distribution of marker enzymes and protein after preparing low-density 'endosome' fractions according to the procedure
described for preparation of liver 'endosome' fractions [13]. The
'post-**ML** supernatant' (Scheme 1) was loaded into a B14 zonal rotor
containing 550 ml of sucrose gradient ranging from 15% to 40% (w/v)
and 150 ml 2.3 M sucrose as the cushion. The rotor was spun at
40,000 rev/min for 3 h, and fractions were assayed for marker
enzymes [7, 11, 14]. Values are expressed on a per-fraction basis:
mg for protein; μmol/min (U) for N-acetyl-β-glucosaminidase (NAG),
5'-nucleotidase (5'AMP'ase), acid β-glycerophosphatase (Acid P'ase)
and alkaline phosphatase (Alk. P'ase); mU for cathepsin B (Cath. B),
tripeptidyl peptidase 'pH 4.0' (TPP pH 4), dipeptidyl peptidase IV
(DPP IV) and leucine β-naphthylamidase (LAP). Positions a, b and c
are densities of 1.10, 1.175 and 1.18 respectively.

5. Andersen, K-J., Haga, H.J. & Dobrota, M. (1980) *Biochem. Soc.
   Trans. 8*, 597-598.
6. Andersen, K-J., Rygh, T. & Josok, R. (1986) *Int. J. Biochem. 18*,
   305-310.
7. Andersen, K-J., Haga, H.J. & Dobrota, M. (1987) *Kidney Internat.
   31*, 886-897.
8. Hinton, R.H. & Dobrota, M. (1976) *Density Gradient Centrifugation*
   in Vol. 6, *Laboratory Techniques in Biochemistry and Molecular
   Biology* (Work, T.S. & Work, E., eds.), North-Holland, Amsterdam,
   pp. 1-290.

9. Dobrota, M., Bonner, F.W. & Carter, B.A. (1982) in
    Nephrotoxicity: Assessment and Pathogenesis (Bach, P.H., Bonner, F.W.,
    Bridges, J.W. & Lock, E.A., eds.), Wiley, Chichester, pp. 320-324.
10. Dobrota, M., Haga, H.J. & Andersen, K-J. (1985) in *Renal
    Heterogeneity and Target Cell Toxicity* (Bach, P.H. & Lock, E.A.,
    eds.), Wiley, Chichester, pp. 241-244.
11. Andersen, K-J. & Dobrota, M. (1987) *Renal Physiol.*, in press.
12. Diment, S. & Stahl, P. (1985) *J. Biol. Chem.* 260, 15311-15317.
13. Evans, W.H. (1985) *Meths. Enzym.* 109, 246-257.
14. Andersen, K-J. & McDonald, J.K. (1987) *Am. J. Physiol.*, in
    press.
15. Andersen, K-J. & Haga, H.J. (1985) *Renal Physiol.* 8, 348-356.

#F-5

INVOLVEMENT OF KIDNEY LYSOSOMES IN THE NEPHROTOXICITY
OF CADMIUM AND OTHER METALS

M. Dobrota

Robens Institute of Industrial and Environmental Health
and Safety, University of Surrey
Guildford GU2 5XH, U.K.

*For studying uptake, redistribution and toxic mechanisms of metals (which associate with macromolecules), subcellular techniques are ideal. Their application to study of the renal handling of thionein labelled both with ^{109}Cd and with ^3H, administered i.v., indicates significant differences between the distributions of the two isotopic labels amongst the subcellular organelles. The rapid appearance and retention of ^{109}Cd (**Cd***) in the cytosol strongly suggests that the endosomal/lysosomal acidification and proteolytic enzymes cause a release of Cd^{2+} from the thionein and transfer to the cytosol where it binds to endogenous Zn-MT, whilst the thionein moiety is rapidly degraded. Thus Cd does not appear to accumulate in lysosomes although the very active endosomes/lysosomes system of the proximal tubule cells appears to be instrumental in the delivery and retention of the metal in the cytosol.*

Preliminary experiments on renal handling of Au, Ni and Pt at the subcellular level suggest that lysosomes may also play a role in the uptake, retention and nephrotoxicity of these metals.

Lysosomal association and accumulation of many heavy metals may lead to their inactivation or alternatively to a potentiation of toxic effects. In the mammalian kidney, Au, Hg, Pb, U, etc. are accumulated in lysosomes of the proximal tubule cells where these and other metals (e.g. Cd, Cu) may cause significant cellular damage and subsequently renal failure [1]. Probably the uptake and accumu-

---

* *Representation adopted by Editor:* **Cd** = ^{109}Cd, **Au** = ^{195}Au, **Ni** = ^{63}Ni. Elements represented by symbols (e.g. Cd, *not* cadmium) even where name used in author's text. MT = metallothionein. **ML** signifies a centrifugal fraction containing mitochondria and lysosomes.

lation of heavy metals into kidney lysosomes occurs because they
are inherently lysosomotropic and are delivered aboard carrier proteins
or directly by virtue of diffusion, lipophilicity or pH. Since toxicity
may arise because of uncoupling or the metal from its ligand, it
is important to embrace in the lysosomal compartment the 'endosome'
where dissociation of receptor and ligand takes place and degradation
of ligand may indeed begin.

The processes of uptake and intracellular transport of metals
are clearly very important in understanding their toxic mechanisms.
Preliminary studies on uptake and renal handling of Cd-thionein [2]
have illustrated the usefulness of carefully developed subcellular
approaches as employed in this study to examine the association of
metals with the various lysosomal populations of the kidney.

EXPERIMENTAL

Male Wistar albino rats were used in the studies described.
The classical subfractions of the kidney cortex homogenate were
prepared by differential pelleting exactly as in Fig. 1 of the
preceding article (#F-4). **ML** was further fractionated by rate-zonal
sedimentation in the HS-type zonal rotor (MSE Scientific Insts.)
in order to resolve the various lysosomal populations and other
organelles. Details are described elsewhere [3].

Dual-labelled Cd-thionein was prepared by incubating purified
hepatic MT-I with $CdCl_2$, isolating the protein, then labelling with
succinimidyl [^{3}H]propionate (Amersham Internatl.) and finally
isolating the pure protein from the reaction mixture by Sephadex
G-25 chromatography. Gold (**Au**) sodium aurothiomalate was prepared
and administered as described by Taylor et al. [4]. **Ni**(HIS)$_2$ was
prepared by mixing **NiCl**$_2$ and histidine at the appropriate molar concentrations.

Sephadex G-75 chromatography was employed to fractionate low
mol. wt. proteins from the kidney cytosols. The ^{3}H and **Cd** isotopes
were counted with an LKB Rackbeta 'Spectral' (Model 1219) counter
using automatic dual label correction.

RESULTS AND DISCUSSION

Cadmium [**Cd** signifies ^{109}Cd]

The time course of uptake and intracellular processing of dual-
labelled Cd-thionein administered i.v. is shown in Table 1 as % of
the dose recovered in the kidney cortex and also as % distribution
of the cortical label found in the classical subfractions. It is
important to note that the % of **Cd** label retained in the cortex remains
constant during the period 10 min–24 h with a steady transfer of
Cd from particles to cytosol over this period. Whilst the proportion

Table 1. Recovery in kidney cortex (% of the i.v. dose) and in classical subfractions (% of homogenate) of thionein labels – ^3H *(italics)* and ^{109}Cd **(Cd)**, at different times after injection of 0.2 ml containing 0.38 mg thionein, 1.92 μCi ^3H and 4.21 μCi **Cd**. Each time point represents 1 expt. with cortices from 5 animals.

Time	% in cortex	Nuclear	**ML**	Microsomal	Supernatant
10 min	*25*, 26	*11*, 12	*26*, 26	*26*, 24	*23*, 27
30 min	*15*, 23	*17*, 23	*32*, 27	*16*, 16	*47*, 50
1.5 h	*3*, 25	*19*, 15	*80*, 11	<*0.1*, 19	<*0.1*, 77
24 h	<*0.1*, 29	<*0.1*, 8	<*0.1*, 5	<*0.1*, 10	<*0.1*, 83

Fig. 1. Distribution patterns of ^3H and ^{109}Cd **(Cd)** after subfractionation of the cortical **ML** fraction by rate-zonal sedimentation at the 4 time points after administration of the dual-labelled Cd-thionein. Radioactivity expressed as cpm × 10³/0.5 ml of fraction. The long bar at centre of gradient represents the region of 'small lysosomes', and the short bar indicates the position of the 'large lysosomes'.

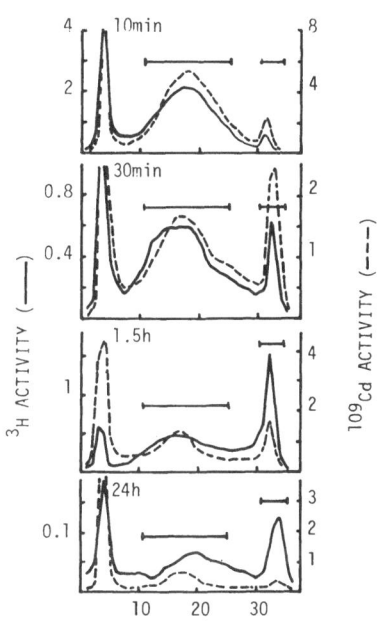

FRACTION No.

of the cortical **Cd** recovered in **ML** steadily decreases with time, the ^3H reaches a peak of 80% at 1.5 h, indicating that a proportion of the thionein moiety is broken down in lysosomes.

When **ML** is subfractionated by a rate-sedimentation method developed specifically to resolve the various lysosomal populations ([3], & preceding article), the distribution patterns of **Cd** and ^3H appear to be closely associated with these lysosomes (Fig. 1). Although no marker enzymes are illustrated in Fig. 1, the patterns for both isotopes closely follow the distribution patterns of lysosomal acid hydrolases as shown in our earlier study [2]. In establishing the association of **Cd** with lysosomes it should firstly be noted that over the period 10-30 min there is an increase in the **Cd** peak in the large lysosomes (Fig. 1) whilst right-hand peak sizes already show a shift of **Cd** from particles to the cytosol. Secondly, the distinct ^{3}H peak in the large lysosomes at 1.5 h suggests that the thionein is being degraded in these lysosomes. When the relative proportions of ^{3}H in the small- and large-lysosomal regions are expressed as % of the total ^{3}H label in **ML** (see Table 1 in the preceding art.), it is apparent that with time there is a distinct shift of MT as indicated by ^{3}H from small to large lysosomes.

The disappearance with time of ^{3}H and **Cd** from the microsomal fraction (Table 1), which most likely contains a significant proportion of endocytic vesicles present in the cortical homogenate, and the relative increase in ^{3}H label in **ML**, suggest that MT is first transferred from endocytic vesicles to small lysosomes.

It was initially surprising to find 23% of the label in the cytosol fraction at the 10 min time-point, especially since subsequent chromatography on Sephadex G-75 showed that the cytosolic ^{3}H label was associated exclusively with the intact MT molecule. Since the exogenously administered MT is not likely to cross cellular membranes it is most likely that its apparent cytosolic location is attributable to disruption of the endosome compartment and the release of the undegraded protein into the cytosol (supernatant fraction). This in turn would suggest that the uptake of Cd-thionein in the proximal tubule occurs mainly via fluid-phase endocytosis.

Fractionation of the cytosols from subsequent time-points by G-75 chromatography shows, from the ^{3}H patterns, that even by 30 min most of the MT is degraded to oligopeptides of low mol. wt. It thus appears that most of the MT is degraded very rapidly in the endosome compartment. Since at these early time-points (and also at 24 h) **Cd** is associated only with intact MT, it appears that Cd is transferred rapidly from the exogenous to the endogenous MT.

Altogether it appears that in the form of Cd-thionein, which is known to be considerably more nephrotoxic than Cd^{2+} [5] and which is also the form in which Cd (after subchronic exposure) is normally delivered to the kidney, it is the acidification and degradation in the endosome and lysosome compartments that are instrumental in the retention and accumulation of Cd and its toxicity in the kidney.

Gold [**Au** signifies ^{195}Au]

The kidney plays an important role in Au metabolism. Its rapid accumulation in the kidney after a single i.p. dose of sodium aurothiomalate leads to extensive tubular necrosis followed by regeneration

Table 2. Recovery in kidney cortex (% of the i.v. dose) and in classical subfractions (% of homogenate) of ^{63}Ni-histidine complex, **Ni**(HIS)$_2$, at different times after injection (200 µg Ni/kg).

Time	% in cortex	Nuclear	**ML**	Microsomal	Supernatant
10 min	5.1	9.2	18.9	14.0	55.0
1.5 h	7.8	11.9	27.4	19.6	56.8
3 h	6.1	11.9	28.3	14.6	48.0
24 h	0.9	18.7	48.3	13.5	32.9

[4] during which the Au levels in the kidney continue to increase. At the subcellular level these changes are characterized by a re-distribution of **Au** radiolabel from the cytosol into particles, notably the lysosome-containing **ML** fraction:
- at 30 min: nuclear 12.7%, **ML** 9.7%, microsomal 14.0%, supernatant 63.4%;
- at 48 h: nuclear 22.7%, **ML** 33.3%, microsomal 5.6%, supernatant 32.6%.

During this Au transfer there is an overall increase of renal Cu [see also R.J. Ward in Vol. 14, this series - *Ed.*] which appears to be attributable to an increase of Cu-thionein. These observations suggest that Au nephrotoxicity may be due to changes in Cu homeostasis and that the transfer of Au from the cytosol into lysosomes may play a role in this process.

Nickel [**Ni** signifies ^{63}Ni]

Although Ni is efficiently excreted via the kidneys it is also reported to cause nephrotoxicity in experimental animals [6]. The results shown in Table 2 represent the renal uptake and subcellular distribution of **Ni**, after administration of **Ni**(HIS)$_2$ complex   (the main physiological form of Ni in the circulation) at a typical pharmaco-logical dose.   The decrease in cortical **Ni** with time illustrates the rapid excretion of Ni via the kidneys. However, over the period 10 min-24 h there is a 'relative' increase in the % of cortical Ni present in **ML**. Possibly the basic nature of Ni(HIS)$_2$ is responsible for this complex being lysosomotropic and thus being retained in renal lysosomes. It is proposed that the association of Ni with the various lysosomal populations may thus be particularly useful for studying Ni nephrotoxicity.

Platinum

From their study on the subcellular distribution of Pt in the kidney, Choie et al. [7] concluded that Pt is associated with nuclei, mitochondria and microsomes. Whilst these authors make no reference to lysosomes, a similar study by Sharma & Edwards [8] indicated that following a single i.p. dose of *cis*-Pt (CDDP) the peak of Pt in the kidney is reached at 24 h. At this time-point the highest proportion of Pt was found in the cytosol, with the second- and third-highest concentrations in the microsomal and lysosomal fractions. As preliminary X-ray microprobe studies also suggest that Pt is associated with kidney lysosomes, it is intended to apply the subcellular approaches adopted to study the renal handling of Cd-thionein to examine in detail the association of Pt with the various populations of kidney cortical lysosomes.

References

1. Sternlieb, I. & Goldfischer, S. (1976) in *Lysosomes in Biology and Pathology*, Vol. 5 (Dingle, J.T. & Dean, R.T., eds.), North Holland/Elsevier, Amsterdam, pp. 185-200.
2. Dobrota, M., Bonner, F.W. & Carter, B.A. (1982) in *Nephrotoxicity: Assessment and Pathogenesis* (Bach, P.H., Bonner, F.W., Bridges, J.W. & Lock, E.A., eds.), Wiley, Chichester, pp. 320-324.
3. Andersen, K-J., Haga, H.J. & Dobrota, M. (1987) *Kidney Internat.*, in press.
4. Taylor, M., Dobrota, M., Hinton, R.H. & Taylor, A. (1985) in *Renal Heterogeneity and Target Cell Toxicity* (Bach, P.H. & Lock, E.A., eds.), Wiley, Chichester, pp. 245-248.
5. Squibb, K.S., Pritchard, J.B. & Fowler, B.A. (1984) *J. Pharmacol. Exp. Ther. 229*, 311-321.
6. Sunderman, F.W. & Horak, E. (1981) in *Organ-Directed Toxicity: Chemical Indices and Mechanisms* (Brown, S.S. & Davies, D.S., eds.), Pergamon, Oxford, pp. 52-64.
7. Choie, D.D., Del Campo, A.F. & Guarino, A.M. (1981) *Toxicol. Appl. Pharmacol. 55*, 245-252.
8. Sharma, R.P. & Edwards, I.R. (1983) *Biochem. Pharmacol. 32*, 2665-2669.

#F-6

LYSOSOMAL POPULATIONS IN KIDNEY CORTEX
OF RATS WITH HEAVY PROTEINURIA

H-J. Haga, K-J. Andersen, B.M. Iversen and M. Dobrota

Section for Clinical Research and Molecular Medicine
Medical Department A, University of Bergen
N-5016 Haukeland Sykehus, Bergen, Norway   and
Robens Institute of Industrial and Environmental Health
and Safety, University of Surrey, Guildford GU2 5XH, U.K.

With a view to elucidating functional relationships among rat lysosomal populations from nephron regions, we induced acute nephritis immunologically, marked by heavy proteinuria and increased urinary activity of acid hydrolases. Acid hydrolases rose too in preparations of large lysosomes (protein droplets) derived from kidney proximal tubules and concerned in protein degradation. Isopycnic centrifugation of small lysosomes yielded a new population of lysosomes with a relatively high membrane content probably resulting from increased membrane turnover induced by the increase in protein uptake.

The lysosomes in rat kidney cortex are more heterogeneous than in liver with respect to size, enzyme content and density ([1], & #F-4, this vol.). This heterogeneity may be due to different lysosomal populations or different cellular origins, or possibly the various lysosomal populations represent quite distinct cellular pathways, e.g. those derived from the autophagic route [2] and those containing exogenous proteins. In order to get a better understanding of the functional relationship between the various populations, they were studied in homogenates of kidney cortices from rats with heavy protein-uria induced immunologically in two ways.

EXPERIMENTAL

The rats were male Wistars of wt. 130-170 g. Acute passive Heyman nephritis was induced by i.v. injection of heterologous antibodies (rabbit) against tubular antigen (FxIA) as described by Iversen et al. [3]. Rats injected with normal rabbit serum served as controls.

Overflow proteinuria was induced as described by Lawrence et al. [4]: 0.87 g bovine serum albumin (BSA; Cohn fraction I, #A-2153) was injected i.p. twice daily for 8 days, and controls received buffered saline instead. The nephritic rats were killed 2 weeks after induction of nephritis, and the BSA rats 8 days after the first injection. Urine was collected over the last 24 h before sacrifice.

Renal cortices (including deep cortex) were homogenized in 0.25 M sucrose/5 mM Tris-HCl pH 7.4, giving a 10% homogenate [1]. The homogenates in each experimental group (8 rats) were pooled and centrifuged ([1], & #F-4 in this vol.) to give fractions: nuclear, **N**; mitochondrial/lysosomal, **ML**; microsomal, **Mic**, and supernatant, **Sup**. **ML** was subfractionated by rate sedimentation in an HS zonal rotor, with pooling of specific regions for further fractionation by equilibrium banding [1]. The various subfractions were assayed for protein, N-acetyl-β-glucosaminidase (NAG) and acid β-galactosidase [1]. The total activities of acid hydrolases were assayed by including digitonin (0.0016% w/v) in the enzyme assays. Protein in urine was measured by the biuret method with BSA as standard [5].

RESULTS

Proteinuria was detectable 4-6 days after induction of nephritis, maximally at 2 weeks after induction (342 mg/24 h). With BSA injections, urinary protein excretion reached a maximum 1-2 days after the start and remained high (263 mg/24 h) during continued i.p. injection of BSA (8 days). Protein excretion in controls never exceeded 20 mg/24 h.

After differential pelleting of the cortical homogenates, the % recovery of NAG and acid β-galactosidase in the **N** fraction was significantly higher in nephritic and BSA rats than in controls (Fig. 1), whilst only minor differences were observed in the other subcellular fractions. After rate-zonal centrifugation of the **ML** fraction the acid hydrolases were recovered in the fast-sedimenting band (fractions 31-35) containing the large lysosomes and in the slower sedimenting broad band (fractions 12-24) containing the small lysosomes [1]. As shown in Fig. 2, the distribution of acid hydrolases in the rate-zonal spin is essentially the same in control, nephritic and BSA rats, while the recovery of acid hydrolases was significantly higher in the large lysosomes from proteinuric rats compared to controls.

The lysosomal nature of the acid hydrolases found in the rate-zonal spin was verified by demonstrating latency in fractions 10-40. The small-lysosome region (fractions 12-24) was pooled and spun to equilibrium: in controls the acid hydrolases banded at three densities which probably represent lysosomal fragments (d 1.11-1.15), light lysosomes (d 1.20) and dense lysosomes (d 1.235). As shown in Fig. 3, the acid hydrolases banded at d = 1.22 in proteinuric rats and also,

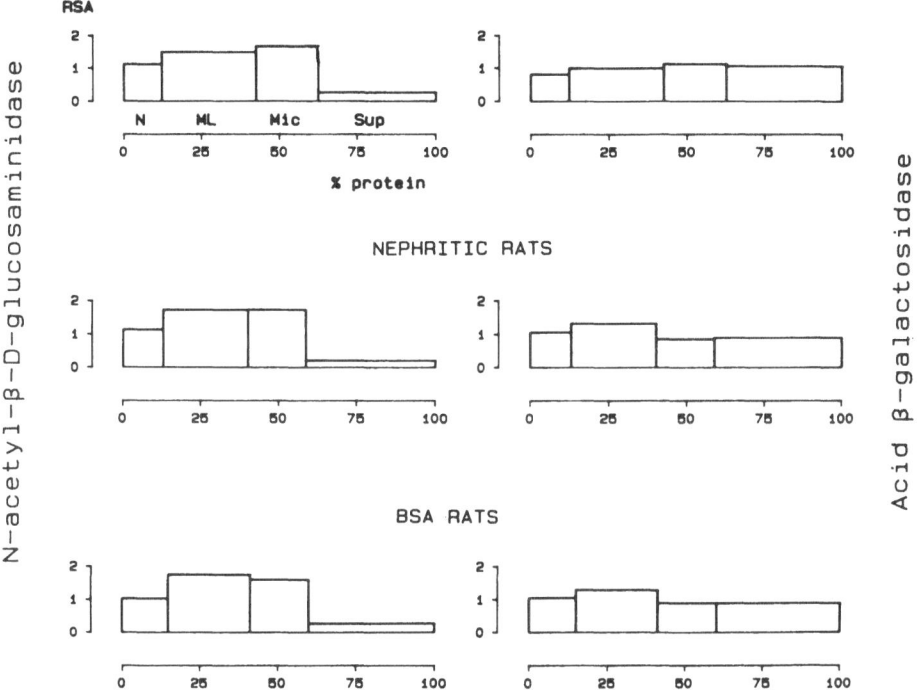

Fig. 1. Distributions of NAG, acid β-galactosidase and protein in subcellular fractions (see text) prepared from cortical homogenates by differential pelleting. Recoveries were >80%. RSA denotes relative specific activity (homogenate = 1).

to a significant extent, at d = 1.20 and 1.185 but only to a slight extent at d = 1.235, in contrast with normal rats. All acid hydrolases recovered in the density region 1.185–1.235 showed latency in protein-uric and control rats, demonstrating the presence of intact lysosomes. Isopycnically, the large lysosomes from the rate-zonal spin (fractions 31–35) banded at the same density (1.235) in proteinuric and control rats.

DISCUSSION

Consistent with the observations made by Straus [6] in proteinuric rats, heavy proteinuria caused by nephritis or overflow-proteinuria induces a significant increase in the recovery of acid hydrolases in the large lysosomes or 'protein droplets' present in the N fraction and in the fast-sedimenting band in the rate-zonal spin. The large lysosomes are derived essentially from the proximal tubule where most of the reabsorption of urinary proteins has been shown to occur [7];

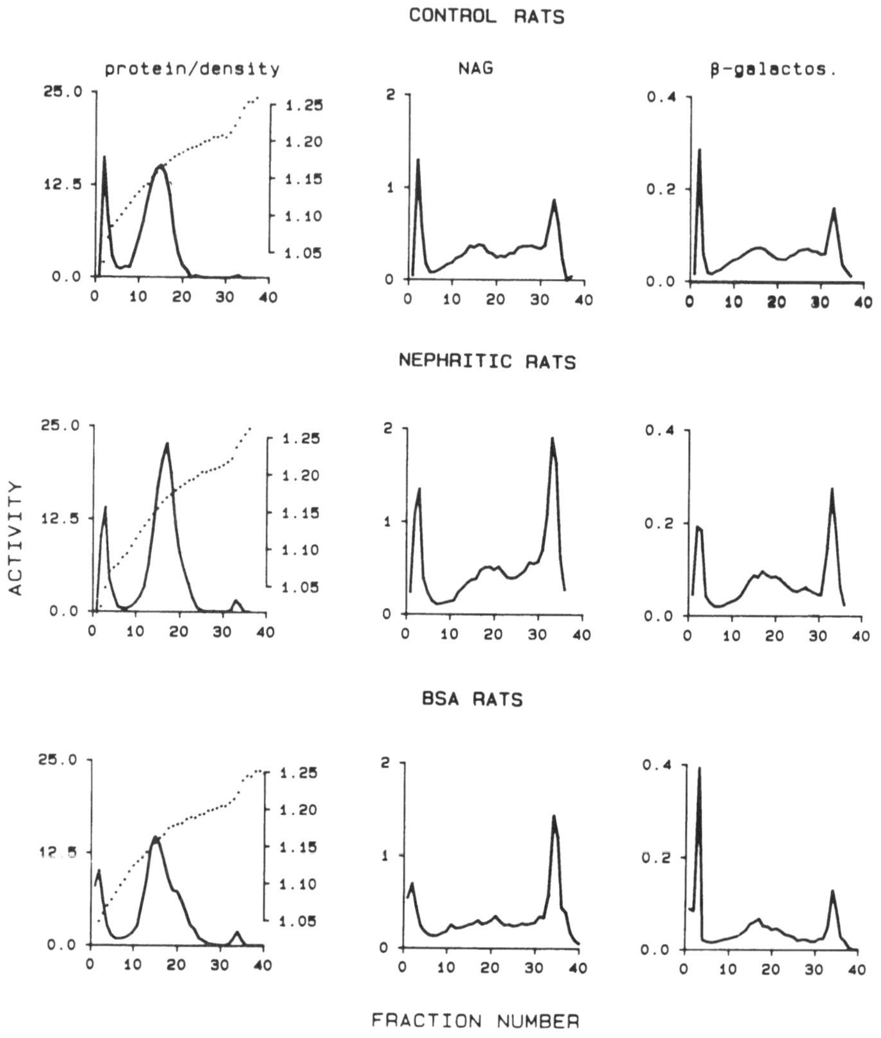

Fig. 2.   Distributions of NAG, acid β-galactosidase and protein
after rate-sedimentation of the **ML** fraction in an HS zonal rotor.
The rotor contained a 550 ml exponential sucrose gradient from
0.5 M to 1.7 M and 150 ml of 2 M sucrose as the cushion. After
loading the resuspended **ML** fraction the rotor was spun at 8000 rev/
min for 1 h.   Protein is presented as mg/fraction, and enzyme
activities are given as μmol/min/fraction.

the association of the 'protein droplets' with reabsorbed labelled
proteins [8] indicates that these lysosomes are actively engaged in
the degradation of reabsorbed proteins.   The populations of small
lysosomes banding at d = 1.20 and 1.235 are also associated with
labelled reabsorbed proteins, and likewise the populations of small

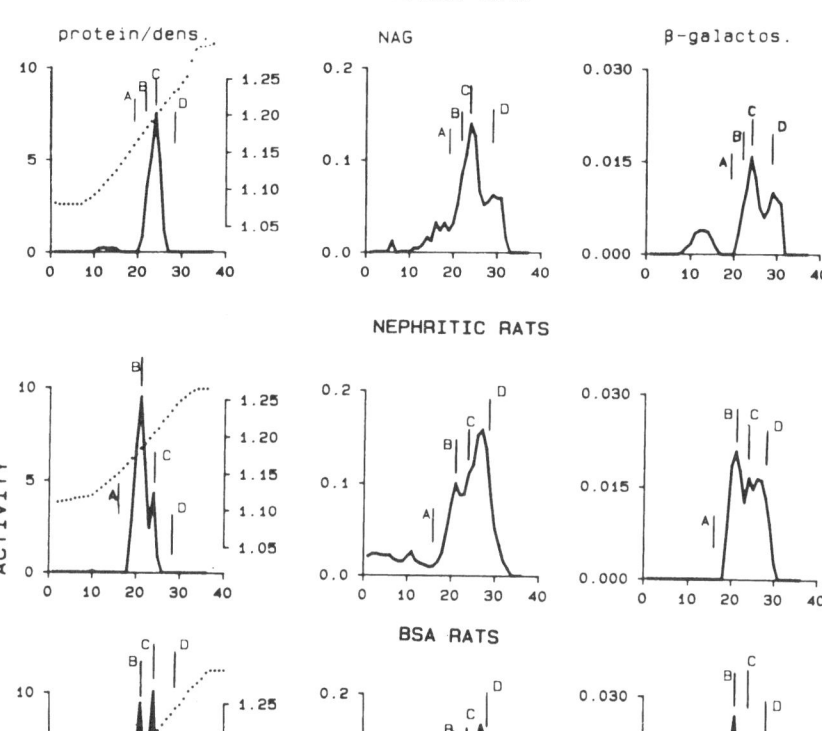

Fig. 3. Distributions (as in Fig. 2) after isopycnic sedimentation of the small lysosomes (fractions 12-24, pooled) from the rate spin in Fig. 1 in a B14 zonal rotor (45,000 rev/min, 16 h). Protein and enzyme activities expressed as in Fig. 2. Density positions: A, 1.16; B, 1.185; C, 1.20; D, 1.235.

lysosomes in proteinuric rats banding at slightly lower densities (1.185 and 1.22) [8].

Possibly the light lysosomes appeared because of higher protein uptake increasing the membrane turnover, which may result in secondary lysosomes having a higher membrane content and hence a fall in overall lysosomal density. Seglen et al. [9] have demonstrated that dense lysosomes are converted to light lysosomes when the lysosomes are activated in rat liver - a feature common to autophagy, fluid

pinocytosis and receptor-mediated endocytosis.   The appearance of
small light lysosomes in rat kidney cortex may indicate that the small
lysosomes are activated by increased reabsorption of urinary proteins,
while the absence of large light lysosomes in proteinuric rats may
indicate that increased membrane turnover has no influence on the
density of large lysosomes, or that the membrane turnover is less
pronounced in large compared to small lysosomes.

Acknowledgement

The work was supported by research grants from the Norwegian
Research Council for Science and the Humanities.

References

1.   Andersen, K-J., Haga, H-J. & Dobrota, M. (1987) *Kidney Int. 31*,
     886-897.
2.   Berkenstam, A., Ahlberg, J. & Glaumann, H. (1983) *Virchows
     Arch. (Cell Path.) 44*, 275-286.
3.   Iversen, B.M., Matre, R. & Ofstad, J. (1982) *Acta Path.
     Microbiol. Immunol. Scand. Sect. C 90*, 241-250.
4.   Lawrence, G.M. & Brewer, D.M. (1981) *Clin.Sci. 60*, 693-702.
5.   Savory, I., Pu, P.H. & Sunderman, F.W. Jr. (1968) *Clin. Chem.
     14*, 1160-1171.
6.   Straus, W. (1957) *J. Biophys. Biochem. Cytol. 3*, 933-947.
7.   Beh, L.V. & Fedynsky, J.N. (1967) *Endocrinology 81*, 475-485.
8.   Dobrota, M., Haga, H-J. & Andersen, K-J. (1985) in *Renal
     Heterogeneity and Target Cell Toxicity* (Bach, P.H. & Lock, E.A.,
     eds.), Wiley, Chichester, pp. 241-244.
9.   Seglen, P.O. & Solheim, A. (1985) *Exp. Cell Res. 157*, 550-555.

#F-7

IMMUNE DERANGEMENTS IN EXPERIMENTAL, TOXIN-INDUCED AUTOIMMUNE NEPHROPATHIES

P. Druet, L. Pelletier, R. Pasquier,
J. Rossert, E. Druet and F. Hirsch

INSERM U28, Hôpital Broussais
96, rue Didot, 75674 Paris Cedex 14, France

Toxin-induced nephropathies represent an increasing cause of renal injury. Injected $HgCl_2$ induces autoimmunity, furnishing a model system, in animals of a susceptible species and strain. With Brown-Norway (BN) rats we have studied features of this action; thus, T-cells exposed to $HgCl_2$ in vivo or in vitro become autoreactive. We have concluded that (1) a toxic agent can act at the cellular level and induce a dysregulation of the immune system, and (2) depending on the genetic background, the same agent may cause either autoimmunity or immunosuppression.

Toxin-induced autoimmune nephropathies have long been recognized, and represent an important problem in nephrology. Mercury and gold salts, D-penicillamine and related drugs are the best known culprits [1]. Although this action is demonstrable with many drugs, the mechanisms await elucidation. A rat model using $HgCl_2$ as an inducing agent has been developed. We had observed that $HgCl_2$ induces in 30% of outbred Wistar rats a nephritis of immune-complex type [2]. We decided to study the susceptibility of various strains, since the induction of immune responses [3] and of autoimmune diseases [4] including drug-induced autoimmunity [5] are well known to be genetically determined. We found that susceptibility indeed greatly depends on the strain tested.

First we describe the autoimmune manifestations observed in Brown Norway (BN) rats after $HgCl_2$ injections, and we summarize our findings on mechanism of action. Then we describe the susceptibility of other strains and the genetic control of susceptibility in BN rats, and consider possible mechanisms for resistance in Lewis (LEW) rats.

THE AUTOIMMUNE DISEASE IN BN RATS

    In most of the experiments reported here, non-toxic doses of
$HgCl_2$, 0.1 mg in 0.1 ml/100 g body wt., were injected s.c. twice weekly
for 2 months. Controls received dist. water equivalent in vol. and
pH (3.8).   The autoimmune abnormalities become manifest from day
7 or 8, reach a peak between days 14 and 21, then decline progressively
even if $HgCl_2$ injections are pursued [6-9]. The abnormalities are
as follows:

(a) lymphoproliferation;
(b) increase in total serum Ig level (mainly IgE);
(c) production of autoantibodies (autoAb's) - anti-glomerular basement
membrane (anti-GBM), anti-DNA;
(d) circulating immune complexes;
(e) autoimmune glomerulonephritis [cf. Haga et al., #F-6, this vol.
-*Ed*.].

    The lymphoproliferation, (a), is characterized by an increased
number of spleen cells and lymph-node cells [9]. The increase in
Ig (b), related to (a), is most marked - although seemingly non-specific -
for IgE [10].   Also (c) amongst the host of autoAb's synthesized,
some are harmful mainly to the kidney; the anti-GBM Ab's recognize
all the basement membranes but *in vivo* are fixed mainly on the
glomerular basement membrane (GBM). The exact determinants recognized
are still unknown. Deposited Ab's are responsible for the typical
smooth, linear pattern of fixation of fluoresceinated anti-rat Ig
conjugates along the GBM [6-8].   Such Ab's are responsible for a
heavy proteinuria and the nephrotic syndrome without renal failure
and for a moderate influx of macrophages to the glomeruli [11].
There is ~50% mortality during the third week, probably because of
intravascular coagulation [12].

    Circulating complexes (d), of unknown composition, probably
account for a disease of immune-complex type, usually observed during
the second month [7, 8].   Granular IgG deposits are found in the
GBM together with subepithelial electron-dense deposits, and also
in the walls of most of the small vessels [7, 13]. Such deposition probably
causes the second phase of a biphasic glomerular disease, the first
phase being due to anti-GBM Ab deposition.

MECHANISMS OF ACTION OF $HgCl_2$ IN BN RATS

    Drugs may act as haptens or they may bind to and modify auto-
antigens resulting in the production of autoAb's against both the
denatured and the native antigen. Although these mechanisms have
been implicated in drug-induced haemolytic anaemia, rarely have they
been demonstrated in drug-induced nephropathies. A quite distinct
way for toxic agents to induce autoimmunity is to disturb the regulation
of the immune system. Autoimmunity may be due to a decrease in

suppressor T-cells, an increase in helper T-cells, to a direct polyclonal activation of B-cells, or a combination of these mechanisms.

(1) HgCl₂ induces a T-dependent polyclonal activation of B-cells

The multiple abnormalities observed during this disease suggested
to us that $HgCl_2$ could disturb immune regulation.-
- Spleen cells exposed to $HgCl_2$ *in vivo* or *in vitro* can produce Ab's
 towards exogenous antigens (without immunization).
- Helper T-cells and B-cells proliferate in lymphoid organs.
- BN rats deprived of T-cells do not develop autoimmunity.
- T-cells exposed to $HgCl_2$ induce syngeneic normal T- and B-cells
to proliferate *in vivo*.
- $HgCl_2$ induces autoreactive T-cells.

To examine this hypothesis we first investigated whether the
number of plaque-forming cells was increased by spleen cells from
rats injected with $HgCl_2$ or normal cells exposed *in vitro* to non-toxic
amounts of $HgCl_2$. We observed [9] that spleen cells from such rats
contained an increased number of B-cells that synthesized IgG and
IgM anti-TNP and anti-sheep-erythrocyte Ab's without prior immunization. In other experiments we showed *in vitro* that T-cells were
required for this polyclonal activation to occur [9]. These data
suggested that helper T-cells were essential.

Another demonstration that $HgCl_2$ induces polyclonal activation
in BN rats was obtained from fusion experiments. When spleen cells
from $HgCl_2$-injected rats were fused with non-secreting myeloma cells,
a number of hybrids were obtained. These hybrids produced a variety
of autoAb's (anti-GBM, anti-DNA, etc.) and also anti-TNP Ab's [14].

(2) T-cells are required for autoimmunity to occur *in vivo*

Using BN rats deprived of T-cells, their role has recently been
confirmed (to be published). The nude mutation was transferred in
BN rats. Such rats (BN rnu/rnu) did not develop autoimmunity, yet
when crossed with BN rats they again became susceptible. In the
same strain, BN 'B' rats (thymectomized, irradiated and replenished
with foetal liver cells) were not susceptible, while autoimmunity
was again observed when BN 'B' rats were replenished with normal
T-cells.

(3) Phenotype of lymphoid cells in HgCl₂-injected BN rats

Another interesting finding underlining the role of T-cells
emerged from study of the phenotype of spleen and lymph-node cells
of $HgCl_2$-injected BN rats. Cell types were enumerated by the classical
technique of membrane immunofluorescence with specific antisera or
monoclonal Ab's. With BN rats given $HgCl_2$ injections, during days
7-14 we found a significant increase in the numbers of B-cells,

total T-cells and helper T-cells but not of suppressor T-cells.
In contrast, the latter increased in number after day 15 while B-
and helper T-cells decreased in number.

These findings suggested that helper T-cells were activated
in this model, resulting in activation of B-cells. This could explain
the activation phase. In contrast, the secondary increase in OX-8+
cells (suppressor/cytotoxic cells) suggested that suppressor cells
are activated as a secondary event and that they could be responsible
for the autoregulation observed. These data accord with other experiments [15] showing that suppressor T-cells collected at the time
the BN rats had recovered were able to attenuate the disease process
when injected into naïve BN rats.

(4) Cellular mechanisms

The role of T-cells and the effect of $HgCl_2$ exposure was then
further assessed. Helper T-cells were purified from $HgCl_2$-injected
rats, or from normal rats by using petri plates coated with goat
anti-rat Ig Ab's [16, 17]. Normal T cells were then exposed *in vitro*
to $HgCl_2$.

In both cases, T-cell injection into the rear footpad of normal
syngeneic BN rats caused an increase in the number of helper T- and
B-cells in the draining popliteal lymph nodes [16], suggesting that
T-cells exposed to $HgCl_2$ can stimulate normal syngeneic lymphocytes
in vivo.

Different cells or cell subsets exposed *in vivo* or *in vitro*
to $HgCl_2$ were irradiated and co-cultured with normal syngeneic
lymphocytes. Thymidine incorporation was measured. With Ia-positive
normal cells also present, helper T-cells exposed to $HgCl_2$ were then
found to stimulate normal helper T-cells to proliferate [17]. Ia
determinants are encoded by immune response genes of the major histocompatibility complex, and are essential for antigen presentation
and for cell cooperation.

These results strongly suggest that in BN rats $HgCl_2$ can generate or expand autoreactive T-cells that recognize Ia determinants.
Such anti-Ia T-cells would then polyclonally stimulate B-cells.

GENETIC CONTROL OF SUSCEPTIBILITY

The response to $HgCl_2$ differed markedly amongst different rat
strains tested (Rt-1 haplotype listed parenthetically): both anti-GBM
Ab's and immune-complex nephritis developed in strain BN (n), but
neither developed in LEW (l), in F,344, AS or BS (l), or in WAG,
WF or LOU (u). The nephritis but not anti-GBM Ab's developed in
PVG/c and AUG (c), in AVN and DA (a), and in BD V (d), BUF (b), OKA (k)
and AS 2 (f). There is some evidence that anti-nuclear Ab's could

play a pathogenic role at least in PVG/c rats [18-20]. In this strain too, $HgCl_2$ could inhibit suppressor T-cells. Evidently, then, a toxic agent can induce autoimmunity in different ways, depending on the strain tested.

Interestingly, we observed that several strains with a given haplotype [specified thus () above] at the major histocompatibility complex (MHC) - RT-1 in the rat - did not develop autoimmune abnormalities. Resistance was not dose-related [18]. This finding allowed us to study the genetic control of susceptibility by crossing the susceptible BN strain with the resistant LEW strain, with testing of segregants obtained between the two strains. The results established that both the occurrence of autoimmune glomerulonephritis [21, 22] and the rise in total serum IgE level are genetically determined [23] and that 3 or 4 genes are involved - one being localized within the MHC. Susceptibility is inherited as an autosomal and dominant trait.

Further studies are in progress to localize the various genes involved. There is already good evidence from studies using congenic and recombinant rats that the MHC-linked gene is an immune-response (Class II) gene. This is of interest because Class II-encoded genes are known to play a major role in autoimmune diseases [4] and because in the present model we have shown the role of Ia (Class II) determinants in the induction of autoimmunity.

MECHANISMS OF THE RESISTANCE IN LEWIS RATS

Experiments similar to those in BN rats were performed in LEW rats [results submitted for publication] with the idea that immune derangements at the cell level could also account for the resistance observed in that strain, wherein the effects of $HgCl_2$ are as follows:

- increase in number of suppressor/cytotoxic T-cells (OX-8+);
- depression of responsiveness to T-cell mitogens and alloantigens;
- abrogation of local graft-*vs.*-host reaction;
- down-modulation of autoimmunity.

Phenotype of cells.- The T-cell increase, shown by our usual methodology, was maximal by day 7 and remained at the same level for at least a month.

Induced non-specific suppression.- We investigated whether the suppressor/cytotoxic T-cells observed in Lewis rats were indeed able to induce suppression. We found that responsiveness to T-cell mitogens (Con A and PHA) was profoundly depressed. Similarly lymphocytes from $HgCl_2$-injected LEW rats were unable to respond to alloantigens in mixed lymphocyte culture or to induce a local graft-*vs.*-host reaction when injected into Fl hybrids. Moreover, the Ab response to sheep erythrocytes was markedly diminished in LEW rats injected with $HgCl_2$.

All these data were suggestive of a non-specific immunosuppression probably mediated by suppressor/cytotoxic T-cells. The precise role of these cells was demonstrated in two ways. (1) Responsiveness to T-cell mitogens and to alloantigens was restored when OX-8+ (suppressor/cytotoxic) T-cells were removed; i.e. helper T-cells were able to respond normally. (2) When lymphoid cells from HgCl$_2$-injected rats were added *in vitro* to normal LEW lymphocytes, they were able to abrogate the response to alloantigens. This shows that it is an active suppression that has been observed.

Down-modulation of autoimmunity in LEW rats.- Since HgCl$_2$ leads to non-specific immunosuppression, we investigated whether LEW rats injected with HgCl$_2$ would still develop autoimmunity. LEW rats are highly susceptible to the induction of an Ab-mediated immune nephritis (Heymann's nephritis) and to a cell-mediated autoimmune disease (experimental allergic encephalomyelitis). We found that these two diseases were either prevented or markedly attenuated when induced in LEW rats injected with HgCl$_2$.

CONCLUDING REMARKS

HgCl$_2$ evidently has quite different effects depending upon the strain tested.-
(1) It induces autoimmunity in BN rats, probably because it generates or expands autoreactive (anti-Ia?) T-cells.
(2) In contrast, the same agent causes non-specific immunosuppression in LEW rats, wherein it generates or augments non-specific suppressor/cytotoxic T-cells by a mechanism that awaits elucidation.

These findings are quite reminiscent of those reported for mice by Gleichmann et al. [24] in the graft-*vs.*-host model. In its chronic form, parental T-cells recognize Ia-positive donor cells; this abnormal cell-cell cooperation leads to a lupus-like syndrome. It was postulated that normal Ia determinants modified by viruses or toxins could be recognized in a similar way by autologous T-cells [24]. Though we too found autoreactive T-cells probably directed towards Ia determinants in the mercury model, our experiments clearly show that it is not a modified but rather a normal Ia determinant that is recognized. The relationship between the two models awaits clarification.

Gleichmann et al. [24] also showed that in other strain combinations in mice, there may appear an acute graft-*vs.*-host disease, wherein suppressor T-cells are induced that may account for the observed immunosuppression. These findings are akin to ours in LEW rats.

Finally we stress potential general interest of the rat mercury model. Some drugs, e.g. D-penicillamine, captopril and phenytoin, are well known to cause autoimmunity or immunosuppression in humans, as can some viruses (e.g. EBV). The cellular or subcellular basis is not yet known. We have demonstrated that, at least for some agents such as $HgCl_2$, the outcome of interaction with genetically defined cellular determinants can vary. That MHC-linked genes are important in drug-induced autoimmunity has long been known. We suggest too that some drugs may have an immunosuppressive effect that is genetically determined.

References

1. Fillastre, J.P., Mery, J.P. & Druet, P. (1984) in *Acute Renal Failure. Correlations between Morphology and Function* (Solez, K. & Whelton, A., eds.), Dekker, New York, pp. 389-407.
2. Bariety, J., Druet, P., Liberte, F. & Sapin, C. (1971) *Am. J. Path. 65*, 293-302.
3. Benacerraf, B. & McDevitt, H.O. (1972) *Science 175*, 273-279.
4. Amos, D.B. & Yunis, E.J. (1979) in *Immune Mechanisms and Disease* (Amos, D.B., Schwartz, R.S. & Janicki, B.W., eds.), Academic Press, New York, pp. 139-161.
5. Wooley, P.H., Griffin, J., Panayi, G.S., Batchelor, J.R., Welsh, K.I. & Gibson, T.J. (1980) *N. Engl. Med. J. 303*, 300-302.
6. Sapin, C., Druet, E. & Druet, P. (1977) *Clin. Exp. Immunol. 28*, 173-179.
7. Druet, P., Druet, E., Potdevin, F. & Sapin, C. (1978) *Ann. Immunol. (Inst. Pasteur) 129C*, 777-792.
8. Bellon, B., Capron, M., Druet, E., Verroust, P., Vial, M.C., Sapin, C., Girard, J.F., Foidart, J.M., Mahieu, P. & Druet, P. (1982) *Eur. J. Clin. Invest. 12*, 127-133.
9. Hirsch, F., Courderc, J., Sapin, C., Fournié, G. & Druet, P. (1982) *Eur. J. Immunol. 12*, 620-625.
10. Prouvost-Danon, A., Abadie, A., Sapin, C., Bazin, H. & Druet, P. (1981) *J. Immunol. 126*, 699-702.
11. Hinglais, N., Druet, P., Grossete, J., Sapin, C. & Bariety, J. (1979) *Lab. Invest. 41*, 150-159.
12. Michaud, A., Sapin, C., Leca, G., Aiach, M. & Druet, P. (1983) *Thrombosis Res. 33*, 77-88.
13. Bernaudin, J.F., Druet, E., Belair, M.F., Pinchon, M.C. & Druet, P. (1979) *Clin. Exp. Immunol. 38*, 264-273.
14. Hirsch, F., Kuhn, J., Ventura, M., Vial, M.C., Fournié, G. & Druet, P. (1986) *J. Immunol. 136*, 3272-3276.
15. Bowman, C., Mason, D.W., Pusey, C.D. & Lockwood, C.M. (1984) *Eur. J. Immunol. 14*, 465-470.
16. Pelletier, L., Pasquier, R., Hirsch, F., Sapin, C. & Druet, P. (1985) *Eur. J. Immunol. 14*, 460-464.
17. Pelletier, L., Pasquier, R., Hirsch, F., Sapin, C. & Druet, P. (1986) *J. Immunol. 136*, 2548-2554.
18. Druet, E., Sapin, C., Fournié, G., Mandet, C., Gunther, E. & Druet, P. (1982) *Clin. Immunol. Immunopath. 25*, 203-212.
19. Weening, J.J., Fleuren, G.J. & Hoedemaeker, Ph.J. (1978) *Lab. Invest 39*, 405-411.

20. Weening, J.J., Hoedemaeker, Ph.J. & Bakker, W.W. (1981) *Clin. Exp. Immunol.* *45*, 64-71.
21. Druet, E., Sapin, C., Gunther, E., Feingold, N. & Druet, P. (1977) *Eur. J. Immunol.* *7*, 348-351.
22. Sapin, C., Mandet, C., Druet, E., Gunther, E. & Druet, P. (1982) *Clin. Exp. Immunol.* *48*, 700-704.
23. Sapin, C., Hirsch, F., Delaporte, J.P., Bazin, H. & Druet, P. (1984) *Immunogenetics* *20*, 227-236.
24. Gleichmann, E., Pals, S.T., Rolink, A.G., Radaszkiewicz, T. (1984) *Immunology Today* *5*, 324-333.

#F-8

GLOMERULAR BASEMENT MEMBRANE AND THE INFLAMMATORY RESPONSE

John D. Williams and Malcolm Davies

Department of Renal Medicine
KRUF Institute, The Royal Infirmary
Cardiff CF2 1SZ, U.K.

In diverse nephritic conditions inflammatory cells exist within the glomeruli and gain direct access to the GBM*. In this situation the phagocytes may become activated, either in response to immune reactants including activated complement components or by direct contact with the GBM. Confirming that human kidney GBM can activate the alternative complement pathway, the activated third component of complement (C3b) was deposited under experimental conditions where the activation of the classical complement pathway was inhibited. In addition the interaction of the GBM with neutrophils in the presence of Ab resulted in the release of granule enzymes and in the generation of reactive oxygen species. These results confirm the potential pathological role of exposed intact or modified membrane resulting in the augmentation of the inflammatory process within the kidney.

The GBM - the principal barrier for the filtration of plasma in the first stage of urine formation - is a specialized basement membrane formed from the fusion of basal laminae of two cell types: epithelial (podocytes) and endothelial [1]. Like all basement membranes it is a complex matrix characterized by unique molecules including type IV collagen, heparan sulphate proteoglycan, laminin and entactin [2, 3]. The type IV collagen serves as a structural matrix for the GBM whereas proteoglycan contributes to the anionic barrier which restricts the passage of negatively charged molecules across the membrane [4]. Further components of the charge barrier may also include sialoglycoproteins [2].

In immune-mediated renal disorders deposits of Ig's and complement components are frequently located adjacent to or within the GBM [5]. Some insight into the pathogenesis of immune-mediated glomeruloneph-

---

* Abbreviations: GBM, glomerular basement membrane; CL, chemiluminescence; PBS, phosphate-buffered saline; PMN, peripheral human neutrophils; Ab, antibody; KRPG, VBS *etc.: see text.*

ritis has been gleaned from animal studies, in particular Heymann nephritis, Masugi nephritis and chronic bovine serum albumin sickness which are models for anti-GBM disease and immune complex glomerulonephritis [6-8]. The exact mechanisms by which immune deposits mediate glomerular damage are poorly understood; but based on data from experimental studies several possibilities exist. These include both humoral-mediated damage via Ig and the complement cascade, as well as cellular damage associated with neutrophil and/or macrophage infiltration and activation [5].

Several studies have demonstrated that in a wide variety of nephritic conditions inflammatory cells are present within the glomeruli and gain direct access to the GBM by inserting themselves between it and the adjacent endothelial cell [9-12]. Under such conditions it is postulated that the phagocytes become activated, either in response to immune reactants or by direct contact with the GBM, resulting in the release of pro-inflammatory products with the potential to damage the GBM. This article reviews work from our own laboratory which investigates the interaction between inflammatory cells and the GBM both in the presence and the absence of humoral mediators.

HUMAN GBM ISOLATION

Our understanding of the function of the GBM in normal physiology, and its role in various forms of renal diseases, has been greatly aided by the ability to isolate this membrane in an acellular form devoid of tubular contamination. The first preparative step is the isolation of glomeruli. The method of choice for human material is the sieving technique of Krakower & Greenspon [13] essentially as described by Spiro [14].

All procedures were carried out in a cold-room at 5°. Human kidneys obtained at post-mortem were processed immediately or stored at -20° until required. Kidneys were dissected free of fat, the capsule removed, and bisected pole-to-pole. The cortex was carefully removed by blunt dissection, washed in ice-cold PBS, finely sliced with a scalpel blade, and passed through a coarse stainless steel mesh (420 µm) [Endecotts (Test Series) Ltd., London] to remove much of the loose connective tissue. The brie obtained was collected in ice-cold PBS and gently passed sequentially through sieves of 180, 120, 105 and 90 µm. Glomeruli were collected on either the 105 or the 90 µm sieve, transferred with ice-cold PBS into Universal containers, and washed (× 5) with ice-cold 0.15 M NaCl by centrifugation at 600 **g** for 5 min until the supernatant was protein-free.

A pure membrane preparation was obtained by sonic disruption. Glomeruli were suspended in 25 ml of 1 M NaCl in a glass beaker which in turn was placed in a 100 ml glass beaker packed with ice chips. The suspension was disrupted by 1 min bursts using a sonicator (Model 500; MSE Instruments, Crawley, U.K.) with a stainless steel

probe tuned to give maximal sonication. Disruption was usually achieved
with 8 bursts, as assessed by phase contrast microscopy. The GBM
was collected by centrifugation at 1800 **g** for 15 min at 4° and then
washed (× 5) with 1 M NaCl by re-centrifugation at 150 **g** until the
supernatant was protein-free. Finally the salt was removed by washing
under the same conditions with distilled water (× 5) and the preparation
lyophilized.

The purity of the final GBM preparation was assessed by several
methods. By phase-contrast and electron microscopy it was adjudged
to be free of cellular and tubular contamination (<1%). Per mg dry wt.
the GBM preparation contained <10 µg each of DNA (diphenylamine assay)
and RNA (Schmidt-Thannhauser procedure). The amino acid composition
was in agreement with published data. Alklaline phosphatase (a plasma-
membrane marker) was undetectable. Proteinase activity at neutral
pH (azocasein as substrate) and at acid pH (Hb) was also absent.
The final preparation was free of endotoxin as assessed by the Limulus
reaction.

APPROACHES BASED ON CHEMILUMINESCENCE (CL)

When phagocytes are incubated with several agents, including
certain bacteria, zymosan particles or immune complexes, they undergo
activation of the hexose monophosphate shunt with an increase in
oxygen consumption and the generation of reactive oxygen products
[15]. This process is associated with the emission of photons which
can be readily and conveniently measured by luminol-enhanced CL [16].
In the experiments here described we have utilized the CL technique
to investigate the interaction between GBM - untreated or pre-treated
with different immune proteins - and PMN's [cf. 17].

Using a Luminometer (LKB Wallach) for measurement, experiments
were carried out in CL cuvettes (holder kept at 37°) into which the
GBM was weighed directly. Human PMN were separated from peripheral
blood, obtained by consent from laboratory staff, by centrifugation
using Ficoll-Hypaque after dextran sedimentation essentially as
described by Boyum [18]. Contaminating erythrocytes were removed
by hypotonic lysis with ice-cold 0.2% NaCl. The final preparation
was washed 3 times with ice-cold PBS, resuspended in the same buffer
to a concentration of 5×10^6 cells/ml, and kept at 4° till required.
Using cytocentrifuged preparations (Cytospin II; Shandon Southern
Products, U.K.) >95% of these cells were identified as PMN by their
morphology.

To determine enhanced CL, 100 µl PMN (5×10^5 cells), 100 µl
luminol (Sigma), 50 µM in KRPG (viz. Krebs-Ringer-phosphate buffer
pH 7.3 with 11 mM glucose, 0.7 mM Ca^{2+}, 1.2 mM Mg^{2+}) and 200 µl KRPG
were pre-incubated (37°) for 6 min to allow temperature equilibrium
and establish background CL, and then transferred to the cuvette
containing GBM. CL was recorded in mV every 2 min over a 30 min

Table 1. CL responses after pre-incubation of GBM with immune serum.

Pre-incubation, h	GBM, mg/ml	CL peak max., mV	Time to attain peak, min
2	1	10	7
4	1	21	8
24	1	24	9
4	1	25	11
4	2	36	6
4	4	50	6

incubation period.   To assess the quality of each batch of PMN, opsonized zymosan (500 µg/100 µg KRPG) was used as control. Elsewhere [16, 19, 20] we describe fully how suitable PMN are prepared and PMN-derived CL is measured.

In experiments now summarized, the GBM was pre-incubated at 37° with non-immune serum or rabbit anti-(human GBM) serum for up to 24 h.   The serum was removed, the GBM washed (× 3) with PBS and resuspended in 100 µl KRPG, and then PMN (5 × 10^5 cells) added. Neither native GBM nor GBM pre-incubated with non-immune serum initiated a CL response, whereas GBM pre-incubated with rabbit antiserum invoked a significant and prompt CL response (Table 1).   It was consistently observed with several batches of GBM prepared from different human kidneys.   Table 1 shows that the peak CL response was concentration-dependent, being decreased and delayed with a lesser amount of GBM, and was increased by prolonging the pre-incubation time, indicating that the GBM/anti-GBM complex was rate-limiting.   A similar type of response has been noted with BSA/anti-BSA immune complexes (unpublished work by M. Davies & M.J. Harber).

The GBM-induced CL response was significantly (>80%) reduced by the inclusion of either catalase (3000 U/ml) or superoxide dismutase (3000 U/ml) in the reaction mixture, indicating that the response involves the release of toxic oxygen metabolites such as superoxide and hydrogen peroxide with a potential tissue-damaging function.

Goodpasture's syndrome is a condition associated with the production of anti-BM Ab and clinical features of glomerulonephritis and pulmonary haemorrhage.   The presence of anti-GBM Ab can be demonstrated as a characteristic linear deposit of IgG along the GBM, and circulating Ab to a BM antigen can be detected in the serum of affected patients.   This interaction of Ig and GBM within the substance of the glomerulus has the potential for complement activation as well as the triggering of inflammatory cells.   Thus the pre-incubation of GBM with serum obtained from a patient with Goodpasture's syndrome should activate PMN to generate the respiratory burst.   This was clearly demonstrated using the CL technique, with a maximum peak activity of 16 mV after 6 min.

The technique can also be extended to study the role of complement in the CL response. It was demonstrated that inactivation of the complement activity of the rabbit antiserum by heating at 56° for 30 min or by chelation with EDTA (12.5 mM) decreased but did not abolish the CL response compared to that given by untreated anti-GBM serum. A second incubation with fresh normal human serum, but not heat-inactivated serum, restored the CL to its original level [19].

COMPLEMENT ACTIVATION BY THE ALTERNATIVE PATHWAY

These results demonstrate that complement proteins play an important role in that process which allows GBM, in the presence of Ab, to activate the inflammatory cell. Indeed, activation of the human complement system is associated with various forms of glomer-ulonephritis. This is evidenced by the decrease in the serum level of complement components [21], the appearance of complement cleavage products in plasma [22], and the demonstration of complement components by immunofluorescence techniques as deposits within glomeruli in biopsies from patients with nephritis [23]. In the majority of these patients the deposition of complement components is also associated with Ig deposition and activation of the complement cascade via the classical pathway. In the absence of detectable deposits of Ig, C1q and C4, however, the deposition within the glomeruli of C3, the C5-9 complex and P in 10% of renal biopsies suggests activation of complement via the alternative pathway [24]. In addition, a percentage of patients with persistent and progressive nephritis will often show the disappearance of deposited Ig but the persistence of complement deposits. In these cases there seems to be a switch from classical pathway to alternative pathway activation.

Triggering of the alternative pathway is associated with a continuous fluid-phase turnover of C3 which involves the hydrolytic cleavage of a thio-ester bond within the C3 molecule [25]. This allows formation of a fluid-phase convertase and a subsequent cleavage of C3 to true C3b [26], which can then bind covalently to any adjacent receptive surface [27]. The discrimination between activating and non-activating surfaces occurs following the deposition of C3, since the former favour the binding of B to C3b, allow its cleavage by D and ultimately protect the C3/C5 convertase from decay dissociation by the regulatory proteins H and I [28]. Thus the activity of the cell-bound convertase depends on a balance between the promoting and stabilizing effects on the one hand and regulatory or inhibitory events on the other. The common feature of activating surfaces such as zymosan, rabbit erythrocytes and desialated sheep erythrocytes is their capacity to provide bound C3b and the amplification convertase C3bBbP with a protected environment [29].

Based on the above histological observations we examined the potential for GBM to serve as an activator of the alternative pathway of complement. Native human GBM was isolated as described above.

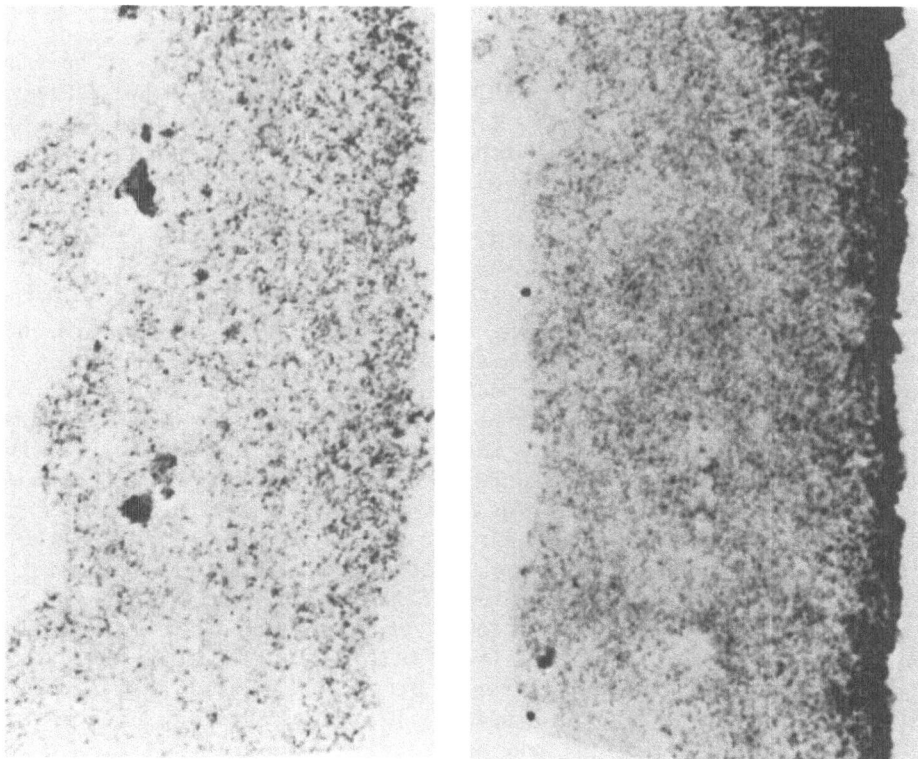

Fig. 1. Electron micrographs (×67,000). *Left.-* Isolated human GBM: the right-hand surface represents the denser smoother epithelial side of the membrane. *Right.-* Human GBM incubated in human serum chelated with Mg^{2+}/EGTA and processed for immunohistochemical localization of C3. The dark reaction product of peroxidase is confined to the epithelial surface and was never seen on the inner endothelial aspect of the membrane. *(Text mention: 2 pp. later.)*

Ultrastructural studies allowed differentiation between the epithelial and endothelial sides of the membrane (cf. Fig. 1). Human complement components, B [30], D [31], P [32], H [33] and I [34] were purified to apparent homogeneity and quantified. Normal human serum was depleted of D by passage through a Sephadex G-75 column equilibrated in veronal-buffered saline, pH 7.5 (VBS). The effluent was then concentrated to its original C3 concentration. Reconstitution of this serum with purified D was monitored by measuring its haemolytic capacity with sheep erythrocytes sensitized with Ab and bearing C4b and C3b [35].

Activation of the human alternative complement pathway by GBM was assessed by the deposition of C3b. This was detected immunologic-ally by the uptake of radiolabelled monoclonal Ab to C3c and immuno-histochemically by peroxidase staining of this Ab.

Complement activation by isolated GBM

In order to examine the capacity of isolated GBM to activate complement, samples were incubated in normal human serum diluted with either VBS containing 8 mM EGTA (a buffer which although chelating Ca^{2+} and blocking the classical pathway will allow the alternative pathway to proceed since its divalent cation requirement is only for Mg^{2+}) or with VBS containing 40 mM EDTA (a buffer which will chelate both Ca^{2+} and Mg^{2+} and totally block complement activation).

Following a pre-determined optimum period of incubation, membrane fragments were washed repeatedly and the second incubation was carried out with ^{125}I-labelled monoclonal anti-human C3c. Mouse ascitic fluid containing monoclonal anti-human C3c (which recognized C3, C3b and C3c) was applied to a protein A-Sepharose column equilibrated to pH 7 with 2 M Tris buffer pH 8.0. The peak protein fraction was quantitated and 100 µg was labelled with ^{125}I by the Iodogen method. The amount of Ab binding to the GBM fragments during the second incubation indicated the capacity of the sample to activate the alternative pathway of complement.

GBM incubated in the presence of serum with Mg^{2+}/EGTA showed an uptake of ^{125}I of 39,000 ±9643 cpm whereas samples incubated in the presence of serum with EDTA had an uptake of only 12,362 ±7497 cpm (mean ±S.D.). This significant increase in the uptake of C3b in the presence of Mg^{2+} indicated direct activation of the alternative pathway by GBM with a mean deposition of 120 ng of C3b/mg of GBM.

Further studies of the deposition of C3b on GBM clearly demonstrated that this was a time-related event. The capacity of the GBM to activate the alternative pathway increased progressively up to 15 min and approached a plateau of C3b deposition by 30 min. In the presence of EDTA-chelated serum the uptake was maximal by 5 min and constant throughout the remainder of the experiment, suggesting a non-enzymatic process or trapping by the GBM.

In order to confirm that the mechanism of deposition of complement was via the alternative pathway, two further studies were carried out. The first demonstrated that the addition of the purified control proteins H and I to twice their normal serum concentration abolished the capacity of GBM to activate the alternative pathway. Secondly, the depletion of factor D from human serum removed its capacity to support activation of the alternative pathway. This was, however, restored when physiological amounts of purified D were added back to that serum.

In a quite separate study [19] we demonstrated, by 2-D immunoelectrophoresis, the fluid-phase conversion of C3 in the presence of GBM. This capacity was increased following enzymatic treatment of the GBM by neutrophil lysosome enzymes.

The presence of sialic acid residues on the surface of GBM has long been established [2], as has the capacity to convert a non-activating surface to an activating surface by the removal of such residues [28]. Neuraminidase was isolated from *Clostridium perfringens* free of detectable protease activity [36] to a specific activity of 38 U/mg; 0.5 U in pH 6.5 acetate buffer and 1 mg of isolated human GBM were incubated for 30 min at 37°, this enzyme concentration being sufficient to remove >50% of the surface sialic acid residues. (The total sialic acid concentration was determined after acid hydrolysis, and free sialic acid residues measured by the thiobarbituric acid method.) Such a procedure increased the capacity of GBM to activate the alternative pathway by up to 50%.

Morphological confirmation of alternative pathway activation by GBM was gained using a second-Ab technique and a horseradish peroxidase label [35]. GBM was incubated as before in serum diluted with appropriate buffer. The Ab anti-C3 was now used unlabelled, and a second Ab (affinity-purified goat anti-mouse IgG) was used, conjugated to the peroxidase. Samples were fixed with 2% glutaraldehyde, processed in 0.05% diaminobenzidine with 0.01% H_2O_2 in 0.1 M phosphate buffer, post-fixed in 2% osmium tetroxide, and examined by electron microscopy. Incubation of GBM in serum diluted in EDTA failed to demonstrate any specific deposition of C3b. In the presence of serum diluted with Mg^{2+}/EGTA, however, the dense reaction product of peroxidase was seen on the outer denser epithelial side of the membrane (Fig. 1, *right portion*). Reaction product was never seen on the endothelial side. Thus it seems that human GBM has the capacity to serve as a surface activator of the alternative complement pathway, a feature which may be augmented by enzymatic modification.

DIRECT ACTIVATION OF INFLAMMATORY CELLS; ENZYME RELEASE STUDIES

The implications of these findings can be amplified by considering a role for GBM not only in the activation of complement but also in its capacity to activate inflammatory cells directly. Human monocytes and neutrophils bear on their surfaces two distinct phagocytic receptors. One recognizes particles bearing IgG, the Fc receptor [37]. The second receptor recognizes target activators of the alternative pathway of complement [38, 39] such as zymosan, rabbit erythrocytes and desialated sheep erythrocytes. Although the latter recognition mechanism is poorly defined and ill-understood, its role in perpetuating the inflammatory response may be of great significance. Thus, having established a potential for GBM to activate the alternative pathway of complement, its capacity to activate inflammatory cells directly needs also to be investigated.

In preliminary studies, rat GBM purified similarly to human GBM was able to promote the release of neutral protease from purified rat peritoneal macrophages as measured with azocasein as substrate.

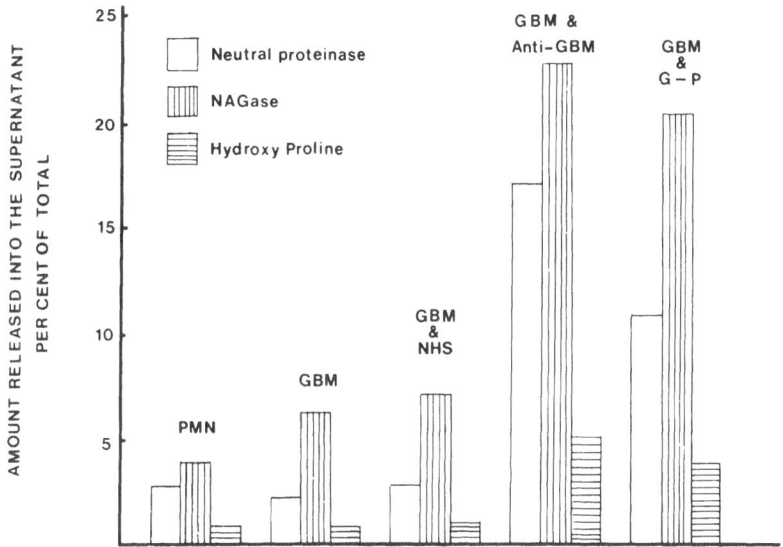

Fig. 2. Release of GBM constituents by PMN. GBM (2 mg) + 250 µl normal serum (NHS), 250 µl rabbit anti-(human GBM) serum (Anti-GBM), 100 µl Goodpasture serum (G-P) **or**, for controls, 250 µl Hanks' Balanced Salts (HBSS) was kept for 60 min at 37°. The serum was removed and the GBM washed (×3) with HBSS and finally suspended in 750 µl HBSS and made up to 1.0 ml with PMN (5 × 10⁵ cells; incubated alone in no-GBM control). After 6 h at 37° the tubes were centrifuged and the supernatants assayed as indicated (NAGase = *N*-acetyl-β-D-glucosaminidase). Details in ref. [19].

Such experiments suggest that under conditions where inflammatory cells may be directly exposed to GBM in a modified or unmodified form the GBM has the potential to activate these cells. Thus GBM *in situ*, when exposed as a result of an inflammatory process or when damaged and modified by lysosomal enzymes, may serve to perpetuate and amplify the inflammatory reaction by its capacity to react either with the alternative pathway of complement or, in the case of the macrophage, directly with the phagocytic cell.

   The release of granule proteins by PMN is thought to be an integral part of the inflammatory reaction and to contribute to tissue destruction. In a previous series of experiments we clearly showed that highly purified PMN elastase and cathepsin G released soluble hydroxyproline-containing material from native insoluble GBM [40]. At 37° degradation of the GBM collagen was extensive and the majority of products appeared as small peptides (M_r <10,000). Evidently neutral proteinases normally located within PMN azurophil granules [41] can extensively damage GBM. GBM pre-treated with immune serum (nephrotoxic or Goodpasture) initiated both neutral proteinase and *N*-acetyl-β-D-glucosaminidase release from human PMN (Fig. 2). Cell integrity was confirmed by absence of lactate dehydrogenase from the medium; hence proteins had been selectively released from primary and tertiary granules.

The above release of lysosomal enzymes was accompanied by the appearance of soluble hydroxyproline material the amount of which indicated that, under the conditions of the experiment, between 3.5% and 5.7% of the hydroxyproline present in the GBM was solubilized (Fig. 2). Thus PMN activation by GBM/anti-GBM is accompanied by membrane damage. More recently these findings have been confirmed and extended to show that the release of reactive oxygen species potentiates GBM damage by lysosomal enzymes [42, 43].

The mechanical stability of the GBM is thought to depend on a large insoluble collagen network. This hypothesis is based on the ability of type IV monomers to form a 'spider-like' framework [44]. Thus the release of a neutral proteinase which cleaves the type IV monomer, albeit at a single site, is capable of distorting the collagen network. Consequently the release from PMN of neutral proteinases such as lysosomal elastase and cathepsin G could potentially initiate considerable damage to the normal GBM.

CONCLUDING COMMENT

The foregoing studies demonstrate that GBM and invading inflammatory cells can interact in the presence or absence of humoral mediators, and suggest mechanisms both for the initiation of acute glomerular damage and for establishing chronic disease.

Acknowledgements

The work was supported by grants from the Kidney Research Unit for Wales Foundation, the Royal College of Physicians, and the Mason Medical Foundation. Acknowledgement is made to Dr. D.D. Abrahamson for carrying out the immunohistochemical studies.

References

1. Farquhar, M.G.(1981) in *Cell Biology pf Extracellular Matrix* (Hay, E.D., ed.), Plenum, New York, pp. 335-378.
2. Rand-Weaver, M. & Price, R.G. (1983) *Biosci. Repts. 3*, [*See also* #F-9, this vol.]
3. Templ, R. (1986) *Kidney Internat. 30*, 293-298.
4. Kanwar, Y.S., Linker, A. & Farquhar, M.G. (1980) *J. Cell Biol. 86*, 688-693.
5. Couser, W.G. (1985) *Kidney Internat. 28*, 569-583.
6. Heymann, W., Hackel, D.B., Harwood, S., Wilson, S.G.F. & Hunter, J.L.P. (1959) *Proc. Soc. Exp. Biol. Med. 100*, 660-669.
7. Masugi, M. (1934) *Beitr. J. Anat. 92*, 429-441.

8. Germuth, F.G. jr. (1953) *J. Exp. Med. 97*, 257–272.

9. Cochrane, C.G., Unanue, E.R. & Dixon, F.J. (1965) *J. Exp. Med. 122*, 99–117.

10. Schrier, G.F., Cotran, R.S., Pardo, V. & Unanue, E. (1978) *J. Exp. Med. 147*, 369–384.

11. Brentjens, J.R., Milgrom, M.L. & Andres, G.A. (1979) in *Immunolog Mechanismsa of Renal Disease* (Wilson, G.B., ed.), Churchill Livingston, London, pp. 215–254.

12. Magil, A.B., Wadsworth, L.D. & Loewen, M. (1981) *Lab. Invest. 44*, 27–33.

13. Krakower, C.A. & Greenspon, S.A. (1951) *Am. Med. Arch. Pathol. 51*, 629–650.

14. Spiro, R.G. (1967) *J. Biol. Chem. 242*, 1915–1922.

15. Allen, R.C. (1982) in *Chemical and Biological Generation of Excited States* (Adam, W. & Cliento, G., eds.), Academic Press, New York, pp. 309–344.

16. Harber, M.J. & Topley, N. (1986) *J. Biolum. Chemilum. 1*, 15–27.

17. Babior, B.M. (1978) *N. Engl. J. Med. 298*, 659–668.

18. Boyum, A. (1968) *Scand. J. Clin. Lab. Invest., 21 (Suppl. 97)*, 77–109.

19. Davies, M., Coles, G.A. & Harber, M.J. (1984) *Immunology 52*, 151–159.

20. Williams, J.D., Topley, N., Alobaidi, H.M. & Harber, M.J. (1986) *Immunology 58*, 117–124.

21. Lewis, E.J., Carpenter, C.B. & Schur, P.H. (1971) *Ann. Int. Med. 75*, 555–560.

22. Perrin, L.H., Lambert, P.H. & Miescher, P.A. (1975) *J. Clin. Invest. 56*, 165–176.

23. Wyatt, R.J., McAdams, A.J., Forristal, J., Snyder, J. & West, C.D. (1979) *Kidney Internat. 16*, 505–512.

24. Verroust, P.J., Wilson, C.B., Cooper, N.R., Edgington, T.S. & Dixon, F.J. (1974) *J. Clin. Invest. 53*, 77–84.

25. Panburn, M.K., Schreiber, R.D. & Müller-Eberhard, H.J. (1981) *J. Exp. Med. 154*, 856–867.

26. Fearon, D.T. & Austen, K.F. (1975) *J. Immunol. 115*, 1357–1361.

27. Law, S.K. & Levine, R.P. (1977) *Proc. Nat. Acad. Sci. 74*, 2701–2705.

28. Kazatchkine, M.D., Fearon, D.T. & Austen, K.F. (1979) *J. Immunol. 122*, 75–81.

29. Fearon, D.T. & Austen, K.F. (1977) *J. Exp. Med. 146*, 22–33.

30. Hunsicker, L.G., Ruddy, S. & Austen, K.F. (1973) *J. Immunol. 110*, 128–138.

31. Davies, A.E., Zault, C., Rosen, F.S. & Alper, C.A. (1979) *Biochemistry 18*, 5802–5807.

32. Fearon, D.T. & Austen, K.F. (1975) *J. Exp. Med. 142*, 856–863.

33. Weiler, J.M., Daha, M.R., Austen, K.F. & Fearon, D.T. (1976) *Proc. Nat. Acad. Sci. 73*, 3268–3272.

34. Fearon, D.T. (1977) *J. Immunol. 119*, 1248–1252.

35. Williams, J.D., Czop, J.K., Abrahamson, D.R. Davies, M. & Austen, K.F. (1984) *J. Immunol. 133*, 394–399.

36. Fearon, D.T. (1978) *Proc. Nat. Acad. Sci. 75*, 1971-1976.
37. Mantovani, B. (1975) *J. Immunol. 115*, 15-17.
38. Czop, J.K. Austen, K.F. (1980) *J. Immunol. 125*, 124-128.
39. Williams, J.D., Lee, T.H., Lewis, R.A. & Austen, K.F. (1985) *J. Immunol. 134*, 3624-3630.
40. Davies, M., Barrett, A.J., Travis, J., Sanders, E. & Coles, G.A. (1978) *Clin. Sci. Mol. Med. 54*, 233-241.
41. Murphy, G., Reynolds, J.J., Bretz, U. & Baggiolini, M. (1979) *Biochem. J. 162*, 195-197.
42. Fligiel, S.E.G., Lee, E.C., McCoy, J.P., Johnson, K.J. & Varani, J. (1984) *Am. J. Path. 115*, 418-425.
43. Shah, S.V., Baricos, W.H. & Basci, A. (1986) *J. Clin. Invest. 79*, 25-31.
44. Yurchenco, P.D., Tsillbary, E.C., Charonis, A.S. & Furthmayr, H. (1986) *J. Histochem. Cytochem. 34*, 93-102.

#F-9

BIOCHEMICAL AND IMMUNOLOGICAL CHARACTERIZATION OF BASEMENT MEMBRANES IN RENAL DISEASE

M. Wong, M. Rand-Weaver, P. Mitchell and R.G. Price

Department of Biochemistry
King's College, Kensington Site
Campden Hill, London W8 7AH, U.K.

Immunological techniques, especially fluorescent staining, aid the diagnosis of immune-based glomerulonephritis. MAb's* enable the isolation and characterization of BM antigens that may be involved in disease. Two MAb's directed against conformational and sequential epitopes in type IV collagen have been investigated using rotary shadowing and immunoblotting techniques: one was specific for BM collagens, while the other recognized an epitope that appeared to be common to collagen types I and V as well. This approach can be used in the immunomapping of connective tissue components.

A different approach was used to study Goodpasture's antigen. The low mol. wt. antigen was released from the BM by collagenase digestion and could be identified by immunoblotting using serum from a patient with Goodpasture's disease. The antigen was non-collagenous in nature and probably resides in the globular domain of type IV collagen. A combination of immunological and biochemical techniques can now be used to characterize the molecular changes in renal BM's in disease.

BM's are specialized forms of extracellular matrices which adjoin the cells that synthesize and secrete them. They have two main biological functions – as a supportive structure or microskeleton to which the cells can adhere, and as selective filters. The latter is best exemplified by the vertebrate kidney GBM, the principal barrier between plasma and its filtrate. It is composed of both collagenous and non-collagenous components as well as other glycoproteins [1]. The principal collagenous component, comprising ~40% of the GBM,

---

*Abbreviations: Ab, antibody: MAb if monoclonal; BM, basement membrane: GBM if glomerular; PAGE, polyacrylamide gel electrophoresis; ELISA, FITC, NEPHGE, etc. – see text.

is type IV collagen [2]. The main non-collagenous BM component
is laminin [3] which may be involved in the adhesion of epithelial
cells to BM's [4]. (Amplification concerning these and other BM
components can be found in reviews.)

In view of the importance of the GBM in renal function and disease,
a more thorough understanding of its structure and function is required.
GBM thickening is the hallmark of diabetic microangiopathy and is
characterized by an increase in the amount of type IV collagen present,
whereas Goodpasture's Syndrome is characterized by the linear
deposition of IgG along the GBM. Further information on the nature
of the macromolecules which contribute to the matrix will aid the
understanding of the underlying mechanism of this disease.

Much effort has been concentrated on establishing the composition
and organization of the BM matrix [5]. Unfortunately, bichemical
investigations on the assembly of BM components and their role in
tissue interactions have been limited by the difficulty in obtaining
homogeneous preparations of different BM's [6]. Another problem
is that small amounts of the components which confer specificity
are present and these may be lost during purification. The differences
between BM's may not be entirely qualitative or quantitative, but
may involve conformational differences in the assembly of the various
components. Some of these inherent difficulties can be overcome
by an immunological approach using MAb's.

BM's are well recognized as important targets in autoimmune
disease in man, and are especially prone to Ab-mediated injury.
Ab's to BM components have been reported to be present in the serum
of some patients [7, 8] and in experimentally produced renal disorders
[9, 10]. However, the antigen(s) involved in the induction of immunity
and the subsequent development of disease have not been identified.

We have adopted immunochemical techniques in the study of the
BM components involved in one such autoimmune disease of the BM,
Goodpasture's Syndrome. This article will review briefly the application and limitations of immunological and biochemical approaches
to the characterization of the GBM using antisera from patients with
this syndrome and MAb's produced against BM components.

MONOCLONAL ANTIBODY PRODUCTION

The antigenic composition of the GBM has been studied by several
investigators using anti-GBM Ab's [11-13]. However, MAb's are specific
for single antigenic determinants (epitopes) and therefore are useful
in localizing a particular antigen in both normal and pathological
BM's. They have the advantage that they are generated by immortal
cell lines.

Several MAb's towards type IV collagen have been reported, most of which are directed against the triple helical domain [14-16]. MAb's towards the 7S collagen domain have also been reported [17-19]. The two MAb's used in this study, MBM4 and MRW4, were raised against citrate-extracted (pH 3.2) renal BM and particulate human GBM respectively. The initial screening of hybridoma-containing culture wells was carried out by enzyme-linked immunosorbent assay (ELISA). If the MAb's are to be used simply for immunohistochemical studies, then screening by immunofluorescence microscopy is a more appropriate method as this excludes any Ab's directed against epitopes that are masked or otherwise inaccessible.

The two MAb's together with Goodpasture's antisera, obtained through plasmapheresis, were used to study collagenase and pepsin digests of BM's from human and bovine kidney [20].

IMMUNOFLUORESCENCE STUDY OF INTACT BASEMENT MEMBRANE SECTIONS

Immunofluorescence microscopy is widely used in the study of renal and skin biopsies, and in the detection of auto-Ab's. This method is most useful in distinguishing between anti-GBM Ab's and immune complex-based renal disease. A granular staining pattern is characteristic of immune-complex disorder while a linear pattern is found in anti-GBM Ab-mediated disease.

Experiments with Goodpasture's serum as the primary Ab (1:20) and fluorescein isothiocyanate- (FITC-)conjugated goat anti-human IgG (1:20) as the secondary Ab demonstrated a characteristic linear staining pattern along the entire length of the BM in renal biopsies (Fig. 1). A heterogeneity and spatial organization of antigens within the GBM which has not been previously appreciated was demonstrated by epifluorescence microscopy at high magnification [12]. In our hands minimal or a complete lack of staining of the GBM or any other glomerular components was observed when bovine sections were used as the antigen source. This could be attributed to either lower amounts of the Goodpasture's antigen or the epitopes being sterically hindered or masked. Enzymatic treatment of the sections with pepsin or collagenase prior to immunostaining failed to produce any positive reactions. However, when the sections are incubated with 6 M urea/ 0.1 M glycine (pH 3.5) at 4° for 60 min [21], the linear deposition of IgG along the BM is established.

Yoshioka et al. [22] demonstrated that nephritogenic GBM antigens, which were not detected by conventional indirect immunofluorescence, were unmasked after denaturation of tissue sections, suggesting that certain epitopes were concealed or restricted in native renal and non-renal BM. Urea is known to interfere with protein interactions by disrupting non-covalent, hydrophobic bonds. When a high molarity of urea is used, changes occur in the protein tertiary structure,

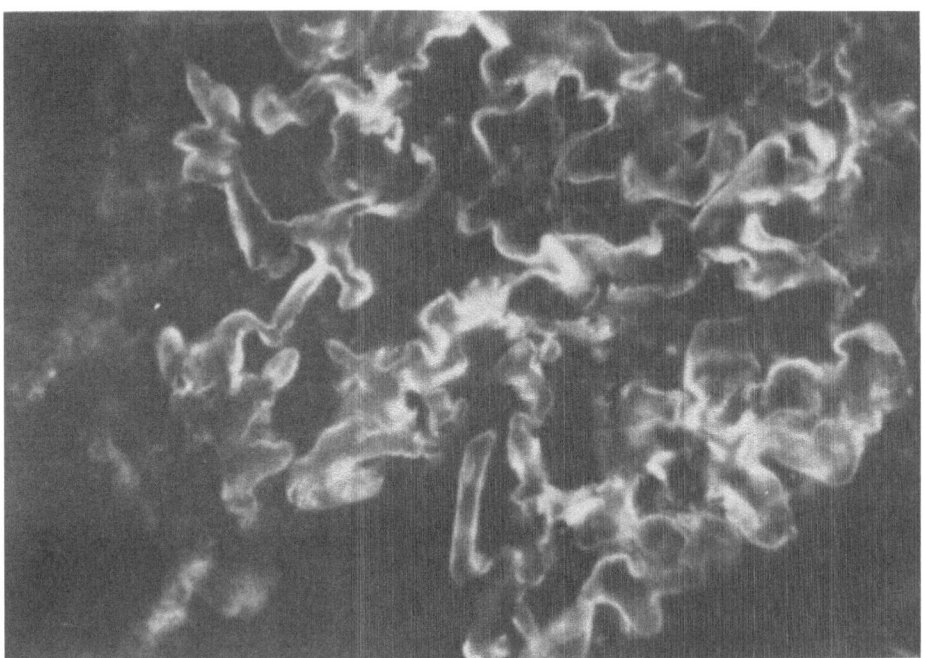

Fig. 1. Immunofluorescent staining of normal adult human kidney sections with Goodpasture serum. Ab's reacted positively along the GBM in a linear manner. Neither the tubular BM nor Bowman's capsule was stained. ×310.

possibly unmasking GBM determinants more fully. The use of pronase and trypsin as well as pepsin for improved localization of tissue antigens in immunochemistry has also been reported [23].

When tissue sections were stained with the MAb's (MBM4 and MRW4) they showed a selective localization in normal and pathological human renal biopsies. Two other MAb's, MBM7 and 15, showed brilliant immuno-fluorescence staining of the GBM in a pattern similar to that observed in anti-GBM disease and were therefore of great interest. However, despite thorough exploration of conditions these Ab's failed to immuno-stain proteins separated by SDS-PAGE and transferred to nitrocellulose sheets, presumably due to low affinity binding characteristics. It is, then, noteworthy that good reactivity in immunofluorescence is not necessarily indicative of suitability of MAb's in other binding systems such as immunoblotting or ELISA.

ROTARY SHADOWING STUDIES

The precise localization of the epitopes recognized by domain-specific MAb's using strictly biochemical approaches is difficult because of the large number of Type IV-like molecular fragments. A non-biochemical approach is required if the macromolecular organization of type IV collagen is to be understood. Recent advances in low-angle metal rotary shadowing casting [24] have made it possible to envisage the individual molecular components of the BM, with resolution down to the level of their domain structure. Direct visualization of Ab-binding sites on type IV collagen has been reported using rotary-shadowing electron microscopy [15, 19, 25]. Thereby we were able to immunomap the epitope recognized by MBM4 to the carboxyl end of the triple helical domain. This powerful procedure has made a crucial contribution to the study of macromolecules and their antigenic composition. However, it is important to be aware that this technique is non-quantitative, as shear forces during sample application can dissociate pre-formed antigen-Ab complexes [15].

M_r OF ANTIGENIC COMPONENTS USING WESTERN BLOTTING AND IMMUNOSTAINING

One of the most widely used approaches to BM study has been immunoblotting [26], entailing PAGE separation of proteins followed by transfer of the protein bands to nitrocellulose membranes which can then be probed with reagents such as Ab's or lectins. Although proteins can be detected on gels the transfer to a solid support has many advantages: antigens are removed from reagents used in electrophoresis which include SDS and reducing agents which are harmful to Ab's, the nitrocellulose sheets are easy to handle, smaller amounts of reagents are required, processing times are generally reduced and transferred patterns can be stored for months prior to being dealt with. Recent excellent reviews [27-30] discuss the conditions affecting elution of proteins from gels and transfer to nitrocellulose, as well as applications of the method.

Our studies on Goodpasture's antigen showed that it was present in the collagenase-resistant portion of human and bovine GBM. Immunoblotting and staining of components solubilized by pepsin with Goodpasture's sera did not react. Antigenic activity was lost when 2-mercaptoethanol was present in the samples, suggesting that the antigen contains intra-chain disulphide bonds that are essential for the activity of the antigen. The antigenic activity was recovered in dimers having apparent M_r's 44,000-49,500 and monomers having apparent M_r's 24,000-29,000 (Fig. 2). The dimer and monomer regions were each shown to be composed of at least two distinct antigenic polypeptides designated, D-1, D-2 and M-1, M-2 respectively.

In contrast to the collagenase digests which reacted with Goodpasture Ab's, both MAb's were reactive with polypeptides obtained from the domains of BM after pepsin digestion. When MBW4 was tested against human GBM solubilized by different methods, only pepsin-digested and guanidine hydrochloride-extracted human GBM were detected by the Ab on immunoblots in both preparations. High mol. wt. material

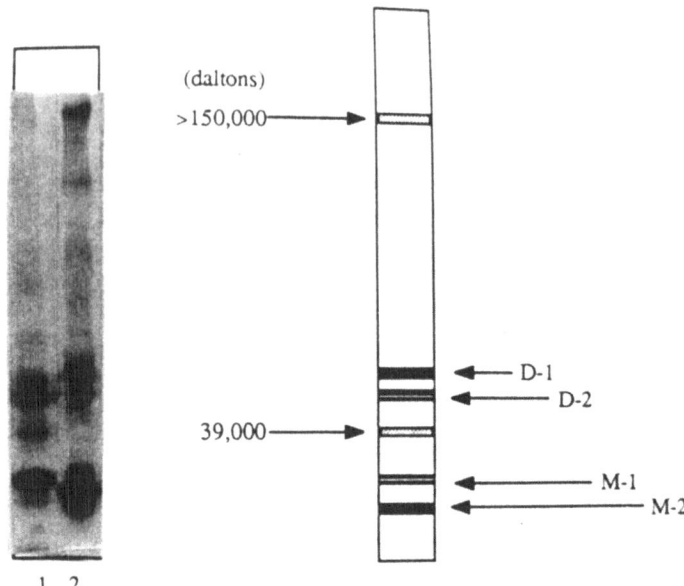

Fig. 2. 'Western' blotting analysis of collagen-digested human and bovine GBM. Proteins from bovine *(lane 1)* and human *(lane 2)* GBM digests were separated by SDS-PAGE (10% gel) and blotted onto nitro-cellulose sheets. Antigenic components were stained using serum from a Goodpasture's patient. Anti-human IgG conjugated to biotin was used as secondary Ab. Colour was developed by adding avidin-horseradish peroxidase and substrate. *On right:* a schematic diagram (not to scale) of the antigenic bands; M and D refer to monomer and dimer regions. Manifesting incomplete breakdown of the antigenic components, lane 2 showed a diffuse band (kM_r >150) in the high mol. wt. region, and lane 1 a lower mol. wt. component (kM_r 39).

of M_r >250,000 was stained.   MBM4 is therefore probably directed against a structural epitope on a protein of collagenous nature, as reduction of particulate human GBM destroyed its antigenicity.

Unlike MBM4, MRW4 is directed towards a sequential determinant on type IV collagen, as both native and reduced collagen are reactive. Three strongly stained bands were observed following immunoblotting of native type IV collagen (M_r = 11,000, 75,000 and 60,000). Following reduction, two new immunoreactive bands appear (M_r = 140,000 and 90,000) in the high mol. wt. region. Because of the molecular size of the components reacting with MRW4 it is believed that the Ab is directed against an epitope in the major triple helix rather than the 7S collagen domain of type IV collagen.

Practical points.- When proteins transferred to nitrocellulose membranes were probed with Ab's, an alternative buffer system for the immunostaining of Goodpasture's antigen was selected because it resulted in a reduced background on the nitrocellulose sheet. Unreacted sites on the sheet were blocked using 1 M glycine containing 1% bovine serum albumin and 5% dried skimmed milk; the washing steps (3 × 5 min) were with pH 6.8 phosphate-buffered saline (PBS) containing 0.1% dried skimmed milk, and all Ab's were diluted in the wash solution. Various brands of skimmed milk can be used, but Marvel (Cadbury Ltd.) dissolves almost instantaneously. The incubation times were 90 and 60 min for primary Ab and secondary Ab respectively. When the second Ab used was coupled to peroxidase, staining was accomplished by immersing the nitrocellulose sheet in a 0.05% solution of 3,3'-diaminobenzidine-tetrahydrochloride in 0.1 M sodium phosphate/citric acid pH 6.0. An alternative substrate is O-chloronaphthol (BDH Chemicals) which dissolves more rapidly and gives a lower background.

It is important to detect the total proteins transferred to the nitrocellulose sheet, as direct comparison between the immuno-stained proteins and a replica gel is unsatisfactory because the gel swells during the staining/destaining procedure. Nitrocellulose sheets could be stained with Coomassie brilliant blue G to detect transferred proteins, but the sheets tended to shrink, making comparison with immunostained sheets difficult. This problem was overcome using Pelikan fount india drawing ink for fountain pens (Pelikan AG, Hanover 1, FRG), at a concentration of 10 µl/ml PBS, as it does not alter the size of the sheet [31].

ANALYSIS OF HETEROGENEITY OF ANTIGENIC COMPONENTS

Two-dimensional SDS-PAGE is becoming the method of choice for the study of complex mixtures of proteins. However, many of the proteins resolved remain unidentified. In our laboratory, antigenic protein spots were identified using the immunoblotting technique in an attempt to characterize whether the antigenic components involved in Goodpasture's Syndrome were heterogeneous.

The first-dimensional separation according to charge, by non-equilibrium pH gradient electrophoresis (NEPHGE), was carried out by Pollard's method [26], and the second by the SDS-PAGE system intro-duced by Laemmli. The immunostaining pattern obtained (Fig. 3) demonstrated that the antigenic components were localized towards the cationic region of the gel (pH 8.0 to 9.0) and were species-specific. Pathological sera were grouped into 3 sets depending on the immuno-staining pattern. The relationship of these sets to the clinical features of the patients has not been determined. However, the results obtained by this combination of both NEPHGE and SDS-PAGE and immuno-blotting have demonstrated for the dimer and monomer regions that considerable complexity of proteins exists which could not be fully appreciated by 1-D gel immunoblots.

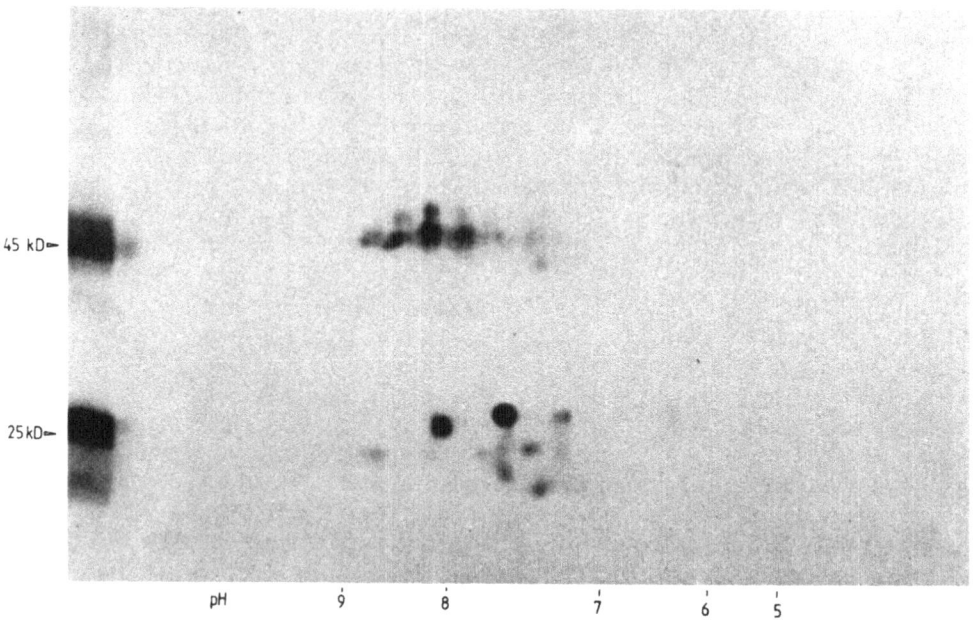

Fig. 3. Two-dimensional NEPHGE/SDS-PAGE and immunoblot analysis of collagenase-digested bovine GBM using Goodpasture's serum as the primary Ab.

SCOPE OF MONOCLONAL ANTIBODIES IN MEDICAL RESEARCH

Ab's have served to probe complicated biosystems and to detect, quantify and localize trace constituents. Such studies, and the diagnosis and treatment of disease, are now powerfully aided by MAb's, some reacting with hitherto unidentified antigens, and others made to antigens already characterized with conventional antisera and offering valuable supplementation to their polyclonal counterparts in routine serology. The trend towards preferential use of MAb's will continue, in view of their purity, specificity and availability in quantity from effectively immortal cell lines. Yet conventional antisera, being readily and inexpensively produced, will remain in use.

Acknowledgements

This work was supported by the Wellcome Trust, the National Kidney Research Fund, the Central Research Fund of London University, and the Royal Society.

References

1. Kefalides, N.A. (1973) *Int. Rev. Conn. Tissue Res. 6*, 63-72.
2. Timpl, R.,   Glanville, R.W., Wick, G. & Martin, G.R. (1979) *Immunology 38*, 109-116.
3. Risteli, L. & Timpl, R. (1981) *Biochem. J. 193*, 749-755.
4. Johansson, S., Kjellen, L., Hook, M. & Timpl, R. (1981) *J. Cell Biol. 90*, 260-264.
5. Kefalides, N.A., ed. (1978) *Biology and Chemistry of Basement Membranes*, Academic Press, New York, 615 pp.- See pp. 215-228.
6. Price, R.G. (1979) in *Cell Populations* [Vol. 8, this series] (Reid, E., ed.), Horwood, Chichester, pp. 105-110.
7. Mahieu, P.M., Lambert, P.H. & Maghuin-Rogister, A.R. (1973) *Eur. J. Biochem. 40*, 599-603.
8. McIntosh, R.M. & Griswold, W. (1971) *Arch. Pathol. 92*, 329-332.
9. Lerner, R.A. & Dixon, F.J. (1966) *J. Exp. Med. 124*, 431-434.
10. Steblay, R.W. (1962) *J. Exp. Med. 116*, 253-256.
11. Henriksen, K. & Jensen, H. (1975) *Scand. J. Immunol. 4*, 699-706.
12. Fish, A.J., Cardomy, K.M. & Michael, A.F. (1979) *J. Lab. Clin. Med. 94*, 447-457.
13. Scheinman, J.I., Foidart, J.M., Gehron-Robey, P., Fish, A.J. & Michael, F. (1980) *Clin. Immunol. Immunopath. 15*, 175-189.
14. Sakai, L.Y., Engvall, E., Hollister, D.W. & Burgerson, R.E. (1982) *Am. J. Pathol. 108*, 310-318.
15. Foellmer, H.G., Madri, J.A. & Furthmayr, H. (1983) *Lab. Invest. 48*, 639-649.
16. Odermatt, B.F., Lang, A.B., Rutter, J.R., Winterhalter, K.H. & Trueb, B. (1984) *Proc. Nat. Acad. Sci. 81*, 7343-7347.
17. Fitch, J.M., Gibney, E., Sanderson, R.D., Mayne, R. & Linsenmayer, T.F. (1982) *J. Cell Biol. 95*, 641-647.
18. Scheinman, J.I. & Tsai, C. (1984) *Lab. Invest. 50*, 101-112.
19. Mayne, R., Wiederman, H., Irwin, M.H., Sanderson, R.D., Fitch, J.M., Linsenmayer, T.F. & Kuhn, K. (1984) *J. Cell Biol. 98*, 1637-1644.
20. Wong, M., Lockwood, C.M. & Price, R.G. (1987) in *Renal Basement Membranes in Health and Disease* (Price, R.G. & Hudson, B.G., eds.), eds.), Academic Press, London, in press.
21. Bene, M.C., Fauré, G., de Ligny, H., Kessler, M. & Duheille, J. (1983) *J. Clin. Invest. 71*, 1342-1347.
22. Yoshioka, K., Kleppel, M. & Fish, A.J. (1985) *J. Immunol. 134*, 3831-3837.
23. Finley, J.C.W. & Petrusz, P. (1982) in *Techniques in Immunocytochemistry* (Bullock, G.R. & Petrusz, P., eds.), Academic Press, New York, pp. 239-249.
24. Furthmayr, H. & Madri, J.A. (1982) *Collagen Rel. Dis. 2*, 349-354.
25. Dieringer, H., Hollister, D.W., Glanville, R.W., Sakai, L.Y. & Kuhn, K. (1985) *Biochem. J. 227*, 217-222.
26. Pollard, J. (1984) in *Methods in Molecular Biology*, Vol. 1:- Proteins (Walker, J.M., ed.), Humana Press, Clifton, NJ, pp. 81-96

27. Towbin, H., Staehlin, T. & Gordon, J. (1979) *Proc. Nat. Acad. Sci. 76*, 4350-4354.
28. Gershoni, J.M. & Palade, G.E. (1983) *Anal. Biochem. 131*, 1-15.
29. Gershoni, J.M. (1985) *Trends Biochem. Sci. 10*, 103-106.
30. Lin, W. & Kasamatu, H. (1983) *Anal. Biochem. 128*, 302-311.
31. Hancock, K. & Tsang, V.C.W. (1983) *Anal. Biochem 133*, 157-162.

#NC(F)

COMMENTS related to

RENAL INVESTIGATIONS

Comments on #F-1: J.K. Nicholson et al. – NEPHROTOXICITY STUDIED BY NMR

R.G. Price asked about the instrument price, relevant to possible use for toxicity studies. **Reply:** £250,000 at present, but prices will come down in time. **Question by G. Land.–** Is it necessary to have a reference compound to be able to quantitate? – Could a drug metabolite that is not yet synthesized be assayed using theoretical assumptions? **Nicholson's reply.–** In theory one can calculate approx. concentrations using only theoretical assumptions, taking into account signal-to-noise ratios, scanning times, sample volumes, etc. In practice this is often not possible. However, unknown compounds can still be quantified by a comparative integration with a known standard. The only assumption necessary is the number of protons (per mol of unknown) contributing to a given resonance (from the unknown). This is often quite easy as the functionality giving rise to an NMR resonance, e.g. CH_2 or CH_3, can often be identified from chemical-shift and coupling-constant data alone.

Comments on #F-2: R.G. Price et al. – RENAL ENZYME RELEASE
    and #F-3: P.H. Bach – CELL CHANGES IN NEPHROTOXICITY

Remarks (to Price) by J.K. McDonald.– Concerning terminology, I assume that your use of the term 'LAP' does not include or refer to the (cytosolic) classical LAP, but rather to the enzyme generally known as aminopeptidase M (or N). For the sake of specificity and sensitivity the assay substrate could advantageously be an alanyl arylamide. **Reply.–** The context is indeed a brush-border marker enzyme. We agree that a substrate for alanine aminopeptidase would be preferable, and intend to switch to this. **Iona Pratt asked** whether the demonstrated functional difference in the proximal tubule of diabetic kidney might account for the relatively greater resistance to damage by nephrotoxins such as gentamicin. **Price's reply.–** Little is known about the functionality of the diabetic kidney, but the findings may indicate an alteration in membrane properties resulting in a decreased uptake of toxins into proximal tubular cells.

Iona Pratt asked Bach whether the Nile Red-positive lipid in epithelial cells derived from glomerular cultures is involved in the redox cycling of adriamycin. **Reply.–** This lipid is not demonstrable by conventional lipid stains (Oil Red O, Sudan Black; lipid has not been previously demonstrated in glomeruli). It may represent a sub-cellular pool of lipid material; we don't believe that it is involved in adriamycin redox cycling (via lipid peroxidation). Although interstitial cells contain similar lipid material and are exquisitely sensitive to BEA, they are not damaged by adriamycin, and similarly glomerular cells are resistant to BEA. **Bach, answering Price.**- We did meet with success in efforts, at present suspended, to culture tubular (proximal) cells; the preparation was not pure, but did respond to toxins in the expected way.

Comments on #**F-4**: K-J. Andersen - RENAL LYSOSOMES; #**F-5**: M. Dobrota
          and #**F-6**: H-J. Haga et al. - LYSOSOMES IN DAMAGED KIDNEY

T. **Berg** asked **Andersen** about the cellular localization of the tripeptidyl peptidase for which he had shown tremendously high activity histochemically. **Reply**: not known as yet; we like to think that it is lysosomal but some may also be in the brush border. **Remark by B.A. Fowler**, concerning the study of degradation where the CdMT dose (≤ 2 mg/rat) was probably above nephrotoxic level: the uptake of CdMT by the proximal tubule cells is saturable, and the % degradation is a function of dose.

Dobrota, answering T. Berg: MT indeed appears first in the small and then in the dense lysosomes. **Comments by Fowler.**- With the nephro-toxic CdMT dose used, normal lysosome biogenesis might be affected and be manifest in an alteration in sedimentation pattern (**Dobrota**: none found). In animals exposed to Cd, MT is mainly cytosolic but there may be adsorption onto organelles, and release from organelles by sonication. **Remark by R. Wattiaux.**- Trial of different homogeni-zation procedures might show differences in the proportion of Cd associated with the cytosol fraction. **Dobrota's reply.**- We used conventional conditions (a Potter-type homogenizer, at 1000 rev/min) as in our studies of normal-kidney lysosomes where marker distribution is well established. Milder conditions might indeed help keep the endosomal compartment intact, but we have not yet devised ideal condi-tions. **Remark by R.G. Price** relating to the model system used being such as to entail proteinuria (agreed by **Dobrota** and **Fowler**): the simplest explanation is an overload situation, but conceivably the cause could be feedback as a result of damaged lysosomes.

Haga, answering T. Berg.- Gradient separation has to contend with the droplets (large lysosomes) and the small, dense lysosomes being similar in density. The association of Cd with MT is pH-depen-dent, hence dissociation might occur in endosomes if they are acidic. **Answer to R.G. Price** relating to the onset of the urinary enzyme increase: increased latency was evident at 2 weeks. **Points raised by A. Schram.**- Concerning examination of urine for the neutral cyto-solic β-glucosidase.- It needs to be distinguished, e.g. by using detergent, from the acid lysosomal glucosidase: but (**Haga's reply**) this was a non-problem because the assay was at pH 7. A urinary rise in lysosomal enzymes in nephrosis could be due to active excretion or to tissue damage, but (**Haga**) the IEF patterns suggest excretion rather than derivation from plasma or damaged cells; the presence of β-glucosidase in normal amounts argues against mere leakage. It was agreed that the picture could be complicated if there were different molecular forms, possibly with pathological variations; enzyme precursors of high mol. wt. might be present, or conceivably degraded lysosomal enzymes.

Comments on #**F-7**: P. Druet et al. - AUTO-IMMUNE NEPHROPATHIES
        and #**F-8**: J.D. Williams & M. Davies - GLOMERULI & INFLAMMATION

Druet, answering C.A. Pasternak.- Various mechanisms are involved in the effects of Hg^{2+} on different lymphocytic cells; (**J. Williams**) one wonders what role macrophages might play in the activation of helper/suppressor cells. **Davies, answering K. Donaldson.-** The chemotactic stimulus which attracts neutrophils to the glomeruli might involve complement (C5a). **Reply to S. Zucker** who asked what role mesangial-cell gelatinase might play in glomerular disease: it can attack type IV collagen in the GBM. **J.P. Luzio asked** whether the GBM and anti-GBM effect on PMN luminescence is a complement-dependent phenomenon; **answer** - to some extent (as amplified in art.# F-8).

Comment on #**F-9**: *see overleaf.*

Supplementary refs. contributed by Senior Editor, on Renal studies

Zonal-rotor methods for isolating p.m., glomeruli and tubular fragments were described in Vol. 3 of this series (Robinson, D., Price, R.G. & Taylor, D.G.). Rat proximal-tubular coated pits, microvilli, endosomes and lysosomes have been compared in respect of membrane composition.- Rodman, J.S., Seidman, L. & Farquhar, M.G. (1986) *J. Cell Biol. 102*, 77-87.

Membrane recycling has also been studied in proximal tubule cells, by e.m. cytochemistry after ferritin infusion; in contrast with recycling between Golgi apparatus and p.m. in other cell types, there is fast recycling that reflects the large demand for membrane material posed by resorption of large fluid volumes.- Christensen, E.I. (1982) *Eur. J. Cell Biol. 29*, 43-49.

In the context of aminoglycoside-induced tubule injury, the distribution of gentamycin (administered or added *in vitro*) was examined in rat cortical homogenates subjected to density-gradient centrifugation (cf. earlier studies with fibroblasts): the main locus was a light-membrane fraction enriched in NAG, a lysosomal marker.- Weinberg, J.M., Hunt, D. & Humes, H.D. (1985) *Biochem. Pharmacol. 34*, 1779-1787.

Immunofluorescence methods have been applied to study the distribution of glomerular basement membrane antigens in various human nephropathies (e.g. with Goodpasture Ab), which manifested character-

istic abnormalities.- Schiffer, M.S., Michael, A.F., Kim, Y. & Fish, A.J.
(1981) *Lab. Invest. 44*, 234-240.

Differences from liver, and enzyme-assay problems, feature in
a study of the regulation of fatty acid metabolism in normal and
streptozotocin- or alloxan-diabetic rats (cf. #F-2): Δ-6 desaturase
activity was unaffected by diabetes, but acyl-CoA formation (a rate-
limiting step) was enhanced. Whole homogenates were used for assay
of desaturase activity; it was low in p.m. and other membrane fractions.- Clark, D.L. & Queener, S.F. (1985) *Biochem. Pharmacol. 34*,
4305-4310.

Human kidney medulla, from which microsomes were prepared, was
studied in respect of Na^+-H^+ exchange by LaBelle, E.F. (1987) *Am. J.
Physiol. 250*, F232-237.

Amplified chemiluminescence, as in #F-8, was used to investigate antirheumatic drugs as scavengers of oxygen radicals in phagocytes.- Müller-Peddinghaus, R., & Wurl, M. (1987) *Biochem. Pharmacol.
36*, 1125-1132.

===========

Comment on #**F-9**: M. Wong et al. - RENAL BASEMENT MEMBRANES

P.K.C. Austwick asked whether the controls manifested a normal
degree of BM thickening. **Reply by R.G. Price.-** The thickness of
the GBM depends on the age of the animals: e.g. in 9-month old rats
the normal is found to be 206 ±14, compared with 356 ±19 in diabetic
and dietary model animals.

Subject Index

This Index, patterned on those in previous vols. to aid back-consultation, focuses on cellular phenomena and processes (including abnormalities), and on bioconstituents (including organelles) and perturbing influences.   As alerting marks (°) indicate, for some aspects a novel collation has to be consulted: **an informative Sub-Index which gives complete contexts and so repays the trouble of use.**-
Thus, kidney studies are fully listed, enabling (say) Hg effects on kidney lysosomes to be tracked down (not readily achievable with a conventional Index).   Indexing is comprehensive for peptidases and heavy metals, but not for other enzymes or agents, nor for investigative approaches.   For major citations the page is represented (e.g.) '25-', the ensuing pp. being relevant too.

° signifies consult Sub-Index, p. 490
  - Im (e.g.) means look for Im

° *signifies see Sub-Index, p. 490*

° *signifies see Sub-Index, p. 490*

---

* *Heavy metals – cell handling, e.g.*
  Cd, *or effects, e.g.* **Cd**, *appear in:-*
°*Sub-Index*, *p. 490* (*& a few in main Index*)

° signifies see Sub-Index, p. 490

° *signifies see Sub-Index,* p. *490*

°signifies see Sub-Index, p. 490

OVERLEAF: Sub-Index
denoted ° above

SUB-INDEX
covering kidney, digestive system and, for aspects
marked * below, other tissues and cells
listed by article, in 4 'blocks' as below

SAMPLE TYPE (& see later headings)

H = human
T = tumour or cancer-related
C = cell incubate - CL if cell line
wc = white cells

M (or +M if not the main approach)
= microscopy:
Mm, morphometry;
ImM, immuno;
CyM, cytochem.

⋯ = a NIL entry.
[ ] = a minor entry, not in-depth.
Order of listing within each of the groups below: INITIALLY collagen (Cg)/ proteolysis/invasiveness; THEN ions.
On right: 293- (e.g.) = p. 293 onwards; (F-2) = art. F-2.
The MAIN INDEX may give more detail, especially for unasterisked items.

CELLULAR FOCUS

Nuc = nucleus
Mt = mitochondria
Ls = lysosomes
Px = peroxisomes
Es = endosomes/ coated vesicles
Mc = microsomes
er = endoplasmic reticulum
pm = plasma memb.
mv = microvilli
Ga = Golgi complex
So = cytosol
BOLD, e.g. **pm**, IF ISOLATED
mx = matrix
⋯

CONSTITUENTS & PHENOMENA

*EcsB = cathepsin B (similarly D, &c)
*En = various endopeptidases
*Ex = exopeptidases
Ez = various enzymes incl. markers
ag = antigen(icity)
FP = fibrous proteins
*Cg = collagens
Pr = various proteins
ℛ = O_2 species/metabolism; peroxidation
*IONS, e.g. H = H^+:
← = ingress, → = exit; ↔, exchange
(Covers ATPases; also
Enzyme egress, →)
MT = metallothionein

*Dg = degradation; Dg if a structure
*Tr = translocation; Tr " "
 # = auto/endogenous; thus #PrDg = autoproteolysis, #Im = autoimmunity.

AGENT/PARTICULAR 'SLANT'

*Heavy metals denoted **Hg** &c.
Dr = drug/other chemical agent
Tx = toxin
*Cp = complement, effects &c
*Im = other immuno effects
*Iv = invasiveness
Id = indicator for diagnosis/therapy
Ij = injury/pathology, various
Mb = membrane perturbations/shifts incl. fluidity changes

KIDNEY (usually cortex): G = glomerulus, P = proximal tubule, D = distal; Md = medulla; U = urine; BM = basement membrane, BB = brush border (pm) or BM

G (&**wc**)| **pm**(GBM); +CyM(457-); mx| ctd. in next line
ag(incl. anti-BM); *Cg(453, 461-), EcsG &c, → (461); [Ez→, Pr→] [ℛ: 456, 462]|
Incl. H: G| BM, +ImM| *ag; Cg; Ex*| Id |465- (F-9)[& 475] Cp, #Im; Dg| 453-(F-8)
[BM: 446; **wc**]| +ImM| *ag*| **Hg**; #Im| 445- (F-7)
P| Nuc, Ls(L03); Mm| *CdMT| [EcsD]*| Cd [**Pb**: 101]| 99- (A-9)
[P&D]| Ls[**Mc**; Es (430)]| *CdMT, Tr & Dg[EcsB, EcsD: 428]* | - | 427- (F-4)
- | Ls [So] | *Pr & CdMT, Dg & Tr; Au, Tr(436); Ni & Pt, Tr(437)*| [**Au, Hg, Pb,** &c: 433]| 433- (F-5)
- | Ls | *Ez*| Im| 439- (F-6)
G & U| **pm**(BB; mv) [Mt, Ls: 410] [+M: 412] | *Ex & Ez →U); [NaK↔: 414]*| Id; **Hg**, Ij| 409- (F-2)
C: P(419-), Md (421-, 150), G(422-)|Px &c| *Ez; ℛ(419-)*|Dr, Tx |419- (F-3)[& 149]
U (vs. kidney)| [Mt, M: 403] | e.g. #citrate, #alanine| Dr, **Hg**; Id| 397- (F-1)

Md| **pm**| – | – | 150   **KIDNEY,**

P| Ls, Px; M| – | Dr| 72, 76   continued

G| BM| *ag*| Ij| 477

H: Md| **Mc**| *NaH*↔|– | 478

G &c [& T, CL: brain: 391] | mx; ImM|
  Cg [#Cg, Dg: 389] | [Iv: 391] | 387-

C: P (& liver, **pc**)| +M|
  Ca, Tr; Na◄| **Hg**, Dr| 233

P| **mv, Es, Ls**; +CyM| – | Tr| 477
– | Px | *Ez* |– |

....

DIGESTIVE SYSTEM: g = gastric parietal cells; **pa** = pancreas (acini);
  i = intestinal mucosa (enterocytes); BB = brush border, **BB** if isolated

i| **BB**; mv [Ls: 258]; **Es** (258, 263–, 285) [Ga: 258]; +ImM (260, 264) |*Ez*; *Ex(256,259)*;
Ez,Tr(258)| Tr| 255- (D-2)

g / i | **pm**, (mv), **Mc /pm** (BB); [+M: 247] | *Na, H*↔; *H, K*↔; *H*◄| [Dr: 244, 253][Mb:
i [**g**: 150] | **BB**: **Mt,pm** (basolateral)|*Ex* |– | 149-   250] | 243- (D-1) [& 285]

rectum (biopsy)[& **wc**] | **Ls**, So &c | *Ez [Ca, Fe*↔:*19]* | Id| 14- (A-2)

T, CL: **pa** & lung| **pm**, So| *Cg,FP:* **En**| Dg, Iv| 373-[& 372]

pa| **pm** [er: 274]|*Ca*↔: *273–, 281, 286] [Mg*↔: *274]*| – | 273- (D-4)

i, BB| **pm**| *Fe*◄| – | 269-   colon| **pm**| – | – | 150
i |mv|*En* | – | 372   H: **i**| **BB**: **mv** &c| *Ez*| – | 286

....

LIVER: pc (or, if isolated, **pc**), parenchymal cells (hepatocytes);
     npc (or **npc**), non-parenchymal cells (See also p. 233 entry above)

C: **pc** | (Start or final:) **Nuc, Mt, Ls, Mc, So** | *Pr, Dg (301 & 372)*; *Ez, Dg (302–)*;
[Ecs(B, D, H, L), **pc** *vs.* **npc**: *371] [Pr, Tr: 304 & 371]* |– | 299- (E-2)

pc & **npc**| **Es**; **Ls** &c| *PrTr*; *PrDg(321) [EzTr]* | Mb, Tr| 315- (E-4)

pc| **Ls**| *Ez* | Dg & Tr [Diet: 383]| 383- | – |**Ls**|*Fe* ; ♔|Ij| 109- (A-10)

– | (Final or, for **Es**, start) **pm, Es** [Ga, Ls: 309] | *Pr, Dg & Tr*|– |307- (E-3)

– | **Es, Ls, pm** | *[Ez: 294] Pr, Tr*; *[Pr, Dg: 293, 295]* ; *K*↔*[H*↔: *290]*| Mb,Tr
– | **Ls** &c| *EcsC, Ez* | Ij | 7 (A-1) [& 122]   [Tx:290]| 289- (E-1)

pc| **Nuc, Mt**; er, **Mc, Ls, So**;+Mm|*Ez* | **In, Tl** [Hg, 101 & 140][As, 99, 101, 134] | 99- (A-9)

....

OTHER SAMPLE TYPES (besides foregoing citations for **wc**, brain & lung)
                rc, red cells; **fb**, fibroblasts

CL: lung alveolar epith. & **wc**| – | *Cg, Dg*| 379- (Cf. lung in p. 273- entry above)

rc incl. ghosts; H: muscle; fat cells; **wc** [T, CL: 206] | – | *Ca*◄| Cp; Dg| 199- (C-2)

CL &c| – | *Na* &c↔*[Na, K*↔: *195]; Pr*→ |Cp, Tx[Hg, Cu, Cd, Zn: 196]; Mb |189- (C-1)

H: **rc**, sickled| **pm**| *Ca*◄; *K & Cl*→; *lipids, Dg(213)*; *[#FP, Dg: 212, 216]* | Mb| 211-
wc (phago), **fb** (transformed)| **pm**, phagosomes, So; +CyM (228)|*Ez*; *FP*|Mb| (C-3)

connect. tissues, glands, sperm, &c [**wc**: 342; T: 341] | **Ls,mx**; +CyM|ctd. 219- (C-4)
Ex [EcsB &c: 336; C: *338];Cg*; *En(343–);Dg*| remodelling [Iv:336]; Dg| 335- (E-6)

T, CL: brain (glioma)| **Ls**, mx| *En (Cg), Ex & Ez:*→|Iv, Dg | 351- (E-7)

T, CL: lung, **wc**, &c|mx; M, ImM|– | Iv(Dg), Id | 359- (E-8) |– |mx|*Cg, En*|– | 336, 371